Lecture Notes in Artificial Intelligence 5306

Edited by R. Goebel, J. Siekmann, and W. Wahlster

Subseries of Lecture Notes in Computer Science

T0223364

Chien-Chung Chan Jerzy W. Grzymala-Busse
Wojciech P. Ziarko (Eds.)

Rough Sets
and Current Trends
in Computing

6th International Conference, RSCTC 2008
Akron, OH, USA, October 23-25, 2008
Proceedings

 Springer

Series Editors

Randy Goebel, University of Alberta, Edmonton, Canada
Jörg Siekmann, University of Saarland, Saarbrücken, Germany
Wolfgang Wahlster, DFKI and University of Saarland, Saarbrücken, Germany

Volume Editors

Chien-Chung Chan
The University of Akron
Department of Computer Science
Akron, OH 44325-4003, USA
E-mail: chan@uakron.edu

Jerzy W. Grzymala-Busse
The University of Kansas
Department of Electrical Engineering and Computer Science
Lawrence, KS 66045, USA
E-mail: jerzy@eecs.ku.edu

Wojciech P. Ziarko
University of Regina
Department of Computer Science
Regina, SK S4S 0A2, Canada
E-mail: ziarko@cs.uregina.ca

Library of Congress Control Number: 2008936101

CR Subject Classification (1998): I.2, F.4.1, F.1, I.5.1, I.4, H.2.8, H.3, H.4

LNCS Sublibrary: SL 7 – Artificial Intelligence

ISSN   0302-9743
ISBN-10   3-540-88423-8 Springer Berlin Heidelberg New York
ISBN-13   978-3-540-88423-1 Springer Berlin Heidelberg New York

Springer is a part of Springer Science+Business Media

springer.com

© Springer-Verlag Berlin Heidelberg 2008
Printed in Germany

Typesetting: Camera-ready by author, data conversion by Scientific Publishing Services, Chennai, India
Printed on acid-free paper   SPIN: 12538836   06/3180   5 4 3 2 1 0

Preface

The articles in this volume were selected for presentation at the Sixth International Conference on Rough Sets and Current Trends in Computing (RSCTC 2008), which took place on October 23–25 in Akron, Ohio, USA.

The conference is a premier event for researchers and industrial professionals interested in the theory and applications of rough sets and related methodologies. Since its introduction over 25 years ago by Zdzislaw Pawlak, the theory of rough sets has grown internationally and matured, leading to novel applications and theoretical works in areas such as data mining and knowledge discovery, machine learning, neural nets, granular and soft computing, Web intelligence, pattern recognition and control. The proceedings of the conferences in this series, as well as in Rough Sets and Knowledge Technology (RSKT), and the Rough Sets, Fuzzy Sets, Data Mining and Granular Computing (RSFDGrC) series report a variety of innovative applications of rough set theory and of its extensions. Since its inception, the mathematical rough set theory was closely connected to application fields of computer science and to other areas, such as medicine, which provided additional motivation for its further development and tested its real-life value. Consequently, rough set conferences emphasize the interactions and interconnections with related research areas, providing forums for exchanging ideas and mutual learning. The latter aspect is particularly important since the development of rough set-related applications usually requires a combination of often diverse expertise in rough sets and an application field. This conference was not different in that respect, as it includes a comprehensive collection of research subjects in the areas of rough set theory, rough set applications as well as many articles from the research and application fields which benefit from the results of rough set theory. To be more specific, major topics of the papers presented at RSCTC 2008 included theoretical aspects of rough set theory, rough set methodology enhanced by probability theory, fuzzy set theory, rough mereology, rule induction, rough set approaches to incomplete data, dominance-based rough set approaches, rough clustering, evolutionary algorithms, granular computing and applications of rough set theory to analysis of real-world data sets.

We would like to express our gratitude to Lotfi Zadeh, Lakhmi Jain and Janusz Kacprzyk for accepting our request to present keynote talks.

This conference was partially supported by the University of Akron, especially the Office of the Vice President for Research, the Buchtel College of Arts and Sciences, and the Department of Computer Science. The conference Web hosting was provided by the Computer Science Department of the University of Akron. The submissions, reviews, and conference proceedings were made through the EasyChair Conference System (http://www.easychair.org). The Infobright Inc. and ZL Technologies Inc. provided support for industrial speakers.

The International Rough Set Society provided technical and publicity support. We express our thanks to these organizations and the EasyChair system development team.

We would like to express our gratitude to Alfred Hofmann, Editor at Springer, and to Ursula Barth, Anna Kramer, and Brigitte Apfel, all from Springer.

Finally, our special thanks go to George R. Newkome, Ronald F. Levant, Wolfgang Pelz, Kathy J. Liszka, Timothy W. O'Neil, Peggy Speck, and Anthony W. Serpette for their help in organizing the conference and registration.

October 2008

Chien-Chung Chan
Jerzy W. Grzymala-Busse
Wojciech Ziarko

Organization

Honorary Chair	Lotfi A. Zadeh
General Conference Chair	Chien-Chung Chan
Program Committee Chairs	Jerzy W. Grzymala-Busse Wojciech Ziarko
Publicity Chairs	Jianchao (Jack) Han Guilong Liu
	Marcin Szczuka JingTao Yao
Local Committee Chairs	Kathy J. Liszka Timothy W. O'Neil
Finance Chair	Wolfgang Pelz

Steering Committee

Ganpiero Cataneo   Andrzej Skowron
Juan-Carlos Cubero   Roman Słowiński
Masahiro Inuiguchi   Shusaku Tsumoto
Tsau Young Lin   Guoyin Wang
James F. Peters   Yiyu Yao
Lech Polkowski

Program Committee

Aijun An	Ryszard Janicki	Wojtek Michalowski
Mohua Banerjee	Jouni Jarvinen	Sushmita Mitra
Jan Bazan	Janusz Kacprzyk	Sadaaki Miyamoto
Malcolm Beynon	Halina Kwasnicka	Mikhail Moshkov
Nicholas Cercone	Jacek Koronacki	Tetsuya Murai
Mihir K. Chakraborty	Bozena Kostek	Michinori Nakata
Davide Ciucci	Vladik Kreinovich	Hung Son Nguyen
Chris Cornelis	Marzena Kryszkiewicz	Sankar K. Pal
Martine De Cock	Yasuo Kudo	Witold Pedrycz
Jitender Deogun	Tianrui Li	Georg Peters
Didier Dubois	Yuefeng Li	Vijay Raghavan
Ivo Düntsch M.-C.	Churn-Jung Liau	Sheela Ramanna
Fernandez-Baizan	Pawan Lingras	Zbigniew Raś
Anna Gomolinska	Jan Małuszyński	Leszek Rutkowski
Salvatore Greco	Victor Marek	Henryk Rybinski
Jianchao Han	Benedetto Matarazzo	Hiroshi Sakai
Aboul E. Hassanien	Lawrence Mazlack	Arul Siromoney
Shoji Hirano	Ernestina	Władysław Skarbek
Tzung-Pei Hong	Menasalvas-Ruiz	Dominik Ślęzak
Xiaohua (Tony) Hu	Duoqian Miao	Jerzy Stefanowski

Jarosław Stepaniuk   Gwo-Hshiung Tzeng   Huanglin Zeng
Piotr Synak   Julio V. Valdes   Justin Zhan
Andrzej Szałas   Alicja Wakulicz-Deja Bo Zhang
Marcin Szczuka   Hui Wang   Wen-Xiu Zhang
Zbigniew Suraj   Anita Wasilewska   Ning Zhong
Soe Than   Junzo Watada   Constantin Zopou-nidis
Li-Shiang Tsay   Arkadiusz Wojna
I. Burhan Turksen   Jing Tao Yao

Additional Reviewers

Wolfram Kahl   Steven Schockaert   Yi Zeng
Xiaobing Liu   Jinhui Yuan

Table of Contents

Data Mining

Decision Support Systems

Clustering

Pattern Recognition and Image Processing

Bioinformatics

Special Sessions

Neuroeconomics: Yet Another Field Where Rough Sets Can Be Useful?

Janusz Kacprzyk*

Systems Research Institute, Polish Academy of Sciences
ul. Newelska 6, 01–447 Warsaw, Poland
kacprzyk@ibspan.waw.pl
www.ibspan.waw.pl/kacprzyk
Google: kacprzyk

Abstract. We deal with neuroeconomics which may be viewed as a new emerging field of research at the crossroads of economics, or decision making, and brain research. Neuroeconomics is basically about neural mechanisms involved in decision making and their economic relations and connotations. We briefly review first the traditional formal approach to decision making, then discuss some experiments of real life decision making processes and point our when and where the results prescribed by the traditional formal models are not confirmed. We deal with both decision analytic and game theoretic type models. Then, we discuss results of brain investigations which indicate which parts of the brain are activated while performing some decision making related courses of action and provide some explanation about possible causes of discrepancies between the results of formal models and experiments. We point out the role of brain segmentation techniques to determine the activation of particular parts of the brain, and point out that the use of some rough sets approaches to brain segmentation, notably by Hassanien, Ślęzak and their collaborators, can provide useful and effective tool.

1 Introduction

First, we wish to briefly introduce the concept of neuroeconomics which is emerging as a new field of science at the crossroads of economics, or maybe more generally decision making, and brain research.

The first question is: what is economics? For our purposes the following classic definition by Robbins [18] should be appropriate:

" ...economics is the science which studies human behavior as a relationship between ends and scarce means which have alternative uses ..."

We can see at the first glance that in that classic definition of economics decision making plays a central and pivotal role. The definition, which emphasizes

---

* Fellow of IEEE.

C.-C. Chan et al. (Eds.): RSCTC 2008, LNAI 5306, pp. 1–12, 2008.

the decision making aspect, has expressed what people have always been aware of, i.e. of the importance of acting rationally. This crucial problem has clearly become a subject of interest of thinkers, scholars and scientists for many centuries, and even millennia. Basically, the developments of science have always been motivated to a decisive extent by practical needs. A natural consequence of this interest has finally been attempts at some formal analyzes which should provide the analysts and decision makers with more objective tools and techniques. Mathematics has been considered crucial in this respect. This trend has gained momentum in the period between World War I and World War II, and in particular after World War II.

In this paper we will often speak about decision making but our analysis will apply to a large extent to broadly perceived economics since for our purposes the very essence of decision making and economics is to choose a choice of action that would make the best use of some scarce means or resources that can be used in various ways leading to different outcomes.

Basically, the point of departure for virtually all decision making models in the formal direction is simple:

- There is a set of options, $X = x_1, x_2, \ldots, x_n$, which represent possible (normally all) choices of a course of action like.
- There is some preference structure over the above set of options which can be given in different ways exemplified by: (1) preferences over pairs of options, for instance: $x_1 \succeq x_2$, $x_2 = x_3$, $x_3 \preceq x_4$, etc. (2) a preference ordering exemplified by $x_1 \geq x_3 \geq \ldots x_k$, or (3) a utility function $f : X \longrightarrow R$ (R is the real line but may be some other set which is naturally ordered.
- a natural rationality is assumed which in the context of the utility function is to find an optimal option $x^* \in X$ such that $f(x^*) = \max_{x \in X} f(x)$.

An agent operating according to such simple and intuitively appealing rules has been named a *homo economicus*, and virtually all traditional approaches to decision making and economics are in fact about various forms of behavior of a homo economicus. These simple conditions have been considered so natural and obvious that only a few reserachers have been considering tricky issues related to what can happen if they do not hold.

This simple model has been a point of departure of a plethora of models accounting for: multiple criteria, multiple decision makers, dynamics, etc. On the other hand, it has triggered the emergence of many distinct areas as: optimization and mathematical programming, optimal control, mathematical game theory, etc. which have shown their strength in so many areas and applications.

Unfortunately, these successes of mathematical models of decision making have mostly happened in *inanimate systems* in which a human being is not a key element like in missile control. The situation changes drastically when a human perception or valuation becomes essential, when we cannot neglect human characteristics like inconsistency and variability of judgments, imprecise preferences, etc. as in in virtually all *animate systems*. Economics is clearly concerned with such systems.

Let us now mention two classes of decision making problems, viewed from the perspective of rational choice theory, which will crucial for us: *decision theory* (*analysis*) and *game theory*. They provide formal tools for determining optimal decisions in the context of individual and multiperson decision making; the former concerns situation with individuals (agents) and/or their groups operating without interaction, and the latter concerns in which there are at least two agents involved but operating with an interaction like, for instance, a sequence of proposals and responses, i.e. decision making is the selection of a strategy meant as a set of rules that govern the possible actions (options) together with their related payoffs to all participating agents.

For example, in the famous prisoners dilemma (cf. Poundstone [16]), two individuals, A and B, are criminals suspected of having jointly committed a crime but there is not enough evidence to convict them. They are into two separate cells in prison, and the police offer each of them the following deal: the one who implicates the other one will be freed. If none of them agrees, they are seen as cooperating and both will get a small sentence due to a lack of evidence so that they both gain in some sense. However, if one of them implicates the other one by confessing, the defector will gain more, since he or she is freed, while the one who remained silent will receive a longer sentence for not helping the police, and there is enough evidence now because of the testimony of the defector. If both betray, both will be punished, but get a smaller sentence. Each individual (agent) has two options but cannot make a good decision without knowing what the other one will do.

If we employed traditional game theoretic tools to find the optimal strategy, we would find that the players would never cooperate as the traditionally rational decision making means that an agent makes decision which is best for him/her without taking into account what the other agents may choose.

So far we have discussed decision making in the sense of what is obviously rational which boils down to the maximization of some utility function. This rationality is clearly a wishful thinking, probablytoo primitive for real life, but results in solvable models in both the analytic and computational sense. This important aspect is clearly reflected in all mathematical models that should be a compromise between complexity, and adequacy and tractability.

However, the formal mathematical direction in decision making (economics) is not the only one, and many other directions have appeared with roots in psychology, sociology, cognitive sciences, and recently brain research. An example can be experimental and behavioral economics, and recently *neuroeconomics*. Basically, as opposed to approaches of the rational choice type mentioned above which focus on normative or prescriptive issues, virtually all those social and brain science related approaches to decision making are rather concerned with the descriptive aspects. They study how subjects make decisions, and which mechanisms they employ. For instance, well known works of Tversky and Kahneman and their collaborators (Tversky and Kahneman, 1991; Kahneman and Tversky, 2003, Kahneman, Slovic and Tversky, 1982) showed that decision makers judgments and behavior deviate to a large extent from results derived by normative theories as agents tend to make decisions due to their so called "framing" of a decision situation (the way they represent the situation as, e.g., a gain or a loss), and often exhibit "strange" loss aversion, risk aversion, and ambiguity aversion. So, their choices do not follow "obvious" results of traditional normative theories.

Moreover, many psychological studies have also showed that people are not as selfish and greedy as the solutions obtained using tools of rational choice approaches may suggest. For instance, subjects cooperate massively in prisoners dilemma and in other similar games.

The *ultimatum game* concerns a one move bargaining (cf. Gŭth W., Schmitberger R. and Schwarze B. [4]. There is a proposer, A, who makes an offer to a responder, B, who can either accept it or not. Suppose that A is to propose to split some amount of money between himself or herself and B. If B accepts the offer of A, B keeps the amount offered and A keeps the rest. If B rejects it, both A and B receive nothing. According to game theory, rational agents should: A should offer the smallest possible amount, and B should accept any amount just to have anything which is clearly better than nothing. Unfortunately, this is not the solution adopted by human agents in real life. Basically, most experiments show that a purely rational game theoretic strategy is rarely played and people tend to make more just and fair offers. In general, proposers tend to offer about 50% of the amount, and responders tend to accept these offers, rejecting most of the unfair offers, experimentally shown to be less than about 20%. So, agents seem to have a tendency to cooperate and to value fairness as opposed to some greedy behavior of traditional game theoretic approaches.

In the *trust game*, A has an initial amount of money he or she could either keep or transfer to B. If A transfers it to B, the amount is tripled. B could keep this amount, or transfer it (partially or totally) to A. Following the solutions given by game theory, A should keep everything, or if A transfers any amount to B, then B should keep all without transferring it back to A. Once again, unfortunately, experimental studies have shown that agents tend to transfer about 50% of their money and get more or less what they invest (cf. Camerer [1]), and this tendency towards fairness and cooperation holds for all cultures, sexes, etc.

To summarize, experimental approaches to rationality and how decisions are really made can thus be informative for the theory of decision making as they clearly indicate that our practical reasoning does not fully obey the axioms of either decision theory or game theory, and that the traditional approaches which somehow neglect morality, fairness and consideration for other people might be inadequate. Thus, we can argue that specific features of a human being should be taken into account in decision analytic and game theoretic models in order to obtain solutions that would be human consistent and hence would be presumably easier acceptable and implementable.

In recent years, however, there is another big boost to such deeper analyses of decision making in various context, both strategic and not, and this comes from *brain research*. In the next section we will discuss how brain research can contribute to the development of economics. This new field, still at its infancy, is called *neuroeconomics* and seems to be able to open new perspectives and vistas.

2   Towards Neuroeconomics

Both the descriptive and prescriptive approach to decision making and economics may be viewed from the point of view of what is being observed and mimicked or what is being rationalized or even optimized as they both concern the behavior of an agent(s) in the sense of "externally visible" choices, courses of action, etc. However, it is quite obvious that this externally visible behavior is just an implication or consequence of some more general mental processes that happen in the brain. One can therefore argue that what really matters is what happens in the brain not what is externally visible as a resulting behavior or resulting testimonies. It should therefore make much sense to look deeply into brain processes while investigating decision making and economics. Clearly, this concerns both the decision analytic and game theoretic aspects. This is basically the motivation the very essence of neuroeconomics that has been initiated in the beginning of the 2000s, cf. Glimcher [3], McCabe [12] or Zak![27] for some pioneering works, cf. also Kenning and Plassmann [11].

First: what is actually neuroeconomics? An often cited definition, which is obviously related to Robbins' [18] definition of economics cited before, is attributed to Ross [19]:

> "... neuroeconomics ... is the program for understanding the neural basis of the behavioral response to scarcity ... ".

In neuroeconomics one can briefly outline the methodology employed as follows:

- Choosing a formal model of decision making and its related rationality, whether in a decision analytic or a game theoretic form, and then deducing what decisions the rational agents should make;
- Testing the model behaviorally, i.e. with respect to externally visible characteristics, to see if agents follow those courses of actions determined in the first stage;
- Identifying the brain areas and neural mechanisms that underlie the particular choice behavior;
- Explaining why agents follow or not the normative courses of actions mentioned.

Neuroeconomics proceeds therefore basically by comparing formal models with behavioral data, and by identifying neural structures causally involved in (maybe underlying) economic, or decision making related behavior.

In neuroeconomics attempts are made to explain decision making as an implication or consequence of brain processes which occur in the representation, anticipation, valuation, selection, and implementation of courses of action (options). It breaks down the whole process of decision making into separate components which are related to specific brain areas. Namely, certain brain areas may perform (or maybe just decisively contribute to?) the representation of the value of an outcome of a course of action before decision, other brain areas may perform the representation of the value of a course of action chosen, and yet other brain areas may perform the representation of these values at the time when a decision is determined and is to be implemented.

The remarks given above are valid both for the decision analytic type and game theoretic type decision processes and we will now consider the consecutively from the neuroeconomic perspective.

Another class of tools needed by neuroeconomics is related to being able to discover what is happening in specific areas of the brain while an activity is being performed. This includes the tools and techniques for: brain imaging, single-neuron measurement, electrical brain stimulation, psychopathology and brain damage in humans, psychophysical measurements, diffusion tensor imaging, etc.

Brain imaging is currently the most popular neuroscientific tool. Basically, the main procedure is to obtain and then compare two brain images: when an agent performs a specific task or not. The difference detected can indicate that a specific area of the brain is activated during the performance of that particular task. There are many methods for brain imaging, but the following three are basic:

- the *electro-encephalogram* (or EEG), which is the oldest, boils down to the attachment of some electrodes to the scalp and then to the measuring of induced electrical currents after some stimulus,
- the *positron emission topography* (PET) scanning, an old technique but still useful, which measures blood flow in the brain which can be considered as an equivalent to neural activities,
- the *functional magnetic resonance imaging* (fMRI), the newest and most often used, which measures blood flow in the brain using changes in magnetic properties due to blood oxygenation.

but though fMRI is the most popular and often considered to be the best, each of those methods has some pros and cons, cf. Camerer, Loewenstein and Prelec [2].

Clearly, brain imaging mentioned above does not allow to see what is happening at the level of single neurons but this will not be considered here. Moreover, we will not study what happens when some part of the brain is damaged or an individual suffers from a mental disease (e.g. schizophrenia) or a developmental disorder (e.g., autism) though by observing differences between healthy and ill people one can draw many interesting conclusions relevant in our context.

2.1   Decision Analysis and Neuroeconomics

The division of the decision making process into stages (set of options, a preference structure and it related utility, and a rational choice) is quite convincing as it is related to some results obtained in the studies of the very essence of rational behavior. For instance, Kahneman, Wakker and Sarin [10] have advocated that the utility should be divided into: (1) *decision utility* which is maybe the most important, and refers to expected gains and losses, or cost and benefits, (2) *experienced utility* which has to do with the pleasant or unpleasant, or even a hedonic aspect implied by a decision, (3) *predicted utility* which is related to the anticipation of experienced utility, and (4) *remembered utility* which boils down to how experienced utility is remembered after a decision, like regretting or rejoicing. Such a *distributed utility* has relations to some structures and processes in the brain, and plays a very important role in the field of neuroeconomics. For instance, the distributed perspective of utility can help explain more precisely why human agents exhibit loss aversion. To be more specific, agents usually pay much more attention to a loss of EUR 10 than to a gain of EUR 10, and Tversky and Kahneman [23] attribute this loss aversion to a bias in the representation of the values of gain and loss.

On the other hand, neuroeconomics explains loss aversion as an interaction of neural structures in the brain which are involved in the anticipation, registration and computation of the hedonic affect of a risky decision. To be more specific, the amygdalae which are are almond shaped groups of neurons located deep within the medial temporal lobes of the brain play a primary role in the processing and memorizing of emotional reactions, and are involved in fear, emotional learning and memory modulation. The amygdalae register the emotional impact of the loss. The ventromedial prefrontal cortex, which is a part of the prefrontal cortex, is usually associated with the processing of risk and fear. In our context, the ventromedial prefrontal cortex predicts that a loss will result in a given affective impact. The midbrain dopaminergic neurons compute the probability and magnitude of the loss, etc.

Agents are therefore loss averse because they have a negative response to losses (experienced utility). When they expect a loss to occur (decision utility), they anticipate their affective reaction (predicted utility). They might be also attempting to minimize their post decision feeling of regret (remembered utility). They anticipate their affective reaction (predicted utility). They might be also attempting to minimize their post-decision feeling of regret (remembered utility).

One may say that the midbrain dopaminergic systems are where the human natural rationality resides, or at least one of its major component. These systems compute utility, stimulate motivation and attention, send reward prediction error signals, learn from these signals and devise behavioral policies.

Similar investigations have referred to other phenomena as the *ambiguity aversion*, i.e. that the human agents exhibit a strong preference for risky prospects, whose whose occurrence is uncertain but probabilities of occurrence are known, over ambiguous prospects, that is those for which the probabilities of occurrence are not known or are very imprecisely known.

And, if we continue, we can see that one of the most robust finding in neuroeconomics concerns the decision utility which is related to the calculation of cost and benefits (or gains and losses). Results of many investigations strongly suggest that this process is realized by dopaminergic systems. They refer to neurons that make and release a neurotransmitter called the dopamine. The dopaminergic system is involved in the pleasure response, motivation and valuation. The dopaminergic neurons respond in a selective way to prediction errors, either the presence of unexpected rewards or the absence of expected rewards. Therefore they detect the discrepancy between the predicted and experienced utility. Moreover, dopaminergic neurons learn from own mistakes: they learn to predict future rewarding events from prediction errors, and the product of this learning process can then be a bias in the process of choosing a course of action; these learning processes can be modeled using temporal difference reinforcement learning algorithms (cf. Sutton and Barto [22]).

So far, the main contribution of neuroeconomics to decision theory may be viewed as giving justifications to the fact that decision makers are adaptive and affective agents, i.e. a *homo neuroeconomicus* is a fast decision maker who relies less on logic and more on a complex collection of flexible neural circuits associated with affective responses. The utility maximization in real life and by human agents is more about feelings and emotions and less about careful deliberations. This is in a sharp contrast to a highly deliberative, cold blooded and greedy type process of traditional, formal decision analysis.

2.2 Game Theory and Neuroeconomics

Now, in the game theoretic decision making context, we will basically be concerned with the strategic rationality. And again, the paradigm of neuroeconomics mentioned in the previous section clearly suggest that strategic decision making is again a highly affection centered activity.

For instance, brain scans of human agents playing the ultimatum game indicate that unfair offers by A trigger in the brain of B a "moral disgust". To be more specific, the anterior insula, which is associated with emotional experience, including anger, fear, disgust, happiness and sadness, is activated in such situations of a moral disgust resulting from an unfair offer. What is interesting is that such activation is proportional to the degree of unfairness and correlated with the decision to reject unfair offers.

In the ultimatum game not only the anterior insula is involved but also two other areas of the brain. First, this is the dorsolateral prefrontal cortex which serves as the highest cortical area responsible for motor planning, organization and regulation and plays an important role in the integration of sensory and mnemonic information and the regulation of intellectual function, goal maintenance and action. It should however be noticed that the dorsolateral prefrontal cortex is not exclusively responsible for the executive functions because virtually all complex mental activities require additional cortical and subcortical circuits which it is connected with. Second, it is the anterior cingulate cortex which is the frontal part of the cingular cortex that relays neural signals between the right and left cerebral hemispheres of the brain. The anterior cingulate cortex seems to play a role in a wide variety of autonomic functions (for instance, regulation of blood pressure or heart beat) as well as some rational cognitive functions exemplified by reward anticipation, decision making, conflict recognition and empathy and emotions. In our context, when an offer is fair, it seems normal to accept it: there is a monetary gain and no aversive feelings. When the offer is unfair, however, the brain faces a dilemma: punish the unfair proposer, or get a little money? The final decision depends on whether the dorsolateral prefrontal cortex or the anterior cingulate cortex dominates. It has been found that anterior cingulate cortex is more active in rejections, while the dorsolateral prefrontal cortex is more active in acceptance. Thus, the anterior cingulate cortex, which is more active itself when an offer is unfair, behaves as a moderator between the cognitive goal (to have more money) and the emotional goal (punishing).

Some other strange types of behavior can be observed in strategic games when cooperation is really needed, and occurs in real life, but is not taken into account. For instance, in the prisoners dilemma, players who initiate and players who experience mutual cooperation display activation in nucleus accumbens (accumbens nucleus or nucleus accumbens septi) which are a collection of neurons within the forebrain and are thought to play an important role in reward, laughter, pleasure, addiction and fear. Some other reward related areas of the brain are also strongly activated.

On the other hand, in the trust game, where cooperation is common but again not prescribed by game theory, players are ready to lose money for punishing untrustworthy players or cheaters. And here again, both the punishing of cheaters and even anticipating such a punishment activate the nucleus accumbens suggesting that a revenge implies some pleasure.

To put it simply, all these results suggest that fairness, trust and cooperation are common because they have some generally accepted values. This is well reflected by activations of some specific areas of the brain but is beyond the scope of the traditional game theoretic approaches.

3   Some Remarks on a Possible Usefulness of Rough Sets Theory for Neuroeconomics

In this section, duen to lack of space, we will only point out so,e potential contributions of rough sets theory to the development of neuroeconomics. One should however notice that this is the view of the author only and has a very general form, of an *ad hoc* type.

Basically, looking at what proponents of neuroeconomics advocate and how they proceed one can notice that emphasis is on relating brain functions. maybe areas, to some courses of actions or behaviors of human decision makers. However, to discover those brain areas and functions brain imaging should be performed to discover them. Brain imaging is currently the most popular neuroscientific tool and the main procedure is to obtain and then compare two brain images: when an agent performs a specific task or not or exhibiting a special behavior or nor. The difference detected can indicate that a specific area of the brain is activated during the performance of that particular task. The functional magnetic resonance imaging (fMRI) can notably be employed.

The data obtained through fMRI undergoes some processing, for instance segmentation which is a process of assigning proper labels to pixels (in 2D) or voxels (in 3D) given in various modalities to distinuish different tissues: white matter, grey matter, cerebrospinal fluid, fat, skin, bones, etc. This information can help properly differentiate parts of the brain responsible for brain functions that are relevant to neuroeconomics.

It is quite obvious that the analysis of fMRI images is difficult as it is plagued by various uncertainties, noise, subjective human judgment, etc. It is quite natural that the use of many computational intelligence tools has been proposed (cf. Ramirez, Durde and Raso [17]. However, it seems that a new impetus in this respect can be given, and new vistas can be opened by using rough sets which would make possible to provide much insight and, for instance, to reduce the attribute set by using a reduct based analysis.

It seems that a proper approach would be to proceed along the lines of Hassianen and Ślęzak [7], and Hassianen [6], and also Widz, Revett and Ślęzak [24], [25], and Widz and Ślęzak [26]. In fact, some concluding remarks and future research directions. For our purposes the most interesting seem to be analyses related to comparisons of various brain images, notably for a healthy and ill person. In our context more importatnt would be the comparison of brain images with and without some behavior, activity or emotion. Moreover, their intended extension to unsupervised classification should be very relevant too. Yet another issue they intend to tackle, an extension towards more complex structure of dependencies (multi-attribute ones), and then the reformulation of the segmentation problem in terms of tissue distributions instead of tissue labels should give more strength.

In general, it seems that rough sets can provide very much for brain imaging as needed for neuroeconomic purposes. The papers cited above seem to provide a very good point of departure.

4   Conclusions

We have presented a very brief account of a new nascent field of neuroeconomics, mainly from the perspective of decision making. Then, we have presented results of some experiments with the real human decision makers and shown how these results deviate from those prescribed by the traditional formal decision making and game theoretic models. Finally, we have presented some results obtained by brain researchers which have shown relations between a stronger activation of some parts of the brain in real situations in agents participating in the decision making and games considered. One could clearly see that some effects which have not been prescribed by traditional formal models but can clearly be viewed as results of human features imply the activation of corresponding parts of the brain involved in or maybe responsible for the particular cognitive, emotional, etc. activities. Finally, we have briefly mentioned that rough sets theory can be useful by providing new insights and richer tools for brain imaging needed in neuroeconomic analyses.

References

1. Camerer, C.F.: Psychology and economics. Strategizing in the brain 300, 1673–1675 (2003)
2. Camerer, C.F., Loewenstein, G., Prelec, D.: Neuroeconomics: How Neuroscience Can Inform Economics. Journal of Economic Literature XLIII, 9–64 (2005)

3. Glimcher, P.W.: Decisions, Uncertainty, and the Brain: The Science of Neuroeconomics. MIT Press, Cambridge (2003)
4. Gŭth W., Schmittberger, R., Schwarze, B.: An Experimental Analysis of Ultimatum Bargaining. Journal of Economic Behavior and Organization 3(4), 367–388 (1982)
5. Hardy-Vallée, B.: Decision-Making: A Neuroeconomic Perspective. Philosophy Compass 2(6), 939–953 (2007)
6. Hassanien, A.E.: Fuzzy-rough hybrid scheme for breast cancer detection. Image and Computer Vision 25(2), 172–183 (2007)
7. Hassanien, A.E., Ślęzak, D.: Rough Neural Intelligent Approach for Image Classification: A Case of Patients with Suspected Breast Cancer. International Journal of Hybrid Intelligent Systems 3/4, 205–218 (2006)
8. Kahneman, D.: A Perspective on Judgment and Choice: Mapping Bounded Rationality. American Psychologist 58(9), 697–720 (2003)
9. Kahneman, D., Slovic, P., Tversky, A. (eds.): Judgment under Uncertainty: Heuristics and Biases. Cambridge University Press, Cambridge (1982)
10. Kahneman, D., Wakker, P.P., Sarin, R.: Back to Bentham? Explorations of Experienced Utility. The Quarterly Journal of Economics 112(2), 375–397 (1997)
11. Kenning, P., Plassmann, H.: NeuroEconomics: An Overview from an Economic Prespective. Brain Reserach Nulletin 67, 343–354 (2005)
12. McCabe, K.: Neuroeconomics. In: Nadel, L. (ed.) Encyclopedia of Cognitive Science, pp. 294–298. Wiley, New York (2005)
13. Montague, R., King-Casas, B., Cohen, J.D.: Imaging Valuation Models in Human Choice. Annual Review of Neuroscience 29, 417–448 (2006)
14. Montague, R., Berns, G.S.: Neural Economics and the Biological Substrates of Valuation. Neuron 36(2), 265–284 (2002)
15. Naqvi, N., Shiv, B., Bechara, A.: The Role of Emotion in Decision Making: A Cognitive Neuroscience Perspective. Current Directions in Psychological Science 15(5), 260–264 (2006)
16. Poundstone, W.: Prisoner's Dilemma. Doubleday, New York (1992)
17. Ramirez, L., Durdle, N.G., Raso, V.J.: Medical image registration in computational intelligence framework: a review. In: Proceedings of IEEE–CCECE 2003: Canadian Conference on Electrical and Computer Engineering, vol. 2, pp. 1021–1024 (2003)
18. Robbins, L.: An Essay on the Nature and Significance of Economic Science. Macmillan, London (1932)
19. Ross, D.: Economic Theory and Cognitive Science: Microexplanation. MIT Press, Cambridge (2005)
20. Rubinstein, A.: Comments on Behavioral Economics. In: Blundell, Newey, W.K., Persson, T. (eds.) Advances in Economic Theory (2005 World Congress of the Econometric Society), vol. II, pp. 246–254. Cambridge University Press, Cambridge (2006)
21. Samuelson, L.: Economic Theory and Experimental Economics. Journal of Economic Literature 43, 65–107 (2005)
22. Sutton, R.S., Barto, A.G. (eds.): Reinforcement Learning: An Introduction. Adaptive Computation and Machine Learning. MIT Press, Cambridge (1998)
23. Tversky, A., Kahneman, D.: Loss Aversion in Riskless Choice: A Reference- Dependent Model. The Quarterly Journal of Economics 106(4), 1039–1061 (1991)
24. Widz, S., Revett, K., Ślęzak, D.: A Hybrid Approach to MR Imaging Segmentation Using Unsupervised Clustering and Approximate Reducts. In: Ślęzak, D., Yao, J.T., Peters, J.F., Ziarko, W., Huo, X. (eds.) RSFDGrC 2005. LNCS (LNAI), vol. 3642, pp. 372–382. Springer, Heidelberg (2005)

25. Widz, S., Revett, K., Ślęzak, D.: A Rough Set-Based Magnetic Resonance Imaging Partial Volume Detection System. In: Pal, S.K., Bandyopadhyay, S., Biswas, S. (eds.) PReMI 2005. LNCS, vol. 3776, pp. 756–761. Springer, Heidelberg (2005)
26. Widz, S., Slezak, D.: Approximation Degrees in Decision Reduct-Based MRI Segmentation. In: FBIT 2007: Proceedings of the 2007 Frontiers in the Convergence of Bioscience and Information Technologies, pp. 431–436. IEEE Computer Society Press, Los Alamitos (2007)
27. Zak, P.J.: Neuroeconomics. Philosophical Transactions of the Royal Society of London, Series B, Biological Science 359(1451), 1737–1748 (2004)

Research Directions in the KES Centre

Lakhmi Jain and Jeffrey Tweedale*

School of Electrical and Information Engineering,
Knowledge Based Intelligent Engineering Systems Centre,
University of South Australia, Mawson Lakes, SA 5095, Australia
{Lakhmi.Jain,Jeff.Tweedale}@unisa.edu.au

Abstract. The ongoing success of the Knowledge-Based Intelligent Information and Engineering Systems (KES) Centre has been stimulated via collaborated with industry and academia for many years. This Centre currently has adjunct personnel and advisors that mentor or collaborate with its students and staff from Defence Science and Technology Organisation (DSTO), BAE Systems (BAE), Boeing Australia Limited (BAL), Ratheon, Tenix, the University of Brighton, University of the West of Scotland, Loyola College in Maryland, University of Milano, Oxford University, Old Dominion University and University of Science Malaysia. Much of our research remains unpublished in the public domain due to these links and intellectual property rights. The list provided is non-exclusive and due to the diverse selection of research activities, only those relating to Intelligent Agent developments are presented.

Keywords: Computational Intelligence, Intelligent Agents, Multi-Agent Systems.

1 Introduction

The KES Centre held its first conference in 1997. This marked a new beginning in Knowledge-Based Engineering Systems, as this conference brought together researchers from around the world to discuss topics relating to this new and emerging area of engineering. Due to its great success the KES conference has now attained full international status. A full history of the KES International Conferences can be found on our web site[1]. The research directions of the KES are focused on modelling, analysis and design in the areas of Intelligent Information Systems, Physiological Sciences Systems, Electronic commerce and Service Engineering. The KES Centre aims to provide applied research support to the Information, Defence and Health Industries. The overall goal will be to synergies contributions from researchers in the diverse disciplines of Engineering, Information Technology, Science, Health, Commerce and Security Engineering. The research projects undertaken in the Centre include adaptive mobile robots, aircraft landing support, learning paradigms and teaming in Multi-Agent System (MAS).

Compare this with the 1st recorded conference relating to the science of Intelligent Agent (IA) itself, which dates back to Dartmouth in 1958. This is 39 years prior to

---

[1] http://www.unisa.edu.au/kes/International_conference/default.asp

C.-C. Chan et al. (Eds.): RSCTC 2008, LNAI 5306, pp. 13–20, 2008.

KES, although microcomputers only started to appear on desktops in the mid-eighties which is when KES was founded. Most of this research involved the progressive use of technology in science to collect and interpret data, prior to representing this as knowledge using "folk law" or "symbology". The wealth of data became unwieldy, forcing researchers to explore data-mining, warehousing and Knowledge Based System (KBS), however the key research domains remained focused on problem solving using formal/structured or reasoning systems [1]. This era was accompanied with an expansion in research into a variety of intelligent decision support systems that where created to derive greater confidence in the decision being generated [2]. The growing density of data had an overall effect on the efficiency of these systems. Conversely a series of measures where created to report on the performance of Decision Support System (DSS). Factors such as; accuracy, response time and explain-ability were raised as constraints to be considered before specifying courses of action [3]. Since the eighties Artificial Intelligence (AI) applications have concentrated on problem solving, machine vision, speech, natural language processing/translation, common-sense reasoning and robot control [4]. In the nineties there was a flurry of activity using "firmware" solutions to overcome speed and compiler complexities, however around the turn of the century, a return to distributed computing techniques has prevailed. Given his time over, John McCarthy would have labelled AI as "Computational Intelligence" [5]. Today the Windows/Mouse interface currently still dominates as the predominant Human Computer Interface (HCI), although it is acknowledged as being impractical for use with many mainstream AI applications.

2   Research Projects

Several of the projects undertaken by the KES Centre involved the use of intelligent paradigms as below:

- Coordination and Cooperation of Unmanned Air Vehicle Swarms in Hostile Environment,
- Communication and Learning in Multi-Agent Systems,
- Simulation of Pulsed Signals above 100 MHz in a Knowledge-Based Environment,
- Using Artificial Intelligence and Fusion Techniques in Target Detection,
- Intelligent decision support feedback using MAS in a Defence maintenance environment, and
- Improving Agent Communication in a Distributed Application Environment.

2.1   Coordination and Cooperation of Unmanned Air Vehicle Swarms in Hostile Environment

The aim of this research is to integrate coordination abilities into agent technology using the concepts of cooperation, collaboration and communication [6]. An example could include the coordination of a swarm of Unmanned Air Vehicles (UAVs) in a hostile environment. There has been substantial research conducted in this area, however the coordination aspect that have been implemented are either specific to an application or difficult to implement. As a result, a rigid and complex architecture is required. Implementing the concepts of cooperation, collaboration and communication in coordination may enhance performance; reduce complexity, and assist in applying coordination in a simple manner. The link between agent coordination and cooperation has been established and two principles have been developed from this link: Coordinative Cooperation and Cooperative Coordination. An architecture, known as the Agent Coordination and Cooperation Cognitive Model (AC^{3}M), is being developed which incorporates these principles into a MAS. It uses the concepts of Coordinative and Cooperative "events" to allow for each to be realized as a cognitive loop. The next approach is to incorporate the Beliefs, Desires, Intentions (BDI) at a physical level for control and link this to the Observe Orient Decide and Act (OODA) loop at a cognitive level for situation awareness and cooperation [7, 8].

2.2   Communication and Learning in Multi-agent Systems

This research involves encompassing communication and learning in multi-agent systems [9]. Firstly, we develop a hybrid agent teaming framework and analyze how to adapt the simulation system for investigating agent team architecture, learning abilities, and other specific behaviors. Secondly, we adopt the reinforcement learning algorithms to verify goal-oriented agents' competitive and cooperative learning abilities for decision-making. In doing so, a simulation test-bed is applied to test the learning algorithms in the specified scenarios. In addition, the function approximation technique, known as Tile Coding (TC), is used to generate value functions, which can avoid the value function growing exponentially with the number of the state values. Thirdly, Bayesian parameter learning algorithms in conjunction with reinforcement learning techniques are proposed for inferencing and reasoning in the cooperative learning. Finally, we will integrate the learning techniques with an agent teaming architecture with the abilities of coordination, cooperation, and dynamic role assignment. The ultimate goal of our research is to investigate the convergence and efficiency of the learning algorithms and then develop a hybrid agent teaming architecture.

2.3   Multiple UAV Communication in an Intelligent Environment

This research concentrates on using coordination and collaboration within UAV teams to accomplish better communication amongst UAV entities [10]. As the UAVs have limited sensor capabilities, cooperative control relies heavily on communication with appropriate neighbors. The advantages of coordinating and collaborating UAV teams include accomplishing the missions in a shorter period and successfully completing many goals simultaneously. This problem is of interest in UAV applications, as communication is often required between nodes that would not otherwise be able to communicate for instance because of range constraints or line-of-site obstructions. Efficient, reliable, low latency communication is required to fully realize and utilize the benefits of multi-vehicle teams. Achieving the leashing goal for instance in a more optimal way by knowledge sharing is one of the research goals. The design and development of an intelligent communication routing protocol for UAV applications that use heterogeneous networks is another goal of this research. The medium access layer will be modified for accessing the medium and for sensor scheduling. The intelligent routing protocol will be modified to accommodate the network layer. Reinforcement learning will be applied in order to add intelligence to the electronic leashing.

2.4   Simulation of Pulsed Signals Above 100 MHz in a Knowledge-Based Environment

The determination of the nature and identity of a pulsed electromagnetic radiation source has been a evolving for some decades [11]. To date the use of Knowledge Based techniques has not been examined to the same extent as in some other aspects. This research topic, considers the feasibility of the use of Artificial Intelligence techniques as support to the traditional techniques for extraction of data. To achieve this, analysis is performed of the nature of pulsed radiation sources and receiving system characteristics. The model used for traditional simulation is examined and used to generate selected key performance indicators. A more precise, temporal based model, which is considered more applicable to AI techniques evaluation, is generated and the same performance indicators are generated and subsequently compared with the traditional model, thus enabling conclusions to be drawn as to their respective merits. Finally, changes to the respective models, if appropriate, are examined and evaluated. Once a model has been developed, AI techniques will be used to assess the suitability of the model in the future evaluation of AI algorithms as a supplement, and as an alternative, to the traditional DSP methods. Techniques currently considered as suitable for this assessment include Fuzzy Logic, Neural Networks, Expert Systems and Evolutionary Computing.

2.5   Using Artificial Intelligence and Fusion Techniques in Target Detection

Automatic Target Recognition (ATR) is a problem which involves extraction of critical information from complex and uncertain data for which the traditional approaches of signal processing, pattern recognition, and rule based artificial intelligence (AI) techniques have been unable to provide adequate solutions. Target recognition of fixed signatures in stationary backgrounds is a straightforward task for which numerous effective techniques have been developed [12, 13]. If the target signatures and the background are variable in either a limited or known manner, more complex techniques such as using rule-based AI (i.e., expert systems) methods can be effective.

However, rule based AI systems exhibit brittle than robust behavior (i.e., there is great sensitivity to the specific assumptions and environments). When the target signatures or backgrounds vary in an unlimited or unknown manner, the traditional approaches have not been able to furnish appropriate solutions.

The aim of this project is to employ multiple sensors concurrently for detection and recognition using a suitable neural network paradigms. The data is fused in this processing scheme to exploit the spectral and geometric differences and arrive at a more reliable decision. However, there are any instances for multi-sensor fusion such as method for correlating non-simultaneous data from multiple, independent sensors and determination of correct classification of targets when there are conflicting reports. Fusion techniques are be used to improve the detection of man-made/artificial targets in multi-spectral or SAR images (taken on two different platforms with different view

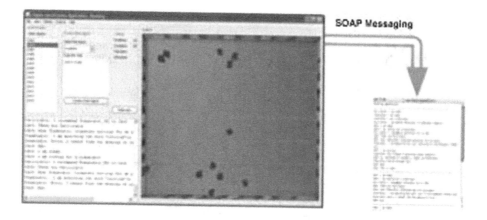

Fig. 1. Example of Agent Communications Demonstrator

angles), using spectral signature, shape/texture, a priori information, and surrounding geography.

2.6 Agent Communication in a Distributed Application Environment

As technology advances, humans are increasingly introducing delays and errors by the lack of response within system time limits. Human intervention must be minimized to avoid system conflicts while enabling the operators avoid repetitive, dull or dangerous tasks. Automation has become necessary in various applications. Since agents are not generally intelligent, they need to posses a capability to interoperate. They also need to interact, communicate, and share knowledge in order to successfully achieve their goal(s). We have found the dynamic nature of the Interface Description Language (IDL), invoked by Simple Object Access Protocol (SOAP) at run time, enables the application to adaptively configure its functionality in real time making the development of intelligent agent applications easier [14, 15]. This research aims to develop improved communication between distributed systems as shown in Figure 1.

This concept demonstrator is developed using Java to simulate this scenario and investigate the interaction, communication, and knowledge-sharing activities among agents within MASs. With the introduction of distributed computing, the problem of inter system communicated created a wide range of solution. The relationship between Web-Services Description Language (WSDL) and OWL-Services (OWL-S) is implied. Both languages are NOT covered in the same domain, however the overlap is obvious. The "service descriptions provide a powerful way of assembling information resources in contexts [16]". Threads, Agents and Distributed computing and reconfigurable silicon designs have attracted serious attention, forcing both industry and developers to reflect on existing paradigms in order to rethink the future. SOA makes it easier to product functional designs with limited functionality. The research conducted so far by KES has developed a blackboard design upon which segregated functions can be integrated into an application of aimed at achieving this goal. More research is required to

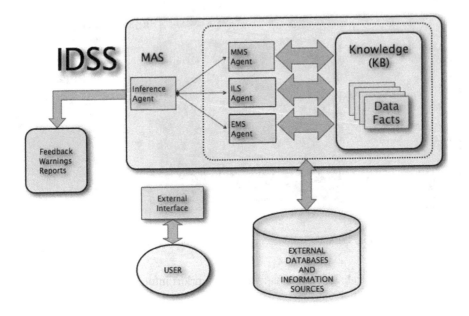

Fig. 2. Example of the Agent Communications Demonstrator

enable agents to communicate and cooperate and self organize in order to maximize the efficiency of any MAS.

2.7   Intelligent Decision Support Feedback Using MAS in a Defence Maintenance Environment

Safety and airworthiness of airborne platforms rest heavily on maintainability and reliability to maximize availability and reduce logistics down time. Maintenance data from test results rely heavily on paper trails and generally fail to provide preventive analysis. An expert system using intelligent agents could be employed to create an expert system in the form of an Information Management Systems (IMS). This concept would develop into an Intelligent Decision Support System (IDSS) that extrapolates forecasts and warnings as shown in Figure 2.

An Intelligent Decision Support System (IDSS) is required to provide adaptive automated responses for provisioning and maintenance of an increasing number of Defence platforms. Many are now emerging to support an increasing number of long-term maintenance contracts from within the private sector. Traditional methods of repair are already being modified to include automated testing although legacy platforms which still rely heavily on manual maintenance techniques. The conceptual development of a multi-agent expert system; referred to above as an IDSS. This system should be able to provide cognitive feedback to support reliability predictions and informed decision making to proactively minimize the issues of logistics down time, obsolescence and any associated risks [17, 18].

3   Future

The KES Centre has entered a new era fuelled by technology that has surpassed a milestone that has enabled renewed vigor into research activities that had previously stalled. It is widely acknowledged that current computer architectures have limited the wide spread implementation of many large scale, commercial quality applications in the artificial intelligence arena. The terms: automation, dynamic reconfiguration, learning, inference and self directed (intelligent) team behavior are all approaching a maturity level upon which large scaled, distributed applications, will become interoperable, spurred on by the technology leap required to surpass the barriers currently being experienced in many of these fields.

References

1. Bigus, J.P., Bigus, J.: Constructing Intelligent Agents Using Java. Professional Developer's Guide Series. John Wiley & Sons, Inc., New York (2001)
2. Yen, J., Fan, X., Sun, S., Hanratty, T., Dumer, J.: Agents with shared mental models for enhancing team decision makings. Decis. Support Syst. 41(3), 634–653 (2006)
3. Dhar, V., Stein, R.: Intelligent decision support methods: the science of knowledge work. Prentice-Hall, Inc., Upper Saddle River (1997)
4. Rich, E., Knight, K.: Artificial Intelligence. McGraw-Hill College, New York (1991)
5. Andresen, S.L.: John mccarthy: Father of ai. IEEE Intelligent Systems 17(5), 84–85 (2002)
6. Consoli, A., Tweedale, J., Jain, L.C.: An architecture for agent coordination and cooperation. In: Apolloni, B., Howlett, R.J., Jain, L.C. (eds.) KES 2007, Part III. LNCS (LNAI), vol. 4694, pp. 934–940. Springer, Heidelberg (2007)
7. Consoli, A., Tweedale, J., Jain, L.C.: An overview of agent coordination and cooperation. In: Gabrys, B., Howlett, R.J., Jain, L.C. (eds.) KES 2006. LNCS (LNAI), vol. 4253, pp. 497–503. Springer, Heidelberg (2006)
8. Consoli, A., Tweedale, J., Jain, L.C.: Cooperative agents in a hostile environment. In: India?. IEEE, Los Alamitos (in press, 2008)
9. Leng, J., Fyfe, J., Jain, L.C.: Reinforcement learning of competitive skills with soccer agents. In: Apolloni, B., Howlett, R.J., Jain, L.C. (eds.) KES 2007, Part I. LNCS (LNAI), vol. 4692, pp. 572–579. Springer, Heidelberg (2007)
10. Leng, J., Sathyaraj, B.M., Jain, L.C.: Temporal difference learning and simulated annealing for optimal control: A case study. In: Nguyen, N., Jo, G., Howlett, R., Jain, L. (eds.) KES-AMSTA 2008. LNCS (LNAI), vol. 4953, pp. 495–504. Springer, Heidelberg (2008)
11. Fitch, P.: Is there a role for artificial intelligence in future electronic support measures? In: Gabrys, B., Howlett, R.J., Jain, L.C. (eds.) KES 2006. LNCS (LNAI), vol. 4252, pp. 523–530. Springer, Heidelberg (2006)
12. Filippidis, A., Jain, L.C., Martin, N.S.: Multisensor data fusion for surface land-mine detection. IEEE Transactions on Systems, Man, and Cybernetics, Part C 30(1), 145–150 (2000)
13. Filippidis, A., Jain, L.C., Martin, N.M.: Using genetic algorithms and neural networks for surface land mine detection. IEEE Transactions on Signal Processing 47(1), 176–186 (1999)
14. Tweedale, J., Jain, L.C.: Interoperability with Multi-Agent Systems. Journal of Intelligence and Fuzzy System 19(6) (in press, 2008)
15. Tweedale, J., Jain, L.C.: Designing agents with dynamic capability. In: Breuker, J., de Mántaras, R.L., Dieng-Kuntz, R., Mizoguchi, R., Guarino, N., Musen, M., Kok, J., Pal, S., Liu, J., Zhong, N. (eds.) 8th Joint Conference on Knowledge Based Software Engineering (JCKBSE 2008), Piraeus, Greece. Frontiers in Artificial Intelligence and Applications. IOS Press, Amsterdam (in press, 2008)

16. Gibbins, N., Harris, S., Shadbolt, N.: Agent-based semantic web services. In: WWW 2003: Proceedings of the 12th international conference on World Wide Web, pp. 710–717. ACM, New York (2003)
17. Haider, K., Tweedale, J., Urlings, P., Jain, L.: Intelligent decision support system in defense maintenance methodologies. In: International Conference of Emerging Technologies (ICET 2006), Peshawar, pp. 560–567. IEEE Press, New York (2006)
18. Haider, K., Tweedale, J., Jain, L.C., Urlings, P.: Intelligent decision support feedback using multi-agent system in a defence maintenance environment. IJIIDS 1(3/4), 311–324 (2007)

On Irreducible Descriptive Sets of Attributes for Information Systems

Mikhail Moshkov[1], Andrzej Skowron[2], and Zbigniew Suraj[3]

[1] Institute of Computer Science, University of Silesia
Będzińska 39, 41-200 Sosnowiec, Poland
moshkov@us.edu.pl
[2] Institute of Mathematics, Warsaw University
Banacha 2, 02-097 Warsaw, Poland
skowron@mimuw.edu.pl
[3] Chair of Computer Science, University of Rzeszów
Rejtana 16A, 35-310 Rzeszów, Poland
zsuraj@univ.rzeszow.pl

Abstract. The maximal consistent extension $Ext(S)$ of a given information system S consists of all objects corresponding to attribute values from S which are consistent with all true and realizable rules extracted from the original information system S. An irreducible descriptive set for the considered information system S is a minimal (relative to the inclusion) set B of attributes which defines exactly the set $Ext(S)$ by means of true and realizable rules constructed over attributes from the considered set B. We show that there exists only one irreducible descriptive set of attributes. We also present a polynomial algorithm for this set construction. The obtained results will be useful for the design of concurrent data models from experimental data.

Keywords: rough sets, information systems, maximal consistent extensions, irreducible descriptive sets.

1 Introduction

Let $S = (U, A)$ be an information system [11], where U is a finite set of objects and A is a finite set of attributes defined on U. We identify objects and tuples of values of attributes on these objects. The information system S can be considered as a representation of a concurrent system: attributes are interpreted as local processes of the concurrent system, values of attributes – as states of local processes, and objects – as global states of the considered concurrent system. This idea is due to Pawlak [10].

Let $Rul(S)$ be the set of all true realizable rules in S of the kind

$$a_1(x) = b_1 \wedge \ldots \wedge a_{t-1}(x) = b_{t-1} \Rightarrow a_t(x) = b_t ,$$

where $a_1, \ldots, a_t \in A$ and $b_1, \ldots, b_t$ are values of attributes $a_1, \ldots, a_t$. *True* means that the rule is true for any object from U. *Realizable* means that the left hand

C.-C. Chan et al. (Eds.): RSCTC 2008, LNAI 5306, pp. 21–30, 2008.
© Springer-Verlag Berlin Heidelberg 2008

side of the rule is true for at least one object from U. Let $V(S)$ be the Cartesian product of ranges of attributes from A.

The knowledge encoded in a given information system S can be represented by means of rules from $Rul(S)$. Besides "explicit" global states, corresponding to objects from U, the concurrent system generated by the considered information system can also have "hidden" global states, i.e., tuples of attribute values from $V(S)$ not belonging to U but consistent with all rules from $Rul(S)$. Such "hidden" states can also be considered as realizable global states. This was a motivation for introducing in [15] the maximal consistent extensions of information systems with both "explicit" and "hidden" global states. More exactly, the maximal consistent extension of U is the set $Ext(S)$ of all objects from $V(S)$ for which each rule from $Rul(S)$ is true. The maximal consistent extensions of information systems were considered in [1,14,15,20,21].

In this paper, we study the problem of construction of an irreducible descriptive set of attributes. A set of attributes $B \subseteq A$ is called a descriptive set for S if there exists a set of rules $Q \subseteq Rul(S)$ constructed over the attributes from B only such that $Ext(S)$ coincides with the set of all objects from $V(S)$ for which all rules from Q are true. A descriptive set B for S is called irreducible if each proper subset of B is not a descriptive set for S. We prove that there exists only one irreducible descriptive set of attributes for S, and we present a polynomial in time algorithm for construction of this set. Let us recall that there is no polynomial in time algorithm for constructing the set $Ext(S)$ from a given information system S [5].

The obtained results will be useful for study of concurrent systems, generated by information systems [8,16,19,22].

For other issues on information systems and dependencies in information systems the reader is referred to, e.g., [2,3,7,9,12,13,17].

The paper consists of seven sections. Irreducible descriptive sets of attributes are considered in Sects. 2–6. Sect. 7 contains short conclusions.

2   Maximal Consistent Extensions

Let $S = (U, A)$ be an information system [11], where $U = \{u_1, \ldots, u_n\}$ is a set of objects and $A = \{a_1, \ldots, a_m\}$ is a set of attributes (functions defined on U). For simplicity of reasoning, we assume that for any two different numbers $i_1, i_2 \in \{1, \ldots, n\}$ tuples $(a_1(u_{i_1}), \ldots, a_m(u_{i_1}))$ and $(a_1(u_{i_2}), \ldots, a_m(u_{i_2}))$ are different. Hence, for $i = 1, \ldots, n$ we identify object $u_i \in U$ and corresponding tuple $(a_1(u_i), \ldots, a_m(u_i))$.

For $j = 1, \ldots, m$ let $V_{a_j} = \{a_j(u_i) : u_i \in U\}$. We assume that $|V_{a_j}| \geq 2$ for $j = 1, \ldots, m$.

We consider the set $V(S) = V_{a_1} \times \ldots \times V_{a_m}$ as the universe of objects and study extensions U^* of the set U such that $U \subseteq U^* \subseteq V(S)$. We assume that for any $a_j \in A$ and any $u \in V(S)$ the value $a_j(u)$ is equal to the j-th component of u.

Let us consider a rule

$$a_{j_1}(x) = b_1 \wedge \ldots \wedge a_{j_{t-1}}(x) = b_{t-1} \Rightarrow a_{j_t}(x) = b_t \ , \tag{1}$$

where $t \geq 1$, $a_{j_1}, \ldots, a_{j_t} \in A$, $b_1 \in V_{a_{j_1}}, \ldots, b_t \in V_{a_{j_t}}$, and numbers $j_1, \ldots, j_t$ are pairwise different. The rule (1) is called *true for an object* $u \in V(S)$ if there exists $l \in \{1, \ldots, t-1\}$ such that $a_{j_l}(u) \neq b_l$, or $a_{j_t}(u) = b_t$. The rule (1) is called *true* if it is true for any object from U. The rule (1) is called *realizable* if there exists an object $u_i \in U$ such that $a_{j_1}(u_i) = b_1, \ldots, a_{j_{t-1}}(u_i) = b_{t-1}$.

By $Rul(S)$ we denote the set of all rules each of which is true and realizable. By $Ext(S)$ we denote the set of all objects from $V(S)$ for which each rule from $Rul(S)$ is true. The set $Ext(S)$ is called the *maximal consistent extension of* U *relative to the set of rules* $Rul(S)$.

3   On Membership to $Ext(S)$

First, we recall a polynomial algorithm $\mathcal{B}_1$ from [4] which for a given information system $S = (U, A)$ and an element $u \in V(S)$ recognizes if this element belongs to $Ext(S)$ or not. Let $U = \{u_1, \ldots, u_n\}$ and $A = \{a_1, \ldots, a_m\}$.

Algorithm 1. Algorithm $\mathcal{B}_1$

Input   : Information system $S = (U, A)$,
          where $U = \{u_1, \ldots, u_n\}$, $A = \{a_1, \ldots, a_m\}$, and $u \in V(S)$.
Output: Return Yes if $u \in Ext(S)$, and No, otherwise.
for $i = 1, \ldots, n$ **do**
  | $M_i(u) \leftarrow \{j \in \{1, \ldots, m\} : a_j(u) = a_j(u_i)\}$;
end
for $i \in \{1, \ldots, n\}$ *and* $j \in \{1, \ldots, m\} \setminus M_i(u)$ **do**
  | $P_i^j(u) \leftarrow \{a_j(u_t) : u_t \in U$ *and* $a_l(u_t) = a_l(u)$ *for each* $l \in M_i(u)\}$;
end
if $|P_i^j(u)| \geq 2$ *for any* $i \in \{1, \ldots, n\}$ *and* $j \in \{1, \ldots, m\} \setminus M_i(u)$ **then**
  | return "Yes";
else
  | return "No";
end

Let us observe that using the indiscernibility relation $IND(A_i(u))$ [11], where $A_i(u) = \{a_l : l \in M_i(u)\}$, we obtain that $P_i^j(u) = a_j([u]_{IND(A_i(u))})$, i.e., $P_i^j(u)$ is equal to the image under a_j of the $A_i(u)$-indiscernibility class $[u]_{IND(A_i(u))}$ defined by u.

The considered algorithm is based on the following criterion.

Proposition 1. [4] *The relation* $u \in Ext(S)$ *holds if and only if* $|P_i^j(u)| \geq 2$ *for any* $i \in \{1, \ldots, n\}$ *and* $j \in \{1, \ldots, m\} \setminus M_i(u)$.

4   Separating Sets of Attributes

A set of attributes $B \subseteq A$ is called a *separating set for* $Ext(S)$ if for any two objects $u \in Ext(S)$ and $v \in V(S) \setminus Ext(S)$ there exists an attribute $a_j \in B$ such that $a_j(u) \neq a_j(v)$ or, which is the same, tuples u and v are different in the j-th component. A separating set for $Ext(S)$ is called *irreducible* if each its proper subset is not a separating set for $Ext(S)$.

It is clear that the set of irreducible separating sets for $Ext(S)$ coincides with the set of decision reducts for the decision system $D = (V(S), A, d)$, where for any $u \in V(S)$

$$d(u) = \begin{cases} 1, \text{ if } u \in Ext(S) \ , \\ 0, \text{ if } u \notin Ext(S) \ . \end{cases}$$

Let us show that the core for this decision system is a reduct. It means that D has exactly one reduct coinciding with the core. We denote by $C(Ext(S))$ the set of attributes $a_j \in A$ such that there exist two objects $u \in Ext(S)$ and $v \in V(S) \setminus Ext(S)$ which are different only in the j-th component. It is clear that $C(Ext(S))$ is the core for D, and $C(Ext(S))$ is a subset of each reduct for D.

Proposition 2. *The set $C(Ext(S))$ is a reduct for the decision system $D = (V(S), A, d)$.*

Proof. Let us consider two objects $u \in Ext(S)$ and $v \in V(S) \setminus Ext(S)$. Let us show that these objects are different on an attribute from $C(Ext(S))$. Let u and v be different in p components $j_1, \ldots, j_p$. Then there exists a sequence $u_1, \ldots, u_{p+1}$ of objects from $V(S)$ such that $u = u_1$, $v = u_{p+1}$, and for $i = 1, \ldots, p$ the objects u_i and u_{i+1} are different only in the component with the number j_i. Since $u_1 \in Ext(S)$ and $u_{p+1} \in V(S) \setminus Ext(S)$, there exists $i \in \{1, \ldots, p\}$ such that $u_i \in Ext(S)$ and $u_{i+1} \in V(S) \setminus Ext(S)$. Therefore, $a_{j_i} \in C(Ext(S))$. It is clear that u and v are different on the attribute a_{j_i}. Thus, $C(Ext(S))$ is a reduct for D.   □

From Proposition 2 it follows that $C(Ext(S))$ is the unique reduct for the decision system D. Thus, a set $B \subseteq A$ is a separating set for $Ext(S)$ if and only if $C(Ext(S)) \subseteq B$. One can show that $C(Ext(S)) \neq \emptyset$ if and only if $Ext(S) \neq V(S)$.

5   On Construction of $C(Ext(S))$

In this section, we present a polynomial in time algorithm for construction of $C(Ext(S))$. First, we define an auxiliary set $N(Ext(S))$. Next, we present a polynomial in time algorithm for constructing this set and finally we show that this auxiliary set $N(Ext(S))$ is equal to $C(Ext(S))$.

Let us define the set $N(Ext(S))$. An attribute $a_j \in A$ belongs to $N(Ext(S))$ if and only if there exist objects $u \in U$ and $v \in V(S) \setminus Ext(S)$ such that u and v are different only in the j-th component. Notice that the only difference in the definition of $N(Ext(S))$ in comparison with the definition of $C(Ext(S))$ is that the first condition for u. In the former case we require $u \in U$ and in the latter case $u \in Ext(S)$.

We now describe a polynomial algorithm $\mathcal{B}_2$ for the set $N(Ext(S))$ construction.

Algorithm 2. Algorithm $\mathcal{B}_2$

Input : Information system $S = (U, A)$, where $A = \{a_1, \ldots, a_m\}$.
Output: Set $N(Ext(S))$.
$N(Ext(S)) = \emptyset$;
for $u \in U$ **do**

    for $j \in \{1, \ldots, m\}$ and $b \in V_{a_j} \setminus \{b_j\}$, where $u = (b_1, \ldots, b_m)$ **do**
        $v \leftarrow (b_1, \ldots, b_{j-1}, b, b_{j+1}, \ldots, b_m)$;
        Apply algorithm $\mathcal{B}_1$ to v;
        if algorithm $\mathcal{B}_1$ returns "No" **then**
          | $N(Ext(S)) \leftarrow N(Ext(S)) \cup \{a_j\}$;
        end
    end
end

Theorem 1. $C(Ext(S)) = N(Ext(S))$.

Proof. Let $a_r \in A$. It is clear that if $a_r \in N(Ext(S))$ then $a_r \in C(Ext(S))$. We now show that if $a_r \notin N(Ext(S))$ then $a_r \notin C(Ext(S))$. To this end we must prove that for any two objects u and v from $V(S)$, if $u \in Ext(S)$ and v is different from u only in the r-th component then $v \in Ext(S)$.

Let us assume that $u \in Ext(S)$ and $v \in V(S)$ is different from u only in the r-th component. We now show that $v \in Ext(S)$.

Taking into account that $u \in Ext(S)$ and using Proposition 1 we conclude that $|P_i^j(u)| \geq 2$ for any $i \in \{1, \ldots, n\}$ and $j \in \{1, \ldots, m\} \setminus M_i(u)$.

We now show that $|P_i^j(v)| \geq 2$ for $i \in \{1, \ldots, n\}$ and $j \in \{1, \ldots, m\} \setminus M_i(v)$. Let us consider four cases.

1. Let $r \notin M_i(u)$ and $a_r(v) = a_r(u_i)$. Then $M_i(v) = M_i(u) \cup \{r\}$ and $j \neq r$. Since $|P_i^j(u)| \geq 2$, there exists an object $u_t \in U$ such that $a_l(u_t) = a_l(u)$ for each $l \in M_i(u)$ and $a_j(u_t) \neq a_j(u_i)$. If $a_r(v) = a_r(u_t)$ then $|P_i^j(v)| \geq 2$. Let $a_r(v) \neq a_r(u_t)$. We denote by w an object from $V(S)$ which is different from u_t only in the r-th component and for which $a_r(w) = a_r(v)$. Since $a_r \notin N(Ext(S))$, we have $w \in Ext(S)$. Let us assume that

$$K_i = \{s \in \{1, \ldots, m\} : a_s(w) = a_s(u_i)\} \ .$$

It is clear that $M_i(v) \subseteq K_i$ and $j \notin K_i$. Taking into account that $w \in Ext(S)$ and using Proposition 1 we conclude that there exists an object $u_p \in U$ such that $a_l(u_p) = a_l(u_i)$ for each $l \in K_i$ and $a_j(u_p) \neq a_j(u_i)$. Since $M_i(v) \subseteq K_i$, we obtain $|P_i^j(v)| \geq 2$.

2. Let $r \notin M_i(u)$ and $a_r(v) \neq a_r(u_i)$. Then $M_i(v) = M_i(u)$. Since $|P_i^j(u)| \geq 2$, there exists an object $u_t \in U$ such that $a_l(u_t) = a_l(u)$ for each $l \in M_i(u)$ and $a_j(u_t) \neq a_j(u_i)$. Taking into account that $M_i(v) = M_i(u)$ and $a_l(u_t) = a_l(v)$ for each $l \in M_i(u)$ we obtain $|P_i^j(v)| \geq 2$.

3. Let $r \in M_i(u)$ and $r \neq j$. Then $M_i(v) = M_i(u) \setminus \{r\}$. Since $|P_i^j(u)| \geq 2$, there exists an object $u_t \in U$ such that $a_l(u_t) = a_l(u)$ for each $l \in M_i(u)$ and $a_j(u_t) \neq a_j(u_i)$. It is clear that $a_l(u_t) = a_l(v)$ for each $l \in M_i(v)$ and $a_j(u_t) \neq a_j(u_i)$. Therefore, $|P_i^j(v)| \geq 2$.

4. Let $r \in M_i(u)$ and $r = j$. Then $M_i(v) = M_i(u) \setminus \{r\}$. By w we denote an object from $V(S)$ which is different from u_i only in the r-th component. Since $a_r \notin N(Ext(S))$, we have $w \in Ext(S)$. Using Proposition 1, one can show that there exists an object $u_p \in U$ which is different from u_i only in the r-th component. It is clear that $a_l(u_p) = a_l(v)$ for each $l \in M_i(v)$, and $a_r(u_p) \neq a_r(u_i)$. Therefore, $|P_i^j(v)| \geq 2$.

Using Proposition 1, we obtain $v \in Ext(S)$. Thus, $a_r \notin C(Ext(S))$. $\qquad\square$

6   Descriptive Sets of Attributes

In this section, we show that the maximal consistent extension $Ext(S)$ of a given information system S cannot be defined by any system of true and realizable rules in S constructed over a set of attributes not including $C(Ext(S))$.

Proposition 3. *Let Q be a set of true realizable rules in S such that the set of objects from $V(S)$, for which any rule from Q is true, coincides with $Ext(S)$, and let B be the set of attributes from A occurring in rules from Q. Then $C(Ext(S)) \subseteq B$.*

Proof. Let us assume the contrary, i.e., $a_j \notin B$ for some attribute $a_j \in C(Ext(S))$. Since $a_j \in C(Ext(S))$, there exist objects $u \in Ext(S)$ and $v \in V(S) \setminus Ext(S)$ which are different only in the component with the number j. Let us consider a rule from Q which is not true for the object v. Since this rule does not contain the attribute a_j, the considered rule is not true for u which is impossible. $\qquad\square$

Now, we will show that using true realizable rules in S with attributes from $C(Ext(S))$ only it is possible to describe exactly the set $Ext(S)$.

Proposition 4. *There exists a set Q of true realizable rules in S such that the set of objects from $V(S)$, for which any rule from Q it true, coincides with $Ext(S)$, and rules from Q use only attributes from $C(Ext(S))$.*

Proof. Let us consider an arbitrary rule from the set $Rul(S)$. Let, for the definiteness, this will be the rule

$$a_1(x) = b_1 \wedge \ldots \wedge a_{t-1}(x) = b_{t-1} \Rightarrow a_t(x) = b_t \ . \tag{2}$$

We show that $a_t \in C(Ext(S))$. Let us assume the contrary, i.e., $a_t \notin C(Ext(S))$. Since (2) is realizable, there exists an object $u_i \in U$ such that

$$a_1(u_i) = b_1, \ldots, a_{t-1}(u_i) = b_{t-1} .$$

Since (2) is true, $a_t(u_i) = b_t$. Using Theorem 1, we conclude that $a_t \notin N(Ext(S))$. Let w be an object from $V(S)$ which is different from u_i only in the component with the number t. Since $a_t \notin N(Ext(S))$, we have $w \in Ext(S)$. Using Proposition 1, we conclude that there exists an object $u_p \in U$ which is different from u_i only in the component with the number t. It is clear that the rule (2) is not true for u_p which is impossible. Thus, $a_t \in C(Ext(S))$.

Let us assume that there exists $j \in \{1, \ldots, t-1\}$ such that $a_j \notin C(Ext(S))$. Now, we consider the rule

$$\bigwedge_{l \in \{1, \ldots, t-1\} \setminus \{j\}} a_l(x) = b_l \Rightarrow a_t(x) = b_t . \tag{3}$$

We show that this rule belongs to $Rul(S)$. Since (2) is realizable, (3) is realizable too. We now show that (3) is true. Let us assume the contrary, i.e., there exists object $u_i \in U$ for which (3) is not true. It means that $a_l(u_i) = b_l$ for any $l \in \{1, \ldots, t-1\} \setminus \{j\}$, and $a_t(u_i) \neq b_t$. Since (2) is true, $a_j(u_i) \neq b_j$. Let us consider the object $w \in V(S)$ such that w is different from u_i only in the j-th component, and $a_j(w) = b_j$. Taking into account that $a_j \notin C(Ext(S))$ we obtain $w \in Ext(S)$, but this is impossible. Since (2) is true, (2) must be true for any object from $Ext(S)$. However, (2) is not true for w.

Thus, if we remove from the left hand side of a rule from $Rul(S)$ all conditions with attributes from $A \setminus C(Ext(S))$ we obtain a rule from $Rul(S)$ which uses only attributes from $C(Ext(S))$. We denote by $Rul^*(S)$ the set of all rules from $Rul(S)$ which use only attributes from $C(Ext(S))$.

It is clear that the set of objects from $V(S)$, for which each rule from $Rul^*(S)$ is true, contains all objects from $Ext(S)$. Let $u \in V(S) \setminus Ext(S)$. Then there exists a rule from $Rul(S)$ which is not true for u. If we remove from the left hand side of this rule all conditions with attributes from $A \setminus C(Ext(S))$ we obtain a rule from $Rul^*(S)$ which is not true for u. Therefore, the set of objects from $V(S)$, for which each rule from $Rul^*(S)$ is true, coincides with $Ext(S)$. Thus, as the set Q we can take the set of rules $Rul^*(S)$. □

We will say that a subset of attributes $B \subseteq A$ is a *descriptive set for S* if there exists a set of rules $Q \subseteq Rul(S)$ that uses only attributes from B, and the set of objects from $V(S)$, for which each rule from Q is true, coincides with $Ext(S)$. A descriptive set B will be called *irreducible* if each proper subset of B is not a descriptive set for S. Next statement follows immediately from Propositions 3 and 4.

Theorem 2. *The set $C(Ext(S))$ is the unique irreducible descriptive set for S.*

From Theorem 1 it follows that $C(Ext(S)) = N(Ext(S))$. The algorithm $\mathcal{B}_2$ allows us to construct the set $N(Ext(S))$ in polynomial time.

7 Descriptions of $Ext(S)$ and $Rul(S)$

In this section, we outline some problems of more compact description of sets $Ext(S)$ and $Rul(S)$ which we would like to investigate in our further study.

Let us start from a proposal for (approximate) description of maximal extensions.

We consider an extension of the language of boolean combinations of descriptors [13] of a given information system by taking instead of descriptors of the form $a = v$ over a given information system $S = (U, A)$, where $a \in A$, $v \in V_a$, and V_a is the set of values of a, their generalization to $a \in W$ where W is a nonempty subset of V_a. Such new descriptors are called generalized descriptors. The semantics of the generalized descriptor $a \in W$ relative to a given information system $S = (U, A)$ is defined by the set $\|a \in W\|_{V(S)} = \{u \in V(S) : a(u) \in W\}$ or by $\|a \in W\|_S \cap U$, if one would like to restrict attention to the set U only. This semantics can be extended, in the standard way, on boolean combination of descriptors defined by classical propositional connectives, i.e., conjunction, disjunction, and negation. Let us consider boolean combinations of generalized descriptors defined by conjunctions of generalized descriptors only. We call them as *templates*. Now, we define decision systems with conditional attributes defined by generalized descriptors. Let us consider a sample U' of objects from $V(S) \setminus U$ and the set GD of all binary attributes $a \in W$ such that $(a \in W)(u) = 1$ if and only if $a(u) \in W$, where $u \in V(S)$. Next, we consider decision systems of the form $DS_B = (U \cup U', B, d)$, where $B \subseteq GD$ and $d(u) = 1$ if and only if $u \in Ext(S)$. Using such decision systems one can construct classifiers for the set $Ext(S)$. The problem is to search for classifiers with the high quality of classification. Searching for such classifiers can be based on the minimal length principle. For example, for any DS_B one can measure the size of classifier by the size of the generated set of decision rules. The size of a set of decision rules can be defined as the sum of sizes of the left hand sides of decision rules from the set. Observe that the left hand sides of the considered decision rules are templates, i.e., conjunctions of generalized descriptors. In this way, some approximate but compact descriptions of $Ext(S)$ by classifiers can be obtained. Another possibility is to use lazy classifiers for $Ext(S)$ based on DS_B decision systems.

Dealing with all rules of a given kind, e.g., all realizable and true deterministic rules [6], one may face problems related to the large size of the set of such rules in a given information system. Hence, it is necessary to look for more compact description of such sets of rules. It is worthwhile mentioning that this problem is of great importance in data and knowledge visualization.

A language which can help to describe the rule set $Rul(S)$ in a more compact way can be defined by dependencies, i.e., expressions of the form $B \longrightarrow C$, where $B, C \subseteq A$ (see, e.g., [13]). A dependency $B \longrightarrow C$ is true in S, in symbols $B \longrightarrow_S C = 1$, if and only if there is a functional dependency between B and C in S what can be expressed using the positive region by $POS_B(C) = U$. Certainly, each true in S dependency $B \longrightarrow C$ in S is representing a set of deterministic, realizable and true decision rules in S. The aim is to select dependencies true in S which are representing as many as possible rules from the given rule set

$Rul(S)$. For example, in investigating decompositions of information systems [16,19,20] some special dependencies in a given information system called as *components* were used. One could also use dependencies called as association reducts [18]. The remaining rules from $Rul(S)$ set which are not represented by the chosen functional dependencies can be added as *links* between components. They are interpreted in [16,19,20] as constraints or interactions between modules defined by components. The selected dependencies and links create a covering of $Rul(S)$. Assuming that a quality measure for such coverings was fixed, one can consider the minimal exact (or approximate) covering problem for $Rul(S)$ set by functional dependencies from the selected set of dependencies and some rules from $Rul(S)$.

Yet another possibility is to search for minimal subsets of a given $Rul(S)$ from which $Rul(S)$ can be generated using, e.g., some derivation rules.

8 Conclusions

We proved that for any information system S there exists only one irreducible descriptive set of attributes, and we proposed a polynomial in time algorithm for this set construction. We plan to use the obtained results in applications of information systems to analysis and design of concurrent systems.

Acknowledgements

The research has been supported by the grant N N516 368334 from Ministry of Science and Higher Education of the Republic of Poland and by the grant Innovative Economy Operational Programme 2007-2013 (Priority Axis 1. Research and development of new technologies) managed by Ministry of Regional Development of the Republic of Poland.

References

1. Delimata, P., Moshkov, M., Skowron, A., Suraj, Z.: Inhibitory Rules in Data Analysis. A Rough Set Approach. Springer, Heidelberg (in press, 2008)
2. Düntsch, I., Gediga, G.: Algebraic Aspects of Attribute Dependencies in Information Systems. Fundamenta Informaticae 29(1-2), 119–134 (1997)
3. Marek, W., Pawlak, Z.: Rough Sets and Information Systems. Fundamenta Informaticae 7(1), 105–116 (1984)
4. Moshkov, M., Skowron, A., Suraj, Z.: On Testing Membership to Maximal Consistent Extensions of Information Systems. In: Greco, S., Hata, Y., Hirano, S., Inuiguchi, M., Miyamoto, S., Nguyen, H.S., Słowinski, R. (eds.) RSCTC 2006. LNCS (LNAI), vol. 4259, pp. 85–90. Springer, Heidelberg (2006)
5. Moshkov, M., Skowron, A., Suraj, Z.: On Maximal Consistent Extensions of Information Systems. In: Conference Decision Support Systems, Zakopane, Poland, December 2006, vol. 1, pp. 199–206. University of Silesia, Katowice (2007)

6. Moshkov, M., Skowron, A., Suraj, Z.: Maximal Consistent Extensions of Information Systems Relative to Their Theories. Information Sciences 178(12), 2600–2620 (2008)
7. Novotný, J., Novotný, M.: Notes on the Algebraic Approach to Dependence in Information Systems. Fundamenta Informaticae 16, 263–273 (1992)
8. Pancerz, K., Suraj, Z.: Synthesis of Petri Net Models: A Rough Set Approach. Fundamenta Informaticae 55, 149–165 (2003)
9. Pawlak, Z.: Information Systems: Theoretical Foundations. WNT, Warsaw (1983) (in Polish)
10. Pawlak, Z.: Concurrent Versus Sequential – The Rough Sets Perspective. Bulletin of the EATCS 48, 178–190 (1992)
11. Pawlak, Z.: Rough Sets – Theoretical Aspects of Reasoning about Data. Kluwer Academic Publishers, Dordrecht (1991)
12. Pawlak, Z., Rauszer, C.: Dependency of Attributes in Information Systems. Bull. Polish. Acad. Sci. Math. 9-10, 551–559 (1985)
13. Pawlak, Z., Skowron, A.: Rudiments of Rough Sets. Information Sciences 177(1), 3–27 (2007); Rough Sets: Some Extensions. Information Sciences 177(1), 28–40 (2007); Rough Sets and Boolean Reasoning. Information Sciences 177(1), 41–73 (2007)
14. Rząsa, W., Suraj, Z.: A New Method for Determining of Extensions and Restrictions of Information Systems. In: Alpigini, J.J., Peters, J.F., Skowronek, J., Zhong, N. (eds.) RSCTC 2002. LNCS (LNAI), vol. 2475, pp. 197–204. Springer, Heidelberg (2002)
15. Skowron, A., Suraj, Z.: Rough Sets and Concurrency. Bulletin of the Polish Academy of Sciences 41, 237–254 (1993)
16. Skowron, A., Suraj, Z.: Discovery of Concurrent Data Models from Experimental Tables: A Rough Set Approach. In: First International Conference on Knowledge Discovery and Data Mining, pp. 288–293. AAAI Press, Menlo Park (1995)
17. Skowron, A., Stepaniuk, J., Peters, J.F.: Rough Sets and Infomorphisms: Towards Approximation of Relations in Distributed Environments. Fundamenta Informaticae 54(1-2), 263–277 (2003)
18. Ślęzak, D.: Association Reducts: A Framework for Mining Multi-attribute Dependencies. In: Hacid, M.S., Murray, N.V., Ras, Z.W., Tsumoto, S. (eds.) ISMIS 2005. LNCS (LNAI), vol. 3488, pp. 354–363. Springer, Heidelberg (2005)
19. Suraj, Z.: Discovery of Concurrent Data Models from Experimental Tables: A Rough Set Approach. Fundamenta Informaticae 28, 353–376 (1996)
20. Suraj, Z.: Some Remarks on Extensions and Restrictions of Information Systems. In: Ziarko, W., Yao, Y.Y. (eds.) RSCTC 2000. LNCS (LNAI), vol. 2005, pp. 204–211. Springer, Heidelberg (2001)
21. Suraj, Z., Pancerz, K.: A New Method for Computing Partially Consistent Extensions of Information Systems: A Rough Set Approach. In: 11th International Conference on Information Processing and Management of Uncertainty in Knowledge-Based Systems, Editions E.D.K., Paris, vol. III, pp. 2618–2625 (2006)
22. Suraj, Z., Pancerz, K.: Reconstruction of Concurrent System Models Described by Decomposed Data Tables. Fundamenta Informaticae 71, 121–137 (2006)

Dominance-Based Rough Set Approach and Bipolar Abstract Rough Approximation Spaces

Salvatore Greco[1], Benedetto Matarazzo[1], and Roman Słowiński[2]

[1] Faculty of Economics, University of Catania,
Corso Italia, 55, 95129 Catania, Italy
[2] Institute of Computing Science, Poznań University of Technology,
60-965 Poznań, and Systems Research Institute,
Polish Academy of Sciences, 01-447 Warsaw, Poland

Abstract. We take into consideration Dominance-based Rough Set Approach and its recently proposed algebraic modeling in terms of bipolar de Morgan Brower-Zadeh distributive lattice. On this basis we introduce the concept of bipolar approximation space and we show how it can be induced from a bipolar quasi Brower-Zadeh lattice.

1 Introduction

In order to handle monotonic relationships between premises and conclusions, such as "the greater the mass and the smaller the distance, the greater the gravity", "the more a tomato is red, the more it is ripe" or "the better the school marks of a pupil, the better his overall classification", Greco, Matarazzo and Słowiński [3,4,5,6,7] have proposed the Dominance-based Rough Set Approach (DRSA), where dominance relation is used instead of indiscernibility relation originally proposed for the classical rough set approach [9,10]. Recently, an algebraic model of DRSA in terms of bipolar de Morgan Brower-Zadeh distributive lattice has been proposed in [8]. It is a generalization of the de Morgan Brower-Zadeh distributive lattice [2], proposed to characterize the classical rough set approach in [1]. In this paper, we go further in this direction, introducing the concept of bipolar approximation space being for DRSA the counterpart of the approximation space proposed for the classical rough set approach [1]. We prove that a bipolar approximation space can be induced from any bipolar quasi Brower-Zadeh lattice of which we investigate the properties. The paper is organized as follows. The next section presents the DRSA approximations. The third section recalls the de Morgan Brower-Zadeh distributive lattice and the modeling of the classical rough set approach in its terms. The fourth section introduces the bipolar approximation space and shows how it can be induced from a bipolar quasi Brower-Zadeh lattice. The last section contains conclusions.

2 Dominance-Based Rough Set Approach

In this section, we recall the Dominance-based Rough Set Approach [4], taking into account, without loss of generality, the case of rough approximation of fuzzy sets [7].

C.-C. Chan et al. (Eds.): RSCTC 2008, LNAI 5306, pp. 31–40, 2008.

A *fuzzy information base* is the 3-tuple $\mathbf{B} =< U, F, \varphi >$, where U is a finite set of *objects* (universe), $F=\{f_1, f_2, ..., f_m\}$ is a finite set of *properties*, and $\varphi : U \times F \to [0,1]$ is a function such that $\varphi(x, f_h) \in [0,1]$ expresses the degree in which object x has property f_h. Therefore, each object x from U is described by a vector

$$Des_F(x) = [\varphi(x, f_1), \ldots, \varphi(x, f_m)]$$

called *description* of x in terms of the evaluations of the properties from F; it represents the available information about x. Obviously, $x \in U$ can be described in terms of any non-empty subset $G \subseteq F$, and in this case we have

$$Des_G(x) = [\varphi(x, f_h), f_h \in G].$$

Let us remark that the concept of fuzzy information base can be considered as a generalization of the concept of property system [11]. Indeed, in a property system an object may either possess a property or not, while in the fuzzy information base an object may possess a property in some degree between 0 and 1.

With respect to any $G \subseteq F$, we can define the *dominance relation* D_G as follows: for any $x, y \in U$, x dominates y with respect to G (denoted as xD_Gy) if, for any $f_h \in G$,

$$\varphi(x, f_h) \geq \varphi(y, f_h).$$

For any $x \in U$ and for each non-empty $G \subseteq F$, let

$$D_G^+ (x) = \{y \in U : yD_Gx\}, \quad D_G^- (x) = \{y \in U : xD_Gy\}.$$

Given $G \subseteq F$, for any $X \subseteq U$, we can define its *upward lower approximation* $\underline{G}^{(>)}(X)$ and its *upward upper approximation* $\overline{G}^{(>)}(X)$ as:

$$\underline{G}^{(>)}(X) = \left\{x \in U : D_G^+(x) \subseteq X\right\},$$

$$\overline{G}^{(>)}(X) = \left\{x \in U : D_G^-(x) \cap X \neq \emptyset\right\}.$$

Analogously, given $G \subseteq F$, for any $X \subseteq U$, we can define its *downward lower approximation* $\underline{G}^{(<)}(X)$ and its *downward upper approximation* $\overline{G}^{(<)}(X)$ as:

$$\underline{G}^{(<)}(X) = \left\{x \in U : D_G^-(x) \subseteq X\right\},$$

$$\overline{G}^{(<)}(X) = \left\{x \in U : D_G^+(x) \cap X \neq \emptyset\right\}.$$

Let us observe that in the above definition of rough approximations $\underline{G}^{(>)}(X)$, $\overline{G}^{(>)}(X)$, $\underline{G}^{(<)}(X)$, $\overline{G}^{(<)}(X)$, the elementary sets, which in the classical rough set theory are equivalence classes of the indiscernibility relation, are the sets $D_G^+(x)$ and $D_G^-(x)$, $x \in U$.

The rough approximations $\underline{G}^{(>)}(X), \overline{G}^{(>)}(X), \underline{G}^{(<)}(X), \overline{G}^{(<)}(X)$ can be used to analyze data relative to gradual membership of objects to some concepts representing properties of objects and their assignment to decision classes. This analysis takes into account the following monotonicity principle: "the greater the degree to which an object has properties from $G \subseteq F$, the greater its degree of membership to a considered class". This principle can be formalized as follows. Let us consider a fuzzy set X in U, characterized by the membership function $\mu_X : U \to [0,1]$. This fuzzy set represents a class of interest, such that function μ specifies a graded membership of objects from U to considered class X. For each cutting level $\alpha \in [0,1]$, we can consider the following sets

- weak upward cut of fuzzy set X:

$$X^{\geq \alpha} = \{x \in U : \mu(x) \geq \alpha\},$$

- strict upward cut of fuzzy set X:

$$X^{> \alpha} = \{x \in U : \mu(x) > \alpha\},$$

- weak downward cut of fuzzy set X:

$$X^{\leq \alpha} = \{x \in U : \mu(x) \leq \alpha\},$$

- strict upward cut of fuzzy set X:

$$X^{< \alpha} = \{x \in U : \mu(x) < \alpha\}.$$

Let us remark that, for any fuzzy set X and for any $\alpha \in [0,1]$, we have that

$$U - X^{\geq \alpha} = X^{< \alpha}, \quad U - X^{\leq \alpha} = X^{> \alpha},$$

$$U - X^{> \alpha} = X^{\leq \alpha}, \quad U - X^{< \alpha} = X^{\geq \alpha}.$$

Given a family of fuzzy sets $\mathbf{X} = \{X_1, X_2, ..., X_p\}$ on U, whose respective membership functions are $\mu_1, \mu_2, ..., \mu_p$, let $P^{>}(\mathbf{X})$ be the set of all the sets obtained through unions and intersections of weak and strict upward cuts of fuzzy sets from $\mathbf{X}$. Analogously, let $P^{<}(\mathbf{X})$ be the set of all the sets obtained through unions and intersections of weak and strict downward cuts of fuzzy sets from $\mathbf{X}$.

$P^{>}(\mathbf{X})$ and $P^{<}(\mathbf{X})$ are closed under set union and set intersection operations, i.e. for all $Y_1, Y_2 \in P^{>}(\mathbf{X})$, $Y_1 \cup Y_2$ and $Y_1 \cap Y_2$ belong to $P^{>}(\mathbf{X})$, as well as for all $W_1, W_2 \in P^{<}(\mathbf{X})$, $W_1 \cup W_2$ and $W_1 \cap W_2$ belong to $P^{<}(\mathbf{X})$. Observe, moreover, that the universe U and the empty set $\emptyset$ belong both to $P^{>}(\mathbf{X})$ and to $P^{<}(\mathbf{X})$ because, for any fuzzy set $X_i \in \mathbf{X}$,

$$U = X_i^{\geq 0} = X_i^{\leq 1}$$

and

$$\emptyset = X_i^{> 1} = X_i^{< 0}.$$

3   Bipolar de Morgan Brower-Zadeh Distributive Lattices

A system $\langle \Sigma, \Sigma^+, \Sigma^-, \wedge, \vee, '^+, '^-, {}^{\sim +}, {}^{\sim -}, 0, 1 \rangle$ is a *bipolar quasi Brower-Zadeh distributive lattice* if the following properties (1b)-(4b) hold:

(1b) Σ is a distributive lattice with respect to the join and the meet operations $\vee$ and $\wedge$

(1b') $\Sigma^+, \Sigma^- \subseteq \Sigma$ are distributive lattices with respect to the join and the meet operations $\vee$ and $\wedge$. Σ is bounded by the least element 0 and the greatest element 1, which implies that also Σ^+ and Σ^- are bounded.

(2b) The unary operations $'^+ : \Sigma^+ \to \Sigma^-$ and $'^- : \Sigma^- \to \Sigma^+$ are Kleene (also Zadeh or fuzzy) bipolar complementation, that is, for arbitrary $a, b \in \Sigma^+$ and $c, d \in \Sigma^-$,

(K1b) $a'^{+'-} = a$, $\quad c'^{-'+} = c$,

(K2b) $(a \vee b)'^+ = a'^+ \wedge b'^+$, $\quad (c \vee d)'^- = c'^- \wedge d'^-$,

(K3b) $a \wedge a'^+ \le b \vee b'^+$, $\quad c \wedge c'^- \le d \vee d'^-$.

(3b) The unary operations $^{\sim +} : \Sigma^+ \to \Sigma^-$ and $^{\sim -} : \Sigma^- \to \Sigma^+$ are Brower (or intuitionistic) bipolar complementations, that is, for arbitrary $a, b \in \Sigma^+$ and $c, d \in \Sigma^-$,

(B1b) $a \wedge a^{\sim + \sim -} = a$, $\quad c \wedge c^{\sim - \sim +} = c$

(B2b) $(a \vee b)^{\sim +} = a^{\sim +} \wedge b^{\sim +}$, $\quad (c \vee d)^{\sim -} = c^{\sim -} \wedge d^{\sim -}$,

(B3b) $a \wedge a^{\sim +} = 0$, $\quad c \wedge c^{\sim -} = 0$.

(4b) Complementation $'^+$ and complementation $^{\sim +}$ on one hand, and complementation $'^-$ and complementation $^{\sim -}$ on the other hand, are linked by the interconnection rule, that is, for arbitrary $a \in \Sigma^+$ and arbitrary $b \in \Sigma^-$:

(in-b) $a^{\sim +} \le a'^+$, $\quad b^{\sim -} \le b'^-$.

A structure $\langle \Sigma, \Sigma^+, \Sigma^-, \wedge, \vee, '^+, '^-, {}^{\sim +}, {}^{\sim -}, 0, 1 \rangle$ is a *bipolar Brower-Zadeh distributive lattice* if it is a bipolar quasi Brower-Zadeh distributive lattice satisfying the stronger interconnection rule, that is, for arbitrary $a \in \Sigma^+$ and arbitrary $b \in \Sigma^-$:

(s-in-b) $a^{\sim + \sim -} = a^{\sim + '-}$, $\quad b^{\sim - \sim +} = b^{\sim - '+}$.

A bipolar Brower-Zadeh distributive lattice is a*bipolar de Morgan Brower-Zadeh distributive lattice*, if it satisfies also the $\vee$ de Morgan property, that is, for arbitrary $a, b \in \Sigma^+$ and $c, d \in \Sigma^-$:

(B2a-b) $(a \wedge b)^{\sim +} = a^{\sim +} \vee b^{\sim +}$, $\quad (c \wedge d)^{\sim -} = c^{\sim -} \vee d^{\sim -}$.

The bipolar de Morgan Brower-Zadeh distributive lattice is an algebraic structure which can be given to the collection of all rough approximations within the Dominance-based Rough Set Approach as follows. Fixed $G \subseteq F$, for any $X \subseteq U$, let us consider the pairs $\left\langle \underline{G}^{(\le)}(X), U - \overline{G}^{(\le)}(X) \right\rangle$ and $\left\langle \underline{G}^{(\ge)}(X), U - \overline{G}^{(\ge)}(X) \right\rangle$, and the sets

$$B = \{(I, E) : I, E \subseteq U \text{ such that } I \cap E = \emptyset\},$$

$$B^- = \left\{ (I, E) : \exists X \subseteq U \text{ for which } I = \underline{G}^{(\leq)}(X) \text{ and } E = U - \overline{G}^{(\leq)}(X) \right\},$$

$$B^+ = \left\{ (I, E) : \exists X \subseteq U \text{ for which } I = \underline{G}^{(\geq)}(X) \text{ and } E = U - \overline{G}^{(\geq)}(X) \right\}.$$

The following result holds.

Theorem 1 [8]. The structure $\langle B, B^+, B^-, \sqcap, \sqcup, {}^{--}, {}^{-+}, {}^{\approx-}, {}^{\approx+}, \langle \emptyset, U \rangle, \langle U, \emptyset \rangle \rangle$, where for any $\langle I_1, E_1 \rangle, \langle I_2, E_2 \rangle \in B$, $\langle I_3, E_3 \rangle \in B^-$, $\langle I_4, E_4 \rangle \in B^+$,

$$\langle I_1, E_1 \rangle \sqcap \langle I_2, E_2 \rangle := \langle I_1 \cap I_2, E_1 \cup E_2 \rangle,$$

$$\langle I_1, E_1 \rangle \sqcup \langle I_2, E_2 \rangle := \langle I_1 \cup I_2, E_1 \cap E_2 \rangle,$$

$$\langle I_3, E_3 \rangle^{--} := \langle E_3, I_3 \rangle, \quad \langle I_4, E_4 \rangle^{-+} = \langle E_4, I_4 \rangle,$$

$$\langle I_3, E_3 \rangle^{\approx-} := \langle E_3, U - E_3 \rangle, \quad \langle I_4, E_4 \rangle^{\approx+} = \langle E_4, U - E_4 \rangle,$$

is a bipolar de Morgan Brower-Zadeh distributive lattice. □

4 Bipolar Approximation Space Induced from Bipolar Quasi Brower-Zadeh Lattice

Generalizing to the context of DRSA the concept of an abstract generalized approximation space [1], we introduce the concept of *bipolar generalized approximation space* as a structure

$$\langle \Sigma, \leq, \Sigma^+, \Sigma^-, \mathcal{O}^+(\pm^+), \mathcal{O}^-(\pm^-), \mathcal{C}^+(\pm^+), \mathcal{C}^-(\pm^-), i^+, i^-, o^+, o^- \rangle,$$

where

1. $\langle \Sigma, \leq, 0, 1 \rangle$ is an abstract lattice with respect to the partial order relation $\leq$, bounded by the first (or minimum) element 0 ($\forall x \in \Sigma, 0 \leq x$) and the last (maximum) element 1 ($\forall x \in \Sigma, x \leq 1$).
2. $\Sigma^+, \Sigma^- \subseteq \Sigma$ such that $0, 1 \in \Sigma^+$ and $0, 1 \in \Sigma^-$, $\langle \Sigma^+, \leq, 0, 1 \rangle$ and $\langle \Sigma^-, \leq, 0, 1 \rangle$ are abstract lattices with respect to the partial order relation $\leq$.
3. $\mathcal{O}^+(\pm^+)$ is a sublattice of Σ^+ consisting of all available *open* (also, *inner*) *definable* elements of Σ^+. Analogously, $\mathcal{O}^-(\pm^-)$ is a sublattice of Σ^- consisting of all available *open* (also, *inner*) *definable* elements of Σ^-.
4. $\mathcal{C}^+(\pm^+)$ is a sublattice of Σ^+ consisting of all available *closed* (also, *outer*) *definable* elements of Σ^+. Analogously, $\mathcal{C}^-(\pm^-)$ is a sublattice of Σ^- consisting of all available *closed* (also, *outer*) *definable* elements of Σ^-.
5. $i^+ : \Sigma^+ \to \mathcal{O}^+(\pm^+)$ is the *upward inner approximation mapping* associating to any approximable elements $x \in \Sigma^+$ the *lower* (or *inner*) approximation $i^+(x) \in \mathcal{O}^+(\pm^+)$, i.e. an upward open definable element such that

$$i^+(x) \leq x,$$

$$\forall \alpha \in \mathcal{O}^+(\pm^+), (\alpha \leq x \Rightarrow \alpha \leq i^+(x)),$$

i.e.,
$$i^+(x) := max\left\{\alpha \in \mathcal{O}^+(\pm^+) : \alpha \le x\right\}.$$

Analogously, $i^- : \Sigma^- \to \mathcal{O}^-(\pm^-)$ is the *downward inner approximation mapping* associating to any approximable elements $x \in \Sigma^-$ the *lower* (or *inner*) approximation $i^-(x) \in \mathcal{O}^-(\pm^-)$, i.e. a downward open definable element such that
$$i^-(x) \le x,$$
$$\forall \alpha \in \mathcal{O}^-(\pm^-), (\alpha \le x \Rightarrow \alpha \le i^-(x)),$$

i.e.,
$$i^-(x) := max\left\{\alpha \in \mathcal{O}^-(\pm^-) : \alpha \le x\right\}.$$

6. $o^+ : \Sigma^+ \to \mathcal{C}^+(\pm^+)$ is the *upward outer approximation mapping* associating to any approximable elements $x \in \Sigma^+$ the *upper* (or *outer*) approximation $o^+(x) \in \mathcal{C}^+(\pm^+)$, i.e. an upward open definable element such that
$$x \le o^+(x),$$
$$\forall \gamma \in \mathcal{C}^+(\pm^+), (x \le \gamma \Rightarrow o^+(x) \le \gamma),$$

i.e.,
$$o^+(x) := min\left\{\gamma \in \mathcal{C}^+(\pm^+) : x \le \gamma\right\}.$$

Analogously, $o^- : \Sigma^- \to \mathcal{C}^-(\pm^-)$ is the *downward outer approximation mapping* associating to any approximable elements $x \in \Sigma^-$ the *upper* (or *outer*) approximation $o^-(x) \in \mathcal{C}^-(\pm^-)$, i.e. a downward closed definable element such that
$$x \le o^-(x),$$
$$\forall \gamma \in \mathcal{C}^-(\pm^-), (x \le \gamma \Rightarrow o^-(x) \le \gamma),$$

i.e.,
$$o^-(x) := min\left\{\gamma \in \mathcal{C}^-(\pm^-) : x \le \gamma\right\}.$$

The rough approximation of any $x \in \Sigma^+$ is the pair $r^+(x) := (i^+(x), o^+(x))$ while the rough approximation of any $y \in \Sigma^-$ is the pair $r^-(y) := (i^-(y), o^-(y))$. An equivalent way to define a rough approximation is to consider, instead of the interior-closure pair, the interior-exterior pair such that, for all $x \in \Sigma^+$, the rough approximation is defined as

$$r_e(x) := (i^+(x), e^+(x)) = (i^+(x), o^+(x)'^+),$$

while, for all $y \in \Sigma^-$, the rough approximation is defined as

$$r_e^-(y) := (i^-(y), e^-(y)) = (i^-(y), o^-(y)'^-).$$

Let us stress that in the above definitions it is not required that the sets of open and closed definable elements must be the same, and, in general, $\mathcal{O}^+(\Sigma^+) \ne \mathcal{C}^+(\Sigma^+)$ and $\mathcal{O}^-(\Sigma^-) \ne \mathcal{C}^-(\Sigma^-)$. We denote by $\mathcal{CO}^+(\pm^+) = \mathcal{C}^+(\pm^+) \cap \mathcal{O}^+(\pm^+)$ the set of all *upward clopen* (simultaneously closed and open) definable elements.

Analogously, we denote by $\mathcal{CO}^-(\pm^-) = \mathcal{C}^-(\pm^-) \cap \mathcal{O}^-(\pm^-)$ the set of all *downward clopen* definable elements. Only in some particular cases, for instance in case of the basic DRSA based on criteria giving partial or complete preorder on the universe U presented in the above section 2, these two open and closed approximation environments coincide.

The bipolar generalized approximation space is an abstraction of the standard way to deal with roughness within DRSA. Now, we show how a structure of this type can be induced from any bipolar quasi Brower-Zadeh lattice. Making use of the Kleene complementations $'^+$ and $'^-$, and of the two Brower complementations $\sim^+$ and $\sim^-$, it is possible to define the mappings

$$^{b+} : a \in \Sigma^+ \mapsto a^{b+} := a'^{+\sim-'+} \in \Sigma^-$$

and

$$^{b-} : a \in \Sigma^- \mapsto a^{b-} := a'^{-\sim+'-} \in \Sigma^+,$$

which are anticomplementations, i.e. for any $a, b \in \Sigma^+$ and $c, d \in \Sigma^-$:

$$a^{b+b-} \le a, \ c^{b-b+} \le c,$$

$$a \le b \text{ implies } b^{b+} \le a^{b+}, \ c \le d \text{ implies } d^{b-} \le c^{b-},$$

$$a \vee a^{b+} = 1, \ c \vee c^{b-} = 1.$$

For any $a \in \Sigma^+$ and any $b \in \Sigma^-$, we have

$$a^{\sim+} \le a'^+ \le a^{b+}, \ b^{\sim-} \le b'^- \le b^{b-}.$$

The following statements are equivalent for a fixed element $a \in \Sigma^+$:

$$a^{\sim+} = a'^+, \ a = a'^{+\sim-}, \ a^{\sim+\sim-} = a'^{+\sim-}, \ a = a^{\sim+'-}.$$

Moreover, the following logical implication holds for a fixed element $a \in \Sigma^+$:

$$a^{\sim+} = a'^+ \Rightarrow [a = a^{b+b-} = a^{\sim+\sim-}].$$

Analogously, for a fixed element $b \in \Sigma^-$, the following statements are equivalent:

$$b^{\sim-} = b'^-, \ b = b'^{-\sim+}, \ b^{\sim-\sim+} = b'^{-\sim+}, \ b = b^{\sim-'+},$$

and the following logical implication holds

$$b^{\sim-} = b'^- \Rightarrow [b = b^{b-b+} = b^{\sim-\sim+}].$$

We can introduce the following sets:

– the set of all *upward (downward) exact* elements

$$\Sigma_e^+ := \{f \in \Sigma^+ : f'^+ = f^{\sim+}\} \quad (\Sigma_e^- := \{f \in \Sigma^- : f'^- = f^{\sim-}\}),$$

- the set of all *upward (downward) open* elements

$$\Sigma_o^+ := \left\{ f \in \Sigma^+ : f = f^{\flat+\flat-} \right\} \quad (\Sigma_o^- := \left\{ f \in \Sigma^- : f = f^{\flat-\flat+} \right\}),$$

- the set of all *upward (downward) closed* elements

$$\Sigma_c^+ := \{ f \in \Sigma^+ : f = f^{\sim+\sim-} \} \quad (\Sigma_c^- := \{ f \in \Sigma^- : f = f^{\sim-\sim+} \}),$$

- the set of all *upward (downward) clopen* elements

$$\Sigma_{co}^+ = \Sigma_c^+ \cap \Sigma_o^+ \quad (\Sigma_{co}^- = \Sigma_c^- \cap \Sigma_o^+).$$

We also have

$$\Sigma_e^+ \subseteq \Sigma_{co}^+, \; \Sigma_e^- \subseteq \Sigma_{co}^-,$$

$$\Sigma_o^+ = \{ f \in \Sigma^+ : f'^+ \in \Sigma_c^- \}, \; \Sigma_o^- = \{ f \in \Sigma^- : f'^- \in \Sigma_c^+ \},$$

$$\Sigma_c^+ = \{ f \in \Sigma^+ : f'^+ \in \Sigma_o^- \}, \; \Sigma_c^- = \{ f \in \Sigma^- : f'^- \in \Sigma_o^+ \}.$$

Observe that in any bipolar quasi Brower-Zadeh lattice we have that, for any $a \in \Sigma^+$ and for any $b \in \Sigma^-$: $a'^{+\sim-}, a^{\sim+\sim-} \in \Sigma_c^+$, $a^{\flat+\flat-}, a^{\sim+/-} \in \Sigma_o^+$, $b'^{-\sim+}, b^{\sim-\sim+} \in \Sigma_c^-$, and $b^{\flat-\flat+}, a^{\sim-/+} \in \Sigma_o^-$. On the basis of these results, eight further unary operators can be introduced:

$$\nu^+ : a \in \Sigma^+ \mapsto \nu^+(a) := a'^{+\sim-} \in \Sigma_c^+, \text{ (upward necessity)}$$

$$\nu^- : a \in \Sigma^- \mapsto \nu^-(a) := a'^{-\sim+} \in \Sigma_c^-, \text{ (downward necessity)}$$

$$\mathcal{I}^+ : a \in \Sigma^+ \mapsto \mathcal{I}^+(a) := a^{\flat+\flat-} \in \Sigma_o^+, \text{ (upward interior)}$$

$$\mathcal{I}^- : a \in \Sigma^- \mapsto \mathcal{I}^-(a) := a^{\flat-\flat+} \in \Sigma_o^-, \text{ (downward interior)}$$

$$\mathcal{C}^+ : a \in \Sigma^+ \mapsto \mathcal{C}^+(a) := a^{\sim+\sim-} \in \Sigma_c^+, \text{ (upward closure)}$$

$$\mathcal{C}^- : a \in \Sigma^- \mapsto \mathcal{C}^-(a) := a^{\sim-\sim+} \in \Sigma_c^-, \text{ (downward closure)}$$

$$\mu^+ : a \in \Sigma^+ \mapsto \mu^+(a) := a^{\sim+/-} \in \Sigma_o^+, \text{ (upward possibility)}$$

$$\mu^- : a \in \Sigma^- \mapsto \mu^-(a) := a^{\sim-/+} \in \Sigma_o^-. \text{ (downward possibility)}$$

According to the above definitions, the complementations $\sim+$, $\sim-$, $\flat+$ and $\flat-$ can be interpreted as follows: $a^{\sim+} = \mu^+(a)'^+$ (upward impossibility), $a^{\sim-} = \mu^-(a)'^-$ (downward impossibility), $a^{\flat+} = \nu^+(a)'^+$ (upward contingency) and $a^{\flat-} = \nu^-(a)'^-$ (downward contingency).

In any bipolar quasi Brower-Zadeh lattice, the mappings

$$\mathcal{I}^+ : \pm^+ \to \pm_o^+, \; a \mapsto \mathcal{I}^+(a) = a^{\flat+\flat-}$$

and

$$\mathcal{I}^- : \pm^- \to \pm_o^-, \; a \mapsto \mathcal{I}^-(a) = a^{\flat-\flat+}$$

are interior operator since they are normalized, decreasing, idempotent and sub-multiplicative, i.e. the following hold:

$$1 = \mathcal{I}^+(1),\ 1 = \mathcal{I}^-(1),$$

$$\forall a \in \Sigma^+, \mathcal{I}^+(a) \leq a, \forall b \in \pm^-, \mathcal{I}^-(b) \leq b,$$

$$\forall a \in \Sigma^+, \mathcal{I}^+(a) = \mathcal{I}^+(\mathcal{I}^+(a)), \forall b \in \pm^-, \mathcal{I}^-(b) = \mathcal{I}^-(\mathcal{I}^-(b)),$$

$$\forall a, b \in \Sigma^+, \mathcal{I}^+(a \wedge b) \leq \mathcal{I}^+(a) \wedge \mathcal{I}^+(b),\ \forall a, b \in \Sigma^-, \mathcal{I}^-(a \wedge b) \leq \mathcal{I}^-(a) \wedge \mathcal{I}^-(b),$$

such that for all $a \in \Sigma^+$ and for all $b \in \Sigma^-$:

$$\mathcal{I}^+(a) = \vee\left\{f \in \pm_o^+ : f \leq a\right\},\ \mathcal{I}^-(b) = \vee\left\{f \in \pm_o^- : f \leq b\right\}.$$

In any bipolar quasi Brower-Zadeh lattice, the mappings

$$\mathcal{C}^+ : \pm^+ \rightarrow \pm_C^+,\ a \mapsto \mathcal{C}^+(a) = a^{\sim+\sim-}$$

and

$$\mathcal{C}^- : \pm^- \rightarrow \pm_C^-,\ a \mapsto \mathcal{C}^-(a) = a^{\sim-\sim+}$$

are closure operators since they are normalized, increasing, idempotent and sub-additive, i.e. the following hold:

$$0 = \mathcal{C}^+(0),\ 0 = \mathcal{C}^-(0),$$

$$\forall a \in \Sigma^+, a \leq \mathcal{C}^+(a), \forall b \in \pm^-, b \leq \mathcal{C}^-(b),$$

$$\forall a \in \Sigma^+, \mathcal{C}^+(a) = \mathcal{C}^+(\mathcal{C}^+(a)), \forall b \in \pm^-, \mathcal{C}^-(b) = \mathcal{C}^-(\mathcal{C}^-(b)),$$

$$\forall a, b \in \Sigma^+, \mathcal{C}^+(a) \vee \mathcal{C}^+(b) \leq \mathcal{C}^+(a \vee b),\ \forall a, b \in \Sigma^-, \mathcal{C}^-(a) \vee \mathcal{C}^-(b) \leq \mathcal{C}^-(a \vee b)$$

such that for all $a \in \Sigma^+$ and for all $b \in \Sigma^-$:

$$\mathcal{C}^+(a) = \wedge\left\{f \in \pm_C^+ : a \leq f\right\},\ \mathcal{C}^-(b) = \wedge\left\{f \in \pm_C^- : b \leq f\right\}.$$

Let $\langle \Sigma, \Sigma^+, \Sigma^-, \wedge, \vee, '^+, '^-, {}^{\sim+}, {}^{\sim-}, 0, 1\rangle$ be a bipolar quasi Brower-Zadeh distributive lattice. Then, the structure

$$\langle \Sigma, \leq, \Sigma^+, \Sigma^-, \Sigma_o^+, \Sigma_o^-, \Sigma_c^+, \Sigma_c^-, \mathcal{I}^+, \mathcal{I}^-, \mathcal{C}^+, \mathcal{C}^-\rangle$$

is a bipolar approximation space with respect to: the set of upward open definable elements $\mathcal{O}^+(\pm^+) = \Sigma_c^+$; the set of downward open definable elements $\mathcal{O}^-(\pm^-) = \Sigma_c^-$; the set of upward closed definable elements $\mathcal{C}^+(\pm^+) = \Sigma_c^+$; the set of downward closed definable elements $\mathcal{C}^-(\pm^-) = \Sigma_c^-$; the upward inner approximation mapping $i^+(a) = \mathcal{I}^+(a) = a^{b+b-}$; the downward inner approximation mapping $i^-(a) = \mathcal{I}^-(a) = a^{b-b+}$; the upward outer approximation mapping $o^+(a) = \mathcal{C}^+(a) = a^{\sim+\sim-}$; the downward outer approximation mapping $o^-(a) = \mathcal{C}^-(a) = a^{\sim-\sim+}$. Observe that for all $a \in \Sigma^+$, $\mathcal{I}^+(a) = (\mathcal{C}^-(a'^+))'^-$, and for all $b \in \Sigma^-$, $\mathcal{I}^-(b) = (\mathcal{C}^+(b'^-))'^+$.

In any bipolar quasi Brower-Zadeh lattice, the following chains of inclusions hold, for all $a \in \Sigma^+$ and $b \in \Sigma^-$:

$$\nu^+(a) \leq \mathcal{I}^+(a) \leq a \leq \mathcal{C}^+(a) \leq \mu^+(a),$$
$$\nu^-(b) \leq \mathcal{I}^-(b) \leq b \leq \mathcal{C}^-(b) \leq \mu^-(b).$$

In case of a bipolar Brower-Zadeh lattice for all $a \in \Sigma^+$ and $b \in \Sigma^-$,

$$\mathcal{I}^+(a) = \nu^+(a), \quad \mathcal{I}^-(b) = \nu^-(b),$$
$$\mathcal{C}^+(a) = \mu^+(a), \quad \mathcal{C}^-(b) = \mu^-(b).$$

Let us observe that, for Theorem 1, this is the case of basic DRSA as described in the above section 2.

5 Conclusions

In this paper, we introduced a bipolar approximation space as a general model for DRSA, and we showed that it can be induced from a bipolar quasi Brower-Zadeh lattice. Future research will be devoted to model DRSA in terms of other abstract algebras and their comparison with the bipolar quasi Brower-Zadeh lattice.

References

1. Cattaneo, G.: Generalized Rough Sets (Preclusivity Fuzzy-Intuitionistic (BZ) Lattices). Studia Logica 58, 47–77 (1997)
2. Cattaneo, G., Nisticó, G.: Brower-Zadeh Posets and three valued Łukasiewicz posets. Fuzzy Sets and Systems 33, 165–190 (1989)
3. Greco, S., Matarazzo, B., Słowiński, R.: Extension of the rough set approach to multicriteria decision support. INFOR 38, 161–196 (2000)
4. Greco, S., Matarazzo, B., Słowiński, R.: Rough set theory for multicriteria decision analysis. European Journal of Operational Research 129, 1–47 (2001)
5. Greco, S., Matarazzo, B., Słowiński, R.: Rough approximation by dominance relations. International Journal of Intelligent Systems 17, 153–171 (2002)
6. Greco, S., Matarazzo, B., Słowiński, R.: Decision rule approach. In: Figueira, J., Greco, S., Ehrgott, M. (eds.) Multiple Criteria Decision Analysis: State of the Art Surveys, pp. 507–563. Springer, Berlin (2005)
7. Greco, S., Matarazzo, B., Słowiński, R.: Dominance-based Rough Set Approach as a proper way of handling graduality in rough set theory. In: Peters, J.F., Skowron, A., Marek, V.W., Orłowska, E., Słowiński, R., Ziarko, W. (eds.) Transactions on Rough Sets VII. LNCS, vol. 4400, pp. 36–52. Springer, Heidelberg (2007)
8. Greco, S., Matarazzo, B., Słowiński, R.: Algebraic structures for Dominance-Based Rough Set Approach. In: Wang, G., Li, T., Grzymala-Busse, J.W., Miao, D., Skowron, A., Yao, Y. (eds.) RSKT 2008. LNCS (LNAI), vol. 5009, pp. 252–259. Springer, Heidelberg (2008)
9. Pawlak, Z.: Rough Sets. International Journal of Computer and Information Sciences 11, 341–356 (1982)
10. Pawlak, Z.: Rough Sets. Kluwer, Dordrecht (1991)
11. Vakarelov, D.: Information systems, similarity and modal logics. In: Orłowska, E. (ed.) Incomplete Information: Rough Set Analysis, pp. 492–550. Physica, Heidelberg (1998)

Paraconsistent Logic Programs
with Four-Valued Rough Sets*

Jan Małuszyński[1], Andrzej Szałas[2], and Aida Vitória[3]

[1] The College of Economics and Computer Science
10-061 Olsztyn, Poland
janma@ida.liu.se
[2] Institute of Informatics, Warsaw University
02-097 Warsaw, Poland
andsz@mimuw.edu.pl
[3] Department of Science and Technology, Linköping University
S 601 74 Norrköping, Sweden
aidvi@itn.liu.se

Abstract. This paper presents a language for defining four-valued rough sets and to reason about them. Our framework brings together two major fields: rough sets and paraconsistent logic programming. On the one hand it provides a paraconsistent approach, based on four-valued rough sets, for integrating knowledge from different sources and reasoning in the presence of inconsistencies. On the other hand, it also caters for a specific type of uncertainty that originates from the fact that an agent may perceive different objects of the universe as being indiscernible. This paper extends the ideas presented in [9]. Our language allows the user to define similarity relations and use the approximations induced by them in the definition of other four-valued sets. A positive aspect is that it allows users to tune the level of uncertainty or the source of uncertainty that best suits applications.

1 Introduction

We present a language for defining four-valued rough sets and to reason about them. Our framework relates and brings together two major fields: rough sets [8] and paraconsistent logic programming [3]. On the one hand the work discussed here provides a paraconsistent approach, based on four-valued rough sets, for integrating knowledge from different sources and reasoning with possible inconsistent knowledge resulting from this integration. On the other hand, it also caters for a specific type of uncertainty that originates from the fact that an agent may perceive different objects of the universe as being indiscernible. This type of uncertainty has been widely studied in the rough set field. To this end, the proposed language allows the user to define similarity relations modeling indiscernibility and use the similarity-based approximations in definitions of new four-valued sets.

The language discussed in this paper is based on ideas of our previous work [9] that presents a four-valued framework for rough sets. In this approach membership

---

* Supported in part by the MNiSW grant N N206 399334.

C.-C. Chan et al. (Eds.): RSCTC 2008, LNAI 5306, pp. 41–51, 2008.

function, set containment and set operations are four-valued, where logical values are **t** (true), **f** (false), **i** (inconsistent) and **u** (unknown). Moreover, the similarity relations used to define approximations of a set are also four-valued. Consequently, we also define four-value notions of upper and lower approximations that extend the usual notion of approximations in rough set theory [8].

In contrast to the standard rough set framework, our framework allows different types of boundary cases to be identified and, consequently, different degrees of uncertainty.

We now briefly compare our work with some of the work in the field of paraconsistent logic programming[1]. From a syntactic perspective, the logic programs introduced in Section 3.1 correspond to Fitting programs [5] which do not involve $\otimes$ and $\forall$ in the right-hand side (body) of the rules. Although both frameworks are intended to deal with inconsistencies and use a four-valued logic, there are some major differences at the semantic level. First, the Belnap's logic underlies the semantics of Fitting programs (see [5], Def. 18). This contrasts with our approach since we use a different truth ordering. Second, the semantics of Fittings programs allows to derive conclusions from false premises. For instance, rule *danger :– hot.* can be used to derive that there is no danger, i.e. *danger* is **f**, if *hot* is **f**. In contrast to our framework, a rule of a Fitting program is satisfied if and only if the truth values assigned to the head and to the body are equal.

Paper [1] describes a paraconsistent approach, called P-Datalog, for knowledge base integration based on a four-valued logic and the total order of the four logical values presented there coincides with our truth ordering. However, there are several important differences. First, in contrast to [1], we do not follow the closed-world assumption, i.e. a formula $\neg p(\bar{d})$ is **t** only if some agent states it explicitly and no agent claims that $p(\bar{d})$ is **t**. Second, our language allows explicit negation in the head and bodies of the rules while P-Datalog programs only allow negation by default $\sim$ in the rule's bodies. Consequently, knowledge ordering is used in our framework as a more natural way to combine knowledge from different sources while P-Datalog uses the truth ordering presented in Section 2.1. Third, the rules are interpreted differently. In our language a rule is interpreted as the implication $\rightarrow_k$ defined in Table 1, while the implication $\rightarrow$ underlying the rules of P-Datalog is another. For example, in P-Datalog, the truth-value of $\mathbf{t} \rightarrow \mathbf{i}$ is **f** while in our framework $\mathbf{t} \rightarrow_k \mathbf{i}$ is **t**. Finally, the language we propose allows disjunction $\vee_t$ (join under truth ordering) to be used in the body of a rule.

The paper is organized as follows. Section 2 summarizes the main results of [9]. Section 3 gives a formal definition of the language. Section 4 sketches an implementation proposal. Finally, Section 5 summarizes the paper.

2   The Four-Valued Framework

2.1   Logics Reflecting Truth Ordering and Knowledge Ordering

To construct the language we use two orderings on truth vales, namely the *truth ordering* and *knowledge ordering*. Truth ordering is used for calculations within a single information source while knowledge ordering is used for gathering knowledge from different sources. This approach has been considered in [2] and in the framework of

---

[1] A detailed comparison is outside of the scope of this paper.

bilattices, in [4,6]. The *truth ordering* $\leq_t$ and the *knowledge ordering* $\leq_k$ are defined as the smallest reflexive and transitive relations satisfying $\mathbf{f} \leq_t \mathbf{u} \leq_t \mathbf{i} \leq_t \mathbf{t}$, $\mathbf{u} \leq_k \mathbf{f} \leq_k \mathbf{i}$, and $\mathbf{u} \leq_k \mathbf{t} \leq_k \mathbf{i}$. The knowledge ordering above coincides with Belnap's knowledge ordering [2]. However, our truth ordering is different from the Belnap's truth ordering. This change is motivated by the fact that Belnap's truth ordering can give counterintuitive results when used for reasoning, as shown in [7].

Having two orderings on truth values, we also have two logics: L_t based on truth ordering and L_k based on knowledge ordering. We denote by $\wedge_t$, $\vee_t$ and $\rightarrow_t$ the connectives of L_t and by $\wedge_k$, $\vee_k$ and $\rightarrow_k$ the corresponding connectives in L_k. Negation, denoted as $\neg$, in both logics has the same semantics. Let GLB^t (GLB^k) and LUB^t (LUB^k) denote the greatest lower bound and the least upper bound of a set of logical values w.r.t truth (knowledge) ordering, respectively. Then, $(a \wedge_t b) = \mathrm{GLB}^t\{a,b\}$ ($(a \wedge_k b) = \mathrm{GLB}^k\{a,b\}$) and $(a \vee_t b) = \mathrm{LUB}^t\{a,b\}$ ($(a \vee_k b) = \mathrm{LUB}^k\{a,b\}$), where a and b are two logical values. Table 1 provides the semantics for implication in both logics, L_t and L_k. Observe that the implication $\rightarrow_t$, introduced in [9], is a four-valued extension of the usual logical implication, suitable for determining set containment and approximations in the case of four-valued sets.

Table 1. Truth tables for $\rightarrow_t$, $\rightarrow_k$, and $\neg$

$\rightarrow_t$	**f**	**u**	**i**	**t**		$\rightarrow_k$	**f**	**u**	**i**	**t**		$\neg$
f	t	t	t	t		**f**	t	t	t	t		t
u	u	u	i	t		**u**	t	t	t	t		u
i	i	i	i	t		**i**	f	f	t	f		i
t	f	u	i	t		**t**	f	f	t	t		f

The semantics of quantifier $\forall$ and $\exists$ is given below.

$$\forall x[P(x)] \overset{\text{def}}{=} \underset{x \in U}{\mathrm{GLB}^t}\{P(x)\} \quad \text{and} \quad \exists x[P(x)] \overset{\text{def}}{=} \underset{x \in U}{\mathrm{LUB}^t}\{P(x)\}.$$

Intuitively, $P(x)$ denotes whether an element x has a property P (i.e. membership of x in a four-valued set P) and it is evaluated to one of the four logical values.

We have the following important propositions.

Proposition 1. *The disjunction $\vee_t$ is monotonic w.r.t. the knowledge ordering.* ◁

Proposition 2. *The conjunction $\wedge_t$ is not monotonic w.r.t. the knowledge ordering.* ◁

This is because $(\mathbf{f} \wedge_t \mathbf{u}) = \mathbf{f}$ but $(\mathbf{i} \wedge_t \mathbf{u}) = \mathbf{u}$. This shows the lack of monotonicity, since $\mathbf{f} <_k \mathbf{i}$ and $\mathbf{f} >_k \mathbf{u}$. However, we have the following proposition.

Proposition 3. *Let p and q be truth values such that $(p \wedge_t q) \geq_t \mathbf{i}$. If $p' \geq_k p$ then $(p' \wedge_t q) \geq_k (p \wedge_t q)$. If $q' \geq_k q$ then $(p \wedge_t q') \geq_k (p \wedge_t q)$.* ◁

Thus, the conjunction is monotonic w.r.t. knowledge ordering for arguments greater or equal than $\mathbf{i}$.

2.2   Operations on Four-Valued Sets

Let us now formalize the notion of four-valued sets. Given a universe U, we introduce a new set, disjoint with U, denoted by $\neg U$ and defined by $\neg U \overset{\text{def}}{=} \{\neg x \mid x \in U\}$, where $\neg x$ denotes elements in $\neg U$. A *four-valued set* A on U is any subset of $U \cup \neg U$. Intuitively, $x \in A$ represents the fact that there is an evidence that x is in A and $(\neg x) \in A$ represents the fact that there is an evidence that x is not in A.

In our framework, set membership is four-valued and it extends the usual two-valued membership. We assume that $\neg(\neg x)$ is equal to x.

Set membership, denoted as $\epsilon : U \times 2^{U \cup \neg U} \to \{\mathbf{f}, \mathbf{u}, \mathbf{i}, \mathbf{t}\}$, is defined by

$$
x \, \epsilon \, A = \begin{cases} \mathbf{t} & \text{if } x \in A \text{ and } (\neg x) \notin A \\ \mathbf{i} & \text{if } x \in A \text{ and } (\neg x) \in A \\ \mathbf{u} & \text{if } x \notin A \text{ and } (\neg x) \notin A \\ \mathbf{f} & \text{if } x \notin A \text{ and } (\neg x) \in A. \end{cases} \tag{1}
$$

The *complement* $\neg A$ *of a four-valued set* A, is defined by $\neg A \overset{\text{def}}{=} \{\neg x \mid x \, \epsilon \, A\}$ and the four-valued set inclusion is defined by $X \Subset Y \overset{\text{def}}{=} \forall x \in U [x \, \epsilon \, X \to_t x \, \epsilon \, Y]$.

The four-valued operations of intersection and union, defined as

$$x \, \epsilon \, (X \Cap Y) \overset{\text{def}}{=} (x \, \epsilon \, X) \wedge_t (x \, \epsilon \, Y) \quad \text{and} \quad x \, \epsilon \, (X \Cup Y) \overset{\text{def}}{=} (x \, \epsilon \, X) \vee_t (x \, \epsilon \, Y),$$

generalize the respective standard set operations.

A four-valued extension of rough sets is then defined by four-valued set approximations as follows (cf. [9]). Note that (four-valued) relations are (four-valued) sets of tuples.

Definition 1. A *four-valued similarity relation* σ is any four-valued binary relation on a universe U, satisfying the reflexivity condition, i.e., for any element x of the universe $(x, x) \, \epsilon \, \sigma = \mathbf{t}$. The *neighborhood of element* $x \in U$ *w.r.t.* σ, is the four-valued set $\sigma(x)$ such that $y \, \epsilon \, \sigma(x) \overset{\text{def}}{=} (x, y) \, \epsilon \, \sigma$.   ◁

Definition 2. Let A be a four-valued set. Then, the *lower and upper approximations of* A *w.r.t.* σ, denoted by A_σ^+ and $A_\sigma^\oplus$, respectively, are defined by $(x \, \epsilon \, A_\sigma^+) \overset{\text{def}}{=} \sigma(x) \Subset A$ and $(x \, \epsilon \, A_\sigma^\oplus) \overset{\text{def}}{=} \exists y \in U [y \, \epsilon \, (\sigma(x) \Cap A)]$.   ◁

Note that approximations are also four-valued. For example, let $U = \{o_1, o_2\}$, the set $A = \{o_1, \neg o_2\}$, and $\sigma(o_1, o_2) = \mathbf{u}$. Then, we have that membership of o_1 in A_σ^+ is unknown ($\mathbf{u}$). It might later appear that $\sigma(o_1, o_2)$ is $\mathbf{f}$ and we then conclude that $(o_1 \, \epsilon \, A_\sigma^+) = \mathbf{t}$. Or, it might appear that $\sigma(o_1, o_2)$ is $\mathbf{t}$ and we then get that $(o_1 \, \epsilon \, A_\sigma^+) = \mathbf{f}$.

3   A Rule Language for Defining Four-Valued Sets

Our aim is to present a rule language for defining four-valued sets. A rule consists of an head and a body. The head and the body are formulae of the four-valued logic of Section 2. Thus, each of them gets one of the four truth values, under a given interpretation. A rule is satisfied in a given interpretation iff whenever the body is $\mathbf{t}$ or $\mathbf{i}$ then the truth value of the head is greater or equal than the truth value of the body, w.r.t. $\leq_k$. Thus, rules reflect the semantics of implication $\rightarrow_k$ as provided in Table 1. This choice corresponds to the intuition that a rule is to be used for increasing knowledge by drawing conclusions. No conclusions are drawn from false or unknown premises (bodies).

3.1   The Syntax

The rules are constructed from:

- *literals* of the form $P(\bar{d})$, $\neg P(\bar{d})$, where P is a relation symbol and $\bar{d}$ is a tuple of terms (variables or constants denoting objects of the universe).
- *truth symbols* : $false,\ unknown,\ incons,\ true$.

Rules are of the form

$$head :- l_{11}, \ldots, l_{1k_1} ; l_{21}, \ldots, l_{2k_2} ; \ldots ; l_{m1}, \ldots, l_{mk_m}. \qquad (2)$$

where $m, k_i \geq 1$, for $1 \leq i \leq m$, *head* and each l_{ij} ($1 \leq j \leq k_i$) are literals or truth symbols.

A rule of the form $head :- true.$, called a *fact*, is abbreviated as $head.$. A *program* is a finite set of rules. A *ground instance* of a rule is obtained by replacing each variable of the rule by a selected constant occurring in the program.

Example 1. Consider two robots, r_1 and r_2, recognizing similarities between objects on the basis of their shape. Assume that the only shapes are *round, rectangular, square* and *oval*. Due to perceptual limitations r_1 does not recognize the difference between round and oval, and r_2 does not recognize the difference between rectangular and square. The following rules can be used to express (partially) the similarities between objects, as perceived by the robot r_1.

$$sim(x, y) :-\quad shape(x, round), shape(y, oval) ;$$
$$shape(x, oval), shape(y, round).$$
$$\neg sim(x, y) :-\ shape(x, square), shape(y, rectangular) ;$$
$$shape(x, square), shape(y, round).$$

For the robot r_2 one can consider, e.g., the following rule.

$$sim(x, y) :-\ shape(x, square), shape(y, rectangular) ;$$
$$shape(x, rectangular), shape(y, square).$$

As a similarity relation is required to be reflexive (cf. Definition 2), the program also includes the fact $sim(x, x)$. ◁

3.2   The Declarative Semantics

Let $\mathcal{P}$ be a program and L be the set of all constant symbols occurring in $\mathcal{P}$. Then, the *Herbrand base* $\mathcal{H_P}$ is the set of all literals whose relation symbols occur in $\mathcal{P}$ and whose arguments belong to L.

A four-valued *interpretation* $\mathcal{I}$ of a program $\mathcal{P}$ is any subset of $\mathcal{H}_\mathcal{P}$. It associates each ground literal l with a truth value such that $\mathcal{I}(l) \overset{\text{def}}{=} (l \in \mathcal{I})$ (ϵ is defined in (1)).

An interpretation $\mathcal{I}_1$ is *smaller or equal* than an interpretation $\mathcal{I}_2$, denoted by $\mathcal{I}_1 \sqsubseteq \mathcal{I}_2$, iff $\mathcal{I}_1$ is a (classical) subset of $\mathcal{I}_2$. Observe that if $\mathcal{I}_1 \sqsubseteq \mathcal{I}_2$ then, for every literal l of the Herbrand base, $\mathcal{I}_1(l) \leq_k \mathcal{I}_2(l)$. The interpretation $\emptyset$, called the empty interpretation, is the least interpretation in this ordering. It assigns **u** to every literal in $\mathcal{H}_\mathcal{P}$.

The notion of interpretation is extended to rules of the form (2) by interpreting ',' as conjunction $\wedge_t$, ';' is interpreted as the disjunction $\vee_t$, and a rule $H :- B.$ is interpreted as $B \rightarrow_k H$, with the semantics provided in Table 1. The truth symbols are interpreted as **f**, **u**, **i** and **t**. Thus, a given four-valued interpretation determines the truth values of the head and of the body of each rule. If several rules have the same literal H in their head, then H takes the truth value being the disjunction $\vee_k$ of the values assigned to each body's rule. More precisely, if $H :- B_1., \ldots, H :- B_m.$ are all rules with the head H then the value of H is obtained from $(B_1 \vee_k \ldots \vee_k B_m)$. Note that different rules with the same literal H in their head gather knowledge about H according to knowledge ordering, using $\vee_k$. On the other hand, one often needs to define cases using truth ordering and this motivates the need of ; in the bodies of the rules.

Definition 3. An interpretation $\mathcal{I}$ *satisfies* a rule $H :- B.$ if the implication $(B \rightarrow_k H)$ is **t** in $\mathcal{I}$. An interpretation is said to be a four-valued *Herbrand model* of a program $\mathcal{P}$ iff it satisfies each rule of $\mathcal{P}$. ◁

Theorem 1. *The (classical) intersection of four-valued Herbrand models of a program $\mathcal{P}$ is a four-valued Herbrand model.*

Proof. Assume the theorem does not hold and let $\mathcal{M}$ be the intersection of the Herbrand models $\mathcal{M}_1$ and $\mathcal{M}_2$ of $\mathcal{P}$. Then, there is a rule $H :- B. \in \mathcal{P}$ such that $\mathcal{M}(H) <_k \mathcal{M}(B)$ and $\mathcal{M}(B) \in \{\mathbf{i}, \mathbf{t}\}$. If $\mathcal{M}(B) = \mathbf{i}$ then the truth value of B must have been **i** both in $\mathcal{M}_1$ and in $\mathcal{M}_2$. Hence, the truth values of the head must also have been **i** in both $\mathcal{M}_1$ and in $\mathcal{M}_2$, and consequently, in $\mathcal{M}$. If $\mathcal{M}(B) = \mathbf{t}$ then the body is **t** in one of the models, assume $\mathcal{M}_1$, and **t** or **i** in the other, $\mathcal{M}_2$. Thus, $\mathcal{M}_1(H) \geq_k \mathbf{t}$ and $\mathcal{M}_2(H) \geq_k \mathbf{t}$. Consequently, H must be **t** or **i** in $\mathcal{M}$. We can then conclude that there is no case under which $\mathcal{M}(H) <_k \mathcal{M}(B)$. This implies that $\mathcal{M}$ must be a model of $\mathcal{P}$. ◁

Corollary 1. *For every program $\mathcal{P}$ there exists the least (w.r.t. $\sqsubseteq$) four-valued model.* ◁

We denote this model by $\mathcal{M}_\mathcal{P}$ and consider it the declarative semantics of the program.

3.3 The Fixpoint Semantics

We now define the semantics of a program $\mathcal{P}$ as a fixpoint of an operator on interpretations. We consider here variable-free programs $\mathcal{P}$. If the program has variables then we consider instead all ground instances of its rules. The operator will be denoted $T_\mathcal{P}$ and it is a four-valued extension of the classical $T_\mathcal{P}$ operator used in logic programming. The operator formalizes the intuition of drawing conclusions with rules.

$$T_\mathcal{P}(\mathcal{I}) = \{l \mid l :- B. \in \mathcal{P} \text{ and } \mathcal{I}(B) = \mathbf{t}\} \cup \{l, \neg l \mid l :- B. \in \mathcal{P} \text{ and } \mathcal{I}(B) = \mathbf{i}\} .$$

Thus, the operator collects all the heads of the ground rules whose bodies are **t** in a given interpretation $\mathcal{I}$. In addition, it takes the heads and the negations of the heads of the rules whose bodies are inconsistent in $\mathcal{I}$.

The following theorem shows that the operator $T_{\mathcal{P}}$ is monotonic w.r.t. $\leq_k$. It follows from Propositions 1 and 3, as $T_{\mathcal{P}}$ uses only the rules for which the body is **t** or **i**.

Theorem 2. *Given a program* $\mathcal{P}$ *and two four-valued interpretations* $\mathcal{I}_1$ *and* $\mathcal{I}_2$, *if* $\mathcal{I}_1 \sqsubseteq \mathcal{I}_2$ *then* $T_{\mathcal{P}}(\mathcal{I}_1) \sqsubseteq T_{\mathcal{P}}(\mathcal{I}_2)$.
$\triangleleft$

Corollary 2. $T_{\mathcal{P}}$ *has the least fixpoint, denoted* LFP($T_{\mathcal{P}}$), *which can be computed by iterating* $T_{\mathcal{P}}$ *starting from the empty interpretation.*
$\triangleleft$

It can be shown that the least fixpoint of $T_{\mathcal{P}}$ is the least model of the program, wr.t. $\leq_k$.

Example 2. Consider the rules of Example 1 and a database with five objects: o_1 and o_2 are oval, o_2 is also considered to be round, o_3 is square, o_4 is rectangular, and o_5 has unknown shape. Note that object o_2 is associated with different shapes, perhaps, because different robots perceive it differently. The successive iterations of $T_{\mathcal{P}}$ are given below. Note that $\mathcal{I}_2 = T_{\mathcal{P}}(\mathcal{I}_2)$.

$$\mathcal{I}_1 = \emptyset$$
$$\mathcal{I}_2 = T_{\mathcal{P}}(\mathcal{I}_1) = \{sim(o_1, o_1), sim(o_2, o_2), sim(o_3, o_3), sim(o_4, o_4), sim(o_5, o_5),$$
$$sim(o_1, o_2), sim(o_2, o_1), \neg sim(o_3, o_2),$$
$$\neg sim(o_3, o_4), sim(o_3, o_4), sim(o_4, o_3)\}.$$

According to the definition of four-valued interpretation, we have that both $sim(o_1, o_2)$ and $sim(o_2, o_1)$, as well as for any $sim(x, x)$, receive the value **t**; $sim(o_3, o_2)$ receives the value **f** but $sim(o_2, o_3)$ is **u**; $sim(o_3, o_4)$ receives the value **i** but $sim(o_4, o_3)$ is **t**; for all the remaining pairs (x, y), the value of $sim(x, y)$ is **u**.
$\triangleleft$

3.4 Using Approximations

A Hierarchy of Uncertainty. In our framework, lower and upper approximations are also four-valued sets. Figure 1 shows the truth ordering of pairs $(o \in A_\sigma^+, o \in A_\sigma^\oplus)$. Note that $(t_1, t_2) \leq_t (t_3, t_4)$ iff $(t_1 \leq_t t_3)$ and $(t_2 \leq_t t_4)$. In the figure this is indicated by an edge from pair (t_1, t_2) to (t_3, t_4). Moreover, not all pairs of logical values are allowed because $(o \in A_\sigma^+) \leq_t (o \in A_\sigma^\oplus)$ [9], for any object $o \in U$.

The pair $(\mathbf{t}, \mathbf{t})$ corresponds to the case where an object o certainly belongs to a given set A, while $(\mathbf{f}, \mathbf{f})$ indicates that o certainly does not belong to A. The remaining pairs of logical values in Figure 1 correspond to boundary cases where the object may belong to A. In the standard rough set framework [8], boundary cases correspond to the pair $(\mathbf{f}, \mathbf{t})$ since approximations are two-valued sets. In contrast to the standard rough set framework, our framework allows different types of boundary cases to be identified and different degrees of uncertainty. For instance, the pair $(\mathbf{i}, \mathbf{t})$ indicates that we can be more certain that object o has a property A than the pair $(\mathbf{f}, \mathbf{i})$, although both pairs indicate that there is a possibility of object o having property A. Note that $(\mathbf{f}, \mathbf{i}) <_t (\mathbf{i}, \mathbf{t})$. However, as Figure 1 shows, not all pairs are comparable, e.g., pairs $(\mathbf{f}, \mathbf{t})$ and $(\mathbf{i}, \mathbf{i})$. But,

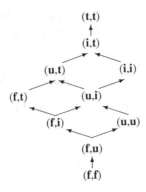

Fig. 1. Truth ordering for $(o \in A_\sigma^+, o \in A_\sigma^\oplus)$

these pairs point then to different sources (types) of uncertainty. For example, the pair $(\mathbf{f}, \mathbf{t})$ indicates that there is at least one object similar to o that does not have property A, i.e. $(o \in A_\sigma^+) = \mathbf{f}$ (see the case (1) of Lemma 1 in Section 4), but there is another object similar to o that has property A, i.e. $(o \in A_\sigma^\oplus) = \mathbf{t}$ (see the case (5) of Lemma 1 in Section 4). Therefore, in the neighborhood of o there are objects that have property A and others that do not have property A. The pair $(\mathbf{i}, \mathbf{i})$ points to a different source of uncertainty, e.g., for all objects in the neighborhood of o there is contradictory evidence about their membership in A (see cases (3) and (6) of Lemma 1 in Section 4).

The informal ideas presented above are reflected in our rule language. Thus, the language allows the user to choose the level of uncertainty or the type of uncertainty that best suits his application.

Extending the Language with Approximations. The rule language makes it possible to define four-valued relations. A defined relation can then be used to specify approximations of another four-valued relation, as discussed in [9] and in Section 2. Such an approximation is itself a four-valued relation. The rule language can thus be extended by allowing approximations of a rough relation (set) to appear in rule bodies. To this end, we need to extend the language with a notation for such symbols. In this paper, the lower approximation (upper approximation) of a relation A w.r.t. a similarity relation σ is denoted A_σ^+ ($A_\sigma^\oplus$). Such approximation symbols can only be used in a program including rules defining A and σ. Programs must also not use recursion through approximations. Intuitively, the relations are not to be defined by referring to their own approximations. Such programs are considered *well-formed*.

Example 3. Consider the rules of Example 1 and the database of objects in Example 2. Based on the accessible knowledge, the robots may be given the task to remove from a given place all round and square objects. Let us introduce an additional unary relation rsq (standing for "round or square") defined as follows.

$$\neg shape(x, y) \;:\text{--}\; shape(x, z), y \neq z.$$
$$rsq(x) \quad\quad\; :\text{--}\; shape(x, round) \;;\; shape(x, square).$$
$$\neg rsq(x) \quad\quad :\text{--}\; \neg shape(x, round), \neg shape(x, square).$$

Observe that $rsq(o_2)$ is **i**, in the declarative semantics of the program, because o_2 is associated with both shapes round and oval.

The required rule expressing the task to be done may be expressed in various non-equivalent ways, according to the intended meaning.

1. $remove(x) :- rsq(x).$ – a traditional formulation where neighborhoods are not taken into account.
2. $remove(x) :- (rsq(x)_{sim}^{+} = true).$ – x is to be removed only when it surely is round or square.
3. $remove(x) :- (rsq(x)_{sim}^{\oplus} = true), rsq(x)_{sim}^{+}.$ – x is to be removed if $(x \in rsq(x)_{sim}^{+}, x \in rsq(x)_{sim}^{\oplus}) \geq_t (\mathbf{i}, \mathbf{t})$.
4. $remove(x) :- rsq(x)_{sim}^{\oplus}.$ – x is to be removed if there is a possibility that it might be round or square.

From the first rule above, we conclude that the truth value of $remove(o_2)$ is **i**, $remove$ (o_3) is **t**, and **u** for all other objects.

The reader can verify[2] that the membership in $rsq(x)_{sim}^{+}$ is **f** for o_1, o_2 and o_4 and it is **u** for o_3 and o_5. The membership in $rsq(x)_{sim}^{\oplus}$ is **i** for o_1 and o_2, **t** for o_3 and o_4, but **u** for o_5. Note that $(o_2 \in rsq(x)_{sim}^{+}) = \mathbf{f}$ because $(o_1 \in sim(o_2)) = \mathbf{t}$ but $rsq(o_1) = \mathbf{f}$, i.e. there is an object similar to o_2 that is neither round nor square, although there is also information indicating that o_2 is round. Consequently, it is not possible to conclude with certainty that o_2 is round (or square).

Using the second rule instead, the truth value of $remove(x)$, is **u** for all objects since for no object in the database it can be proved that it is surely round or square. Note that there is no object o in the database such that $(o \in rsq_{sim}^{+}) = \mathbf{t}$.

The third rule imposes that there must be a quite high believe that an object is round or square in order to remove it, although some uncertainty is acceptable. The rule forces that $(x \in rsq_{sim}^{\oplus}) = \mathbf{t}$. Thus, there must be an object similar to x that is round or square and $(x \in rsq_{sim}^{+}) \geq_t \mathbf{i}$. Remember that no conclusions are drawn from rules with bodies evaluated to false or unknown. If $(x \in rsq_{sim}^{+}) <_t \mathbf{i}$ then the rule does not fire, since the whole body becomes evaluated to **f** or **u**. Consequently with this rule, the truth value of $remove(x)$ is **u**, for all objects x in the database.

With the fourth rule, $remove(o_1)$ and $remove(o_2)$ are **i**, $remove(o_5)$ is **u**, and **t** for all remaining objects. In particular, $remove(o_1)$ is **i** because $(o_1 \in rsq(x)_{sim}^{\oplus}) = \mathbf{i}$. In contrast with the first rule, if this rule is used then o_4 is removed.   ◁

4   Implementation

For a program not using approximations the least model can be computed by iterating the T_P operator, as illustrated in Example 2.

The following lemma, which follows from the definition of approximations, shows how to compute the truth value of an approximation literal under a given interpretation, by consecutive check of simple conditions.

Lemma 1. Let A be a four-valued set on a universe U, σ be a four-valued similarity relation, and $x \in U$.

---

[2] Detailed calculation for the lower and upper approximations are not shown for space reasons.

1. $x \in A_\sigma^+ = \mathbf{f}$ iff $(y \in \sigma(x) = \mathbf{t}$ and $y \in A = \mathbf{f})$, for some $y \in U$.
2. $x \in A_\sigma^+ = \mathbf{u}$ iff (1) does not hold and
   (a) $(y \in \sigma(x) = \mathbf{u}$ and $y \in A \leq_t \mathbf{u})$, for some $y \in U$, or
   (b) $(y \in \sigma(x) = \mathbf{t}$ and $y \in A = \mathbf{u})$, for some $y \in U$.
3. $x \in A_\sigma^+ = \mathbf{i}$ iff 1. and 2. does not hold and
   (a) $(y \in \sigma(x) = \mathbf{i}$ and $y \in A \leq_t \mathbf{i})$, for some $y \in U$, or
   (b) $(y \in \sigma(x) = \mathbf{t}$ and $y \in A = \mathbf{i})$, for some $y \in U$.
4. $x \in A_\sigma^+ = \mathbf{t}$ iff $y \in A = \mathbf{t}$, for all $y \in U$ such that $\sigma(x, y) \geq_t \mathbf{u}$.

Moreover,

5. $x \in A_\sigma^\oplus = \mathbf{t}$ iff $(y \in \sigma(x) = \mathbf{t}$ and $y \in A = \mathbf{t})$, for some $y \in U$.
6. $x \in A_\sigma^\oplus = \mathbf{i}$ iff (1) does not hold and
   (a) $(y \in \sigma(x) = \mathbf{i}$ and $y \in A \geq_t \mathbf{i})$, for some $y \in U$, or
   (b) $(y \in \sigma(x) = \mathbf{t}$ and $y \in A = \mathbf{i})$, for some $y \in U$.
7. $x \in A_\sigma^\oplus = \mathbf{u}$ iff 1. and 2. does not hold and
   (a) $(y \in \sigma(x) = \mathbf{u}$ and $y \in A \geq_t \mathbf{u})$, for some $y \in U$, or
   (b) $(y \in \sigma(x) \geq_t \mathbf{i}$ and $y \in A = \mathbf{u})$, for some $y \in U$.
8. $x \in A_\sigma^\oplus = \mathbf{f}$ iff $y \in A = \mathbf{f}$ or $y \in \sigma(x) = \mathbf{f}$, for all $y \in U$.   ◁

Observe that any well-formed program has a least model which can be computed by a combined use of the T_P operator and Lemma 1.

5   Conclusions

In the current paper, we proposed a rule language for defining four-valued rough sets. The language allows us to work with four-valued approximations which appear frequently in practice. The techniques discussed in the paper open the space for implementation as well as for a pragmatic use of the language. The language is suitable for applications where some information might be unknown or inconsistent, in particular for databases or expert systems.

References

1. Amo, S., Pais, M.S.: A paraconsistent logic approach for querying inconsistent databases. International Journal of Approximate Reasoning 46, 366–386 (2007)
2. Belnap, N.: A useful four-valued logic. In: Epstein, G., Dunn, J.M. (eds.) Modern Uses of Many Valued Logic, pp. 8–37. Reidel (1977)
3. Damásio, C.V., Pereira, L.M.: A survay of paraconsistent semantics for logic programs. In: Gabbay, D.M., Smets, P. (eds.) Handbook of Defeasible Reasoning and Uncertainty Management Systems, vol. 2, pp. 241–320. Kluwer Academic Publishers, Dordrecht (1998)
4. Fitting, M.C.: Bilattices in logic programming. In: Epstein, G. (ed.) Proc. of the 20th International Symposium on Multiple-Valued Logic, pp. 238–246. IEEE CS Press, Los Alamitos (1990)
5. Fitting, M.C.: Fixpoint semantics for logic programming a survey. Theoretical Computer Science 278(1-2), 25–51 (2002)

6. Ginsberg, M.: Multivalued logics: a uniform approach to reasoning in AI. Computational Intelligence 4, 256–316 (1988)
7. Maluszyński, J., Szałas, A., Vitória, A.: A four-valued logic for rough set-like approximate reasoning. In: Duntsch, I., Grzymala-Busse, J., Orlowska, E., Polkowski, L. (eds.) Transactions on Rough Sets VI. LNCS, vol. 4374, pp. 176–190. Springer, Heidelberg (2007)
8. Pawlak, Z.: Rough Sets. Theoretical Aspects of Reasoning about Data. Kluwer Academic Publishers, Dordrecht (1991)
9. Vitória, A., Szałas, A., Maluszyński, J.: A four-valued extension of rough sets. In: Wang, G., Li, T., Grzymala-Busse, J., Miao, D., Skowron, A., Yao, Y. (eds.) RSKT 2008. LNCS (LNAI), vol. 5009, pp. 106–114. Springer, Heidelberg (2008)

An Equivalent Definition of Rough Sets

Guilong Liu[1,*] and James Kuodo Huang[2]

[1] School of Information Science,
Beijing Language and Culture University,
Beijing 100083, China
[2] Association of International Uncertainty Computing
P.O. Box 3355
Alhambra, California 91803, USA
http://www.aiuc.org

Abstract. Using characteristic function of sets, this paper proposes the concept of linear mappings for the power sets of universal sets. Through this concept, we explain rough set upper approximation as a linear mapping and study the linear properties of rough sets. The relationship between the linear mappings and the upper approximations is established. The results and methods given in this paper will hopefully simplify the theoretical and practical researches of rough sets.

Keywords: Rough sets, Binary relations, Approximations, Composition of relations, Characteristic functions.

1 Introduction

The rough set theory was firstly proposed by Pawlak [8,9,10,11,12] in 1982. It is an extension of set theory and can be seen as a new mathematical approach to deal with uncertainty, vagueness and incomplete information [21,22]. The successful application of rough set theory in a variety of real life problems has amply demonstrated its usefulness.

Pawlak introduced the concept of rough sets via equivalence relation. He showed many interesting properties of the lower and upper approximation operators. The lower and upper approximation operators are the key tools to measure the uncertainty [14,15,16,17]. Let U, V be two universal sets and $P(U)$ be the power set of U. We define the concept of linear mapping from $P(V)$ to $P(U)$ in this paper. We explain rough set upper approximation as a linear mapping from $P(V)$ to $P(U)$. We derive many linear properties of lower and upper approximation operators. We also show that there is a one-to-one correspondence between the set of all binary relations from U to V and the set of all linear mappings from $P(V)$ to $P(U)$.

* This work is supported by the Key Project of the Chinese Ministry of Education(108133). James Kuodo Huang is a visiting professor at Beijing Language and Culture University during this work done.

C.-C. Chan et al. (Eds.): RSCTC 2008, LNAI 5306, pp. 52–60, 2008.

Let U be a non-empty set of objects called the universal set. U can be an infinite set, i.e., we do not restrict the universal set to a finite one.

This paper is organized as follows. In Section 2, we present the basic concepts of binary relation and properties of rough sets and give another representation of upper approximation operator. The main objective of Section 3 is to discuss linear properties of the upper approximation operators. We show that each upper approximation operator coincides with a linear mapping. In Section 4, we study the composition of upper approximation operators. Finally, Section 5 concludes the paper.

2   Preliminaries

Let us recall some definitions and properties of binary relations. Let U be a universal set, and $P(U)$ be the power set of U. If X is a subset of U, the characteristic function of X, still denoted by X, is defined for each $x \in U$ as follows [4,5,6,7]:

$$X(x) = \begin{cases} 1, & x \in X \\ 0, & x \notin X. \end{cases}$$

Let U and V be two universal sets and R be a binary relation from U to V. i.e., R is a subset of $U \times V$. Recall that the standard composition of relation R and subset $Y \subseteq V$, which is denoted by $R \circ Y$, produces a subset of U defined by $(R \circ Y)(x) = \vee_{y \in U}(R(x,y) \wedge Y(y))$ for all $x \in U$, where $R(x,y)$ denotes the membership function with value 1 at xRy and 0 otherwise, where $\wedge$ and $\vee$ denote the minimum and maximum, respectively. This composition is often referred to as the max-min composition.

Let U, V and W be three distinct but related universal sets, R a binary relation from U to V, and S a binary relation from V to W. The composition [1] of R and S is the relation consisting of ordered pairs (x, z), where $x \in U, z \in W$, and for which there exists an intermediate element $y \in V$ such that xRy and ySz. We denote the composition of R and S by $R \circ S$. Consequently, $x(R \circ S)z \Leftrightarrow \exists y(xRy \wedge ySz)$ or $(R \circ S)(x, z) = \vee_{y \in V}(R(x,y) \wedge S(y,z))$. A different type of operation on a relation R from U to V is the formation of inverse, usually written R^{-1}. The relation R^{-1} [1] is a relation from V to U defined by $yR^{-1}x \Leftrightarrow xRy$.

With respect to R, we define right neighborhood $r(x)$ of an element x in U, the R-related set of x in U, to be the set of y in V with the property that x is R-related to y. Thus, in symbols, $r(x) = \{y \in V | xRy\}$. Similarly, the left neighborhood $l(y)$ of an element y in V is $l(y) = \{x \in U | xRy\}$. By using concept of right neighborhoods, we define the lower and upper approximation operators $\underline{R}, \overline{R} : P(V) \rightarrow P(U)$ as follows [2,3,13,18,19,20]:

$$\underline{R}Y = \{x \in U | r(x) \subseteq Y\}, \text{ and } \overline{R}Y = \{x \in U | r(x) \cap Y \neq \emptyset\},$$

respectively. The pair $RY = (\underline{R}Y, \overline{R}Y)$ is referred to as the rough set of $Y \in P(V)$. Similarly, by using left neighborhoods, we can also define another pair of the lower and upper approximation operators $\underline{R_l}, \overline{R_l} : P(U) \rightarrow P(V)$ as follows:

$$\underline{R_l}X = \{y \in V | l(y) \subseteq X\}, \text{ and } \overline{R_l}X = \{y \in V | l(y) \cap X \neq \emptyset\},$$

respectively. It is easy to verify that $\overline{R_l}X = \overline{R^{-1}}X$ and $\underline{R_l}X = \underline{R^{-1}}X$. Because of these results, we only study the lower and upper approximation operators induced by right neighborhoods.

Proposition 2.1. Let U, V be two universal sets and R be an arbitrary binary relation from U to V. Then $\overline{R}Y = R \circ Y$ for all $Y \in P(V)$.

Proof. We only need to show that $\overline{R}Y(x) = 1$ if and only if $(R \circ Y)(x) = 1$. Suppose that $(\overline{R}Y)(x) = 1$, then $x \in \overline{R}Y$, by the definition of the upper approximation, $r(x) \cap Y \neq \emptyset$, this means that there exists some $y \in V$ such that $R(x, y) = Y(y) = 1$, hence $\vee_{y \in V}(R(x,y) \wedge Y(y)) = 1$. In other words, $(R \circ Y)(x) = 1$.

Conversely, if $(R \circ Y)(x) = 1$, then $\exists y \in V$ such that $R(x, y) = Y(y) = 1$, this means that $y \in r(x) \cap Y$, thus $r(x) \cap Y \neq \emptyset$ and $x \in \overline{R}Y$. We obtain $\overline{R}Y(x) = 1$. This shows that $\overline{R}Y(x) = 1$ if and only if $(R \circ Y)(x) = 1$. □

Proposition 2.1, in fact, gives an equivalent definition of rough set upper approximation.

If universal sets U and V are finite. How many the lower and upper approximations are there?

Proposition 2.2. Let U and V be two finite universal sets with $|U| = m$ and $|V| = n$. Then there are 2^{mn} different upper approximations and 2^{mn} different lower approximations.

Proof. Note that if $R \neq S$, then there exists at least one $Y \subseteq V$ such that $R \circ Y \neq S \circ Y$. Thus $R \neq S$ implies $\overline{R}Y \neq \overline{S}Y$ and $\underline{R}Y \neq \underline{S}Y$. In addition, there are 2^{mn} different binary relations from U to V. □

Proposition 2.1 illustrates that $\overline{R}Y = R \circ Y$ can be seen as an equivalent definition of the upper approximation operation. As for the lower approximation operation, we use formula $\underline{R}Y = (\overline{R}(Y^C))^C$, where Y^C denotes the complement of Y in V.

3   Properties of Rough Sets

By using the properties of binary relation, we can obtain corresponding properties of rough sets. We define the linear mapping as follows:

Definition 3.1. Let U and V be two universal sets. A mapping $f : P(V) \rightarrow P(U)$ is called to be linear if it satisfies the following conditions:
   (1) $f(\emptyset) = \emptyset$;
   (2) For any given index set I and $Y_i \in P(V), i \in I$, $f(\cup_{i \in I} Y_i) = \cup_{i \in I} f(Y_i)$.

Proposition 3.1. Let U and V be two universal sets, if $f : P(V) \rightarrow P(U)$ is a linear mapping, then there is a unique binary relation from U to V such that $f(Y) = \overline{R}Y$ for all $Y \in P(V)$.

Proof. By using the mapping f, we construct a binary relation from U to V as follows:

$$l(y) = f(\{y\}), y \in V.$$

Then $\overline{R}\emptyset = \emptyset = f(\emptyset)$, and for $\emptyset \neq Y \in P(V)$

$$\overline{R}Y = \cup_{y \in Y} l(y) = \cup_{y \in Y} f(\{y\})$$

$$= f(\cup_{y \in Y}\{y\}) = f(Y).$$

Thus $\overline{R}Y = f(Y)$ for all $Y \in P(V)$. If there exists another binary relation S such that $f(Y) = \overline{R}Y = \overline{S}Y$, then $R \circ Y = S \circ Y$, hence $R = S$. □

Let $\Omega(P(V), P(U))$ denote the set of all linear mappings from $P(V)$ to $P(U)$. We consider the mapping

$$\alpha : P(U \times V) \to \Omega(P(V), P(U)), \alpha(R) = \overline{R}, \tag{1}$$

for all binary relations R from U to V. We will show that α is a bijection between $P(U \times V)$ and $\Omega(P(V), P(U))$.

Proposition 3.2. Let α as in (1) above, then α is a bijection between $P(U \times V)$ and $\Omega(P(V), P(U))$.

Proof. Suppose that $\alpha(R) = \alpha(S)$, then $\overline{R} = \overline{S}$ and $R \circ Y = S \circ Y$ for all subsets $Y \subseteq V$, this means that $R = S$ and α is an injection. Proposition 3.1 guarantee that α is a surjection. Hence α is a bijection between $P(U \times V)$ and $\Omega(P(V), P(U))$. □

For any given index set I, assume that $R_i(i \in I)$ are binary relations from U to V. Recall that the union $\cup_{i \in I} R_i$ and the intersection $\cap_{i \in I} R_i$ are defined by $\cup_{i \in I} R_i = \{(x, y) | \exists i \in I, (x, y) \in R_i\}$ and $\cap_{i \in I} R_i = \{(x, y) | (x, y) \in R_i, \forall i \in I\}$, respectively.

Proposition 3.3. Let U and V be two universal sets. For any given index set I, $R_i(i \in I)$ are binary relations from U to V. Then
    (H1) $\overline{\emptyset}Y = \emptyset$;
    (H2) $(\overline{\cup_{i \in I} R_i})Y = \cup_{i \in I}\overline{R_i}Y$;
    (H3) If R and S are binary relations from U to V, then $\overline{R}Y \subseteq \overline{S}Y$ for all $Y \subseteq V$ if and only if $R \subseteq S$;
    (H4) If R and S are binary relations from U to V, then $\overline{R}Y = \overline{S}Y$ for all $Y \subseteq V$ if and only if $R = S$;
    (L1) $\underline{\emptyset}Y = U$;
    (L2) $(\underline{\cup_{i \in I} R_i})Y = \cap_{i \in I}\underline{R_i}Y$;
    (L3) If R and S are binary relations from U to V, then $\underline{R}Y \subseteq \underline{S}Y$ for all $Y \subseteq V$ if and only if $S \subseteq R$;
    (L4) If R and S are binary relations from U to V, then $\underline{R}Y = \underline{S}Y$ for all $Y \subseteq V$ if and only if $R = S$.

Proof. (H1) $\overline{\emptyset}Y = \emptyset \circ Y = \emptyset$.

(H2) $(\overline{\cup_{i \in I} R_i})Y = (\cup_{i \in I} R_i) \circ Y = \cup_{i \in I} R_i \circ Y = \cup_{i \in I} \overline{R_i}Y$.

(H3) If $\overline{R}Y \subseteq \overline{S}Y$, then $R \circ Y \subseteq S \circ Y$ for all $Y \subseteq V$, thus $R \subseteq S$ and vice versa.

(H4) If $\overline{R}Y = \overline{S}Y$, then $R \circ Y = S \circ Y$ for all $Y \subseteq V$, thus $R = S$ and vice versa.

Using duality, (L1),(L2), (L3) and (L4) can be proved in a similar way.   □

In general, $(R \cap S) \circ Y \neq R \circ Y \cap S \circ Y$, therefore, we do not guarantee that $(\overline{R \cap S})Y = \overline{R}Y \cap \overline{S}Y$. This can be seen from the following counter example.

Example 3.1. Let $U = V = \{1, 2, 3, 4\}$, $R = \{(1,1), (1,2), (1,4), (2,1), (2,3), (3,3), (3,4), (4,2), (4,4)\}$, $S = \{(1,1), (2,2), (2,3), (3,1), (3,3), (3,4), (4,3)\}$ and $X = \{1, 3, 4\}$. Then $(\overline{R \cap S})X = \{1, 2, 3\} \neq \overline{R}X \cap \overline{S}X$.

If $U = V$, much more can be said. Let U be a universal set and R be a binary relation on U. The diagonal relation on U is denoted by $\triangle = \{(x, x) | x \in U\}$. R is called reflexive if $\triangle \subseteq R$; R is called symmetric if $R^{-1} \subseteq R$; R is called transitive if $R^2 \subseteq R$; R is called idempotent if $R^2 = R$; R is called nilpotent if $R^m = 0$ for some positive integer m; R is called an equivalence relation if R is reflexive, symmetric and transitive. The powers of the lower and upper approximations are inductively defined as follows:

$\overline{R}^1 Y = \overline{R}Y, \overline{R}^2 Y = \overline{R}(\overline{R}Y), \cdots, \overline{R}^{r+1} Y = \overline{R}^r(\overline{R}Y)$, and

$\underline{R}^1 Y = \underline{R}Y, (\underline{R})^2 Y = \underline{R}(\underline{R}Y), \cdots, (\underline{R})^{r+1} Y = (\underline{R})^r(\underline{R}Y)$ for all positive integer r.

Proposition 3.4. Let U be a universal set and R be a binary relation on U.

(1) $\overline{\triangle}X = X$ for all $X \subseteq U$, i.e., $\overline{\triangle}$ is an identity mapping;

(2) R is reflexive if and only if $X \subseteq \overline{R}X$ for all $X \subseteq U$;

(3) R is symmetric if and only if $\overline{R^{-1}}X \subseteq \overline{R}X$ for all $X \subseteq U$;

(4) R is transitive if and only if $\overline{R^n}X \subseteq \overline{R}X$ for all $X \subseteq U$ and for $n = 1, 2, \cdots$;

(5) R is idempotent if and only if $\overline{R}(\overline{R}X) = \overline{R}X$ for all $X \subseteq U$;

(6) R is nilpotent if and only if there exists some positive integer m such that $\overline{R}^n X = \emptyset$ for all integer $n \geq m$ and for all $X \subseteq U$.

Proof. The proof of (1) is straightforward.

(2) R is reflexive $\Leftrightarrow \triangle \subseteq R \Leftrightarrow \overline{\triangle}X \subseteq \overline{R}X \Leftrightarrow X \subseteq \overline{R}X$.

(3) R is symmetric $\Leftrightarrow R^{-1} \subseteq R \Leftrightarrow \overline{R^{-1}}X \subseteq \overline{R}X$.

(4) It is well-known that R is transitive if and only $R^n \subseteq R$ if for $n = 1, 2, \cdots$.

(5) R is idempotent $\Leftrightarrow R^2 = R \Leftrightarrow \overline{R}(\overline{R}Y) = \overline{R}Y$.

(6) R is nilpotent if and only if there exists some positive integer m such that $R^m = \emptyset$, if and only if $\overline{R}^n Y = \emptyset$ for all integer $n \geq m$.

For the lower approximation, we have the similar results and omit it.   □

Here, we give counting upper approximations induced by binary relations.

Proposition 3.5. Let U be a universal set with n elements, then

(1) There are 2^{n^2} different upper approximations induced by binary relations on U.

(2) There are $2^{n(n-1)}$ different upper approximations induced by reflexive binary relations on U.

(3) There are $2^{\frac{1}{2}n(n+1)}$ different upper approximations induced by symmetric binary relations on U.

(4) Let $p(n)$ denote the number of different upper approximations induced by equivalence relations on U, then $p(n)$ satisfies the recurrence relation $p(n) = \sum_{j=0}^{n-1} C(n-1,j)p(n-j-1)$ and the initial condition $p(0) = 1$.

Proof. (1) The number of relations on U with n elements is 2^{n^2}.

(2) The number of reflexive relations on U with n elements is $2^{n(n-1)}$.

(3) The number of symmetric relations on U with n elements is $2^{\frac{1}{2}n(n+1)}$.

(4) The number $p(n)$ of equivalence relations on U with n elements satisfies the recurrence relation $p(n) = \sum_{j=0}^{n-1} C(n-1,j)p(n-j-1)$ and the initial condition $p(0) = 1$. Where $C(n-1,j)$ is the binomial coefficient. $\qquad\square$

However, since there is no known general formula for the number of transitive relations on universal set U with n elements, therefore there is no known general formula for the number of upper approximations induced by transitive relations on U with n elements.

Let U be an arbitrary universal set and R be a binary relation from U to V. Consider the following two sets:

$$K = \{Y | Y \in P(V), \overline{R}Y = \emptyset\}$$

and
$$L = \{X | X \in P(U), X = \overline{R}Y, \text{ for some } Y \in P(V)\}.$$

Proposition 3.6. Let U, V, R, K and L be as above, then

(1) K is a completely distributive lattice;

(2) For any given index set I, if $X_i \in L(i \in I)$, then $\cup_{i \in I} X_i \in L$;

(3) If $U = V$ and $R \subseteq R^2$, then $K \cap L = \emptyset$.

Proof. (1) Since $R \circ \emptyset = \emptyset$, we have $\emptyset \in K$ and $K \neq \emptyset$. For any given index set I, if $Y_i \in K(i \in I)$, then $R \circ (\cup_{i \in I} Y_i) = \cup_{i \in I} R \circ Y_i = \emptyset$ and $R \circ (\cap_{i \in I} Y_i) \subseteq R \circ Y_i = \emptyset$. Hence $\cup_{i \in I} Y_i, \cap_{i \in I} Y_i \in K$. Distributivity is trivial. So K is a completely distributive lattice.

(2) For every $i \in I, X_i \in L$, there exists some $Y_i \in P(V)$ such that $X_i = \overline{R}Y_i$, thus $\cup_{i \in I} X_i = \cup_{i \in I} R \circ Y_i = R \circ (\cup_{i \in I} Y_i) \in L$.

(3) Suppose that $X \in K \cap L$, then there exists some $Y \in P(U)$ such that $X = \overline{R}Y = R \circ Y$. It is noted that $X \in L$ implies $\emptyset = \overline{R}X = R \circ X = R^2 \circ Y \supseteq R \circ Y$. This means that $X = \overline{R}Y = R \circ Y = \emptyset$. $\qquad\square$

4   Composition of Approximation Operations

This section will discuss the composition of approximation operations. Let U, V and W be three distinct but related universal sets. Also let R be a binary relation from U to V, and S be a binary relation from V to W.

Proposition 4.1. Let U, V, W, R and S be as above, then

(1) $\overline{R(\overline{S}Z)} = \overline{R \circ S}Z$ for all $Z \in P(W)$;

(2) $\overline{R \circ (\cup_{i \in I} S_i)}Z = \cup_{i \in I} \overline{R \circ S_i}Z$ for any given index set I, $S_i \in P(V \times W)$, $i \in I$ and $Z \in P(W)$;

(3) $\overline{(\cup_{i \in I} R_i) \circ S}Z = \cup_{i \in I} \overline{R_i \circ S}Z$ for any given index set I, $R_i \in P(U \times V)$, $i \in I$ and $Z \in P(W)$;

(4) $\underline{(R \circ S)}Z = \underline{R(\underline{S}Z)}$;

(5) $\underline{R \circ (\cup_{i \in I} S_i)}Z = \cap_{i \in I} \underline{R \circ S_i}Z$ for any given index set I, $S_i \in P(V \times W)$, $i \in I$ and $Z \in P(W)$;

(6) $\underline{(\cup_{i \in I} R_i) \circ S}Z = \cap_{i \in I} \underline{R_i \circ S}Z$, for any given index set I, $R_i \in P(U \times V)$ and $Z \in P(W)$.

Proof. (1) $\overline{R(\overline{S}Z)} = (R \circ S) \circ Z = (R \circ S) \circ Z = \overline{R \circ S}Z$ for all $Z \in P(W)$.

(2) follows from $R \circ (\cup_{i \in I} S_i) = \cup_{i \in I} R \circ S_i$.

(3) follows from $(\cup_{i \in I} R_i) \circ S = \cup_{i \in I} R_i \circ S$.

By duality, the proof of parts (4)-(6) is analogous to that of parts (1)-(3). $\square$

Proposition 4.2. Let U and V be two universal sets and R be a binary relation from U to V, then

(1) $\overline{R\underline{R^{-1}}X} \subseteq X \subseteq \underline{R\overline{R^{-1}}}X$ for all $X \in P(U)$.

(2) $\overline{R^{-1}\underline{R}Y} \subseteq Y \subseteq \underline{R^{-1}\overline{R}}Y$ for all $Y \in P(V)$.

Proof. (1) Since

$$\overline{R\underline{R^{-1}}X}(x) = \vee_{xRy}\underline{R^{-1}}X(y)$$

$$= \vee_{xRy}(\wedge_{yR^{-1}z}X(z))$$

$$= \vee_{xRy}(\wedge_{zRy}X(z))$$

$$\leq \vee_{xRy}X(x) = X(x)$$

and

$$\underline{R\overline{R^{-1}}X}(x) = \wedge_{xRy}\overline{R^{-1}}X(y)$$

$$= \wedge_{xRy}(\vee_{yR^{-1}z}X(z))$$

$$= \wedge_{xRy}(\vee_{zRy}X(z))$$

$$\geq \wedge_{xRy}X(x) = X(x),$$

we have $\overline{R\underline{R^{-1}}}X \subseteq X \subseteq \underline{R\overline{R^{-1}}}X$ for all $X \in P(U)$. The proof of part (2) is analogous to the proof of part (1) and we omit it. $\square$

Corollary 4.1. Let U be a universal set and R be a symmetric relation on U, then $\overline{R\underline{R}}X \subseteq X \subseteq \underline{R\overline{R}}X$ for all $X \in P(U)$.

Proof. If R is symmetric, then $R_{-1} = R$. Part (1) follows $\overline{R\underline{R}}X \subseteq X \subseteq \underline{R\overline{R}}X$ for all $X \in P(U)$. $\square$

5 Conclusion

This paper systematically discussed rough sets via viewpoint of linear mappings. Many interesting linear properties have been derived. As we know, the upper approximation of rough sets is first explained as a linear mappings. The method of linear mappings hopefully will be useful to theoretical and applied researches of rough sets.

References

1. Grassmann, W.K., Tremblay, J.P.: Logical and discrete mathematics, A computer science perspective. Prentice-Hall, Inc., Englewood Cliffs (1996)
2. Kondo, M.: Algebraic approach to generalized rough sets. In: Ślęzak, D., Wang, G., Szczuka, M.S., Düntsch, I., Yao, Y. (eds.) RSFDGrC 2005. LNCS (LNAI), vol. 3641, pp. 132–140. Springer, Heidelberg (2005)
3. Kondo, M.: On the structure of generalized rough sets. Information Sciences 176, 589–600 (2006)
4. Liu, G.L.: Rough sets over the Boolean algebras. In: Ślęzak, D., Wang, G., Szczuka, M.S., Düntsch, I., Yao, Y. (eds.) RSFDGrC 2005. LNCS (LNAI), vol. 3641, pp. 124–131. Springer, Heidelberg (2005)
5. Liu, G.L.: The axiomatization of the rough set upper approximation operations. Fundamenta Informaticae 69, 331–342 (2006)
6. Liu, G.L.: Generalized rough sets over fuzzy lattices. Information Sciences 178, 1651–1662 (2008)
7. Liu, G.L.: Axiomatic Systems for Rough Sets and Fuzzy Rough Sets. International Journal of Approximate Reasoning 48, 857–867 (2008)
8. Pawlak, Z.: Rough sets. International Journal of Computer and Information Sciences 11, 341–356 (1982)
9. Pawlak, Z.: Rough sets: Theoretical aspects of reasoning about data. Kluwer Academic Publishers, Boston (1991)
10. Pawlak, Z., Skowron, A.: Rudiments of rough sets. Information Sciences 177(1), 3–27 (2007)
11. Pawlak, Z., Skowron, A.: Rough sets: Some extensions. Information Sciences 177(1), 28–40 (2007)
12. Pawlak, Z., Skowron, A.: Rough sets and boolean reasoning. Information Sciences 177(1), 41–73 (2007)
13. Pei, D., Xu, Z.: Transformation of rough set models. Knowledge-Based Systems 20, 745–751 (2007)
14. Qi, G., Liu, W.: Rough operations on Boolean algebras. Information Sciences 173, 49–63 (2005)
15. Radzikowska, A.M., Kerre, E.E.: Fuzzy rough sets based on residuated lattices. In: Peters, J.F., Skowron, A., Dubois, D., Grzymała-Busse, J.W., Inuiguchi, M., Polkowski, L. (eds.) Transactions on Rough Sets II. LNCS, vol. 3135, pp. 278–296. Springer, Heidelberg (2004)
16. Skowron, A., Stepaniuk, J.: Tolerance approximation spaces. Fundamenta Informaticae 27, 245–253 (1996)
17. Slowinski, R., Vanderpooten, D.: A Generalized Definition of Rough Approximations Based on Similarity. IEEE Trans. On Knowledge and Data Engineering 12(2), 331–336 (1990)

18. Yao, Y.Y.: Two views of the theory of rough sets in finite universes. International Journal of Approximate Reasoning 15, 291–317 (1996)
19. Yao, Y.Y.: Relational interpretations of neighborhood operators and rough set approximation operators. Information Sciences 111(1–4), 239–259 (1998)
20. Yao, Y.Y.: Constructive and algebraic methods of theory of rough sets. Information Sciences 109, 21–47 (1998)
21. Yao, Y.Y., Lin, T.Y.: Generalization of rough sets using modal logic. Intelligent Automation and Soft Computing, An International Journal 2, 103–120 (1996)
22. Zhang, H., Liang, H., Liu, D.: Two new operators in rough set theory with applications to fuzzy sets. Information Sciences 166(1–4), 147–165 (2004)

A Note on Attribute Reduction in the Decision-Theoretic Rough Set Model

Y. Zhao, S.K.M. Wong, and Y.Y. Yao

Department of Computer Science, University of Regina
Regina, Saskatchewan, Canada S4S 0A2
yanzhao@cs.uregina.ca, skmwong@rogers.com, yyao@cs.uregina.ca

Abstract. This paper considers two groups of studies on attribute reduction in the decision-theoretic rough set model. Attribute reduction can be interpreted based on either decision preservation or region preservation. According to the fact that probabilistic regions are non-monotonic with respect to set inclusion of attributes, attribute reduction for region preservation is different from the classical interpretation of reducts.

1 Introduction

Attribute reduction is an important problem of rough set theory. For classification tasks, we consider two possible interpretations of the concept of a reduct. The first interpretation views a reduct as a minimal subset of attributes that has the same classification power as the entire set of condition attributes [11]. The second interpretation views a reduct as a minimal subset of attributes that produces positive and boundary decision rules with precision over certain tolerance levels [19,22]. Studies on attribute reduction can therefore be divided into two groups.

One group concentrates on the decision class or classes to which an equivalence class belongs. An equivalence class leads to one decision class in consistent decision tables, and possibly more than one decision class in inconsistent decision tables. In the latter case, for each equivalence class Skowron [13] suggests a *generalized decision* consisting of the set of decision classes to which the equivalence class belongs. Similarly, Slezak [16] proposes the notion of *majority decision* that uses a binary vector for each equivalence class to indicate the decision classes to which it belongs. In general, a *membership distribution function* over decision classes may be used to indicate the degree to which an equivalence class belongs [15]. Zhang *et al.* [8,23] propose the *maximum distribution criterion* based on the membership distribution function. A reduct is a minimal subset of attributes that has the same classification power in terms of generalized decision, majority decision, decision distribution, or maximum distribution for all objects in the universe.

The other group concentrates on positive, boundary and negative regions of decision classes to which an equivalence class belongs. In the Pawlak rough set model [10], each equivalence class may belong to one of the two regions. The

C.-C. Chan et al. (Eds.): RSCTC 2008, LNAI 5306, pp. 61–70, 2008.

positive region is the union of equivalence classes that induce certain classification rules. The boundary region is the union of equivalence classes that induce uncertain classification rules. The negative region is in fact the empty set. In the decision-theoretic rough set model [19,20,21], a probabilistic generalization of Pawlak rough sets, probabilistic regions are defined by two threshold values that, in turn, are determined systematically from a loss function by using the Bayesian decision procedure. In this case, the probabilistic negative region may not be the empty set. It represents the fact that we do not want to make any positive or boundary decision for some equivalence classes [19,22]. The positive and boundary regions induce two different types of decision rules called the positive rules and boundary rules [19,22]. Although both types of rules may be probabilistic and uncertain, they have very different semantics. While a positive rule lead to a definite decision, a boundary rule leads to a "wait-and-see" decision.

Attribute reduction in the decision-theoretic rough set model is based on these types of probabilistic rules. Reduct construction may be viewed as the search of a minimal subset of attributes that produces positive and boundary decision rules satisfying certain tolerance levels of precision.

2   The Decision-Theoretic Rough Set Model

In many data analysis applications, objects are only perceived, observed, or measured by using a finite number of attributes, and are represented as an information table [10].

Definition 1. *An information table is the following tuple:*

$$S = (U, At, \{V_a \mid a \in At\}, \{I_a \mid a \in At\}),$$

where U is a finite nonempty set of objects, At is a finite nonempty set of attributes, V_a is a nonempty set of values of $a \in At$, and $I_a : U \to V_a$ is an information function that maps an object in U to exactly one value in V_a.

For classification problems, we consider an information table of the form $S = (U, At = \mathbf{C} \cup \{D\}, \{V_a\}, \{I_a\})$, where $\mathbf{C}$ is a set of condition attributes describing the objects, and D is a decision attribute that indicates the classes of objects.

Let $\pi_D = \{D_1, D_2, \ldots, D_m\}$ be a partition of the universe U defined by the decision attribute D. Each equivalence class $D_i \in \pi_D$ is called a decision class. Given another partition π_A of U defined by a condition attribute set $A \subseteq \mathbf{C}$, each equivalence class is defined as $[x]_A = \{y \in U \mid \forall a \in A(I_a(x) = I_a(y))\}$. The precision of an equivalence class $[x]_A \in \pi_A$ for predicting a decision class D_i is defined as:

$$p(D_i|[x]_A) = \frac{|[x]_A \cap D_i|}{|[x]_A|},$$

where $|\cdot|$ denotes the cardinality of a set. The precision is the ratio of the number of objects in $[x]_A$ that are correctly classified into the decision class D_i and the number of objects in $[x]_A$. The decision-theoretic rough set model utilizes ideas from Bayesian decision theory and computes two thresholds based on the notion of expected loss (conditional risk). For a detailed description, please refer to papers [19,20,21,22].

In the decision-theoretic rough set model, we can introduce tolerance thresholds for defining probabilistic positive, boundary and negative regions of a decision class and the partition formed by all decision classes. By using the thresholds, one can divide the universe U into three regions of a decision partition π_D based on two thresholds $0 \leq \beta < \alpha \leq 1$:

$$\mathrm{POS}_{(\alpha,\beta)}(\pi_D|\pi_A) = \{x \in U \mid p(D_{\max}([x]_A)|[x]_A) \geq \alpha\},$$
$$\mathrm{BND}_{(\alpha,\beta)}(\pi_D|\pi_A) = \{x \in U \mid \beta < p(D_{\max}([x]_A)|[x]_A) < \alpha\},$$
$$\mathrm{NEG}_{(\alpha,\beta)}(\pi_D|\pi_A) = \{x \in U \mid p(D_{\max}([x]_A)|[x]_A) \leq \beta\}, \tag{1}$$

where $D_{\max}([x]_A) \in \pi_D$ is a dominant decision class of the objects in $[x]_A$, i.e., $D_{\max}([x]_A) = \arg\max_{D_i \in \pi_D}\{\frac{|[x]_A \cap D_i|}{|[x]_A|}\}$. The Pawlak model, as a special case, can be derived by setting a loss function that produces $\alpha = 1$ and $\beta = 0$ [12]. We can also derive the 0.50 probabilistic model [12], the symmetric variable precision rough set model [24], and the asymmetric variable precision rough set model [6].

The three regions are pairwise disjoint, and the union is a covering of U. In the Pawlak rough set model, we can easily prove $\mathrm{POS}(\pi_D|\pi_A) \cup \mathrm{BND}(\pi_D|\pi_A) = U$ and $\mathrm{NEG}(\pi_D|\pi_A) = \emptyset$. In the decision-theoretic model, it may happen that $\mathrm{POS}_{(\alpha,\beta)}(\pi_D|\pi_A) \cup \mathrm{BND}_{(\alpha,\beta)}(\pi_D|\pi_A) \neq U$ and $\mathrm{NEG}_{(\alpha,\beta)}(\pi_D|\pi_A) \neq \emptyset$. The union of the probabilistic positive and boundary regions is called a probabilistic non-negative region.

2.1   Decision Making

Skowron proposes a generalized decision δ as the set of all decision classes an equivalence class takes [13]. For an equivalence class $[x]_A \in \pi_A$ the generalized decision is denoted as:

$$\delta([x]_A) = \{I_D(x) \mid x \in [x]_A\}$$
$$= \{D_i \in \pi_D \mid p(D_i|[x]_A) > 0\}.$$

By introducing precision thresholds, we can separate the Skowron's generalized decision into three parts. The part of *positive decisions* is the set of decision classes with the precision higher than or equal to α. A positive decision may lead to a definite and immediate action. The part of *boundary decisions* is the set of decision classes with the precision lower than α but higher than β. A boundary decision may lead to a "wait-and-see" action. A decision with the precision lower than or equal to β is not strong enough to support any further action. The union of positive decisions and boundary decisions can be called the set of *general decisions* that support actual decision making. Let $D_{\mathrm{POS}_{(\alpha,\beta)}}$, $D_{\mathrm{BND}_{(\alpha,\beta)}}$ and $D_{\mathrm{GEN}_{(\alpha,\beta)}}$ denote the positive, boundary and general decision sets, respectively. For an equivalence class $[x]_A \in \pi_A$,

$$D_{\text{POS}_{(\alpha,\beta)}}([x]_A) = \{D_i \in \pi_D \mid p(D_i|[x]_A) \geq \alpha\},$$
$$D_{\text{BND}_{(\alpha,\beta)}}([x]_A) = \{D_i \in \pi_D \mid \beta < p(D_i|[x]_A) < \alpha\},$$
$$D_{\text{GEN}_{(\alpha,\beta)}}([x]_A) = D_{\text{POS}_{(\alpha,\beta)}}([x]_A) \cup D_{\text{BND}_{(\alpha,\beta)}}([x]_A). \tag{2}$$

In the rest of this paper, we only focus on positive and general decision making, and the corresponding positive and non-negative regions. For other rough set regions, one can refer to Inuiguchi's study [4].

Example 1. Consider a simple information table $S = (U, At = \mathbf{C} \cup \{D\}, \{V_a\}, \{I_a\})$ shown in Table 1. The condition attribute set $\mathbf{C}$ partitions the universe into six equivalence classes: $[o_1]_{\mathbf{C}}, [o_2]_{\mathbf{C}}, [o_3]_{\mathbf{C}}, [o_4]_{\mathbf{C}}, [o_5]_{\mathbf{C}}$ and $[o_7]_{\mathbf{C}}$. Suppose $\alpha = 0.75$ and $\beta = 0.60$, we can reformat the table by including $D_{\text{POS}_{(\alpha,\beta)}}$, $D_{\text{BND}_{(\alpha,\beta)}}$ and $D_{\text{GEN}_{(\alpha,\beta)}}$ for all equivalence classes defined by $\mathbf{C}$.

Table 1. An information table and its reformation

	c_1 c_2 c_3 c_4 c_5 c_6	D
o_1	1 1 1 1 1 1	M
o_2	1 0 1 0 1 1	M
o_3	0 1 1 1 0 0	Q
o_4	1 1 1 0 0 1	Q
o_5	0 0 1 1 0 1	Q
o_6	1 0 1 0 1 1	F
o_7	0 0 0 1 1 0	F
o_8	1 0 1 0 1 1	F
o_9	0 0 1 1 0 1	F

	c_1 c_2 c_3 c_4 c_5 c_6	$D_{\text{POS}_{(\alpha,\beta)}}$	$D_{\text{BND}_{(\alpha,\beta)}}$	$D_{\text{GEN}_{(\alpha,\beta)}}$
$[o_1]_{\mathbf{C}}$	1 1 1 1 1 1	$\{M\}$	$\emptyset$	$\{M\}$
$[o_2]_{\mathbf{C}}$	1 0 1 0 1 1	$\emptyset$	$\{F\}$	$\{F\}$
$[o_3]_{\mathbf{C}}$	0 1 1 1 0 0	$\{Q\}$	$\emptyset$	$\{Q\}$
$[o_4]_{\mathbf{C}}$	1 1 1 0 0 1	$\{Q\}$	$\emptyset$	$\{Q\}$
$[o_5]_{\mathbf{C}}$	0 0 1 1 0 1	$\emptyset$	$\emptyset$	$\emptyset$
$[o_7]_{\mathbf{C}}$	0 0 0 1 1 0	$\{F\}$	$\emptyset$	$\{F\}$

2.2   Monotocity Property of the Regions

By considering the two thresholds separately, we obtain the following observations. For a partition π_A, the decrease of the precision threshold α can result an increase of the probabilistic positive region $\text{POS}_{(\alpha,\beta)}(\pi_D|\pi_A)$. Thus, we can make positive decisions for more objects. The decrease of the precision threshold β can result an increase of the probabilistic non-negative region $\neg\text{NEG}_{(\alpha,\beta)}(\pi_D|\pi_A)$, thus we can make general decision for more objects.

Consider any two subsets of attributes $A, B \subseteq \mathbf{C}$ with $A \subseteq B$. For any $x \in U$, we have $[x]_B \subseteq [x]_A$. In the Pawlak model, if $[x]_A \in \text{POS}(\pi_D|\pi_A)$, then its subset $[x]_B$ also is in the positive region, i.e., $[x]_B \in \text{POS}(\pi_D|\pi_B)$. At the same time, if $[x]_A \in \text{BND}(\pi_D|\pi_A)$, its subset $[x]_B$ may be in the positive region or the boundary region. If $[x]_A \in \text{NEG}(\pi_D|\pi_A)$, its subset $[x]_B$ may also belong to the positive region or the boundary region. We immediately obtain the monotonic property of the Pawlak positive and non-negative regions with respect to set inclusion of attributes:

$$A \subseteq B \Longrightarrow \text{POS}(\pi_D|\pi_A) \subseteq \text{POS}(\pi_D|\pi_B);$$
$$A \subseteq B \Longrightarrow \neg\text{NEG}(\pi_D|\pi_A) \subseteq \neg\text{NEG}(\pi_D|\pi_B).$$

That is, a larger subset of $\mathbf{C}$ induces a larger positive region and a larger non-negative region. The entire condition attribute set $\mathbf{C}$ induces the largest positive and non-negative regions.

The *quality of classification*, or the *degree of dependency of D*, is defined as [11]:

$$\gamma(\pi_D|\pi_A) = \frac{|\text{POS}(\pi_D|\pi_A)|}{|U|}, \tag{3}$$

which is equal to the generality of the positive region. Based on the monotocity of the Pawlak positive region, we can obtain the monotocity of the γ measure. That is, $A \subseteq B \Longrightarrow \gamma(\pi_D|\pi_A) \leq \gamma(\pi_D|\pi_B)$.

In the decision-theoretic model, for a subset $[x]_B$ of an equivalence class $[x]_A$, no matter to which region $[x]_A$ belongs, we do not know to which region $[x]_B$ belongs. Therefore, we cannot obtain the monotocity of the probabilistic regions with respect to set inclusion of attributes. The probabilistic positive and non-negative regions are monotonically increasing with respect to the decreasing of the α and β thresholds, respectively, but are non-monotonic with respect to the set inclusion of attributes. Intuitively, the largest condition attribute set $\mathbf{C}$ may not be able to induce the largest positive and non-negative regions.

In the decision-theoretic model, the quantitative γ measure can be extended to indicate the quality of a probabilistic classification. A straightforward transformation of the γ measure is denoted as follows [24]:

$$\gamma_{(\alpha,\beta)}(\pi_D|\pi_A) = \frac{|\text{POS}_{(\alpha,\beta)}(\pi_D|\pi_A)|}{|U|}. \tag{4}$$

Since the probabilistic positive region is non-monotonic, the $\gamma_{(\alpha,\beta)}$ measure is also non-monotonic with respect to the set inclusion of attributes.

3   Definitions and Interpretations of Attribute Reduction

A reduct $R \subseteq \mathbf{C}$ for positive decision preservation can be defined by requiring that the positive decisions of all objects are unchanged.

Definition 2. *Given an information table $S = (U, At = \mathbf{C} \cup \{D\}, \{V_a \mid a \in At\}, \{I_a \mid a \in At\})$, an attribute set $R \subseteq \mathbf{C}$ is a reduct of $\mathbf{C}$ with respect to the certain decisions of all objects if it satisfies the following two conditions:*

(i) $\forall x \in U(D_{\text{POS}_{(\alpha,\beta)}}([x]_R) = D_{\text{POS}_{(\alpha,\beta)}}([x]_\mathbf{C}));$
(ii)   *for any $R' \subset R$ the condition* (i) *does not hold.*

The definition for general decision preservation can be similarly defined by having the condition (i) stated as: $\forall x \in U(D_{\text{GEN}_{(\alpha,\beta)}}([x]_R) = D_{\text{GEN}_{(\alpha,\beta)}}([x]_\mathbf{C})).$

A reduct $R \subseteq \mathbf{C}$ for positive region preservation can be defined by requiring that the induced positive region is the maximum.

Definition 3. *An attribute set $R \subseteq \mathbf{C}$ is a reduct of $\mathbf{C}$ with respect to the positive region of π_D if $R = \arg\max_{A \subseteq \mathbf{C}}\{POS_{(\alpha,\beta)}(\pi_D | \pi_A)\}$. It can be stated loosely as,*

(i) $POS_{(\alpha,\beta)}(\pi_D | \pi_R) \supseteq POS_{(\alpha,\beta)}(\pi_D | \pi_{\mathbf{C}})$;
(ii)   *for any $R' \subset R$, $POS_{(\alpha,\beta)}(\pi_D | \pi_{R'}) \subset POS_{(\alpha,\beta)}(\pi_D | \pi_{\mathbf{C}})$.*

That is, a reduct R is the global maximum regarding all subsets of $\mathbf{C}$; it is also the local maximum regarding all its own subsets.

The definition for non-negative region preservation can be similarly defined as $R = \arg\max_{A \subseteq \mathbf{C}}\{\neg NEG_{(\alpha,\beta)}(\pi_D | \pi_A)\}$.

For simplicity, the qualitative measure can be replaced by the quantitative measure. For example, the set-theoretic measure of a region can be replaced by the cardinality of the region [10,18], or the entropy of the region [9,15,18].

3.1   An Interpretation of Region Preservation in the Pawlak Model

Pawlak defines a reduct as an attribute set satisfying the following two conditions.

Definition 4. *[10]*

(i) $POS(\pi_D | \pi_R) = POS(\pi_D | \pi_{\mathbf{C}})$;
(ii)   *for any attribute $a \in R$, $POS(\pi_D | \pi_{R-\{a\}}) \neq POS(\pi_D | \pi_R)$.*

Based on the fact that the Pawlak positive region is monotonic with respect to set inclusion of attributes, the attribute set $\mathbf{C}$ must produce the largest positive region. A reduct R produces a positive region as big as what $\mathbf{C}$ does, and all proper subsets of R cannot produce a bigger positive region than R does. Thus, only all proper subsets $R - \{a\}$ for all $a \in R$ need to be checked.

Many authors [1,3,10,18] use an equivalent quantitative definition of a Pawlak reduct, i.e., $\gamma(\pi_D | \pi_R) = \gamma(\pi_D | \pi_{\mathbf{C}})$. In other words, R and $\mathbf{C}$ induce the same quantitative measurement of the Pawlak positive region.

In the Pawlak model, for a reduct $R \subseteq \mathbf{C}$ we have $POS(\pi_D | \pi_R) \cap BND(\pi_D | \pi_R) = \emptyset$, and $POS(\pi_D | \pi_R) \cup BND(\pi_D | \pi_R) = U$. The condition $POS(\pi_D | \pi_R) = POS(\pi_D | \pi_{\mathbf{C}})$ is equivalent to $BND(\pi_D | \pi_R) = BND(\pi_D | \pi_{\mathbf{C}})$. The requirement of the same boundary region is implied in the definition of a Pawlak reduct. It is sufficient to consider only the positive region in the Pawlak model.

3.2   Difficulties with the Interpretations of Region Preservation in Probabilistic Models

Parallel to Pawlak's definition, an attribute reduct in a probabilistic model can be defined by requiring that the probabilistic positive region of π_D is unchanged. Such a definition has been proposed by Kryszkiewicz as a β-reduct [7], and by Inuiguchi as a β-low approximation reduct [4,5] for the variable precision rough set model. A typical definition is defined as follows.

Definition 5. *[7]*

    (i) $POS_{(\alpha,\beta)}(\pi_D|\pi_R) = POS_{(\alpha,\beta)}(\pi_D|\pi_C)$;

    (ii)   *for any attribute* $a \in R$, $POS_{(\alpha,\beta)}(\pi_D|\pi_{R-\{a\}}) \neq POS_{(\alpha,\beta)}(\pi_D|\pi_C)$.

In probabilistic models, many proposals have been made to extend the Pawlak attribute reduction by using the extended and generalized measure $\gamma_{(\alpha,\beta)}$. Accordingly, the condition (i) of the definition can be re-expressed as $\gamma_{(\alpha,\beta)}(\pi_D|\pi_R) = \gamma_{(\alpha,\beta)}(\pi_D|\pi_C)$. Although the definition, especially the definition based on the extended $\gamma_{(\alpha,\beta)}$ measure, is adopted by many researchers [2,3,7,17,24], the definition itself is inappropriate for attribute reduction in probabilistic models. We can make the following three observations.

Table 2. Probabilistic positive and non-negative regions defined by some attribute sets

$A \subseteq C$	$POS_{(\alpha,\beta)}(\pi_D|\pi_A)$	$\neg NEG_{(\alpha,\beta)}(\pi_D|\pi_A)$
C	$\{o_1, o_3, o_4, o_7\}$	$\{o_1, o_2, o_3, o_4, o_6, o_7, o_8\}$
$\{c_1, c_2, c_5\}$	$\{o_1, o_3, o_4, o_7\}$	$\{o_1, o_2, o_3, o_4, o_6, o_7, o_8\}$
$\{c_1, c_2\}$	$\{o_3\}$	$\{o_2, o_3, o_5, o_6, o_7, o_8, o_9\}$
$\{c_1, c_5\}$	$\{o_4, o_7\}$	$\{o_3, o_4, o_5, o_7, o_9\}$
$\{c_2, c_5\}$	$\{o_1, o_2, o_3, o_4, o_6, o_7, o_8\}$	$\{o_1, o_2, o_3, o_4, o_6, o_7, o_8\}$
$\{c_1\}$	$\emptyset$	$\emptyset$
$\{c_2\}$	$\emptyset$	U
$\{c_5\}$	$\{o_3, o_4, o_5, o_9\}$	$\{o_3, o_4, o_5, o_9\}$

Problem 1. In probabilistic models, the probabilistic positive region is non-monotonic regarding set inclusion of attributes. The equality relation in condition (i) is not enough for verifying a reduct, and may miss some reducts. At the same time, the condition (ii) should consider all subsets of a reduct R, not only the subsets $R - \{a\}$ for all $a \in R$.

Example 2. Suppose $\alpha = 0.75$ and $\beta = 0.60$ for Table 1. Compare the probabilistic positive regions defined by C and all subsets of $\{c_1, c_2, c_5\}$ listed in Table 2. It is clear that $POS_{(\alpha,\beta)}(\pi_D|\pi_C) = POS_{(\alpha,\beta)}(\pi_D|\pi_{\{c_1,c_2,c_5\}})$, and none of the subset of $\{c_1, c_2, c_5\}$ keeps the same positive region. Though, according to the non-monotocity of the probabilistic positive region, we can verify that $POS_{(\alpha,\beta)}(\pi_D|\pi_{\{c_2,c_5\}})$ is a superset of $POS_{(\alpha,\beta)}(\pi_D|\pi_C)$, and thus support positive decision for more objects. We can verify that $\{c_2, c_5\}$ is a reduct regarding the positive region preservation, and $\{c_1, c_2, c_5\}$ is not.

Problem 2. In probabilistic models, both the probabilistic positive region and the probabilistic boundary region, i.e., the probabilistic non-negative region, need to be considered for general decision making. The definition only reflects the probabilistic positive region and does not evaluate the probabilistic boundary region. Inuiguchi's definition for a β-upper approximation reduct also considers the general decision making. However, the equality relation used may be inappropriate [4].

Example 3. We use the same Table 1 to demonstrate this problem. Suppose $\alpha = 0.75$ and $\beta = 0.60$. Compare the probabilistic non-negative regions defined by $\mathbf{C}$ and all subsets of $\{c_1, c_2, c_5\}$ listed in Table 2. The probabilistic non-negative regions are equal regarding the attribute sets $\{c_1, c_2, c_5\}$, $\{c_2, c_5\}$ and $\mathbf{C}$. Furthermore, $\neg\text{NEG}_{(\alpha,\beta)}(\pi_D | \pi_{\{c_2\}}) = U$ is a superset of $\neg\text{NEG}_{(\alpha,\beta)}(\pi_D | \pi_{\mathbf{C}})$, and thus supports general decision for more objects. Therefore, $\{c_2\}$ is a reduct regarding the non-negative region preservation, and none of its superset is.

Problem 3. Based on the condition $\gamma_{(\alpha,\beta)}(\pi_D | \pi_R) = \gamma_{(\alpha,\beta)}(\pi_D | \pi_{\mathbf{C}})$, we can obtain $|\text{POS}_{(\alpha,\beta)}(\pi_D | \pi_R)| = |\text{POS}_{(\alpha,\beta)}(\pi_D | \pi_{\mathbf{C}})|$, but not $\text{POS}_{(\alpha,\beta)}(\pi_D | \pi_R) = \text{POS}_{(\alpha,\beta)}(\pi_D | \pi_{\mathbf{C}})$. This means that the quantitative equivalence of the probabilistic positive regions does not imply the qualitative equivalence of the probabilistic positive regions.

Example 4. Quantitatively, $|\text{POS}_{(\alpha,\beta)}(\pi_D | \pi_{\{c_5\}})| = |\text{POS}_{(\alpha,\beta)}(\pi_D | \pi_{\mathbf{C}})|$ indicates $\gamma_{(\alpha,\beta)}(\pi_D | \pi_{\{c_5\}}) = \gamma_{(\alpha,\beta)}(\pi_D | \pi_{\mathbf{C}})$. Qualitatively, they indicate two different sets of objects. The positive decision will be made for the two different sets of objects. Similarly, the quantitative equivalence of two regions $\neg\text{NEG}_{(\alpha,\beta)}(\pi_D | \pi_{\{c_1, c_2\}})$ and $\neg\text{NEG}_{(\alpha,\beta)}(\pi_D | \pi_{\mathbf{C}})$ does not imply the qualitative equivalence of them. They lead to general decision for two different sets of objects.

3.3   Constructing Reducts in the Decision-Theoretic Model

Constructing a reduct for decision preservation can apply the traditional methods, for example, the methods based on the discernibility matrix [14]. Both the rows and columns of the matrix correspond to the equivalence classes defined by $\mathbf{C}$. An element of the matrix is the set of all attributes that distinguish the corresponding pair of equivalence classes. Namely, the matrix element consists of all attributes on which the corresponding two equivalence classes have distinct values and distinct decision making. A discernibility matrix is symmetric. The elements of a positive decision-based discernibility matrix M_{POS} and a general decision-based discernibility matrix M_{GEN} are defined as follows. For any two equivalence classes $[x]_{\mathbf{C}}$ and $[y]_{\mathbf{C}}$,

$$M_{\text{POS}}([x]_{\mathbf{C}}, [y]_{\mathbf{C}}) = \{a \in \mathbf{C} \mid I_a(x) \neq I_a(y) \wedge D_{\text{POS}_{(\alpha,\beta)}}([x]_{\mathbf{C}}) \neq D_{\text{POS}_{(\alpha,\beta)}}([y]_{\mathbf{C}})\};$$

$$M_{\text{GEN}}([x]_{\mathbf{C}}, [y]_{\mathbf{C}}) = \{a \in \mathbf{C} \mid I_a(x) \neq I_a(y) \wedge D_{\text{GEN}_{(\alpha,\beta)}}([x]_{\mathbf{C}}) \neq D_{\text{GEN}_{(\alpha,\beta)}}([y]_{\mathbf{C}})\}.$$

Skowron and Rauszer showed that the set of attribute reducts are in fact the set of prime implicants of the reduced disjunctive form of the discernibility function [14]. Thus, a certain decision reduct is a prime implicant of the reduced disjunctive form of the discernibility function

$$\bigwedge\{\bigvee(M_{\text{POS}}([x]_{\mathbf{C}}, [y]_{\mathbf{C}})) | \forall x, y \in U \ (M_{\text{POS}}([x]_{\mathbf{C}}, [y]_{\mathbf{C}}) \neq \emptyset)\}. \tag{5}$$

A general decision reduct is a prime implicant of the reduced disjunctive form of the discernibility function

$$\bigwedge\{\bigvee(M_{\text{GEN}}([x]_{\mathbf{C}}, [y]_{\mathbf{C}})) | \forall x, y \in U \ (M_{\text{GEN}}([x]_{\mathbf{C}}, [y]_{\mathbf{C}}) \neq \emptyset)\}. \tag{6}$$

Based on the non-monotocity of the regions, the construction for region-based reduct is not trivial. One needs to exhaustively search all subsets of **C** in order to find the global optimal attribute set that induces the largest positive region or the largest non-negative region.

For local optimization, we know that if the positive decisions are equivalent regarding two attribute sets A and **C** for all objects in the universe, then the positive regions are also equivalent regarding A and **C**. That is,

$$\forall x \in U \ ([D_{\text{POS}_{(\alpha,\beta)}}([x]_A) = D_{\text{POS}_{(\alpha,\beta)}}([x]_\mathbf{C})]) \implies$$
$$[\text{POS}_{(\alpha,\beta)}(\pi_D|\pi_A) = \text{POS}_{(\alpha,\beta)}(\pi_D|\pi_\mathbf{C})].$$

The revised relation may not be true. It means the set of positive decision reducts is actually the set of attribute sets that can keep the same probabilistic positive region as **C** does. According to the above property, for each positive decision reduct R, there exists a subset of R which is a local optimal positive region reduct. Therefore, if we can construct a positive decision reduct, which is a decision problem, then we can check all its subsets for a local optimal positive region reduct, which is an optimization problem. This method can save time for checking all subsets of **C**. Similarly, for each general decision reduct R, there exists a subset of R which is a local optimal non-negative region reduct.

For our running example, $\{c_1, c_2, c_5\}$ is a positive decision reduct and $\{c_2, c_5\}$ is a local optimal positive region reduct; $\{c_2, c_5\}$ is a general decision reduct and $\{c_2\}$ is a local optimal non-negative region reduct.

4   Conclusion

Definitions of attribute reduction in the decision-theoretic rough set model are examined in this paper, regarding both decision preservation and region preservation. While attribute construction for decision preservation can explore the monotonicity, attribute construction for region preservation cannot be done in a similar manner. Decision-based reducts can be constructed by the traditional approaches such as the ones based on the discernibility matrix, while region-based reducts require exhaustive search methods for reduct construction. Heuristics and algorithms need to be studied for constructing global and local optimal region-based reducts.

References

1. Beaubouef, T., Petry, F.E., Arora, G.: Information-theoretic measures of uncertainty for rough sets and rough relational databases. Information Sciences 109, 185–195 (1998)
2. Beynon, M.: Reducts within the variable precision rough sets model: a further investigation. European Journal of Operational Research 134, 592–605 (2001)
3. Hu, Q., Yu, D., Xie, Z.: Information-preserving hybrid data reduction based on fuzzy-rough techniques. Pattern Recognition Letters 27, 414–423 (2006)
4. Inuiguchi, M.: Attribute reduction in variable precision rough set model. International Journal of Uncertainty, Fuzziness and Knowledge-Based Systems 14, 461–479 (2006)

5. Inuiguchi, M.: Structure-based attribute reduction in variable precision rough set models. Journal of Advanced Computational Intelligence and Intelligent Informatics 10, 657–665 (2006)
6. Katzberg, J.D., Ziarko, W.: Variable precision rough sets with asymmetric bounds. In: Ziarko, W. (ed.) Rough Sets, Fuzzy Sets and Knowledge Discovery, pp. 167–177. Springer, London (1994)
7. Kryszkiewicz, M.: Maintenance of reducts in the variable precision rough sets model, ICS Research Report 31/94, Warsaw University of Technology (1994)
8. Mi, J.S., Wu, W.Z., Zhang, W.X.: Approaches to knowledge reduction based on variable precision rough set model. Information Sciences 159, 255–272 (2004)
9. Miao, D.Q., Hu, G.R.: A heuristic algorithm for reduction of knowledge. Chinese Journal of Computer Research and Development 36, 681–684 (1999)
10. Pawlak, Z.: Rough sets. International Journal of Computer and Information Sciences 11, 341–356 (1982)
11. Pawlak, Z.: Rough Sets: Theoretical Aspects of Reasoning About Data. Kluwer Academic Publishers, Boston (1991)
12. Pawlak, Z., Wong, S.K.M., Ziarko, W.: Rough sets: probabilistic versus deterministic approach. International Journal of Man-Machine Studies 29, 81–95 (1988)
13. Skowron, A.: Boolean reasoning for decision rules generation. In: Proceedings of the International Symposium on Methodologies for Intelligent Systems, pp. 295–305 (1993)
14. Skowron, A., Rauszer, C.: The discernibility matrices and functions in information systems. In: Slowiński, R. (ed.) Intelligent Decision Support, Handbook of Applications and Advances of the Rough Sets Theory. Kluwer, Dordrecht (1992)
15. Slezak, D.: Approximate reducts in decision tables. Proceedings of Information Processing and Management of Uncertainty, 1159–1164 (1996)
16. Slezak, D.: Normalized decision functions and measures for inconsistent decision tables analysis. Fundamenta Informaticae 44, 291–319 (2000)
17. Swiniarski, R.W.: Rough sets methods in feature reduction and classification. International Journal of Applied Mathematics and Computer Science 11, 565–582 (2001)
18. Wang, G.Y., Zhao, J., Wu, J.: A comparitive study of algebra viewpoint and information viewpoint in attribute reduction. Fundamenta Informaticae 68, 1–13 (2005)
19. Yao, Y.Y.: Decision-theoretic rough set models. In: Yao, J., Lingras, P., Wu, W.-Z., Szczuka, M.S., Cercone, N.J., Ślęzak, D. (eds.) RSKT 2007. LNCS (LNAI), vol. 4481, pp. 1–12. Springer, Heidelberg (2007)
20. Yao, Y.Y., Wong, S.K.M.: A decision theoretic framework for approximating concepts. International Journal of Man-machine Studies 37, 793–809 (1992)
21. Yao, Y.Y., Wong, S.K.M., Lingras, P.: A decision-theoretic rough set model. In: Ras, Z.W., Zemankova, M., Emrich, M.L. (eds.) Methodologies for Intelligent Systems, vol. 5, pp. 17–24. North-Holland, New York (1990)
22. Yao, Y.Y., Zhao, Y.: Attribute reduction in decision-theoretic rough set models. Information Sciences 178, 3356–3373 (2008)
23. Zhang, W.X., Mi, J.S., Wu, W.Z.: Knowledge reduction in inconsistent information systems. Chinese Journal of Computers 1, 12–18 (2003)
24. Ziarko, W.: Variable precision rough set model. Journal of Computer and System Sciences 46, 39–59 (1993)

An Interpretation of Belief Functions on Infinite Universes in the Theory of Rough Sets

Wei-Zhi Wu[1] and Ju-Sheng Mi[2]

[1] School of Mathematics, Physics and Information Science,
Zhejiang Ocean University, Zhoushan, Zhejiang, 316004, P.R. China
wuwz@zjou.edu.cn
[2] College of Mathematics and Information Science, Hebei Normal University,
Shijiazhuang, Hebei, 050016, P.R. China
mijsh@263.net

Abstract. A general type of belief structure and its inducing dual pair of belief and plausibility functions on infinite universes of discourse are first defined. Relationship between belief and plausibility functions in Dempser-Shafer theory of evidence and the lower and upper approximations in rough set theory is then established. It is shown that the probabilities of lower and upper approximations induced by an approximation space yield a dual pair of belief and plausibility functions. And for any belief structure there must exist a probability approximation space such that the belief and plausibility functions defined by the given belief structure are just respectively the lower and upper probabilities induced by the approximation space. Finally, essential properties of the belief and plausibility functions are examined. The belief and plausibility functions are respective a monotone Choquet capacity and an alternating Choquet capacity of infinite order.

Keywords: Approximation operators, Belief functions, Belief structures, Rough sets.

1 Introduction

As a generalization of Bayesian theory of subjective judgment, the Dempster-Shafer theory of evidence (also called the theory of belief function) is a method used to model and manipulate uncertainty, imprecise, incomplete, and even vague information. It was originated by Dempster's concepts of lower and upper probabilities [2], and extended by Shafer [10] as a theory. The basic representational structure in this theory is a belief structure, which consists of a family of subsets called focal elements, with associated individual positive weights summing to one. The primitive numeric measures derived from the belief structure are a dual pair of belief and plausibility functions. With more than forty years' development, evidential reasoning has been emerging as a powerful methodology for pattern recognition, image analysis, diagnosis, knowledge discovery, information fusion, and decision making.

C.-C. Chan et al. (Eds.): RSCTC 2008, LNAI 5306, pp. 71–80, 2008.

The original concepts of belief and plausibility functions in Dempster-Shafer theory of evidence come from the lower and upper probabilities induced by a multi-valued mapping carrying a probability measure defined over subsets of the domain of the mapping [10], such a multi-valued mapping is in fact a random set [8]. The belief (resp. plausibility) function is a monotone Choquet capacity of infinite order (resp. alternating Choquet capacity of infinite order) [1] satisfying the sub-additive (resp. super-additive) property at any order [10]. The sub-additive and super-additive at any order form the essential properties of belief and plausibility functions respectively.

Another important method used to deal with uncertainty in intelligent systems characterized by insufficient and incomplete information is the theory of rough sets originated by Pawlak [9]. The basic structure of rough set theory is an approximation space consisting of a universe of discourse and a binary relation imposed on it. Using the concepts of lower and upper approximations in rough set theory, knowledge hidden in information systems may be unravelled and expressed in the form of decision rules. The belief and plausibility functions in the Dempster-Shafer theory of evidence seem to have some natural correspondences with the lower and upper approximations in rough set theory. The relationships between the Dempster-Shafer theory of evidence and rough set theory have received wide attention in the research community [3,4,6,11,12,13,15,19]. In finite universes of discourse, it has been demonstrated that different types of belief structures are associated with various rough approximation spaces such that different dual pairs of lower and upper approximation operators induced by the rough approximation spaces may be used to interpret the corresponding dual pairs of belief and plausibility functions derived by the belief structures [15,19]. It can be observed that the belief and plausibility functions in the Dempster-Shafer theory of evidence and lower and upper approximations in rough set theory capture the mechanisms of numeric and non-numeric aspects of uncertain knowledge respectively. The Dempster-Shafer theory of evidence may be used to deal with knowledge acquisition in information systems [7,14,16,20].

The purpose of this paper is to develop a general framework of belief and plausibility functions on infinite universes of discourse under the interpretation of theory of rough sets. By using an arbitrary belief structure, we introduce a dual pair of generalized belief and plausibility functions. We then establish the relationship between belief and plausibility functions in Dempster-Shafer theory of evidence and lower and upper approximations of rough set theory. We will also examine properties of belief functions and prove that the belief and plausibility functions respectively satisfy the essential properties of sub-additive and super-additive at any order.

2   Generalized Rough Set Models in Infinite Universes

Let X be a nonempty set called the universe of discourse. The class of all subsets of X will be denoted by $\mathcal{P}(X)$. For any $A \in \mathcal{P}(X)$, we denote by $\sim A$ the complement of A.

Let U and W be two nonempty universes of discourse. A subset $R \in \mathcal{P}(U \times W)$ is referred to as a binary relation from U to W. The relation R is referred to as serial if for any $x \in U$ there exists $y \in W$ such that $(x, y) \in R$. If $U = W$, $R \in \mathcal{P}(U \times U)$ is called a binary relation on U, $R \in \mathcal{P}(U \times U)$ is referred to as reflexive if $(x, x) \in R$ for all $x \in U$; R is referred to as symmetric if $(x, y) \in R$ implies $(y, x) \in R$ for all $x, y \in U$; R is referred to as transitive if for any $x, y, z \in U$, $(x, y) \in R$ and $(y, z) \in R$ imply $(x, z) \in R$; R is referred to as an equivalence relation if R is reflexive, symmetric and transitive.

Assume that R is an arbitrary binary relation from U to W. We can define a set-valued function $R_s : U \to \mathcal{P}(W)$ by:

$$R_s(x) = \{y \in W : (x, y) \in R\}, \quad x \in U.$$

$R_s(x)$ is called the successor neighborhood of x with respect to R. Obviously, any set-valued function F from U to W defines a binary relation from U to W by setting $R = \{(x, y) \in U \times W : y \in F(x)\}$. From the set-valued function R_s, we can define a basic set assignment [17,18] $j : \mathcal{P}(W) \to \mathcal{P}(U)$,

$$j(A) = \{u \in U : R_s(u) = A\}, \quad A \in \mathcal{P}(W).$$

It is easy to verify that j satisfies the properties (J1) and (J2):

$$(\text{J1}) \ A \neq B \implies j(A) \cap j(B) = \emptyset, \qquad (\text{J2}) \bigcup_{A \in \mathcal{P}(W)} j(A) = U.$$

If R is an arbitrary relation from U to W, then the triple (U, W, R) is referred to as a generalized approximation space. For any set $A \subseteq W$, a pair of lower and upper approximations, $\underline{R}(A)$ and $\overline{R}(A)$, are defined by

$$\underline{R}(A) = \{x \in U : R_s(x) \subseteq A\}, \quad \overline{R}(A) = \{x \in U : R_s(x) \cap A \neq \emptyset\}. \tag{1}$$

The pair $(\underline{R}(A), \overline{R}(A))$ is referred to as a generalized crisp rough set, and $\underline{R}$ and $\overline{R} : \mathcal{P}(W) \to \mathcal{P}(U)$ are called the lower and upper generalized approximation operators respectively.

From the definitions of approximation operators, the following theorem can be easily derived [5,9,17]:

Theorem 1. *For a given approximation space (U, W, R), the lower and upper approximation operators defined by Eq. (1) satisfy the following properties: for all $A, B, A_i \in \mathcal{P}(W), i \in J, J$ is an index set,*

(LD) $\underline{R}(A) = \sim \overline{R}(\sim A)$,   (UD) $\overline{R}(A) = \sim \underline{R}(\sim A)$;

(L1) $\underline{R}(W) = U$,   (U1) $\overline{R}(\emptyset) = \emptyset$;

(L2) $\underline{R}(\bigcap_{i \in J} A_i) = \bigcap_{i \in J} \underline{R}(A_i)$,   (U2) $\overline{R}(\bigcup_{i \in J} A_i) = \bigcup_{i \in J} \overline{R}(A_i)$;

(L3) $A \subseteq B \implies \underline{R}(A) \subseteq \underline{R}(B)$, (U3) $A \subseteq B \implies \overline{R}(A) \subseteq \overline{R}(B)$;

(L4) $\underline{R}(\bigcup_{i \in J} A_i) \supseteq \bigcup_{i \in J} \underline{R}(A_i)$, (U4) $\overline{R}(\bigcap_{i \in J} A_i) \subseteq \bigcap_{i \in J} \overline{R}(A_i)$.

Properties (LD) and (UD) show that $\underline{R}$ and $\overline{R}$ are dual approximation operators. Properties with the same number may be considered as dual properties. It can be easily checked that property (L2) implies properties (L3) and (L4), and dually, property (U2) yields properties (U3) and (U4).

By property (U2) we observe that $\overline{R}(X) = \bigcup_{x \in X} \overline{R}(\{x\})$. If we set

$$h(x) = \overline{R}(\{x\}), \quad x \in W,$$

then it is easy to verify that

$$h(x) = \{u \in U : x \in R_s(u)\}, \quad x \in W.$$

Conversely,

$$R_s(u) = \{y \in W : u \in h(y)\}, \quad u \in U.$$

Obviously,

$$\overline{R}(X) = \bigcup_{x \in X} h(x), \quad X \in \mathcal{P}(W).$$

Hence h is called the upper approximation distributive function [18]. The relationships between the basic set assignment j of R and the approximation operators can be concluded as follows:

$$\begin{aligned}
\text{(JL)} \quad &\underline{R}(X) = \bigcup_{Y \subseteq X} j(Y), & X \subseteq W; \\
\text{(JU)} \quad &\overline{R}(X) = \bigcup_{Y \cap X \neq \emptyset} j(Y), & X \subseteq W; \\
\text{(LJ)} \quad &j(X) = \underline{R}(X) \setminus \bigcup_{Y \subset X} \underline{R}(Y), & X \subseteq W.
\end{aligned}$$

With respect to certain special types, say, serial, reflexive, symmetric, and transitive binary relations, the approximation operators have additional properties [5,9,17,18].

Theorem 2. *Let R be an arbitrary crisp binary relation from U to W, and $\underline{R}$ and $\overline{R}$ the lower and upper generalized crisp approximation operators defined by Eq. (1). Then*

(1) *R is serial*
$$\Longleftrightarrow \text{(L0)} \quad \underline{R}(\emptyset) = \emptyset,$$
$$\Longleftrightarrow \text{(U0)} \quad \overline{R}(W) = U,$$
$$\Longleftrightarrow \text{(LU0)} \quad \underline{R}(A) \subseteq \overline{R}(A), \forall A \in \mathcal{P}(W).$$

If R is a binary relation on U, then

(2) *R is reflexive*
$$\Longleftrightarrow \text{(L5)} \quad \underline{R}(A) \subseteq A, \forall A \in \mathcal{P}(U),$$
$$\Longleftrightarrow \text{(U5)} \quad A \subseteq \overline{R}(A), \forall A \in \mathcal{P}(U).$$

(3) *R is symmetric*
$$\Longleftrightarrow \text{(L6)} \quad \overline{R}(\underline{R}(A)) \subseteq A, \forall A \in \mathcal{P}(U),$$
$$\Longleftrightarrow \text{(U6)} \quad A \subseteq \underline{R}(\overline{R}(A)), \forall A \in \mathcal{P}(U).$$

(4) *R is transitive*
$$\Longleftrightarrow \text{(L7)} \quad \underline{R}(A) \subseteq \underline{R}(\underline{R}(A)), \forall A \in \mathcal{P}(U),$$
$$\Longleftrightarrow \text{(U7)} \quad \overline{R}(\overline{R}(A)) \subseteq \overline{R}(A), \forall A \in \mathcal{P}(U).$$

If R is an equivalence relation on U, then the pair (U, R) is a Pawlak approximation space and more interesting properties of lower and upper approximation operators can be derived [9].

3 Belief Structures and Belief Functions on Infinite Universes

Definition 1. *Let W be a nonempty universe of discourse which may be infinite. A set function $m : \mathcal{P}(W) \to [0, 1]$ is referred to as a basic probability assignment or mass distribution if*

$$\text{(M1)} \ m(\emptyset) = 0, \quad \text{(M2)} \ \sum_{X \subseteq W} m(X) = 1.$$

A set $X \in \mathcal{P}(W)$ with $m(X) > 0$ is referred to as a focal element of m. We denote by $\mathcal{M}$ the family of all focal elements of m. The pair $(\mathcal{M}, m)$ is called a belief structure.

Associated with the belief structure $(\mathcal{M}, m)$, a pair of belief and plausibility functions can be derived.

Definition 2. *Let $(\mathcal{M}, m)$ be a belief structure on W. A set function $\text{Bel} : \mathcal{P}(W) \to [0, 1]$ is referred to as a belief function on W if*

$$\text{Bel}(X) = \sum_{M \subseteq X} m(M), \quad \forall X \in \mathcal{P}(W). \tag{2}$$

A set function $\text{Pl} : \mathcal{P}(W) \to [0, 1]$ is referred to as a plausibility function on W if

$$\text{Pl}(X) = \sum_{M \cap X \neq \emptyset} m(M), \quad \forall X \in \mathcal{P}(W). \tag{3}$$

Lemma 1. *Let $(\mathcal{M}, m)$ be a belief structure on W, then the focal elements of m constitute a countable set.*

Proof. For every $n \in \mathbf{N}$ (where $\mathbf{N}$ is the set of positive integer numbers), denote $\mathcal{D}_n = \{A \in \mathcal{P}(W) : m(A) > 1/n\}$. Since the sum of the masses of all focal element is 1, $\mathcal{D}_n$ is finite for every $n \in \mathbf{N}$ and therefore the set of the focal elements, that coincides with $\bigcup_{n=1}^{\infty} \mathcal{D}_n$, is countable.

4 Relationship between Belief Functions and Rough Sets on Infinite Universes

The following theorem shows that any belief structure can associate with a probability approximation space such that the probabilities of lower and upper approximations induced from the approximation space yield respectively the corresponding belief and plausibility functions derived from the given belief structure.

Theorem 3. *Let $(\mathcal{M}, m)$ be a belief structure on W which may be infinite. If* Bel $: \mathcal{P}(W) \to [0, 1]$ *and* Pl $: \mathcal{P}(W) \to [0, 1]$ *are respectively the belief and plausibility functions defined in Definition 2, then there exists a countable set U, a serial relation R from U to W, and a probability measure P on U such that*

$$\text{Bel}(X) = P(\underline{R}(X)), \quad \text{Pl}(X) = P(\overline{R}(X)), \quad \forall X \in \mathcal{P}(W). \tag{4}$$

Proof. Since $\sum\limits_{A \in \mathcal{P}(W)} m(A) = 1$, by Lemma 1 we know that the focal elements of m constitute a countable set, with no loss of generality, we assume that $\mathcal{M}$ has infinite countable elements and we denote

$$\mathcal{M} = \{A_i \in \mathcal{P}(W) : i \in \mathbf{N}\},$$

where $\sum\limits_{i \in \mathbf{N}} m(A_i) = 1$. Let $U = \{u_i : i \in \mathbf{N}\}$ be a set having infinite countable elements, we define a set function $P : \mathcal{P}(U) \to [0, 1]$ as follows:

$$P(\{u_i\}) = m(A_i), \quad i \in \mathbf{N},$$

$$P(X) = \sum_{u \in X} P(\{u\}), \quad X \in \mathcal{P}(U).$$

Obviously, P is a probability measure on U.

We further define a binary relation R from U to W as follows:

$$(u_i, w) \in R \iff w \in A_i, \quad i \in \mathbf{N}, w \in W.$$

From R we can obtain a mapping $j : \mathcal{P}(W) \to \mathcal{P}(U)$ as follows:

$$j(A) = \{u \in U : R_s(u) = A\}, \quad A \in \mathcal{P}(W).$$

It is easy to see that $j(A) = \{u_i\}$ for $A = A_i$ and $\emptyset$ otherwise. Consequently, $m(A) = P(j(A)) > 0$ for $A \in \mathcal{M}$ and 0 otherwise. Note that $j(A) \cap j(B) = \emptyset$ for $A \neq B$ and $\bigcup_{A \in \mathcal{P}(W)} j(A) = U$. Then, by property (JL), we can conclude that for any $X \in \mathcal{P}(W)$,

$$P(\underline{R}(X)) = P(\bigcup_{A \subseteq X} j(A)) = \sum_{A \subseteq X} P(j(A)) = \sum_{A \subseteq X} m(A) = \text{Bel}(X).$$

On the other hand, by property (JU), we have

$$P(\overline{R}(X)) = P(\bigcup_{A \cap X \neq \emptyset} j(A)) = \sum_{A \cap X \neq \emptyset} P(j(A)) = \sum_{A \cap X \neq \emptyset} m(A) = \text{Pl}(X).$$

Theorem 4. *Assume that (U, W, R) is a serial approximation space, U is a countable set, and $(U, \mathcal{P}(U), P)$ is a probability space. For $X \in \mathcal{P}(W)$, define*

$$m(X) = P(j(X)), \quad \text{Bel}(X) = P(\underline{R}(X)), \quad \text{Pl}(X) = P(\overline{R}(X)). \tag{5}$$

Then $m : \mathcal{P}(W) \to [0, 1]$ is a basic probability assignment on W and Bel $: \mathcal{P}(W) \to [0, 1]$ *and* Pl $: \mathcal{P}(W) \to [0, 1]$ *are respectively the belief and plausibility functions on W.*

Proof. Let

$$j(A) = \{x \in U : R_s(x) = A\}, \quad A \in \mathcal{P}(W).$$

It can be easily checked that j satisfies properties (J1) and (J2), i.e.,

(J1) $A \neq B \Longrightarrow j(A) \cap j(B) = \emptyset$, (J2) $\bigcup_{A \in \mathcal{P}(W)} j(A) = U.$

Since R is serial, we can observe that $j(\emptyset) = \emptyset$, consequently,

$$m(\emptyset) = P(j(\emptyset)) = P(\emptyset) = 0$$

and

$$\sum_{A \in \mathcal{P}(W)} m(A) = \sum_{A \in \mathcal{P}(W)} P(j(A)) = P(\bigcup_{A \in \mathcal{P}(W)} j(A)) = P(U) = 1.$$

Hence m is a basic probability assignment on W. And for any $X \in \mathcal{P}(W)$, according to properties (JL) and (J1) we have

$$\text{Bel}(X) = P(\underline{R}(X)) = P(\bigcup_{A \subseteq X} j(A)) = \sum_{A \subseteq X} P(j(A)) = \sum_{A \subseteq X} m(A).$$

Therefore, we have proved that Bel is a belief function. Similarly, by properties (JU) and (J1) we can conclude that

$$\text{Pl}(X) = P(\overline{R}(X)) = P(\bigcup_{A \cap X \neq \emptyset} j(A)) = \sum_{A \cap X \neq \emptyset} P(j(A)) = \sum_{A \cap X \neq \emptyset} m(A).$$

Therefore, Pl is a plausibility function.

5 Properties of Belief and Plausibility Functions on Infinite Universes

The following theorem presents the properties of belief and plausibility functions.

Theorem 5. *Let W be a nonempty set which may be infinite and $(\mathcal{M}, m)$ a belief structure on W. If Bel, Pl $: \mathcal{P}(W) \to [0, 1]$ are respectively the belief and plausibility functions induced from the belief structure $(\mathcal{M}, m)$. Then*

(1) $\text{Pl}(X) = 1 - \text{Bel}(\sim X), \quad X \in \mathcal{P}(W),$
(2) $\text{Bel}(X) \leq \text{Pl}(X), \quad X \in \mathcal{P}(W),$
(3) $\text{Bel}(X) + \text{Bel}(\sim X) \leq 1, \quad X \in \mathcal{P}(W),$
(4) $\text{Bel} : \mathcal{P}(W) \to [0, 1]$ *is a monotone Choquet capacity of infinite order on W, i.e., it satisfies the axioms (MC1)–(MC3) as follows:*
(MC1) $\text{Bel}(\emptyset) = 0,$
(MC2) $\text{Bel}(W) = 1.$
(MC3) *For any $n \in \mathbf{N}$ and $\forall X_i \in \mathcal{P}(W)$, $i = 1, 2, \ldots, n,$*

$$\text{Bel}(\bigcup_{i=1}^{n} X_i) \geq \sum_{\emptyset \neq J \subseteq \{1,2,\ldots,n\}} (-1)^{|J|+1} \text{Bel}(\bigcap_{j \in J} X_j).$$

(5) $\mathrm{Pl} : \mathcal{P}(W) \to [0,1]$ *is an alternating Choquet capacity of infinite order on* W, *i.e., it satisfies the axioms* (AC1)–(AC3) *as follows:*

(AC1) $\mathrm{Pl}(\emptyset) = 0$,

(AC2) $\mathrm{Pl}(W) = 1$,

(AC3) *For any* $n \in \mathbf{N}$ *and* $\forall X_i \in \mathcal{F}(W)$, $i = 1, 2, \dots, n$,

$$\mathrm{Pl}(\bigcap_{i=1}^{n} X_i) \leq \sum_{\emptyset \neq J \subseteq \{1,2,\dots,n\}} (-1)^{|J|+1} \mathrm{Pl}(\bigcup_{j \in J} X_j).$$

Proof. By Theorem 3, there exists a countable set U, a serial relation R from U to W, and a probability measure P on U such that

$$\mathrm{Bel}(X) = P(\underline{R}(X)), \quad \mathrm{Pl}(X) = P(\overline{R}(X)), \quad \forall X \in \mathcal{P}(W).$$

Then for any $X \in \mathcal{P}(W)$, by the dual property (UD) in Theorem 1, we have

$$\mathrm{Pl}(X) = P(\overline{R}(X)) = P(\sim \underline{R}(\sim X)) = 1 - P(\underline{R}(\sim X)) = 1 - \mathrm{Bel}(\sim X).$$

Thus property (1) holds.

(2) Notice that R is serial, then, according to property (LU0) in Theorem 2, we have

$$\mathrm{Bel}(X) = P(\underline{R}(X)) \leq P(\overline{R}(X)) = \mathrm{Pl}(X).$$

(3) follows immediately from (1) and (2).

(4) By property (L0) in Theorem 2, we have

$$\mathrm{Bel}(\emptyset) = P(\underline{R}(\emptyset)) = P(\emptyset) = 0,$$

that is, (MC1) holds. On the other hand, by property (L1) in Theorem 1, we have

$$\mathrm{Bel}(W) = P(\underline{R}(W)) = P(U) = 1,$$

thus (MC2) holds.

For any $n \in \mathbf{N}$ and $\forall X_i \in \mathcal{P}(W)$, $i = 1, 2, \dots, n$, by properties (L4) and (L2) in Theorem 1, we have

$$\begin{aligned}
\mathrm{Bel}(\bigcup_{i=1}^{n} X_i) = P(\underline{R}(\bigcup_{i=1}^{n} X_i)) &\geq P(\bigcup_{i=1}^{n} \underline{R}(X_i)) \\
&= \sum_{\emptyset \neq J \subseteq \{1,2,\dots,n\}} (-1)^{|J|+1} P(\bigcap_{j \in J} \underline{R}(X_j)) \\
&= \sum_{\emptyset \neq J \subseteq \{1,2,\dots,n\}} (-1)^{|J|+1} P(\underline{R}(\bigcap_{j \in J} X_j)) \\
&= \sum_{\emptyset \neq J \subseteq \{1,2,\dots,n\}} (-1)^{|J|+1} \mathrm{Bel}(\bigcap_{j \in J} X_j).
\end{aligned}$$

Thus (MC3) holds. Therefore, we have proved that Bel is a monotone Choquet capacity of infinite order on W.

(5) Similar to (4), by Theorems 1 and 2, we have

(AC1) $\mathrm{Pl}(\emptyset) = P(\overline{R}(\emptyset)) = P(\emptyset) = 0$.

(AC2) $\text{Pl}(W) = P(\overline{R}(W)) = P(U) = 1$.

(AC3) For any $n \in \mathbf{N}$ and $\forall X_i \in \mathcal{F}(W)$, $i = 1, 2, \ldots, n$, by properties (U4) and (U2), we have

$$
\begin{aligned}
\text{Pl}(\bigcap_{i=1}^{n} X_i) = P(\overline{R}(\bigcap_{i=1}^{n} X_i)) &\leq P(\bigcap_{i=1}^{n} \overline{R}(X_i)) \\
&= \sum_{\emptyset \neq J \subseteq \{1,2,\ldots,n\}} (-1)^{|J|+1} P(\bigcup_{j \in J} \overline{R}(X_j)) \\
&= \sum_{\emptyset \neq J \subseteq \{1,2,\ldots,n\}} (-1)^{|J|+1} P(\overline{R}(\bigcup_{j \in J} X_j)) \\
&= \sum_{\emptyset \neq J \subseteq \{1,2,\ldots,n\}} (-1)^{|J|+1} \text{Pl}(\bigcup_{j \in J} X_j).
\end{aligned}
$$

Thus we have concluded that Pl is an alternating Choquet capacity of infinite order on W.

From Theorem 5 we can see that semantics of the original Dempster-Shafer theory of evidence is still maintained.

6   Conclusion

We have investigated a general type of belief and plausibility functions on infinite universes of discourse. We have obtained the relationship between Dempster-Shafer theory of evidence and rough set theory on infinite universes of discourse. We have shown that the belief and plausibility functions defined by a belief structure on an infinite universe can be represented as lower and upper probabilities in a countable set induced by an approximation space. We have also examined properties of the belief and plausibility functions. The essential properties are that the belief and plausibility functions are respectively the monotone Choquet capacity and alternating Choquet capacity of infinite order.

Acknowledgement

This work was supported by grants from the National Natural Science Foundation of China (No. 60673096 and No. 60773174) and the Natural Science Foundation of Zhejiang Province in China (No. Y107262).

References

1. Choquet, G.: Theory of capacities. Annales de l'institut Fourier 5, 131–295 (1954)
2. Dempster, A.P.: Upper and lower probabilities induced by a multivalued mapping. Annals of Mathematical Statistics 38, 325–339 (1967)
3. Grzymala-Busse, J.W.: Rough-set and Dempster-Shafer approaches to knowledge acquisition under uncertainty—a comparison, Dept. Computer Science, University of Kansas, Lawrence, KS (unpublished manuscript, 1987)

4. Klopotek, M.A., Wierzchon, S.T.: A new qualitative rough-set approach to modeling belief function. In: Polkowski, L., Skowron, A. (eds.) RSCTC 1998. LNCS (LNAI), vol. 1424, pp. 346–353. Springer, Heidelberg (1998)
5. Kondo, M.: Algebraic approach to generalized rough sets. In: Slezak, D., Yao, J.T., Peters, J.F., Ziarko, W., Yao, Y.Y. (eds.) RSFDGrC 2005. LNCS (LNAI), vol. 3641, pp. 132–140. Springer, Heidelberg (2005)
6. Lin, T.Y.: Granular computing on binary relations II: rough set representations and belief functions. In: Polkowski, L., Skowron, A. (eds.) Rough Sets in Knowledge Discovery: 1. Methodolodgy and Applications, pp. 122–140. Physica, Heidelberg (1998)
7. Lingras, P.J., Yao, Y.Y.: Data mining using extensions of the rough set model. Journal of the American Society for Information Science 49, 415–422 (1998)
8. Nguyen, H.T.: On random sets and belief functions. Journal of Mathematical Analysis and Applications 65, 531–542 (1978)
9. Pawlak, Z.: Rough Sets: Theoretical Aspects of Reasoning about Data. Kluwer Academic Publishers, Boston (1991)
10. Shafer, G.: A Mathematical Theory of Evidence. Princeton University Press, Princeton (1976)
11. Skowron, A.: The relationship between the rough set theory and evidence theory. Bulletin of Polish Academic of Sciences: Mathematics 37, 87–90 (1989)
12. Skowron, A.: The rough sets theory and evidence theory. Fundamenta Informatica XIII, 245–262 (1990)
13. Skowron, A., Grzymala-Busse, J.: From rough set theory to evidence theory. In: Yager, R.R., Fedrizzi, M., Kacprzyk, J. (eds.) Advances in the Dempster-Shafer Theory of Evidence, pp. 193–236. Wiley, New York (1994)
14. Wu, W.-Z.: Attribute reduction based on evidence theory in incomplete decision systems. Information Sciences 178, 1355–1371 (2008)
15. Wu, W.-Z., Leung, Y., Zhang, W.-X.: Connections between rough set theory and Dempster-Shafer theory of evidence. International Journal of General Systems 31, 405–430 (2002)
16. Wu, W.-Z., Zhang, M., Li, H.-Z., Mi, J.-S.: Knowledge reduction in random information systems via Dempster-Shafer theory of evidence. Information Sciences 174, 143–164 (2005)
17. Yao, Y.Y.: Generalized rough set model. In: Polkowski, L., Skowron, A. (eds.) Rough Sets in Knowledge Discovery 1. Methodology and Applications, pp. 286–318. Physica, Heidelberg (1998)
18. Yao, Y.Y.: Relational interpretations of neighborhood operators and rough set approximation operators. Information Sciences 111, 239–259 (1998)
19. Yao, Y.Y., Lingras, P.J.: Interpretations of belief functions in the theory of rough sets. Information Sciences 104, 81–106 (1998)
20. Zhang, M., Xu, L.D., Zhang, W.-X., Li, H.-Z.: A rough set approach to knowledge reduction based on inclusion degree and evidence reasoning theory. Expert Systems 20, 298–304 (2003)

Some Remarks on Approximations of Arbitrary Binary Relations by Partial Orders

Ryszard Janicki*

Department of Computing and Software,
McMaster University,
Hamilton, ON, L8S 4K1 Canada
janicki@mcmaster.ca

Abstract. When a non-numerical ranking is created using Pairwise Comparisons paradigm, its first estimation is a binary relation which may not even be a partial order. In the paper four different partial order approximations of an arbirary binary relation are introduced and discussed.

1 Introduction and Motivation

While ranking the importance of *several* objects is often problematic (as the "perfect ranking" often does not exists [1]), it is often much easier when to do restricted to *two* objects. The problem is then reduced to constructing a global ranking from the set of partially ordered pairs. The method could be traced to the Marquis de Condorcet's 1795 paper (see [1]). At present the numerical version of pairwise comparisons based ranking is practically identified with the controversial Saaty's Analytic Hierarchy Process (AHP, [10]). On one hand AHP has respected practical applications, on the other hand it is still considered by many (see [2]) as a flawed procedure that produces arbitrary rankings. We believe that most of the problems with AHP stem mainly from the following two sources:

1. The final outcome is always expected to be totally ordered (i.e. for all a, b, either $a < b$ or $b > a$),
2. Numbers are used to calculate the final outcome.

An alternative, non-numerical method was proposed in [7] and refined in [5,6]. It is based on the concept of partial order and the concept of *partial order approximation of an arbitrary binary relation*. In [6] the non-numerical approach has been formalised as follows.

A *ranking* is just a partial order $Rank = (X, <^{rank})$, where X is the set of objects to be ranked and $<^{rank}$ is a ranking relation. We assume that $<^{rank}$ is a weak or total order. The ranking relation $<^{rank}$ is unknown and the *ranking problem* is to construct $<^{rank}$ on the basis of *ranking data*.

A *pairwise comparisons ranking data* is a tuple $PCRD = (X, R_0, R_1, ..., R_k)$, where $k \geq 1$, and R_i's are relations satisfying $R_0 \cup R_1 \cup ... \cup R_k = X \times X$ and

---

* Partially supported by NSERC grant of Canada.

$R_k \subseteq R_{k-1} \subseteq ... \subseteq R_1$. The relation R_0, interpreted as *indifference*, is symmetric and reflexive, the relations $R_1, ..., R_k$, interpreted as *preferences*, are asymmetric and irreflexive. In [5] the case $PCRD = (X, \approx, \sqsubseteq, \subset, <, \prec)$, with the following interpretation $a \approx b$: a and b are *indifferent*, $a \sqsubseteq b$: *slightly in favour of* b, $a \subset b$: *in favour of* b, $a < b$: b is *strongly better*, $a \prec b$: b is *extremely better*, was considered in some details. The list $\sqsubseteq, \subset, <, \prec$ may be shorter or longer, but not empty and not much longer (due to limitations of the human mind [7,10]).

We may now state the *ranking problem* more precisely as follows: "*derive the ranking relation $<^{rank}$ from a given pairwise comparison ranking data PCRD*". Note that in a general case, *none* of the relations R_i, $i = 1, ..., k$, could be even a partial order. The problem is that X is believed to be partially or weakly ordered by the ranking relation $<^{rank}$ but the data acquisition process may be so influenced by informational noise, imprecision, randomness, or expert ignorance that the collected data $R_1, R_2, ..., R_k$ are only some relations on X. We may say that they give a fuzzy picture of ranking, and to focus it, we must do some pruning and/or extending.

The methods of finding $<^{rank}$ presented in [6,7] are in principle based on the following three concepts

- partial order approximation $<_R$ of an arbitrary relation R,
- partial order approximation $<_{(R,\lhd)}$ of a pair of relations R and $\lhd$, where R is an arbitrary relation, $\lhd$ is a partial order included in R, and $\lhd \subseteq <_{(R,\lhd)}$ (the relation $\lhd$ represents the part of R that already is a precise ranking),
- weak order approximation of a given partial order.

Approximations of relations (sets, numbers, etc.) are usually defined as follows, a relation R^{up} is an (upper) approximation of R if R^{up} has a desired property and $R \subseteq R^{up}$, or, a relation R^{low} is an (lower) approximation of R if R^{low} has a desired property and $R^{low} \subseteq R$. This idea is behind many *closures* definitions [9] and Pawlak's Rough Sets [8]. Weak order approximations of partial orders follow this scheme [3], but partial order approximations of arbitrary relations do not have to. It appears that for partial order approximations of arbitrary relations the concepts "least" and "greatest" approximations are of limited use and we may have several different approximations, each of them could be considered as "the best" in some circumstances.

The approximation of R proposed in [6,7], denoted $(R^+)^\bullet$ in this paper, can be described as follows: "compute first the transitive closure of R, and next remove all cycles from it". The technique could be traced to E. Schröder's 1895 paper [11]. It seems to work nicely in many cases [5,6,7], but *not always*.

Consider the following example. Suppose we have four objects a, b, c, d, each of them is characterised by a vector of real numbers $(x_1, ..., x_4)$, so $a = (x_1^a, x_2^a, x_3^a, x_4^a)$, etc. Suppose that the measurements have errors so each x_i is only an estimation. We define the relation $<_{(1)}$ on real numbers as follows $x <_{(1)} y \iff y - x \geq 1$. The relation $<_{(1)}$ is a partial order, in fact it is a semi-order [3] (semi-orders are often used to model cases when errors of data are taken into account). We now define:

$$(x_1, x_2, x_3, x_4) \leftarrow (y_1, y_2, y_3, y_4) \iff (\exists i. \, x_i <_{(1)} y_i) \wedge (\forall i. \, \neg(y_i <_{(1)} x_i)).$$

In other words, if either $x_i <_{(1)} y_i$ or x_i and y_i are incomparable w.r.t. $<_{(1)}$ and at least for one j, $x_j <_{(1)} y_j$. This looks like a reasonable way of comparing *two* objects. Let

$$a = (1.0, 0.5, 0.5, 0.1), b = (0, 1.0, 0.5, 0.5), c = (0.5, 0, 1.0, 0.5), d = (0.9, 0.5, 0, 0.5).$$

We now have: $d \leftarrow c \leftarrow b \leftarrow a$, but the relation $\leftarrow$ is *not* transitive, as we have $\neg(c \leftarrow a)$, $\neg(d \leftarrow a)$, $\neg(d \leftarrow b)$. Using the technique of [5,7] we obtain the following totally ordered ranking: $d <^{rank} c <^{rank} b <^{rank} a$. The same result we will get by using AHP [10]. However, since all numerical values are only estimates and we can say that one is bigger than another only if the difference between them is greater or equal 1, the rank $<^{rank} = \emptyset$, i.e. a, b, c and d are *incomparable*, is what we would intuitively expect! In this paper we will propose a solution to this problem. Note that for $d = (0.5, 0.5, 0, 1.1)$ and the same a, b, c, we have $a \leftarrow d \leftarrow c \leftarrow b \leftarrow a$, so the technique of [5,7] produces $<^{rank} = \emptyset$, as expected (but AHP does not!).

In this paper we will introduce and analyse four different kinds of partial order approximations denoted $R^{\subseteq \wedge \bullet}$, $(R^{\bullet})^{\subseteq}$, $(R^{\bullet})^{+}$, $(R^{+})^{\bullet}$, respectively.

2 Relations and Partial Orders

In this section we recall some fairly known concepts and results that will be used in the following sections [3,9].

Let X be a finite set, fixed for the rest of this paper. For every relation $R \subseteq X \times X$, let $R^{+} = \bigcup_{i=1}^{\infty} R^i$, denote the *transitive closure* of R, $id = \{(x,x) \mid x \in X\}$ denote the identity relation, and let $R^{\circ} = R \cup id$ denote the *reflexive closure* of R (see [9] for details).

For each relation R and each $a \in X$ we define:

$$Ra = \{x \mid xRa\} \qquad aR = \{x \mid aRx\}.$$

A relation $< \in X \times X$ is a *(sharp) partial order* if it ir irreflexive and transitive, i.e. if $\neg(a < a)$ and $a < b < c \implies a < c$, for all $a, b, c \in X$.

We write $a \sim_< b$ if $\neg(a < b) \wedge \neg(b < a)$, that is if a and b are either *distinctly incomparable* (w.r.t. $<$) or *identical* elements. We also write

$$a \equiv_< b \iff (\{x \mid a < x\} = \{x \mid b < x\} \wedge \{x \mid x < a\} = \{x \mid x < b\}).$$

The relation $\equiv_<$ is an *equivalence relation* (i.e. it ir reflexive, symmetric and transitive) and it is called *the equivalence with respect to* $<$, since if $a \equiv_< b$, there is nothing in $<$ that can distinguish between a and b (see [3] for details). We always have $a \equiv_< b \implies a \sim_< b$.

A partial order is

- *total* or *linear*, if $\sim_<$ is empty, i.e., for all $a, b \in X$. $a \neq b \implies (a < b \vee b < a)$.
- *weak* or *stratified*, if $a \sim_< b \sim_< c \implies a \sim_< c$, i.e. if $\sim_<$ is an equivalence relation.

If a partial order $<$ is weak than $a \equiv_< b \iff a \sim_< b$ (see [3]).

The sets $R^{\circ}a$ and aR° allow some characterisation of relations in terms of set theory inclusion. We have two folklore results.

Lemma 1. *For every relation R:*

1. $bR^\circ \subset aR^\circ \implies aRb$,
2. $R^\circ a \subset R^\circ b \implies aRb$.

Proof. (1) Let $bR^\circ \subset aR^\circ$. Since $b \in bR^\circ$, then $b \in aR^\circ$, i.e. $aRb \vee a = b$. But $a = b$ implies $bR^\circ = aR^\circ$, so aRb.

(2) Dually to (1).   □

Lemma 2. *If R is a partial order then the following three statements are equivalent:*

1. aRb,
2. $bR^\circ \subset aR^\circ$,
3. $R^\circ a \subset R^\circ b$.

Proof. (2) $\implies$ (1) and (3) $\implies$ (1) follow from Lemma 1.

(1) $\implies$ (2): Let aRb and $x \in bR^\circ$. If $x = b$ then aRb implies $b \in aR^\circ$. If $x \neq b$ then aRb and bRx, which implies aRx, i.e. $x \in aR^\circ$. Hence $bR^\circ \subseteq aR^\circ$. But $aRb \implies a \neq b \wedge \neg bRa$, so $a \notin bR^\circ$, which means $bR^\circ \subset aR^\circ$.

(1) $\implies$ (3): Similarly to (1) $\implies$ (2).   □

Lemma 2 simply says that "*a is smaller that b if and only the set of all elements smaller than a is included in the set of all elements smaller than b, and if and only if the set of all element bigger that b is included in the set of all elements bigger than a*".

We will call the properties (2) and (3) of Lemma 2 *inclusion properties*, and say that the relation R has *inclusion properties* if it satisfies $bR^\circ \subset aR^\circ \wedge R^\circ a \subset R^\circ b$. Lemma 2 just says that R has inclusion properties if and only if it is a partial order.

A relation R is *acyclic* if and only if $\neg x R^+ x$ for all $x \in X$.

For every relation R, define the relations R^{cyc}, R^{cyc}_{id} and $R^\bullet$ as

- $aR^{cyc}b \iff aR^+b \wedge bR^+a$,
- $aR^{cyc}_{id}b \iff aR^{cyc}b \vee a = b$,
- $aR^\bullet b \iff aRb \wedge \neg(aR^{cyc}b)$,

We will call $R^\bullet$ an *acyclic refinement of R*. If $aR^{cyc}b$ we will say that a and b belong to some cycle(s).

Corollary 1

1. R^{cyc}_{id} is an equivalence relation and R is acyclic if and only if $R^{cyc} = \emptyset$,
2. $R^\bullet \subseteq R$, $R^\bullet$ is acyclic (i.e. also irreflexive), and $aR^\bullet b \iff aRb \wedge \neg(bR^+a)$,
3. if R is a partial order then $R = R^+ = R^\bullet$.   □

In this paper expressions like $(R^\bullet)^+$ are interpreted as $(R^\bullet)^+ = Q^+$ where $Q = R^\bullet$. Also for each equivalence relation $E \subseteq X \times X$, $[x]_E$ will denote the equivalence class of E containing x and X/E will donote the set of all equivalence classes of E.

Lemma 3 (Schröder [11]). *For every relation $R \subseteq X \times X$, let $\prec_R \subseteq (X/R^{cyc}_{id}) \times (X/R^{cyc}_{id})$ be the following relation:*

$$[x]_{R^{cyc}_{id}} \prec_R [y]_{R^{cyc}_{id}} \iff xR^+y \wedge \neg yR^+x.$$

The relation $\prec_R$ is a partial order on X/R^{cyc}_{id}.   □

Fig. 1. An example of R such that $aR^c b \wedge aR^{cyc}b$ for some a and b. We have $R^{\circ}a = \{a,c\} \subset \{a,b,c\} = R^{\circ}b$ and $bR^{\circ} = \{b,e\} \subset \{a,b,c\} = aR^{\circ}$, so $aR^c b$ and clearly $aR^{cyc}b$.

3   Inclusion Property and Equivalence w.r.t. a Given Relation

In this section two concepts initially introduced for partial orders will be extended to arbitrary relations. The first one is *inclusion property*.

For every relation R, define the relation R^c as follows :

- $aR^c b \iff bR^{\circ} \subset aR^{\circ} \wedge R^{\circ}a \subset R^{\circ}b.$

We will call R^c the *inclusion property kernel* of R.

Corollary 2.   *1.* $R^c \subseteq R$ and R^c is a partial order.
2. If R is a partial order then $R = R^c$. □

It may however happen that $aR^c b \wedge aR^{cyc}b$, see Figure 1, hence R^c alone can hardly be considered as a ranking derived from R. However it can be used as one of the tools that could be used for such a derivation.

The second concept is the relation $\equiv_<$ which can easily be extended to an arbitrary relation R.

For every relation R, define the relation $\equiv_R$ as follows :

- $a \equiv_R b \iff aR = bR \wedge Ra = Rb.$

The relation $\equiv_R$ is an *equivalence relation* (i.e. it ir reflexive, symmetric and transitive) and it is called *the equivalence with respect to R*, since if $a \equiv_R b$, there is nothing in R that can distinguish between a and b. Note also that:

$$a \equiv_R b \iff \forall x. (xRa \iff xRb) \wedge (aRx \iff bRx).$$

Lemma 4. *For every two relations R and Q:* $a \equiv_R b \wedge a \equiv_Q b \implies a \equiv_{R \cap Q} b.$
Proof. $a \equiv_R b \wedge a \equiv_Q b \implies aR = bR \wedge Ra = Rb \wedge aQ = bQ \wedge Qa = Qb \implies a(R \cap Q) = b(R \cap Q) \wedge (R \cap Q)a = (R \cap Q)b \iff a \equiv_{R \cap Q} b.$ □

It turns out that such operations as transitive closure, acyclic refinement and inclusion property kernel, preserve the equivalence with respect to R.

Lemma 5. *For every relation R we have:*
1. $a \equiv_R b \implies a \equiv_{R^+} b,$
2. $a \equiv_R b \implies a \equiv_{R^{\bullet}} b,$
3. $a \equiv_R b \implies a \equiv_{R^c} b.$

Proof. (1) $xR^+a \iff xRx_1R...Rx_nRa$. But $a \equiv_R b \implies (x_nRa \iff x_nRb)$, so $xR^+a \iff xRx_1R...Rx_nRb \iff xR^+b$. Similarly we show $aR^+x \iff bR^+x$, hence $a \equiv_{R^+} b$.

(2) Since $xR^\bullet a \iff xRa \land \neg aR^+x$ and $xRa \iff xRb$, then $xR^\bullet a \implies xRb$. Suppose bR^+x, i.e. $bRx_1R...x_kRx$. But $bRx_1 \iff aRx_1$, so $bR^+x \iff aR^+x$, a contradiction as $xR^\bullet a \implies \neg aR^+x$. Hence $xR^\bullet a \implies xR^\bullet b$. By replacing a with b, we immediately get $xR^\bullet b \implies xR^\bullet a$, i.e. $xR^\bullet a \iff xR^\bullet b$. In an almost identical manner we show $aR^\bullet x \iff bR^\bullet x$, so $a \equiv_{R^\bullet} b$.

(3) Note that if $a = b$ then clearly $a \equiv_{R^c} b$, so assume $a \neq b$.

First we show that $a \equiv_R b$ implies $\forall x.\ R^\circ x \subset R^\circ a \iff R^\circ x \subset R^\circ b$. Suppose $R^\circ x \subset R^\circ a$, i.e. $Rx \cup \{x\} \subset Ra \cup \{a\}$. Since $Ra = Rb$, then $R^\circ x = Rx \cup \{x\} \subseteq Rb \cup \{a\}$.

We now have to consider two cases:

Case 1: $a \in Rb$. Since $Ra = Rb$ then $a \in Ra$, so we have $R^\circ x \subset R^\circ a \cup \{a\} = Ra = Rb \subseteq Rb \cup \{b\} = R^\circ b$, so $R^\circ x \subset R^\circ b$.

Case 2: $a \notin Rb$. First we show that $a \in Rx \implies a \in Rb$. We have $a \in Rx \iff aRx \iff bRx \iff b \in Rx$ and $b \in Rx \subseteq R^\circ x \subset Ra \cup \{a\} \implies bRa \lor a = b$. Since $a \neq b$ then bRa. Because $a \equiv_R b$ the we have $Ra = Rb$ and $aR = bR$, so $bRa \land Ra = Rb \implies bRb$, while $bRb \land aR = bR \implies aRb$, i.e. $a \in Rb$. This means $a \notin Rb$ implies $a \notin Rb \land a \notin Rx$. Hence we have: $R^\circ a = R^\circ a \setminus \{a\} \subset (Rb \cup \{a\}) \setminus \{a\} = Rb \subseteq R^\circ b$, so $R^\circ x \subset R^\circ b$. In this way we have proved $\forall x.\ R^\circ x \subset R^\circ a \implies R^\circ x \subset R^\circ b$. Similarly we prove that $a \equiv_R b$ implies $\forall x.\ xR^\circ \subset aR^\circ \implies xR^\circ \subset bR^\circ$, which means that $a \equiv_R b$ implies

$$\forall x.\ (R^\circ x \subset R^\circ a \land xR^\circ \subset aR^\circ) \implies (R^\circ x \subset R^\circ b \land xR^\circ \subset bR^\circ).$$

By replacing a with b we get an inverse inclusion, so in fact we proved:

$$\forall x.\ (R^\circ x \subset R^\circ a \land xR^\circ \subset aR^\circ) \iff (R^\circ x \subset R^\circ b \land xR^\circ \subset bR^\circ),$$

i.e. $\forall x.\ (xR^c a \iff xR^c b)$. In almost identical way we can prove $\forall x.\ (aR^c x \iff bR^c x)$. Hence $a \equiv_{R^c} b$. $\square$

4   Approximating Relations by Partial Orders

We will start with a formal definition of a *partial order approximation* of a relation R.

Definition 1. *A partial order* $< \subseteq X \times X$ *is a* partial order approximation *of a relation* $R \subseteq X \times X$ *if it satisfies the following three conditions:*

1. $a < b \implies aR^+b$,
2. $a < b \implies \neg aR^{cyc}b$ *(or, equivalently $a < b \implies \neg bR^+a$)*,
3. $aR^c b \land aR^\bullet b \implies a < b$,
4. $a \equiv_R b \implies a \equiv_< b$. $\square$

Since R^+ is the smallest transitive relation containing R (see [9]), and due to informational noise, imprecision, randomness, etc., some parts of R might be missing, it is reasonable to assume that R^+ is the upper bound of $<$.

If R is interpreted as an estimation of a ranking, then $aR^{cyc}b$ is interpreted that as far as ranking is concerned, a and b are indifferent, so $aR^{cyc}b \implies (\neg a < b \land \neg b < a)$, which is expressed by (2) of the above definition. When $a < b \implies aR^+b$, then $\neg aR^{cyc}b$ can be replaced by $\neg bR^+a$.

The condition (3) defines the lower bound. Note that the greatest partial order included in R often does not exist, however if R is interpreted as an estimation of a ranking, it is reasonable to assume that the inclusion property refinement in included in the ranking order. However, as Figure 1 shows, it may happen that $aR^C b$ and $aR^{cyc}b$ so we need to add $aR^\bullet b$ to avoid a contradiction.

The condition (4) ensures preservation of the equivalence with respect to R.

Since R is constructed on the basis of pairwise comparisons paradigm, it may happen that aRb makes sense only locally, when the domain is restricted to $\{a,b\}$, and it needs to be pruned in global setting (as the relation $\leftarrow$ from Section 1). In such cases we may require $a <^{rank} b \implies aRb$, which leads to the following definition.

Definition 2. *A partial order $< \subseteq X \times X$ is an inner partial order approximation of a relation $R \subseteq X \times X$, if it is a partial order approximation of R, and satisfies:*
$$a < b \implies aRb.$$ □

Every partial order is transitive, acyclic and equal to its inclusion property kernel. An arbitrary relation R may not have these properties but we may try to refine R using transive closure, acyslic refinement and finding inclusion property kernel, in various orders or simultaneously (i.e. using set theory intersection). We will show that there are exactly four partial order approximations that can be derived in this way.

Let us first define the relation $R^{C \wedge \bullet}$ as follows:

$$aR^{C \wedge \bullet} b \iff aR^C b \wedge aR^\bullet b.$$

We can now formulate the main result of this paper.

Theorem 1

1. *The relations $R^{C \wedge \bullet}$, $(R^\bullet)^C$, $(R^\bullet)^+$, $(R^+)^\bullet$ are partial order approximations of R.*
2. *The relations $R^{C \wedge \bullet}$ and $(R^\bullet)^C$ are inner partial order approximations of R.*
3. $R^{C \wedge \bullet} \subseteq (R^\bullet)^C \subseteq (R^\bullet)^+ \subseteq (R^+)^\bullet$.
4. *If R is transitive, i.e. $R = R^+$, then $R^{C \wedge \bullet} = (R^\bullet)^C = (R^\bullet)^+ = (R^+)^\bullet$.*
5. *If R is a partial order, then $R = R^{C \wedge \bullet} = (R^\bullet)^C = (R^\bullet)^+ = (R^+)^\bullet$.*
6. *If R is acyclic, i.e. $R = R^\bullet$, then $R^C = R^{C \wedge \bullet} = (R^\bullet)^C$ and $(R^\bullet)^+ = (R^+)^\bullet$.*
7. *A partial order $<$ is a partial order approximation of R if and only if*
$$aR^{C \wedge \bullet} b \implies a < b \implies a(R^+)^\bullet b.$$
8. $aR^{cyc}b \implies a \equiv_{(R^+)^\bullet} b.$
9. *The realtions $R^{C \wedge \bullet}$, $(R^\bullet)^C$, $(R^\bullet)^+$, $(R^+)^\bullet$ are the only partial order approximations of R that can be derived from R by using operations '$\cap$', '^{C}', '$^+$' and '$^\bullet$'.* □

With an exception of (8), the above theorem is practically self-explained. The assertion (8) says that if a and b belong to a cycle in R then they are equivalent with respect to $(R^+)^\bullet$. This indicate that if we have a reason to believe that all cycles result from errors, informational noise, etc., and all elements of a cycle should be interpreted as indifferent, then $(R^+)^\bullet$ is most likely the best partial order approximation of R.

Proof of Theorem 1. First we show that the relations $R^{C \wedge \bullet}$, $(R^\bullet)^C$, $(R^\bullet)^+$, $(R^+)^\bullet$ are partial orders. Consider $R^{C \wedge \bullet}$. Clearly $aR^{C \wedge \bullet} b \iff aR^C b \wedge aR^\bullet b \iff aR^C b \wedge \neg bR^+ a$. By Corollary 1(2) the relation $R^{C \wedge \bullet}$ is irreflexive so we need only to prove its transitivity. Suppose that $aR^{C \wedge \bullet} b$ and $bR^{C \wedge \bullet} c$. This means $aR^C b$, $bR^C c$, $\neg bR^+ a$ and $\neg cR^+ b$. By

Corollary 1(3), R^C is transitive, so $aR^C c$, and by Lemma 1, aRb, bRc and aRc. Hence we only need to show that $\neg cR^+ a$. Suppose $cR^+ a$. Then $cR^+ a$ and aRb implies $cR^+ b$, a contradition as $aR^{C \wedge \bullet} c$ implies $\neg cR^+ b$. Therefore $R^{C \wedge \bullet}$ is a partial order.

Consider $(R^\bullet)^C$. From Corollary 1(3) it immediately follows that the relation $(R^\bullet)^C$ is a partial order.

Consider $(R^\bullet)^+$. By Corollary 1(2), we have $aR^\bullet b \iff aRb \wedge \neg(bR^+ a)$. The relation $(R^\bullet)^+$ is clearly transitive, we need only to show $\neg(a(R^\bullet)^+ a)$ for all $a \in X$. Since $aRb \wedge \neg(bR^+ a) \implies a \neq b$, then $\neg aR^\bullet a$. Suppose $a(R^\bullet)^+ a$. Since $\neg aR^\bullet a$, this means $aR^\bullet b(R^\bullet)^+ a$, for some $b \neq a$. But $aR^\bullet b \implies aRb$ and $b(R^\bullet)^+ a \implies bR^+ a$, so we have $aRb \wedge bR^+ a$, contradicting $aR^\bullet b$. Hence $\neg(a(R^\bullet)^+ a)$, i.e. $(R^\bullet)^+$ is a partial order.

Consider $(R^+)^\bullet$. Notice that $a(R^+)^\bullet b \iff aR^+ b \wedge \neg bR^+ a \iff [x]_{R_{id}^{cyc}} \prec_R [y]_{R_{id}^{cyc}}$, where $\prec_R$ is the relation from Lemma 3. Hence, by Lemma 3, the relation $(R^+)^\bullet$ is a partial order.

We will now prove (3), i.e. $R^{C \wedge \bullet} \subseteq (R^\bullet)^C \subseteq (R^\bullet)^+ \subseteq (R^+)^\bullet$.
Suppose $aR^{C \wedge \bullet} b$, i.e. $aR^C b \wedge \neg bR^+ a$. Then aRb and $\neg bR^+ a$, so $a \in (R^\bullet)^\circ a \cap (R^\bullet)^\circ b$. Assume that $x \in R^\bullet a$ and $x \notin R^\bullet b$. Since $aR^{C \wedge \bullet} b \implies aR^C b$, then we have $Ra \subset Rb$. But $R^\bullet a \subseteq Ra$, so $x \in Rb$. We now have $x \in Rb$ and $x \notin R^\bullet b$, i.e. $bR^+ x$. Since $x \in R^\bullet a$ means xRa, then $bR^+ ax$ and xRa give us $bR^+ a$, a contradiction as, $aR^{C \wedge \bullet} b \implies \neg bR^+ a$. Hence $R^\bullet a \subseteq R^\bullet b$. Since $a \neq b$ then $R^\bullet a \neq R^\bullet b$, so $(R^\bullet)^\circ a \subset (R^\bullet)^\circ b$. Similarly we show $b(R^\bullet)^\circ \subset a(R^\bullet)^\circ$, hence $a(R^\bullet)^C b$. Therefore $R^{C \wedge \bullet} \subseteq (R^\bullet)^C$.

By Lemma 1 we have $(R^\bullet)^C \subseteq R^\bullet$, and clearly $R^\bullet \subseteq (R^\bullet)^+$, hence $(R^\bullet)^C \subseteq (R^\bullet)^+$. Suppose $a(R^\bullet)^+ b$. Recall that $x(R^+)^\bullet y \iff xR^+ y \wedge \neg yR^+ x$. By Corollary 1(2), we have $R^\bullet \subset R$. Hence $a(R^\bullet)^+ + b \implies aR^+ b$. Suppose $bR^+ a$. Then $aR^{cyc} b$, i.e. $\neg aR^\bullet b$, a contradiction. Hence $a(R^+)^\bullet b$, i.e. $(R^\bullet)^+ \subseteq (R^+)^\bullet$. Therefore we have proved the assertion (3).

Note that (3) together with the fact that all $R^{C \wedge \bullet}$, $(R^\bullet)^C$, $(R^\bullet)^+$, $(R^+)^\bullet$ are partial orders imply that $R^{C \wedge \bullet}$, $(R^\bullet)^C$, $(R^\bullet)^+$, $(R^+)^\bullet$ satisfy (1),(2) and (3) of Definition 1. By Lemma 5, $(R^\bullet)^+$ and $(R^+)^\bullet$ satisfy (4) of Definition 1; and by Lemmas 5 and 4, $R^{C \wedge \bullet}$ and $(R^\bullet)^C$ satisfy satisfy (4) of Definition 1. Therefore the assertion (1) of the above theorem does hold.

The assertion (1) and Corollary 2(2) yield the assertion (2).

Hence (1), (2) and (3) hold. We will now prove (4). It suffices to show that if $R = R^+$ then $(R^+)^\bullet \subseteq R^{C \wedge \bullet}$. Note that in this case $a(R^+)^\bullet b \iff aRb \wedge \neg bRa$. If $R = R^+$ then $(R^+)^\bullet = R^\bullet$, so we only need to show $(R^+)^\bullet \subseteq R^C$. Let $a(R^+)^\bullet b$. This means $a \neq b$ and $\neg bRa$. Furthermore $\neg bRa$ implies $a \notin bR \wedge b \notin Ra$. Assume $x \in bR^\circ$. If $x = b$ then aRb implies $b \in aR$, i.e. $x \in aR^\circ$. If $x \neq b$ then $x \in bR^\circ \implies bRx$. Since R is transitive $aRb \wedge bRx \implies aRx \implies x \in Ra \implies x \in R^\circ a$. Hence $bR^\circ \subseteq aR^\circ$. Since $a \neq b$ and $a \notin bR$, then $a \notin bR^\circ$, which means $bR^\circ \subset aR^\circ$. Dually we show $R^\circ a \subset R^\circ b$, i.e. $aR^C b$, so we have proved (4).

The assertion (5) follows from (4) and Lemma 2.

If $R = R^\bullet$ then clearly $(R^\bullet)^C = R^C$. We also heve $R^{C \wedge \bullet} = R^C \cap R^\bullet = R^C \cap R = R^C$ as, by Lemma 1, $R^C \subseteq R$. From (3) it follows $(R^\bullet)^+ \subseteq (R^+)^\bullet$. If $R = R^\bullet$, then $(R^+)^\bullet \subseteq R^+ = (R^\bullet)^+$, i.e. $(R^\bullet)^+ = (R^+)^\bullet$, so we have proved (6).

The assertion (7) follows from (1), (3) and Definition 1.

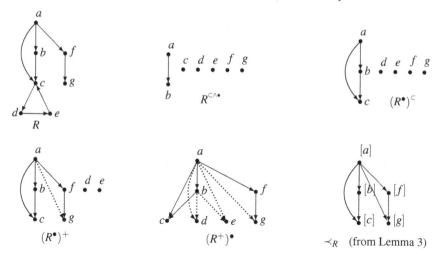

Fig. 2. An example of a relation R, its partial order approximations $R^{\subset\wedge\bullet}$, $(R^\bullet)^\subset$, $(R^\bullet)^+$, $(R^+)^\bullet$, and its relation $\prec_R$ from Lemma 3. Dotted lines in $(R^\bullet)^+$ and $(R^+)^\bullet$ indicate the ralationship that is not in R and was added by transitivity operation. For the relation $\prec_R$, $[x]$ denotes $[x]_{R^{cyc}_{id}}$ for $x \in \{a,b,c,d,e,f,g\}$, and $[a] = \{a\}$, $[b] = \{b\}$, $[c] = \{c,d,e\}$, $[f] = \{f\}$, $[g] = \{g\}$.

The assertion (8) is a consequence of Lemma 3. Recall that we have
$$a \equiv_{(R^+)^\bullet} b \iff \{x \mid x(R^+)^\bullet a\} = \{x \mid x(R^+)^\bullet b\} \wedge \{x \mid a(R^+)^\bullet x\} = \{x \mid b(R^+)^\bullet x\}.$$
If $aR^{cyc}b$ then $[a]_{R^{cyc}_{id}} = [b]_{R^{cyc}_{id}}$. Hence we have
$$x(R^+)^\bullet a \iff [x]_{R^{cyc}_{id}} \prec_{(R^+)^\bullet} [a]_{R^{cyc}_{id}} \iff [x]_{R^{cyc}_{id}} \prec_{(R^+)^\bullet} [b]_{R^{cyc}_{id}} \iff x(R^+)^\bullet b,$$
which means $\{x \mid x(R^+)^\bullet a\} = \{x \mid x(R^+)^\bullet b\}$. Similarly we can prove $\{x \mid a(R^+)^\bullet x\} = \{x \mid b(R^+)^\bullet x\}$. Thus the assertion (8) does hold as well.

To show (9) first notice that, $(R^\subset)^\bullet = (R^\subset)^+ = R^\subset$ (as $R^\subset$ is a partial order), $R^+ \cap R^\bullet = (R^+)^\bullet$ (from the definition of acyclic refinement), and $R^+ \cap R^\subset = R^\subset$ (since $R^\subset \subseteq R \subseteq R^+$). Since $R^+ = (R^+)^+$, from (4) we have $(R^+)^\bullet = ((R^+)^+)^\bullet = ((R^+)^\bullet = (R^+)^\bullet$. From (1), (3) and (5) it follows that additional applications of '$\cap$', '$\subset$', '$+$' and '$\bullet$' do not produce new realtions.   □

5   Approximation with Partially Ordered Kernel

Even if R may in general be imprecise, in most cases *some parts* of R describe the precise ranking. For instance if R is the result of expert voting, if all experts agree that aRb, then we may assume that $a <^{rank} b$ (see *Pereto's principle* [4]). In this section we will formally treat such case.

Let R be a relation and let $\lhd$ be a partial order satisfying $\lhd \subseteq R$. We are looking for a partial order approximation of R that includes $\lhd$. The relation $\lhd$ will be called a *partially ordered kernel* of R. In general it may happen that $\lhd$ is not included in any partial order approximation discussed in the previous section (Figure 1 in [7] shows the case of $a \lhd b$ and $\neg a(R^+)^\bullet b$). In general the union of partial orders may not be a partial order at all, however we may use the following lemma.

Lemma 6. *Let R be a relation, $<_1$ and $<_2$ be partial orders satisfying:*

1. *$a <_1 b \implies aR^+b$, and*
2. *$a <_2 b \implies aR^+b \wedge \neg(bR^+a)$.*

Then $(<_1 \cup <_2)^+$ is the smallest partial order containing $<_1 \cup <_2$.

Proof. $(<_1 \cup <_2)^+$ is evidently the smallest transitive relation containing $<_1 \cup <_2$. It suffices to show that $(<_1 \cup <_2)^+$ is irreflexive. Suppose it is not irreflexive, i.e. there exoists x_0 such that $x_0(<_1 \cup <_2)^+x_0$. This means $x_0Q_1x_1Q_2x_2...x_{n-1}Q_nx_n$, with $x_n = x_0$, where Q_i is either $<_1$ or $<_2$. Since $<_1$ and $<_2$ are sharp partial orders, then at least one of Q_i's, sat Q_k, must be equal to $<_2$. Since $<_1 \subseteq R^+$ and $<_2 \subseteq R^+$, then for each $i, j \leq n$, we have $x_iR^+x_j \wedge x_jR^+x_i$. In particular $x_kR^+x_{k-1}$, a contradiction as $x_{k-1} <_2 x_k \implies \neg x_{k-1}R^+x_k$. Hence $(<_1 \cup <_2)^+$ is irreflexive. □

Corollary 3. *For each $< \in \{R^{c\wedge\bullet}, (R^\bullet)^c, (R^\bullet)^+, (R^+)^\bullet\}$, and each partial order $\triangleleft \subseteq R$, $(\triangleleft \cup <)^+$ is the smallest partial order containing $\triangleleft \cup <$.* □

The special case of Corollary 3 was used in the ranking algorithms proposed in [6,7].

6 Final Comment

A systematic approach to finding partial order approximations of arbitrary relations has been proposed. It is usually assumed that ranking is a *weak order* [4], and none of the relations $R^{c\wedge\bullet}$, R^c, $(R^\bullet)^c$, $(R^\bullet)^+$ and $(R^+)^\bullet$ guarantees it, so they must eventually be extended to appropriate weak orders using one of the method proposed in [3]. This process is not discussed in this paper, an interested reader is referred to [6,7]. By modifying an example from the Introduction one may show that each of the four partial order approximations of R is better than the others in given circumstances, however some experiments made to justify some claims of [5] indicate that often $(R^+)^\bullet$ could be interpreted as the "best" partial order approximation. This appears to be especially true when cycles of R are naturally interpreted as indifference (see Theorem 1(8)). The solution presented in this paper uses classical relational calculus [9]. The problem of partial order approximation of R does not seem to fit well to standard "lower/upper bound" approach, which poses an interesting question "Can we apply *Rough Sets paradigm* [8] to solve this problem?".

References

1. Arrow, K.J.: Social Choice and Individual Values. J. Wiley, Chichester (1951)
2. Dyer, J.S.: Remarks on the Analytic Hierarchy Process. Management Sci. 36, 244–258 (1990)
3. Fishburn, P.C.: Interval Orders and Interval Graphs. J. Wiley, New York (1985)
4. French, S.: Decision Theory. Ellis Horwood, New York (1986)
5. Janicki, R.: Pairwise Comparisons, Incomparability and Partial Orders. In: Proc. of ICEIS 2007 (Int. Conf. on Enterprise Information Systems), Funchal, Portugal, vol. 2, pp. 297–302 (2007)

6. Janicki, R.: Ranking with Partial Orders and Pairwise Comparisons. In: Wang, G., Li, T., Grzymała-Busse, J.W., Miao, D., Skowron, A., Yao, Y. (eds.) RSKT 2008. LNCS (LNAI), vol. 5009, pp. 442–451. Springer, Heidelberg (2008)
7. Janicki, R., Koczkodaj, W.W.: Weak Order Approach to Group Ranking. Computers Math. Applic. 32(2), 51–59 (1996)
8. Pawlak, Z.: Rough Sets. Kluwer, Dordrecht (1991)
9. Rosen, K.H.: Discrete Mathematics and Its Applications. McGraw-Hill, New York (1999)
10. Saaty, T.L.: A Scaling Methods for Priorities in Hierarchical Structure. Journal of Mathematical Psychology 15, 234–281 (1977)
11. Schröder, E.: Algebra der Logik, 2nd edn., Teuber, Leipzig, vol. 1895. Chelsea (1966)

On Rough Equalities and Rough Equivalences of Sets

B.K. Tripathy[1], Anirban Mitra[2], and J. Ojha[3]

[1] School of Computing Sciences, V.I.T. University, Vellore, T.N. 632 014, India
tripathybk@rediffmail.com
[2] Department of Computer Science, Berhampur University, Berhampur,
Orissa 760 007, India
mitra.anirban@gmail.com
[3] Department of Mathematics, Khallikote College, Berhampur - 760001,
Orissa, India

Abstract. The different types of rough equalities [4,5,6] of sets deal with approximate equalities of sets which may not be equal in the usual sense of classical set theory. In this article, we make further study of the properties of rough equalities. A more general kind of equality of sets (we call it rough equivalence) shall be introduced, which captures equalities of sets at a higher level than rough equalities. Many properties of this new notion and its comparison with rough equalities shall be dealt with. We shall make use of the concepts of rough inclusions of sets in this sequel.

Keywords: bottom R-equal, top R-equal, R-equal, bottom R-equivalent, top R-equivalent and R-equivalent.

1 Introduction

The notion of rough sets was introduces by Pawlak [7] as an extension of the concept of crisp sets and to capture impreciseness. Imprecision in this approach is expressed by a boundary region of a set. In fact, the idea of rough set is based upon approximation of a set by a pair of sets, called the lower and upper approximations of the set [9,10,11].

Let U be a universe of discourse and R be an equivalence relation over U. By U/R we denote the family of all equivalence classes of R, referred to as *categories* or *concepts* of R and the equivalence class of an element $x \in U$ is denoted by $[x]_R$. By a knowledge base we understand a relational system $\mathbf{K} = (U, \mathbf{R})$, where U is as above and $\mathbf{R}$ is a family of equivalence relations over U.

For any subset $P(\neq \phi) \subseteq \mathbf{R}$, the intersection of all equivalence relations in P is denoted by $IND(P)$ and is called the *indiscernibility relation* over P. By $IND(K)$ we denote the family of all equivalence relations defined in K, that is $IND(K) = \{IND(P) : P \subseteq \mathbf{R}, P \neq \phi\}$.

Given any $X \subseteq U$ and $R \in IND(K)$, we associate two subsets, $\underline{R}X = \bigcup\{Y \in U/R : Y \subseteq X\}$ and $\bar{R}X = \bigcup\{Y \in U/R : Y \cap X \neq \phi\}$, called the R-*lower* and

C.-C. Chan et al. (Eds.): RSCTC 2008, LNAI 5306, pp. 92–102, 2008.
© Springer-Verlag Berlin Heidelberg 2008

R-*upper approximations* of X respectively. The R-*boundary* of X is denoted by $BN_R(X)$ and is given by $BN_R(X) = \bar{R}X - \underline{R}X$. The elements of $\underline{R}X$ are those elements of U which can certainly be classified as elements of X and elements of $\bar{R}X$ are those elements of U which can possibly be classified as elements of X, employing the knowledge of R. We say that X is rough with respect to R if and only if $\underline{R}X \neq \bar{R}X$, equivalently $BN_R(X) \neq \phi$. X is said to be R-*definable* if and only if $\underline{R}X = \bar{R}X$, or $BN_R(X) = \phi$.

2 Rough Equality of Sets

Extending the idea of equality of sets in crisp set theory, where two sets are said to be equal if and only if they have the same elements, three types of rough or approximate equalities have been introduced by Novotny and Pawlak [4,5,6]. We state these definitions.

2.1 Definitions

Definition 1. Let $K = (U, \boldsymbol{R})$ be a knowledge base, $X, Y \subseteq U$ and $R \in IND(K)$. We say that
(i) Two sets X and Y are *bottom R-equal* ($X=_BY$) if $\underline{R}X = \underline{R}Y$;
(ii) Two sets X and Y are *top R-equal* ($X=_TY$) if $\bar{R}X = \bar{R}Y$;
(iii) Two sets X and Y are *R-equal* ($X=Y$) if ($X=_BY$) and ($X=_TY$). Equivalently, $\underline{R}X = \underline{R}Y$ and $\bar{R}X = \bar{R}Y$.

For simplicity, we drop the suffix R in the above notations. It can be easily verified that the relations bottom R-equal, top R-equal and R-equal are equivalence relations over $P(U)$, the power set of U. The concept of approximate equality of sets refers to the topological structure of the compared sets but not the elements they consist of. Thus sets having significantly different elements may be rough equal. In fact, if $X =_B Y$ then $\underline{R}X = \underline{R}Y$ and as $X \supseteq \underline{R}X, Y \supseteq \underline{R}Y$, X and Y can differ only in elements of $X - \underline{R}X$ and $Y - \underline{R}Y$. However, it is easy to check that two sets X and Y may be R-equal in spite of $X \cap Y = \phi$.

As noted by Pawlak ([8], p.26), rough equality of sets is of relative character, that is things are equal or not equal from our point of view depending on what we know about them. So, in a sense the definition of rough equality refers to our knowledge about the universe. Some more related work on rough equalities can be found in [1,2,3].

2.2 Properties

The following properties of rough equalities are well known (see for instance [8]):
The following properties of rough equalities are well known [8].

(1) $X=_BY$ if and only if $X \cap Y=_BX$ and $X \cap Y=_BY$.
(2) $X=_TY$ if and only if $X \cap Y=_TX$ and $X \cap Y=_TY$.
(3)   If $X=_TX'$ and $Y=_TY'$ then $X \cup Y=_TX' \cup Y'$.

(4)   If $X=_B X'$ and $Y=_B Y'$ then $X \cap Y =_B X' \cap Y'$.
(5)   If $X=_T Y$ then $X \cup -Y =_T U$.
(6)   If $X=_B Y$ then $X \cap -Y =_B \phi$.
(7)   If $X \subseteq Y$ and $Y =_T \phi$ then $X =_T \phi$.
(8)   If $X \subseteq Y$ and $X =_T U$ then $Y =_T U$.
(9) $X =_T Y$ if and only if $-X =_B -Y$.
(10) If $X =_B \phi$ or $Y =_B U$ then $X \cap Y =_B \phi$.
(11) If $X =_T U$ or $Y =_T U$ then $X \cup Y =_T U$.

In the following two properties of lower and upper approximations of rough sets, we find that inclusions hold and equalities do not hold true in general:

(12) $\underline{R}X \cup \underline{R}Y \subseteq \underline{R}(X \cup Y)$ and
(13) $\bar{R}(X \cap Y) \subseteq \bar{R}X \cap \bar{R}Y$.

The following results [12] provide necessary and sufficient conditions for equation to hold in (12) and (13). In these results we take $\{E_1, E_2, ..., E_n\}$ as a partition of the universe U with respect to an equivalence relation R and $\{X_1, X_2, ..., X_m\}$ are subsets of U.

Theorem 1. We have
(14) $\bigcup_{i=1}^{m} \underline{R}(X_i) \subset \underline{R}(\bigcup_{i=1}^{m} X_i)$
   if and only if there exists at least one E_j such that $X_i \cap E_j \subset E_j$, for $i = 1, 2, ..., m$ and $\bigcup_{i=1}^{m} X_i \supseteq E_j$.

Corollary 1. Equality holds in (14) if and only if there exist no E_j such that $X_i \cap E_j \subset E_j, i = 1, 2, ..., m$ and $\bigcup_{i=1}^{m} X_i \supseteq E_j$.

Theorem 2. We have
(15) $\bar{R}(\bigcap_{i=1}^{m} X_i) \subset \bigcap_{i=1}^{m} \bar{R}(X_i)$
   if and only if there exists at least one E_j such that
   $X_i \cap E_j \neq \phi$ for $i = 1, 2, ..., m$ and $(\bigcap_{i=1}^{m} X_i) \cap E_j = \phi$.

Corollary 2. Equality holds in (14) if and only if there exist no E_j such that $X_i \cap E_j \neq \phi, i = 1, 2, ..., m$. and $(\bigcap_{i=1}^{m} X_i) \cap E_j = \phi$.
   It has been noted that (see for instance [8]) the properties (1) to (11) fail to hold if $=_B$ is replaced by $=_T$ or vice versa. However, we have the following observations with regards to their interchange. We omit the proofs to accommodate space.

(i) The properties (7) to (11) hold true under the interchange.
(ii) The properties (5) and (6) holds true under interchange if $BN_R(Y) = \phi$.
(iii) (A) The properties (1) and (4) hold under interchange if conditions of Corollary 2 hold with m = 2.
   (B) The properties (2) and (3) hold if conditions of Corollary 1 hold with m = 2.

3   Rough Equivalence of Sets

In this section we introduce three concepts of rough equivalence of sets.

3.1 Definitions

(i) We say that two sets X and Y are *bottom R-equivalent* if and only if both $\underline{R}X$ and $\underline{R}Y$ are ϕ or not ϕ together (we write, X is b_eqv. to Y). We put the restriction here that for bottom R-equivalence of X and Y either both $\underline{R}X$ and $\underline{R}Y$ are equal to U or none of them is equal to U.

(ii) We say that two sets X and Y are *top R-equivalent* if and only if both $\bar{R}X$ and $\bar{R}Y$ are U or not U together (we write, X is t_eqv. to Y). We put the restriction here that for top R-equivalence of X and Y either both $\bar{R}X$ and $\bar{R}Y$ are equal to ϕ or none of them is equal to ϕ.

(iii) We say that two sets X and Y are *R-equivalent* if and only if X and Y are bottom R-equivalent and top R-equivalent (we write, X is eqv. to Y). We would like to note here that when two sets X and Y are R-equivalent, the restrictions in **(i)** and **(ii)** become redundant.

For example, in case **(i)**, if one of $\underline{R}X$ and $\underline{R}Y$ is equal to U then the corresponding upper approximation must be U and for rough equivalence it is necessary that the other upper approximation must also be U. Similarly, the other case.

3.2 Elementary Properties

(i) It is clear from the definitions above that in all cases (bottom,top,total) R-equality implies R-equivalence and the converses are not true.

(ii) Bottom R-equivalence, top R-equivalence and R-equivalence are equivalence relations on P(U).

(iii) If two sets are roughly equivalent then by using our present knowledge, we may not be able to say whether two sets are approximately equal as described above, but, we can say that they are approximately equivalent. That is both the sets have or not have positive elements with respect to R and both the sets have or not have negative elements with respect to R.

3.3 Example

Let us consider all the cattle in a locality as our universal set C. We define a relation R over C by xRy if and only if x and y are cattle of the same kind. Suppose for example, this equivalence relation decomposes the universe into disjoint equivalence classes as given below.

$C = \{Cow, Buffalo, Goat, Sheep, Bullock\}.$

Let P_1 and P_2 be two persons in the locality having their set of cattle represented by X and Y.

We cannot talk about the equality of X and Y in the usual sense as the cattle can not be owned by two different people.

Similarly we can not talk about the rough equality of X and Y except the trivial case when both the persons do not own any cattle.

We find that rough equivalence is a better concept which can be used to decide the equality of the sets X and Y in a very approximate and real sense.

There are four different cases in which we can talk about equivalence of P_1 and P_2.

Case I. $\bar{R}X, \bar{R}Y$ are not U and $\underline{R}X, \underline{R}Y$ are ϕ. That is P_1 and P_2 both have some kind of cattle but do not have all cattle of any kind in the locality. So, they are equivalent.

Case II. $\bar{R}X, \bar{R}Y$ are not U and $\underline{R}X, \underline{R}Y$ are not ϕ. That is P_1 and P_2 both have some kind of cattle and also have all cattle of some kind in the locality. So, they are equivalent.

Case III. $\bar{R}X, \bar{R}Y$ are U and $\underline{R}X, \underline{R}Y$ are ϕ. That is P_1 and P_2 both have all kinds of cattle but do not have all cattle of any kind in the locality. So, they are equivalent.

Case IV. $\bar{R}X, \bar{R}Y$ are U and $\underline{R}X, \underline{R}Y$ are not ϕ. That is P_1 and P_2 both have all kinds of cattle and also have all cattle of some kind in the locality. So, they are equivalent.

There are two different cases under which we can talk about the non - equivalence of P_1 and P_2.

Case V. One of $\bar{R}X$ and $\bar{R}Y$ is U and the other one is not. Then, out of P_1 and P_2 one has cattle of all kinds and other one dose not have so. So, they are not equivalent. Here the structures of $\underline{R}X$ and $\underline{R}Y$ are unimportant.

Case VI. Out of $\underline{R}X$ and $\underline{R}Y$ one is ϕ and other one is not. Then, one of P_1 and P_2 does not have all cattle of any kind, whereas the other one has all cattle of some kind. So, they are not equivalent. Here the structures of $\bar{R}X$ and $\bar{R}Y$ are unimportant.

It may be noted that we have put the restriction for top rough equivalence that in the case when $\bar{R}X$ and $\bar{R}Y$ are not equal to U, it should be the case that both are ϕ or not ϕ together. It will remove the cases when one set is ϕ and the other has elements from all but one of the equivalence classes but does not have all the elements of any class completely being rough equivalent. Taking the example into consideration it removes cases like when a person has no cattle being rough equivalent to a person, who has some cattle of every kind except one.

Similarly, for bottom rough equivalence we have put the restriction that when $\underline{R}X$ and $\underline{R}Y$ are not equal to ϕ, it should be the case that both are U or not U together.

3.4   General Properties

In this section we establish some properties of rough equivalences of sets. These properties are similar to those for rough equalities. Some of these properties which do not hold in full force, sufficient conditions have been obtained. Also, we shall verify the necessity of such conditions. We need the concepts of different rough inclusions [8] and rough comparisons, which are introduced below.

Definition 2

Let $K = (U, \boldsymbol{R})$ be a knowledge base, $X, Y \subseteq U$ and $R \in IND(K)$. Then
(i) We say that X is *bottom R-included* in $Y (X \sqsubseteq_{BR} Y)$ if and only if $\underline{R}X \subseteq \underline{R}Y$.
(ii) We say that X is *top R-included* in $Y (X \sqsubseteq_{TR} Y)$ if and only if $\bar{R}X \subseteq \bar{R}Y$.

(iii) We say that X is *R-included* in $Y(X \sqsubseteq_R Y)$ if and only if $X \sqsubseteq_{BR} Y$ and $X \sqsubseteq_{TR} Y$.

We shall drop the suffixes R from the notations above in their use of make them simpler.

Definition 3

(i) We say $X, Y \subseteq U$ are *bottom rough comparable* if and only if $X \sqsubseteq_B Y$ or $Y \sqsubseteq_B X$ holds.

(ii) We say $X, Y \subseteq U$ are *top rough comparable* if and only if $X \sqsubseteq_T Y$ or $Y \sqsubseteq_T X$ holds.

(iii) We say $X, Y \subseteq U$ are *rough comparable* if and only if X and Y are both top rough comparable and bottom rough comparable.

Property 1

(i) If $X \cap Y$ is *b_eqv* to X and $X \cap Y$ is *b_eqv* to Y then X is *b_eqv* to Y .

(ii) The converse of (i) is not necessarily true.

(iii) The converse is true if in addition X and Y are bottom rough comparable.

(iv) The condition in (iii) is not necessary.

Proof

(i) The proof is trivial.

(ii) The cases when $\underline{R}X$ and $\underline{R}Y$ are both not ϕ but $\underline{R}(X \cap Y) = \phi$ the converse is not true.

(iii) We have $\underline{R}(X \cap Y) = \underline{R}X \cap \underline{R}Y = \underline{R}X$ or $\underline{R}Y$, as the case may be, since X and Y are bottom rough comparable.

So, $X \cap Y$ is *b_eqv* to X and $X \cap Y$ is *b_eq* to Y.

(iv) We provide an example to show that this condition is not necessary. Let us take $U = \{x_1, x_2, .., x_8\}$ and the partition induced by an equivalence relation R be $\{\{x_1, x_2\}, \{x_3, x_4\}, \{x_5, x_6\}, \{x_7, x_8\}\}$.

Now, for $X = \{x_1, x_2, x_3, x_4\}$ and $Y = \{x_3, x_4, x_5, x_6\}$, we have $\underline{R}X = X \neq \phi, \underline{R}Y = Y \neq \phi, X \cap Y = \{x_3, x_4\}$ and $\underline{R}(X \cap Y) = \{x_3, x_4\} \neq \phi$. So, $X \cap Y$ is *b_eqv* to both X and Y. But X and Y are not bottom rough comparable.

Property 2

(i) If $X \cup Y$ is *t_eqv* to X and $X \cup Y$ is *t_eqv* to Y then X is *t_eqv* to Y.

(ii) The converse of (i) may not be true.

(iii) A sufficient condition for the converse of (i) to be true is that X and Y are top rough comparable.

(iv) The condition in (iii) is not necessary.

Proof. The proof is similar to that of property 1 and hence omitted.

Property 3

(i) If X is *t_eqv* to X' and Y is *t_eqv* to Y' then it may or may not be true that $X \cup Y$ is *t_eqv* to $X' \cup Y'$.

(ii) A sufficient condition for the result in (i) to be true is that X and Y are top rough comparable and X' and Y' are top rough comparable.

(iii) The condition in (ii) is not necessary for result in (i) to be true.

Proof

(i) The result fails to be true when all of $\bar{R}(X)$, $\bar{R}(X')$, $\bar{R}(Y)$ and $\bar{R}(Y')$ are not U and exactly one of $X \cup Y$ and $X' \cup Y'$ is U.

(ii) We have $\bar{R}(X) \neq U$, $\bar{R}(X') \neq U$, $\bar{R}(Y) \neq U$ and $\bar{R}(Y') \neq U$. So, under the hypothesis, $\bar{R}(X \cup Y) = \bar{R}X \cup \bar{R}Y = \bar{R}(X)$ or $\bar{R}(Y)$, which is not equal to U. Similarly, $\bar{R}(X' \cup Y') \neq U$. Hence, $X \cup Y$ is t_eqv to $X' \cup Y'$.

(iii) Continuing with the same example, taking $X = \{x_1, x_2, x_3\}$, $X' = \{x_1, x_2, x_4\}$, $Y = \{x_4, x_5, x_6\}$ and $Y' = \{x_3, x_5, x_6\}$, we find that $\bar{R}X = \{x_1, x_2, x_3, x_4\} = \bar{R}X' \neq U$ and $\bar{R}Y = \{x_3, x_4, x_5, x_6\} = \bar{R}Y' \neq U$. So, X and Y are not top rough comparable. X' and Y' are not top rough comparable. But, $\bar{R}(X \cup Y) = \{x_1, x_2, x_3, x_4, x_5, x_6\} = \bar{R}(X' \cup Y')$. So, $X \cup Y$ is top equivalent to $X' \cup Y'$.

Property 4

(i) X is b_eqv to X' and Y is b_eqv to Y' may or may not imply that $X \cap Y$ is b_eqv to $X' \cap Y'$.

(ii) A sufficient condition for the result in (i) to be true is that X and Y are bottom rough comparable and X' and Y' are bottom rough comparable.

(iii) The condition in (ii) is not necessary for result in (i) to be true.

Proof. The proof is similar to that of property 3 and hence omitted.

Property 5

(i) X is t_eqv to Y may or may not imply that $X \cup (-Y)$ is t_eqv to U.

(ii) A sufficient condition for result in (i) to hold is that $X =_B Y$.

(iii) The condition in (ii) is not necessary for the result in (i) to hold.

Proof

(i) The result fails to hold true when $\bar{R}(X) \neq U$, $\bar{R}(Y) \neq U$ and still $\bar{R}(X \cup (-Y)) = U$.

(ii) As $X =_B Y$, we have $\underline{R}X = \underline{R}Y$. So, $-\underline{R}X = -\underline{R}Y$. Equivalently, $\bar{R}(-X) = \bar{R}(-Y)$. Now, $\bar{R}(X \cup -Y) = \bar{R}(X) \cup \bar{R}(-Y) = \bar{R}(X) \cup \bar{R}(-X) = \bar{R}(X \cup -X) = \bar{R}(U) = U$. So, $X \cup -Y$ is t_eqv to U.

(iii) Continuing with the same example and taking $X = \{x_1, x_2, x_3\}$, $Y = \{x_2, x_3, x_4\}$ we get $-Y = \{x_1, x_5, x_6, x_7, x_8\}$. So that $\underline{R}X = \{x_1, x_2\}$ and $\underline{R}Y = \{x_3, x_4\}$. Hence, it is not true that $X =_B Y$. But, $X \cup -Y = \{x_1, x_2, x_3, x_5, x_6, x_7, x_8\}$. So, $\bar{R}(X \cup -Y) = U$. That is, $X \cup -Y$ t_eqv to U.

Property 6

(i) X is b_eqv to Y may or may not imply that $X \cap (-Y)$ is b_eqv to ϕ.

(ii) A sufficient condition for the result in (i) to hold true is that $X =_T Y$.

(iii) The condition in (ii) is not necessary for the result in (i) to hold true.

Proof. The proof is similar to that of property 5 and hence omitted.

Property 7. If $X \subseteq Y$ and Y is b_eqv to ϕ then X is b_eqv to ϕ .
Proof. As Y is b_eqv to ϕ, we have $\underline{R}(Y) = \phi$. So, if $X \subseteq Y$, $\underline{R}(X) \subseteq \underline{R}(Y) = \phi$.

Property 8. If $X \subseteq Y$ and X is t_eqv to U then Y is t_eqv to U.

Proof. The proof is similar to that of Property 7.

Property 9. X is t_eqv to Y if and only if $-X$ is b_eqv to $-Y$.

Proof. The proof follows from the property, $\underline{R}(-X) = -\bar{R}(X)$.

Property 10. X is b_eqv to ϕ, Y is b_eqv to $\phi \Rightarrow X \cap Y$ is b_eqv to ϕ.

Proof. The proof follows directly from the fact that under the hypothesis the only possibility is $\underline{R}(X) = \underline{R}(Y) = \phi$.

Property 11. If X is t_eqv to U or Y is t_eqv to U then $X \cup Y$ is t_eqv to U.

Proof. The proof follows directly from the fact that under the hypothesis the only possibility is $\bar{R}(X) = \bar{R}(Y) = U$.

3.5 Properties with Interchanges

Like the case of rough equalities, it is curious to know the result of replacing bottom rough equivalence with top rough equivalence and vice versa, in the properties established in the previous section. In this section we shall establish such properties whenever these are valid. Whenever the properties do not hold, we shall provide sufficient conditions under which it can be true. In addition, we shall test if such conditions are necessary also for the validity of the properties. Invariably, it has been found that such conditions are not necessary. We shall show it by providing suitable examples.

Property 12

(i) If $X \cap Y$ is t_eqv to X and $X \cap Y$ is t_eqv to Y then X is t_eqv Y.
(ii) The converse of (i) is not necessarily true.
(iii) A sufficient condition for the converse of (i) to hold true is that conditions of corollary 2 hold with m = 2.
(iv) The condition in (iii) is not necessary.

Proof

(i) The proof is trivial.
(ii) The result fails when $\bar{R}X$ and $\bar{R}(X) = U\bar{R}(Y)$ and $\bar{R}(X \cap Y) \neq U$.
(iii) Under the hypothesis, we have $\bar{R}(X \cap Y) = \bar{R}(X) \cap \bar{R}(Y)$. If X is t_eqv to Y then both $\bar{R}X$ and $\bar{R}Y$ are equal to U or not equal to U together. So,

accordingly we get $\bar{R}(X \cap Y)$ equal to U or not equal to U. Hence the conclusion follows.

(iv) We see that the sufficient condition for the equality to hold when m = 2 in Corollary 2 is that there is no E_j such that $X \cap E_j \neq \phi$, $Y \cap E_j \neq \phi$ and $X \cap Y \cap E_j = \phi$.

Let us take U and the relation as above. Now, taking $X = \{x_1, x_3, x_6\}$, $Y = \{x_3, x_5, x_6\}$. The above sufficiency conditions are not satisfied as $\{x_5, x_6\} \cap X \neq \phi$, $\{x_5, x_6\} \cap Y \neq \phi$ and $\{x_5, x_6\} \cap X \cap Y = \phi$. However, $\bar{R}X = \{x_1, x_2, x_3, x_4, x_5, x_6\} \neq U$.

Property 13

(i) $X \cup Y$ is b_eqv to X and $X \cup Y$ is b_eqv to Y then X is b_eqv to Y.
(ii) The converse of (i) is not necessarily true.
(iii) A sufficient condition for the converse of (i) to hold true is that the condition of corollary 1 holds for m = 2.
(iv) The condition in (iii) is not necessary.

Proof

(i) The proof is trivial.
(ii) The converse is not true when $\underline{R}X = \phi = \underline{R}Y$ but $\underline{R}(X \cup Y) \neq \phi$.
(iii) Suppose X is b_eqv to Y. Then $\underline{R}X$ and $\underline{R}Y$ are ϕ or not ϕ together. If the conditions are satisfied then $\underline{R}(X \cup Y) = \underline{R}X \cup \underline{R}Y$. So, if both $\underline{R}X$ and $\underline{R}Y$ are ϕ or not ϕ together then $\underline{R}(X \cup Y)$ is ϕ or not ϕ accordingly and the conclusion holds.
(iv) Let us take U as above. The classification corresponding to the equivalence relation be given by $\{\{x_1, x_2\}, \{x_3, x_4, x_5\}, \{x_6\}, \{x_7, x_8\}\}$. Let $X = \{x_1, x_3, x_6\}$, $Y = \{x_2, x_5, x_6\}$. Then $\underline{R}(X) \neq \phi$, $\underline{R}(Y) \neq \phi$ and $\underline{R}(X \cup Y) \neq \phi$. The condition in (iii) is not satisfied as taking $E = \{x_1, x_2\}$ we see that $X \cap E \subset E$, $Y \cap E \subset E$ and $X \cup Y \supseteq E$.

Property 14

(i) X is b_eqv to X' and Y is b_eqv to Y' may not imply $X \cup Y$ is b_eqv to $X' \cup Y'$.
(ii) A sufficient condition for the conclusion of (i) to hold is that the conditions of corollary 2 are satisfied for both X, Y and X' , Y' separately with m = 2.
(iii) The condition in (ii) is not necessary for the conclusion in (i) to be true

Proof

(i) When $\underline{R}X$, $\underline{R}Y$, $\underline{R}X'$, $\underline{R}Y'$ are all ϕ and out of $X \cup Y$ and $X' \cup Y'$ one is ϕ but the other one is not ϕ, the result fails to be true.
(ii) Under the additional hypothesis, we have $\underline{R}(X \cup Y) = \underline{R}X \cup \underline{R}Y$ and $\underline{R}(X' \cup Y') = \underline{R}X' \cup \underline{R}Y'$. Here both $\underline{R}X$ and $\underline{R}X'$ are ϕ or not ϕ together and both $\underline{R}Y$ and $\underline{R}Y'$ are ϕ or not ϕ together. If all are ϕ then both $\underline{R}(X \cup Y)$ and

$\underline{R}(X' \cup Y')$ are ϕ. So, they are b_eqv. On the other hand, if at least one pair is not ϕ then we get both $\underline{R}(X \cup Y)$ and $\underline{R}(X' \cup Y')$ are not ϕ and so they are b_eqv.

(iii) The condition is not satisfied means there is E_i with $X \cap E_i \subset E_i, Y \cap E_i \subset E_i$ and $X \cup Y \supseteq E_i$; there exists E_j (not necessarily different from E_i) such that $X' \cap E_j \subset E_j, Y' \cap E_j \subset E_j$ and $X' \cup Y' \supseteq E_j$.

Let us consider the example, $U = x_1, x_2, ..., x_8$ and the partition induced by an equivalence relation R be $\{\{x_1, x_2\}, \{x_3, x_4\}, \{x_5, x_6\}\{x_7, x_8\}\}$. $X = \{x_1, x_5\}$, $Y = \{x_3, x_6\}$, $X' = \{x_1, x_4\}$ and $Y' = \{x_3, x_7\}$. Then $\underline{R}X = \underline{R}X' = \underline{R}Y = \underline{R}Y' = \phi$. Also, $\underline{R}(X \cup Y) \neq \phi, \underline{R}(X' \cup Y') \neq \phi$. So, X is b_eqv to X', Y is b_eqv to Y' and $X \cup Y$ is b_eqv to $X' \cup Y'$. However, $X' \cap \{x_3, x_4\} \subset \{x_3, x_4\}$, $Y' \cap \{x_3, x_4\} \subset \{x_3, x_4\}$ and $X' \cup Y' \supseteq \{x_3, x_4\}$. So, the condition are not satisfied.

Property 15

(i) X is t_eqv to X' and Y is t_eqv to Y' may not necessarily imply that $X \cap Y$ is t_eqv to $X' \cap Y'$.

(ii) A sufficient condition for the conclusion in (i) to hold is the conditions of corollary 1 are satisfied for both X, Y and X' , Y' separately with m = 2.

(iii) The condition in (ii) is not necessary for the conclusion in (i) to hold.

Proof

(i) When $\bar{R}X = \bar{R}X' = \bar{R}Y = \bar{R}Y' = U$ and out of $\bar{R}(X \cap Y), \bar{R}(X' \cap Y')$ one is U whereas the other one is not U the result fails to be true.

(ii) If the conditions of corollary 1 are satisfied for X, Y and X' , Y' separately then the case when $\bar{R}X = \bar{R}X' = \bar{R}Y = \bar{R}Y' = U$, we have $\bar{R}(X' \cap Y') = \bar{R}X' \cap \bar{R}Y' = U$ and $\bar{R}(X \cap Y) = \bar{R}X \cap \bar{R}Y = U$. In other cases, if $\bar{R}X$ and $\bar{R}X'$ not U or $\bar{R}Y$ and $\bar{R}Y'$ not U then as $\bar{R}(X' \cap Y') \neq U$ and $\bar{R}(X \cap Y) \neq U$. So, in any case $X \cap Y$ and $X' \cap Y'$ are t_eqv to each other.

(iii) We continue with the same example. The conditions are not satisfied means there is no E_j such that $X \cap E_j \neq \phi, Y \cap E_j \neq \phi$ and $X \cap Y \cap E_j = \phi$ or $X' \cap E_j \neq \phi, Y' \cap E_j \neq \phi$ and $X' \cap Y' \cap E_j = \phi$. Taking $X = \{x_1, x_5\}$, $Y = \{x_3, x_5\}, X' = \{x_1, x_4\}$ and $Y' = \{x_2, x_4\}$ we have $X \cap \{x_5, x_6\} \neq \phi, Y' \cap \{x_5, x_6\} \neq \phi$ and $X \cap Y \cap \{x_5, x_6\} = \phi$.$X' \cap \{x_3, x_4\}, Y' \cap \{x_3, x_4\} \neq \phi$ and $X' \cap Y' \cap \{x_3, x_4\}$. So, the conditions are violated. But $\bar{R}X \neq U, \bar{R}X' \neq U, \bar{R}Y \neq U, \bar{R}Y' \neq U$. So, X is t_eqv and Y is t_eqv Y'. Also, $\bar{R}(X \cap Y) \neq U$ and $\bar{R}(X' \cap Y') \neq U$. Hence, $X \cap Y$ is t_eqv to $X' \cap Y'$.

Property 16. X is b_eqv to Y may or may not imply that $X \cup -Y$ is b_eqv to U.

We note that all the properties from 7 to 11 hold true under the replacement of t_eqv by b_eqv and vice versa.

4   Conclusions

In this article, study on the concept of rough equalities is carried out further and the validity of their properties are checked when $t\_eqv.$ and $b\_eqv.$ are interchanged. Some of them are found to hold good under the changes, while sufficient conditions are provided for other cases. A new concept of rough equivalence has been introduced and many properties which are parallel to those of rough equality are established. These results include both direct properties and those obtained after interchange of the symbols of bottom rough equivalence and top rough equivalence. An example is provided to show better applicability of rough equivalence over rough equality in representation of approximate knowledge.

References

1. Banerjee, M., Chakraborty, M.K.: Rough Consequence and Rough Algebra in Rough Sets. In: Ziarko, W.P. (ed.) Fuzzy Sets and Knowledge Discovery, Proc. Int. Workshop on Rough Sets and Knowledge Discovery (RSKD 1993), Banf, Canada, pp. 196–207. Springer, London (1994)
2. Bonikowski, Z.: A Certain Conception of the Calculus of Rough sets. Notre Dame Journal of Formal Logic 33, 412–421 (1992)
3. Chakraborty, M.K., Banerjee, M.: Rough Dialogue and Implication Lattices. Fundamenta Informetica 75(1-4), 123–139 (2007)
4. Novotny, M., Pawlak, Z.: Characterization of Rough Top equalities and Rough Bottom Equalities. Bullten of Polish Academy of Scieces and Math 33, 91–97 (1985)
5. Novotny, M., Pawlak, Z.: On Rough Equalities. Bulliten of Polish Academy of Sciences and Math. 33, 99–104 (1985)
6. Novotny, M., Pawlak, Z.: Black Box Analysis and Rough Top Equality. Bulliten of Polish Academy of Sciences and Math. 33, 105–113 (1985)
7. Pawlak, Z.: Rough Sets. Int. Jour. Of Info. and Computer Sc. 11, 341–356 (1982)
8. Pawlak, Z.: Rough Sets, Theoretical aspect of Reasoning about Data. Kluwer Academic Publishers, Dordrecht (1991)
9. Pawlak, Z., Skowron, A.: Rudiments of rough sets. Information Sciences - An International Journal 177(1), 3–27 (2007)
10. Pawlak, Z., Skowron, A.: Rough sets: Some extensions. Information Sciences - An International Journal 177(1), 28–40 (2007)
11. Pawlak, Z., Skowron, A.: Rough sets and Boolean reasoning. Information Sciences - An International Journal 177(1), 41–73 (2007)
12. Tripathy, B.K., Mitra, A.: Topological Properties of Rough Sets and Applications. Communicated to International Journal of Granular Computing, Rough Sets and Intellegent Systems

Statistical Independence of Multi-variables from the Viewpoint of Linear Algebra

Shusaku Tsumoto and Shoji Hirano

Department of Medical Informatics, Faculty of Medicine,
Shimane University
89-1 Enya-cho Izumo 693-8501 Japan
{tsumoto,hirano,abe}@med.shimane-u.ac.jp

Abstract. This paper focuses on statistical independence of three variables from the viewpoint of linear algebra. While information granules of statistical independence of two variables can be viewed as determinants of 2×2- submatrices, those of three variables consist of linear combination of odds ratios.

1 Introduction

Statistical independence between two attributes is a very important concept in data mining and statistics. The definition $P(A, B) = P(A)P(B)$ show that the joint probability of A and B is the product of both probabilities. This gives several useful formula, such as $P(A|B) = P(A)$, $P(B|A) = P(B)$. In a data mining context, these formulae show that these two attributes may not be correlated with each other. Thus, when A or B is a classification target, the other attribute may not play an important role in its classification.

Although independence is a very important concept, it has not been fully and formally investigated as a relation between two attributes.

In this paper, a statistical independence in a contingency table is focused on from the viewpoint of granular computing.

The first important observation is that a contingency table compares two attributes with respect to information granularity. It is shown from the definition that statistifcal independence in a contingency table is a special form of linear depedence of two attributes. Especially, when the table is viewed as a matrix, the above discussion shows that the rank of the matrix is equal to 1.0. Also, the results also show that partial statistical independence can be observed.

The second important observation is that matrix algebra is a key point of analysis of this table. A contingency table can be viewed as a matrix and several operations and ideas of matrix theory are introduced into the analysis of the contingency table.

The paper is organized as follows: Section 2 discusses the characteristics of contingency tables. Section 3 shows the conditions on statistical independence for a 2×2 table. Section 4 gives those for a $2 \times n$ table. Section 5 extends

C.-C. Chan et al. (Eds.): RSCTC 2008, LNAI 5306, pp. 103–112, 2008.
© Springer-Verlag Berlin Heidelberg 2008

these results into a multi-way contingency table. Section 6 discusses statistical independence from matrix theory. Section 7 and 8 show pseudo statistical independence. Finally, Section 9 concludes this paper.

2   Contingency Matrix

Definition 1. *Let R_1 and R_2 denote multinominal attributes in an attribute space A which have m and n values. A contingency tables $T(R_1, R_2)$ is a table of a set of the meaning of the following formulas: $|[R_1 = A_j]_A|$, $|[R_2 = B_i]_A|$, $|[R_1 = A_j \wedge R_2 = B_i]_A|$, $|U|$ ($i = 1, 2, 3, \cdots, n$ and $j = 1, 2, 3, \cdots, m$). This table is arranged into the form shown in Table 1, where: $|[R_1 = A_j]_A| = \sum_{i=1}^{m} x_{1i} = x_{\cdot j}$, $|[R_2 = B_i]_A| = \sum_{j=1}^{n} x_{ji} = x_{i\cdot}$, $|[R_1 = A_j \wedge R_2 = B_i]_A| = x_{ij}$, $|U| = N = x_{\cdot\cdot}$. ($i = 1, 2, 3, \cdots, n$ and $j = 1, 2, 3, \cdots, m$).*

Table 1. Contingency Table ($m \times n$)

	A_1	A_2	$\cdots$	A_n	Sum
B_1	x_{11}	x_{12}	$\cdots$	x_{1n}	$x_{1\cdot}$
B_2	x_{21}	x_{22}	$\cdots$	x_{2n}	$x_{2\cdot}$
$\cdots$	$\cdots$	$\cdots$	$\cdots$	$\cdots$	$\cdots$
B_m	x_{m1}	x_{m2}	$\cdots$	x_{mn}	$x_{m\cdot}$
Sum	$x_{\cdot 1}$	$x_{\cdot 2}$	$\cdots$	$x_{\cdot n}$	$x_{\cdot\cdot} = |U| = N$

Definition 2. *A contigency matrix $M_{R_1, R_2}(m, n, N)$ is defined as a matrix, which is composed of $x_{ij} = |[R_1 = A_j \wedge R_2 = B_i]_A|$, extracted from a contigency table defined in definition 1.*
      That is,

$$M_{R_1, R_2}(m, n, N) = \begin{pmatrix} x_{11} & x_{12} & \cdots & x_{1n} \\ x_{21} & x_{22} & \cdots & x_{2n} \\ \vdots & \vdots & \vdots & \vdots \\ x_{m1} & x_{m2} & \cdots & x_{mn} \end{pmatrix}.$$

□

For simplicity, if we do not need to specify R_1 and R_2, we use $M(m, n, N)$ as a contingency matrix with m rows, n columns and N samples.

One of the important observations from granular computing is that a contingency table shows the relations between two attributes with respect to intersection of their supporting sets. When two attributes have different number of equivalence classes, the situation may be a little complicated. But, in this case, due to knowledge about linear algebra, we only have to consider the attribute which has a smaller number of equivalence classes. and the surplus number of equivalence classes of the attributes with larger number of equivalnce classes can be projected into other partitions. In other words, a $m \times n$ matrix or contingency table includes a projection from one attributes to the other one.

3   Statistical Independence in 2 × 2 Contingency Table

Let us consider a contingency table shown in Table 1 ($m = n = 2$). Statistical independence between R_1 and R_2 gives:

$$P([R_1 = 0], [R_2 = 0]) = P([R_1 = 0]) \times P([R_2 = 0])$$
$$P([R_1 = 0], [R_2 = 1]) = P([R_1 = 0]) \times P([R_2 = 1])$$
$$P([R_1 = 1], [R_2 = 0]) = P([R_1 = 1]) \times P([R_2 = 0])$$
$$P([R_1 = 1], [R_2 = 1]) = P([R_1 = 1]) \times P([R_2 = 1])$$

Since each probability is given as a ratio of each cell to N, the above equations are calculated as:

$$\frac{x_{11}}{N} = \frac{x_{11} + x_{12}}{N} \times \frac{x_{11} + x_{21}}{N}$$
$$\frac{x_{12}}{N} = \frac{x_{11} + x_{12}}{N} \times \frac{x_{12} + x_{22}}{N}$$
$$\frac{x_{21}}{N} = \frac{x_{21} + x_{22}}{N} \times \frac{x_{11} + x_{21}}{N}$$
$$\frac{x_{22}}{N} = \frac{x_{21} + x_{22}}{N} \times \frac{x_{12} + x_{22}}{N}$$

Since $N = \sum_{i,j} x_{ij}$, the following formula will be obtained from these four formulae.

$$x_{11}x_{22} = x_{12}x_{21} \text{ or } x_{11}x_{22} - x_{12}x_{21} = 0$$

Thus,

Theorem 1. *If two attributes in a contingency table shown in a 2 × 2 contingency table are statistical indepedent, the following equation holds:*

$$x_{11}x_{22} - x_{12}x_{21} = 0 \tag{1}$$

$\square$

It is notable that the above equation corresponds to the fact that the determinant of a matrix corresponding to this table is equal to 0. Also, when these four values are not equal to 0, the equation 1 can be transformed into:

$$\frac{x_{11}}{x_{21}} = \frac{x_{12}}{x_{22}}.$$

Let us assume that the above ratio is equal to $C(constant)$. Then, since $x_{11} = Cx_{21}$ and $x_{12} = Cx_{22}$, the following equation is obtained.

$$\frac{x_{11} + x_{12}}{x_{21} + x_{22}} = \frac{C(x_{21} + x_{22})}{x_{21} + x_{22}} = C = \frac{x_{11}}{x_{21}} = \frac{x_{12}}{x_{22}}. \tag{2}$$

It is also notable that this equation is the same as the equation on collinearity of projective geometry [1].

4 Statistical Independence in $m \times n$ Contingency Table

Let us consider a $m \times n$ contingency table shown in Table 1. Statistical independence of R_1 and R_2 gives the following formulae:

$$P([R_1 = A_i, R_2 = B_j]) = P([R_1 = A_i])P([R_2 = B_j])$$
$$(i = 1, \cdots, m, j = 1, \cdots, n).$$

According to the definition of the table,

$$\frac{x_{ij}}{N} = \frac{\sum_{k=1}^{n} x_{ik}}{N} \times \frac{\sum_{l=1}^{m} x_{lj}}{N}. \tag{3}$$

Thus, we have obtained:

$$x_{ij} = \frac{\sum_{k=1}^{n} x_{ik} \times \sum_{l=1}^{m} x_{lj}}{N}. \tag{4}$$

Thus, for a fixed j,

$$\frac{x_{i_a j}}{x_{i_b j}} = \frac{\sum_{k=1}^{n} x_{i_a k}}{\sum_{k=1}^{n} x_{i_b k}}$$

In the same way, for a fixed i,

$$\frac{x_{ij_a}}{x_{ij_b}} = \frac{\sum_{l=1}^{m} x_{lj_a}}{\sum_{l=1}^{m} x_{lj_b}}$$

Since this relation will hold for any j, the following equation is obtained:

$$\frac{x_{i_a 1}}{x_{i_b 1}} = \frac{x_{i_a 2}}{x_{i_b 2}} \cdots = \frac{x_{i_a n}}{x_{i_b n}} = \frac{\sum_{k=1}^{n} x_{i_a k}}{\sum_{k=1}^{n} x_{i_b k}}. \tag{5}$$

Since the right hand side of the above equation will be constant, thus all the ratios are constant. Thus,

Theorem 2. *If two attributes in a contingency table shown in Table 1 are statistical indepedent, the following equations hold:*

$$\frac{x_{i_a 1}}{x_{i_b 1}} = \frac{x_{i_a 2}}{x_{i_b 2}} \cdots = \frac{x_{i_a n}}{x_{i_b n}} = const. \tag{6}$$

for all rows: i_a and i_b $(i_a, i_b = 1, 2, \cdots, m)$.

$\square$

4.1 Three-Way Table

Let "$\bullet$" denote as the sum over the row or column of a contingency matrix. That is ,

$$x_{i\bullet} = \sum_{j=1}^{n} x_{ij} \tag{7}$$

$$x_{\bullet j} = \sum_{i=1}^{m} x_{ij}, \tag{8}$$

where (7) and (8) shows marginal column and row sums. Then, it is easy to see that

$$x_{\bullet\bullet} = N,$$

where N denotes the sample size.

Then, Equation (4) is reformulated as:

$$\frac{x_{ij}}{x_{\bullet\bullet}} = \frac{x_{i\bullet}}{x_{\bullet\bullet}} \times \frac{x_{\bullet j}}{x_{\bullet\bullet}} \tag{9}$$

That is,

$$x_{ij} = \frac{x_{i\bullet} \times x_{\bullet j}}{x_{\bullet\bullet}}$$

Or

$$x_{ij}x_{\bullet\bullet} = x_{i\bullet}x_{\bullet j}$$

Thus, statistical independence can be viewed as the specific relations between assignments of i,j and ".". By use of the above relation, Equation (6) can be rewritten as:

$$\frac{x_{i_1 j}}{x_{i_2 j}} = \frac{x_{i_1 \bullet}}{x_{i_2 \bullet}},$$

where the right hand side gives the ratio of marginal column sums.

Equation (9) can be extended into multivariate cases. Let us consider a three attribute case.

Statistical independence with three attributes is defined as:

$$\frac{x_{ijk}}{x_{\bullet\bullet\bullet}} = \frac{x_{i\bullet\bullet}}{x_{\bullet\bullet\bullet}} \times \frac{x_{\bullet j\bullet}}{x_{\bullet\bullet\bullet}} \times \frac{x_{\bullet\bullet k}}{x_{\bullet\bullet\bullet}}, \tag{10}$$

Thus,

$$x_{ijk}x_{\bullet\bullet\bullet}^2 = x_{i\bullet\bullet}x_{\bullet j\bullet}x_{\bullet\bullet k}, \tag{11}$$

which corresponds to:

$$P(A = a, B = b, C = c) = P(A = a)P(B = b)P(C = c), \tag{12}$$

where A,B,C correspond to the names of attributes for i,j,k, respectively.

In statistical context, statistical independence requires hiearchical model. That is, statistical independence of three attributes requires that all the two pairs of three attributes should satisfy the equations of statistical independence. Thus, for Equation (12), the following equations should satisfy:

$$P(A = a, B = b) = P(A = a)P(B = b),$$
$$P(B = b, C = c) = P(B = b)P(C = c), \; and$$
$$P(A = a, C = c) = P(A = a)P(C = c).$$

Thus,

$$x_{ij\bullet}x_{\bullet\bullet\bullet} = x_{i\bullet\bullet}x_{\bullet j\bullet} \tag{13}$$

$$x_{i\bullet k}x_{\bullet\bullet\bullet} = x_{i\bullet\bullet}x_{\bullet\bullet k} \tag{14}$$

$$x_{\bullet jk}x_{\bullet\bullet\bullet} = x_{\bullet j\bullet}x_{\bullet\bullet k} \tag{15}$$

From Equation (11) and Equation (13),

$$x_{ijk}x_{\bullet\bullet\bullet} = x_{ij\bullet}x_{\bullet\bullet k},$$

Therefore,

$$\frac{x_{ijk}}{x_{ij\bullet}} = \frac{x_{\bullet\bullet k}}{x_{\bullet\bullet\bullet}} \tag{16}$$

In the same way, the following equations are obtained:

$$\frac{x_{ijk}}{x_{i\bullet k}} = \frac{x_{\bullet j\bullet}}{x_{\bullet\bullet\bullet}} \tag{17}$$

$$\frac{x_{ijk}}{x_{\bullet jk}} = \frac{x_{i\bullet\bullet}}{x_{\bullet\bullet\bullet}} \tag{18}$$

In summary, the following theorem is obtained.

Theorem 3. *If a three-way contingency table satisfy statistical independence, then the following three equations should be satisfied:*

$$\frac{x_{ijk}}{x_{ij\bullet}} = \frac{x_{\bullet\bullet k}}{x_{\bullet\bullet\bullet}}$$

$$\frac{x_{ijk}}{x_{i\bullet k}} = \frac{x_{\bullet j\bullet}}{x_{\bullet\bullet\bullet}}$$

$$\frac{x_{ijk}}{x_{\bullet jk}} = \frac{x_{i\bullet\bullet}}{x_{\bullet\bullet\bullet}}$$

□

Thus, the equations corresponding to Theorem 2 are obtained as follows.

Corollary 1. *If three attributes in a contingency table shown in Table 1 are statistical indepedent, the following equations hold:*

$$\frac{x_{ijk_a}}{x_{ijk_b}} = \frac{x_{\bullet\bullet k_a}}{x_{\bullet\bullet k_b}}$$

$$\frac{x_{ij_a k}}{x_{ij_b k}} = \frac{x_{\bullet j_a\bullet}}{x_{\bullet j_b\bullet}}$$

$$\frac{x_{i_a jk}}{x_{i_b jk}} = \frac{x_{i_a\bullet\bullet}}{x_{i_b\bullet\bullet}}$$

for all i,j, and k.

□

4.2   Multi-way Table

The above discussion can be easily extedned into a multi-way contingency table.

Theorem 4. *If a m-way contingency table satisfy statistical independence, then the following equation should be satisfied for any k-th attribute i_k and j_k ($k = 1, 2, \cdots, n$) where n is the number of attributes.*

$$\frac{x_{i_1 i_2 \cdots i_k \cdots i_n}}{x_{i_1 i_2 \cdots j_k \cdots i_n}} = \frac{x_{\bullet\bullet \cdots i_k \cdots \bullet}}{x_{\bullet\bullet \cdots j_k \cdots \bullet}}$$

Also, the following equation should be satisfied for any i_k:

$$x_{i_1 i_2 \cdots i_n} \times x_{\bullet\bullet \cdots \bullet}^{n-1}$$

$$= x_{i_1 \bullet \cdots \bullet} x_{\bullet i_2 \cdots \bullet} \times \cdots \times x_{\bullet\bullet \cdots i_k \cdots \bullet} \times \cdots \times x_{\bullet\bullet \cdots \bullet i_n}$$

□

5   Information Granule for Contingency Matrix

5.1   Residual of Contingency Matrix

Tsumoto and Hirano [2] discusses the meaning of pearson residuals from the viewpoint of linear algebra.

The residual is defined as a difference between an observed value for each cell in a contingency matrix and an expected value:

$$\sigma_{ij} = x_{ij} - \frac{x_{i\bullet} \times x_{\bullet j}}{x_{\bullet\bullet}}.$$

And simple calculation leads to the following theorem.

Theorem 5. *The residual of $M_{R_1, R_2}(m, n, N)$ is obtained as:*

$$\sigma_{ij} = \frac{1}{x_{\bullet\bullet}} \{ x_{ij} x_{\bullet\bullet} - x_{i\bullet} \times x_{\bullet j} \}$$

$$= \frac{1}{x_{\bullet\bullet}} \left\{ x_{ij} \sum_{k \neq i} \sum_{l \neq j} x_{kl} - \left(\sum_{l \neq j} x_{il} \right) \left(\sum_{k \neq i} x_{kj} \right) \right\}$$

$$= \frac{1}{x_{\bullet\bullet}} \sum_{\substack{k \neq i \\ l \neq j}} (x_{ij} x_{kl} - x_{kj} x_{il})$$

$$= \frac{1}{x_{\bullet\bullet}} \sum_{\substack{k \neq i \\ l \neq j}} \Delta_{j,l}^{i,k},$$

where $\Delta_{j,l}^{i,k}$ is the determinant of a 2×2 submatrix of $M_{R_1, R_2}(m, n, N)$ with selection of i and k rows and j and l columns.   □

Thus, a 2×2 submatrix in a contingency table can be viewed as a information granule for statistical (in)dependence.

Can we generalize this results into statistical independence of three variables ? This is our main question to be partially answered in this paper.

5.2   Information Granule for $2 \times 2 \times 2$ Data Cube

Let us get back to Equation (11), (13), (14) and (15).

From Equation (11), the residual for x_{ijk} is obtained as:

$$\sigma_{ijk} = x_{ijk} - \frac{x_{i\bullet\bullet} \times x_{\bullet j\bullet} \times x_{\bullet\bullet k}}{x_{\bullet\bullet\bullet}^2}.$$

For simplicity, let us confine to 2×2-data cube. Then, the above residual for x_{111} will be:

$$\sigma_{111} = x_{111} - \frac{x_{1\bullet\bullet} \times x_{\bullet 1\bullet} \times x_{\bullet\bullet 1}}{x_{\bullet\bullet\bullet}^2}$$

$$= \frac{1}{x_{\bullet\bullet\bullet}^2} \left\{ x_{111} \left(x_{\bullet\bullet\bullet}^2 - x_{\bullet 1\bullet} x_{\bullet\bullet 1} \right) \right\} - \sum_{\substack{k \neq i \text{ or} \\ l \neq j}} x_{1jk} x_{\bullet 1\bullet} x_{\bullet\bullet 1} \}$$

$$= \frac{1}{x_{\bullet\bullet\bullet}^2} \{ x_{\bullet 1\bullet}(x_{111}x_{\bullet\bullet 2} - x_{112}x_{\bullet\bullet 1}) + x_{\bullet\bullet 1}(x_{111}x_{\bullet 2\bullet} - x_{121}x_{\bullet 1\bullet})$$

$$+ x_{111}x_{\bullet 2\bullet}x_{\bullet\bullet 2} - x_{122}x_{\bullet 1\bullet}x_{\bullet\bullet 1} \}$$

$x_{111}x_{\bullet\bullet 2} - x_{112}x_{\bullet\bullet 1}$ and $x_{111}x_{\bullet 2\bullet} - x_{121}x_{\bullet 1\bullet}$ are related with Equations (13), (14) because:

$$\Delta_{111} = x_{111}x_{\bullet\bullet\bullet}^2 - x_{1\bullet\bullet}x_{\bullet 1\bullet}x_{\bullet\bullet 1}$$

$$= x_{\bullet\bullet\bullet}(x_{111}x_{\bullet\bullet\bullet} - x_{11\bullet}x_{\bullet\bullet 1})$$

$$= x_{\bullet\bullet\bullet}\{x_{111}x_{\bullet\bullet\bullet} - (x_{111} + x_{112})x_{\bullet\bullet 1}\}$$

$$= x_{\bullet\bullet\bullet}(x_{111}x_{\bullet\bullet 2} - x_{112}x_{\bullet\bullet 1})$$

and in similar way,

$$\Delta_{111} = x_{111}x_{\bullet\bullet\bullet}^2 - x_{1\bullet\bullet}x_{\bullet 1\bullet}x_{\bullet\bullet 1}$$

$$= x_{\bullet\bullet\bullet}(x_{111}x_{\bullet\bullet\bullet} - x_{1\bullet 1}x_{\bullet 1\bullet})$$

$$= x_{\bullet\bullet\bullet}(x_{111}x_{\bullet 2\bullet} - x_{121}x_{\bullet 1\bullet})$$

Thus, when Equations (13), (14) and (15) holds, $x_{111}x_{\bullet\bullet 2} - x_{112}x_{\bullet\bullet 1} = 0$ and $x_{111}x_{\bullet 2\bullet} - x_{121}x_{\bullet 1\bullet} = 0$.

Then, let us proceed to the calculation of the third part: $\delta_{111} = x_{111}x_{\bullet 2\bullet}x_{\bullet\bullet 2} - x_{122}x_{\bullet 1\bullet}x_{\bullet\bullet 1}$.

Since the following equations satisfies:

$$x_{111}x_{\bullet\bullet\bullet} - x_{1\bullet\bullet}x_{\bullet 11} = x_{111}x_{2\bullet\bullet} - x_{211}x_{1\bullet\bullet}$$

$$x_{122}x_{\bullet\bullet\bullet} - x_{1\bullet\bullet}x_{\bullet 22} = x_{122}x_{2\bullet\bullet} - x_{222}x_{1\bullet\bullet},$$

the former and latter part of δ_{111} is simplified to:

$$x_{111}x_{\bullet 2\bullet}x_{\bullet\bullet 2} = \frac{1}{x_{\bullet\bullet\bullet}}\{(x_{111}x_{2\bullet\bullet} - x_{211}x_{1\bullet\bullet}) + x_{1\bullet\bullet}x_{\bullet 11}\}x_{\bullet 2\bullet}x_{\bullet\bullet 2}$$

$$x_{122}x_{\bullet 1\bullet}x_{\bullet\bullet 1} = \frac{1}{x_{\bullet\bullet\bullet}}\{(x_{122}x_{2\bullet\bullet} - x_{222}x_{1\bullet\bullet})$$

$$+ x_{1\bullet\bullet}x_{\bullet 22}\}x_{\bullet 1\bullet}x_{\bullet\bullet 1} \tag{19}$$

When Equation (15) satisfies, $x_{111}x_{\bullet\bullet 2} - x_{112}x_{\bullet\bullet 1}$ and $x_{122}x_{\bullet\bullet 1} - x_{121}x_{\bullet\bullet 2}$ become 0. Then, the remaining part is given as:

$$x_{1\bullet\bullet}x_{\bullet 11}x_{\bullet 2\bullet}x_{\bullet\bullet 2} - x_{1\bullet\bullet}x_{\bullet 22}x_{\bullet 1\bullet}x_{\bullet\bullet 1}$$

$$= x_{1\bullet\bullet}(x_{\bullet 11}x_{\bullet 2\bullet}x_{\bullet\bullet 2} - x_{\bullet 22}x_{\bullet 1\bullet}x_{\bullet\bullet 1})$$

Since:

$$x_{\bullet 11}x_{\bullet\bullet\bullet} - x_{\bullet 1\bullet}x_{\bullet\bullet 1} = \sum_{\substack{k \neq 1 \text{ or} \\ l \neq 1}} (x_{\bullet 11}x_{\bullet kl} - x_{\bullet 1l}x_{\bullet k1})$$

$$x_{\bullet 22}x_{\bullet\bullet\bullet} - x_{\bullet 2\bullet}x_{\bullet\bullet 2} = \sum_{\substack{k \neq 2 \text{ or} \\ l \neq 2}} (x_{\bullet 22}x_{\bullet 2l} - x_{\bullet 2l}x_{\bullet k2}),$$

these values are given as 2×2 subderminants of matrices generated from a 2×2 data cube.

Theorem 6. *The residual sigma$_{111}$ of $2 \times 2 \times 2$-data cube is obtained as:*

$$\sigma_{111}x_{\bullet\bullet\bullet}^2 = x_{\bullet 1\bullet}(x_{111}x_{\bullet\bullet 2} - x_{112}x_{\bullet\bullet 1})$$

$$+ x_{\bullet\bullet 1}(x_{111}x_{\bullet 2\bullet} - x_{121}x_{\bullet 1\bullet})$$

$$+ \frac{1}{x_{\bullet\bullet\bullet}}(x_{111}x_{2\bullet\bullet} - x_{211}x_{1\bullet\bullet})$$

$$+ \frac{1}{x_{\bullet\bullet\bullet}}(x_{122}x_{2\bullet\bullet} - x_{222}x_{1\bullet\bullet})$$

$$+ \frac{x_{1\bullet\bullet}}{x_{\bullet\bullet\bullet}^2}\sum_{\substack{k \neq 1 \text{ or} \\ l \neq 1}}(x_{\bullet 11}x_{\bullet kl} - x_{\bullet 1l}x_{\bullet k1})$$

$$+ \frac{x_{1\bullet\bullet}}{x_{\bullet\bullet\bullet}^2}\sum_{\substack{k \neq 2 \text{ or} \\ l \neq 2}}(x_{\bullet 22}x_{\bullet 2l} - x_{\bullet 2l}x_{\bullet k2}). \tag{20}$$

$\square$

The above formula shows that even in the context of three variables, the concept of 2×2 subdeterminants of the matrices generated from a data cube play an central role in measuring statistical (in)dependence. It is notable that this equation will become 0 when Equation (13), (14) and (15) are satisfied. However, the converse is not trivial.

5.3   Information Granule for $l \times m \times n$ Data Cube

The above results can be generalized, although the results are rather complicated.

Theorem 7. *The residual σ_{ijk} of $l \times m \times n$-data cube is obtained as:*

$$\sigma_{ijk} x_{\bullet\bullet\bullet}^2 = x_{\bullet j\bullet}(x_{ijk} \sum_{n\neq k} x_{\bullet\bullet n} - \sum_{n\neq k} x_{ijn} x_{\bullet\bullet 1})$$

$$+ x_{\bullet\bullet k}(x_{ijk} \sum_{m\neq j} x_{\bullet m\bullet} - \sum_{m\neq j} x_{imk} x_{\bullet j\bullet})$$

$$+ \frac{1}{x_{\bullet\bullet\bullet}}(x_{ijk} \sum_{l\neq i} x_{l\bullet\bullet} - \sum_{l\neq i} x_{ljk} x_{i\bullet\bullet})$$

$$+ \frac{1}{x_{\bullet\bullet\bullet}}(\sum_{\substack{m\neq j \text{ and} \\ n\neq k}} x_{imn} \sum_{l\neq i} x_{l\bullet\bullet} - \sum_{\substack{l\neq i \text{ and} \\ m\neq j \text{ and} \\ n\neq k}} x_{lmn} x_{i\bullet\bullet})$$

$$+ \frac{x_{i\bullet\bullet}}{x_{\bullet\bullet\bullet}^2} \sum_{i,j=1}^{l,m} \sum_{\substack{p\neq i \text{ or} \\ q\neq j}} (x_{\bullet ij} x_{\bullet pq} - x_{\bullet ip} x_{\bullet qj})$$

$$(21)$$

$\square$

As shown in this formula, alternative sum is very important concept for statistical independence of three variables.

6   Conclusion

This paper focuses on statistical independence of three variables from the viewpoint of linear algebra. While information granules of statistical independence of two variables can be viewed as determinants of 2×2- submatrices, those of three variables consist of several combination s of odds ratios from a data cube, although the formula is rather complicated. Thus, in the case of three attributes, odds ratios play an important role in measuring the degree of statistical independence. It will be our future work to search for the corresponding determinants for $2 \times 2 \times 2$-data cube.

References

1. Coxeter, H. (ed.): Projective Geometry, 2nd edn. Springer, New York (1987)
2. Tsumoto, S., Hirano, S.: Meaning of pearson residuals - linear algebra view. In: Proceedings of IEEE GrC 2007. IEEE press, Los Alamitos (2007)

Rough Mereology in Classification of Data: Voting by Means of Residual Rough Inclusions

Lech Polkowski[1] and Piotr Artiemjew[2]

[1] Polish–Japanese Institute of IT
Warsaw, Poland
[2] Department of Mathematics and Computer Science,
University of Warmia and Mazury
Olsztyn, Poland
polkow@pjwstk.edu.pl, artem@matman.uwm.edu.pl

Abstract. In this work, we pursue the theme of applications of rough mereology, presenting a scheme for classifier construction by voting of training objects, exhaustive set of rules, and granules of training objects according to weights assigned by residual rough inclusions. The results show a high effectiveness of this approach as witnessed by the reported tests with some well–known data sets from UCI repository whose results are compared against the standard rough set exhaustive classifier.

Keywords: granulation of knowledge, rough inclusions, residual implications, granular decision systems.

1 Introduction

We formalize data sets as decision systems of the form of a triple (U, A, d), where U is a set of objects, A is a set of attributes, and $d \notin A$ is the decision.

For a pair of objects u, v, we define sets $DIS(u, v) = \{a \in A : a(u) \neq a(v)\}$ and $IND(u, v) = \{a \in A : a(u) = a(v)\}$, and their variants for a given ε, i.e., $DIS_\varepsilon(u, v) = \{a \in A : ||a(u) - a(v)|| \geq \varepsilon\}$ and $IND_\varepsilon(u, v) = \{a \in A : ||a(u) - a(v)|| < \varepsilon\}$, where the standard metric $||x - y||$ in the real line is applied, i.e., we assume from now on that attributes are real–valued.

We apply in this work an approach to granulation proposed in [5], [7], [8], consisting in using rough inclusions; see [1], [2] for some earlier results.

A rough inclusion is a relation $\mu \subseteq U \times U \times [0, 1]$ see, e. g., [9], where $\mu(u, v, r)$ means that u is similar to v to a degree of r.

One can look at rough inclusions as measures of similarity between objects. To understand the nature of rough inclusions, a few lines of introduction can be followed. One of them, probably intuitively most appealing, goes back to the idea of Henri Poincaré of a similarity relation not being an equivalence: consider points in the real line or plane along with a metric ρ bounded by 1, e.g., the Euclidean metric bounded by 1, i.e. $\rho(x, y) = min\{||x - y||, 1\}$, where $||x - y||$ denotes this time the Euclidean metric in the Euclidean space R^k with a finite

C.-C. Chan et al. (Eds.): RSCTC 2008, LNAI 5306, pp. 113–120, 2008.

dimension k. Given a parameter $\delta > 0$, a small positive real number, one says that points x, y are *similar* in case $\rho(x, y) \leq \delta$, in symbols $sim_\delta(x, y)$. Then, clearly, sim_δ is a tolerance relation which is not any equivalence.

A step further consists in introducing a graded counterpart to sim_δ by letting, for a parameter's r value in the interval $[0, 1]$,

$$sim_\delta(x, y, r) \text{ iff } \rho(x, y) \leq 1 - r.$$

Properties of $sim_\delta(x, y, r)$ follow easily from properties of ρ, and among them one finds,

(MON) If $sim_\delta(x, y, 1)$ then for each z, from $sim_\delta(z, x, r)$ it follows that sim_δ (z, y, r).

(ID) $sim_\delta(x, x, 1)$ for each x.

(EXT) If $sim_\delta(x, y, r)$ and $s \leq r$ then $sim_\delta(x, y, s)$.

Properties (MON), (ID), (EXT) can be taken as generic properties of any similarity. In particular each rough inclusion is required to satisfy them.

A standard rough inclusion is the one induced by the Łukasiewicz t–norm, or, equivalently, by the the Hamming distance on objects in a decision system, see [5], [7], [8], given as

$$\mu(v, u, r) \text{ iff } \frac{|IND(u, v)|}{|A|} \geq r. \tag{1}$$

Any continuous t–norm does induce a rough inclusion by means of its residual implication, see [5]; the residual implication $x \Rightarrow_t y$ of a t–norm t, is defined as

$$x \Rightarrow_t y \geq z \text{ iff } t(x, z) \leq y. \tag{2}$$

In [5], it is shown that $\Rightarrow_t$ does induce a rough inclusion on the interval $[0, 1]$:

$$\mu^{\Rightarrow_t}(u, v, r) \text{ iff } x \Rightarrow_t y \geq r. \tag{3}$$

We use three basic t–norms: the minimum min, the product $P(x, y) = x \cdot y$, and the Łukasiewicz $L(x, y) = max\{0, x + y - 1\}$.

In case $x > y$, these t–norms induce implications given by:

(MIN) $x \Rightarrow_{min} y = y$,
(P) $x \Rightarrow_P y = \frac{y}{x}$, and
(L) $x \Rightarrow_L y = min\{1, 1 - x + y\}$ (when $x \leq y$ the value is always 1).

For a given rough inclusion μ and a radius $r \in [0, 1]$, the granule $g^\mu(u, r)$ is the set of those v for which $\mu(v, u, r)$ holds, see [5].

2   Voting by Training Objects

We apply to data sets the CV-5 cross–validation and as a data set we use Wisconsin Diagnostic Breast Cancer [11].

For each test object u, and a training object v, for each attribute a, the factor $q_a(u, v) = \frac{\|a(u) - a(v)\|}{diam\ a}$ is computed where $diam\ a$ is the length of the interval $T(a) = [m(a), M(a)]$, where : $m(a)$ is the minimal value of a on the training set, and $M(a)$ is the maximal value of a on the training set. In case the test value $a(u)$ lies outside $T(a)$ it is projected into $T(a)$ by the mapping $f_a : x \to m_a$ when $x < m_a$, $x \to M_a$ when $x > M_a$, $x \to x$, otherwise.

In case $q_a(u, v) \geq \varepsilon$, a is included into $DIS_\varepsilon(u, v)$, otherwise a is included into $IND_\varepsilon(u, v)$, and these sets yield quotients $dis_\varepsilon(u, v) = \frac{|DIS_\varepsilon(u,v)|}{|A|}$ and $ind_\varepsilon(u, v) = \frac{|IND_\varepsilon(u,v)|}{|A|}$.

The weight $w_u(v, t) = dis_\varepsilon(u, v) \to_t ind\varepsilon(u, v)$ is then computed according to (3) with respect to a chosen t–norm t.

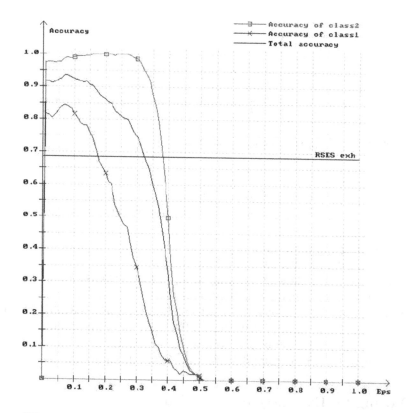

Fig. 1. CV-5; Wisconsin Diagnostic Breast Cancer; Algorithm 5_v1. Granules of training objects. t=min; Best result for $\varepsilon = 0.07$: accuracy=0.936283, coverage=1.0.

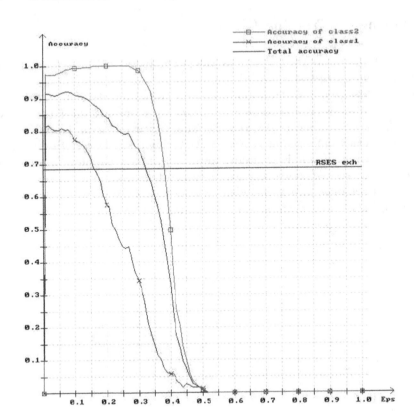

Fig. 2. CV-5; Wisconsin Diagnostic Breast Cancer; Algorithm 6_v1. Granules of training objects. t=P; Best result for $\varepsilon = 0.07$: accuracy=0.922124, coverage=1.0.

For each decision category c, the factor $sel_u(c,t)$
$= \frac{\sum_{v \ in \ training \ set} w_u(v,t)}{size \ c \ in \ training \ set}$ is computed and the test object u is assigned the category with maximal $sel_u(c)$.

The values of the parameter ε are taken every one hundreth, i.e. from 0.0 through 0.01, 0.02, to 0.99, 1.0.

Performance of this classifier was judged against the exhaustive classifier (see, e.g., [10] for a public domain exhaustive classifier).

We show results for Wisconsin data set in which case the exhaustive classifier gave accuracy of 0.6846, and coverage of 0.9928. Fig. 1 shows results in case of the t–norm $t = min$, for Wisconsin data set.

Analogously, Fig.2 shows results in case $t = P$, and Fig. 3 in case $t = L$.

3   Voting by Exhaustive Set of Rules

In this case, rules induced from the training set voted in the manner similar to that in sect. 2 with the difference that instead of the value $a(v)$ the

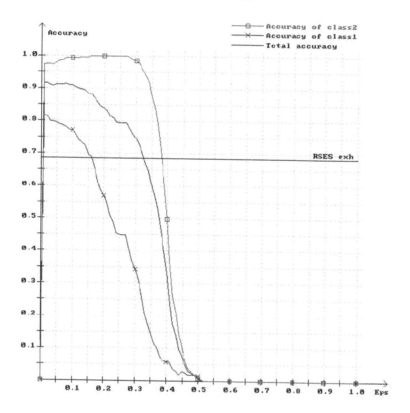

Fig. 3. CV-5; Wisconsin Diagnostic Breast Cancer; Algorithm 7_v1. Granules of training objects. t=L; Best result for $\varepsilon = 0.01$: accuracy=0.916814 , coverage=1.0.

value $a(r)$ assigned to the attribute a in the premise of rule r was inserted. Weights $w_u(r,t)$ took part in voting according to the factor $sel_u(c,t)$ $= \frac{\sum_{rules\ pointing\ to\ c} w_u(r,t)\cdot support\ r}{size\ c\ in\ training\ set}$ computed for each category c, and $c*$ with $sel_u(c*,t) = max_c sel_u(c,t)$ was assigned to u.

Figs. 4, 5, give results for, respectively, $t = min,\ P$.

4   Granules of Granular Reflections of Training Objects

The idea of a granulated data set was proposed in [8]: given a granulation radius r, the set $G(r,\mu)$ of all granules of the radius r is formed. From this set, a covering $C(r,\mu,\mathcal{G})$ of the set of objects U is chosen by means of a strategy $\mathcal{G}$, which is usually a random choice of granules with irreducibility checking.

Given the covering $C(r,\mu,\mathcal{G})$, attributes in the set A are factored through granules to make a new attribute set. For an attribute $a \in A$, and a granule g, the new attribute a_G is defined as

$$a_G(g) = \mathcal{S}(\{a(v) : v \in g\}),\tag{4}$$

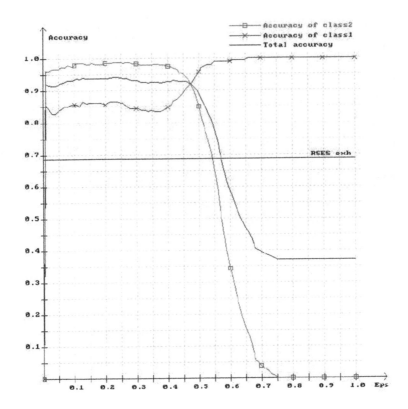

Fig. 4. CV-5; Wisconsin Diagnostic Breast Cancer; Algorithm 5_v2. Granules of rules. t=min; Best result for $\varepsilon = 0.23$: accuracy=0.943363, coverage=1.0.

Table 1. CV-5; Wisconsin Diagnostic Breast Cancer; Algorithm 5_v3. Granular objects. t=min; r_gran=granulation radius, optimal_eps=best optimal epsilon, acc=accuracy, cov=coverage, trn=training set. Best result for $r = 0.1$, $\varepsilon = 0.07$: accuracy=0.938053, coverage=1.0.

r_gran	optimal eps	acc	cov	trn
nil	nil	0.6846	0.9928	456
0.0	0.05	0.631716	0.99469	1.0
0.0333333	0.09	0.782301	1.0	170.0
0.0666667	0.07	0.934513	1.0	438.6
0.1	0.07	0.938053	1.0	446.6
0.133333	0.07	0.938053	1.0	446.6
0.166667	0.07	0.938053	1.0	446.6
0.2	0.07	0.938053	1.0	446.6
0.233333	0.07	0.936283	1.0	455.4
0.266667	0.07	0.936283	1.0	456.0

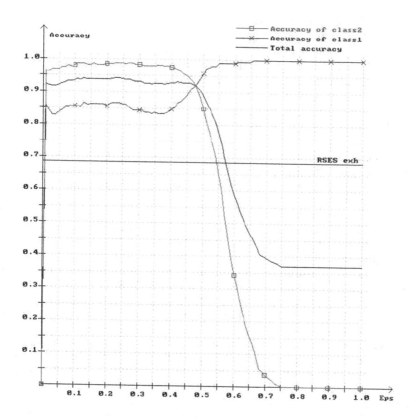

Fig. 5. CV-5; Wisconsin Diagnostic Breast Cancer; Algorithm 6_v2. Granules of rules.
t=P; Best result for $\varepsilon = 0.23$: accuracy=0.943363, coverage=1.0.

where S is a strategy for choosing the value of a_G, and a new information system
is formed: $(U^G = C(r, \mu, \mathcal{G}), A_G = \{a_G : a \in A\})$ called a granular reflection of
the given information system [8],[5]. The new granular object $g = g(u, r)$ with
values $a_G(g)$ of attributes a is called the *granular reflection* of the object u.

In experiments in this case, a granular reflection of the training set is formed,
with the strategy S as majority voting by objects in the granule with random
tie resolution. Then the procedure of sect. 2 is repeated with granular reflections of training objects. Parameters are granulation radius r and ε. Results in
Table 1 are given for each granulation radius in terms of the optimal ε – the
value of ε at which the highest accuracy of classification is obtained. Tests were
performed with Wisconsin data set and CV–5 cross–validation was applied.

5   Conclusions

Results of experiments along with the ones reported earlier [1], [2] witness the
high effectiveness of this approach: quality of classification is much higher than

with standard exhaustive classifier. We observe that optimal classifier obtained with the general radius r and $t = min$ gives better accuracy than the classifier on training objects with $r = 1$ which means that weighting heuristics slightly improve the quality of classification. The best result is obtained with rules and $r = 1$ which implies that passing to rules reduces slightly the noise in data.

References

1. Artiemjew, P.: On classification of data by means of rough mereological granules of objects and rules. In: Wang, G., Li, T., Grzymała-Busse, J.W., Miao, D., Skowron, A., Yao, Y. (eds.) RSKT 2008. LNCS (LNAI), vol. 5009, pp. 221–228. Springer, Heidelberg (2008) (in print)
2. Artiemjew, P.: Rough mereological classifiers obtained from weak variants of rough inclusions. In: Wang, G., Li, T., Grzymała-Busse, J.W., Miao, D., Skowron, A., Yao, Y. (eds.) RSKT 2008. LNCS (LNAI), vol. 5009, pp. 229–236. Springer, Heidelberg (2008) (in print)
3. Bazan, J.G.: A comparison of dynamic and non–dynamic rough set methods for extracting laws from decision tables. In: Polkowski, L., Skowron, A. (eds.) Rough Sets in Knowledge Discovery, vol. 1, pp. 321–365. Physica Verlag, Heidelberg (1998)
4. Pawlak, Z.: Rough Sets: Theoretical Aspects of Reasoning about Data. Kluwer, Dordrecht (1991)
5. Polkowski, L.: On the idea of using granular rough mereological structures in classification of data. In: Wang, G., Li, T., Grzymała-Busse, J.W., Miao, D., Skowron, A., Yao, Y. (eds.) RSKT 2008. LNCS (LNAI), vol. 5009, pp. 213–220. Springer, Heidelberg (2008)(in print)
6. Polkowski, L.: The paradigm of granular rough computing. In: Zhang, D., Wang, Y., Kinsner, W. (eds.) ICCI 2007, pp. 145–163. IEEE Computer Society, Los Alamitos (2007)
7. Polkowski, L.: Formal granular calculi based on rough inclusions (a feature talk). In: Zhang, Y.-Q., Lin, T.Y. (eds.) IEEE GrC 2006, pp. 9–18. IEEE Press, Piscataway (2006)
8. Polkowski, L.: Formal granular calculi based on rough inclusions (a feature talk). In: Hu, X., Liu, Q., Skowron, A., Lin, T.Y., Yager, R.R., Zhang, B. (eds.) IEEE GrC 2005, pp. 57–62. IEEE Press, Piscataway (2005)
9. Polkowski, L., Skowron, A.: Rough mereology: a new paradigm for approximate reasoning. International Journal of Approximate Reasoning 15(4), 333–365 (1997)
10. RSES, http://logic.mimuw.edu.pl/rses
11. UCI Repository, http://www.ics.uci.edu/~mlearn/databases/

Rough Set Approach to Information Tables with Imprecise Decisions

Masahiro Inuiguchi[1] and Bingjun Li[2]

[1] Graduate School of Engineering Science, Osaka University,
Toyonaka, Osaka 560-8531, Japan
`inuiguti@sys.es.osaka-u.ac.jp`
[2] College of Information and Management Science, Henan Agricultural University
Zhengzhou 450002, China
`zzlbjun@163.com`

Abstract. In this paper, we treat information tables with imprecise decisions, for short, imprecise decision tables. In the imprecise decision tables, decision attribute values are specified imprecisely. Under such decision tables, lower and upper object sets for a set of decision attribute values are defined. Their properties are shown. Concepts of reducts of imprecise decision tables are studied. Discernibility matrix methods are investigated for calculations of all reducts.

1 Introduction

Rough set approach proposed by Pawlak [9] provides useful tools for reasoning from data. It is applied to various fields such as medicine, engineering, management and so on. In order to extend the applicability, rough sets have been generalized in various ways [2,3,11,12,13]. Some [6] of them treats imprecise data in decision tables. Nevertheless, decision attribute values have been assumed to be precise, so far. Indeed, precise decision attribute values are usually obtained. In data mining and knowledge discovery, precise data are preferable. On the other hand, imprecise data were treated in incomplete information databases [7] as well as in nondeterministic information systems [8].

In the real world, we come across cases when we only obtain data with imprecise decision attribute values. For example, evaluation of the economic situation would be difficult to tell precisely. Failure diagnosis of complex systems would start from the expert hunch or conjecture. The conjectured source of failure is a decision attribute value and it would be imprecise. Consider a forecast, it would be difficult to be exact and precise. Some tolerance would be necessary. Moreover, evaluations by humans are often imprecise. Even data with imprecise decision attribute values would be useful to induce rough knowledge or to find the condition attributes possibly to effect on the decision attribute value. It is much more informative than ignorance. Utilization and analyzing such data is valuable if a sufficient number of precise data are not available.

Decision tables with imprecise decision attribute values have not yet discussed considerably. In this paper, we study the rough set approach to decision tables

C.-C. Chan et al. (Eds.): RSCTC 2008, LNAI 5306, pp. 121–130, 2008.
© Springer-Verlag Berlin Heidelberg 2008

Table 1. The imprecise decision table

Object	c_1	c_2	$\cdots$	c_m	d
u_1	$c_1(u_1)$	$c_2(u_1)$	$\cdots$	$c_1(u_1)$	$F(u_1)$
u_2	$c_1(u_2)$	$c_2(u_2)$	$\cdots$	$c_2(u_2)$	$F(u_2)$
$\vdots$	$\vdots$	$\vdots$	$\vdots$	$\vdots$	$\vdots$
u_n	$c_1(u_n)$	$c_2(u_n)$	$\cdots$	$c_m(u_n)$	$F(u_n)$

with imprecise decision attribute values. The decision tables are called imprecise decision tables. Because decision attribute values are imprecise, decision classes are often empty. Then we use sets of decision attribute values instead of decision classes. We define upper and lower object sets to a given set of decision attribute values. In this occasion, we combine all information about decision attribute values of objects having common condition attribute values. We also define a set of conflicting objects and sets of boundary objects. The properties of upper and lower object sets as well as a set of conflicting objects and sets of boundary objects are investigated. Using upper and lower object sets, sets of conflicting and boundary objects, condition attribute reduction is discussed. Various kinds of structure preserving reducts [4] are defined and their relationships are shown. As the result, they are consolidated into three kinds of reducts. In order to calculate all reducts, three kinds of discernibility matrices are successfully obtained corresponding is to three kinds of reducts.

This paper organized as follows. In next section, imprecise decision tables are introduced and upper and lower object sets as well as a set of conflicting objects and sets of boundary objects are defined. Their properties are investigated. In Section 3, condition attribute reduction is studied. Various kind of structure preserving reducts [4] are defined and consolidated into three kinds of reducts. Then discernibility matrices corresponding to the three kinds of reducts are shown and all reducts are obtained as prime implicants of three kinds of Boolean functions. In Section 4, concluding remarks are given.

2   Rough Sets Under Imprecise Decision Tables

2.1   Information Tables with Imprecise Decision Values

In this paper, we treat information tables with imprecise decision values shown in Table 1. The information table is a decision table represented by a quadruple $\mathcal{I} = (U, C \cup \{d\}, F, V)$. U is a finite set of objects, $U = \{u_1, u_2, \ldots, u_n\}$. C is a finite set of condition attributes, $C = \{c_1, c_2, \ldots, c_m\}$. Each attribute c_i can be seen as a function from U to V_{c_i}, where V_{c_i} is the domain of condition attribute c_i. The function value $c_i(u_j)$ indicates the attribute value of u_j. d is a decision attribute to which each object u_j takes a unique value $d(u_j)$. F is a set-valued function from U to 2^{V_d}, where 2^{V_d} is a power set of V_d and V_d is the domain of decision attribute d. $F(u_j)$ indicates a set of possible decision attribute values of u_j. Finally, $V = \bigcup_{c \in C} V_c \cup V_d$.

In Table 1, decision attribute values are allowed to take set-values, while in the conventional decision tables, decision attribute values should be singletons, i.e., single values. Each set-value of decision attribute shows possible decision attribute values of the object. Such set-value may be obtained when decision attribute values are not specified precisely. For example, when our knowledge is not complete but partial, we may specify the decision value such as "not d_1". Moreover, when the decision maker hesitates his/her classification of an objects, he/she may specify "d_1 or d_2". In those cases, decision attribute values can be treated as imprecise values. Even imprecise decision values would be more useful than no information and a number of imprecise decision values may collaborate to obtain precise values and useful results. Imprecise decision attribute values can be also regarded as values by conjecture. Then allowing imprecise decision attribute values enables us to analyze data by conjectures.

Information tables with imprecise decisions are called imprecise decision tables in this paper. We propose a rough set approach to imprecise decision tables.

2.2   Rough Sets Under Imprecise Decision Tables

Considering the imprecise nature of decision attribute values of imprecise decision tables, we define generalized decision values $\delta_P(u_j)$ and aggregated decision values $\hat{F}_P(u_j)$ under a given condition attribute set $P \subseteq C$ as follows:

$$\delta_P(u_j) = \{F(u) \mid u \in U, \; c_i(u) = c_i(u_j), \; \forall c_i \in P\}, \tag{1}$$

$$\hat{F}_P(u_j) = \begin{cases} \bigcap \delta_P(u_j), & \text{if } \bigcap \delta_P(u_j) \neq \emptyset, \\ \bigcup \delta_P(u_j), & \text{otherwise.} \end{cases} \tag{2}$$

We note that $\delta_P(u_j)$ is a family of sets of decision attribute values while $\hat{F}_P(u_j)$ is a set of decision attribute values. $\delta_P(u_j)$ collects imprecise decision values $F(u)$ of all objects u taking same condition attribute values with respect to $P \subseteq C$ as u_j takes. Since we assume the true decision attribute value of u_j is in $F(u_j)$ and the same decision attribute value would be assigned for all objects which share same condition attribute values, we may obtain a smaller possible range for decision attribute value of u_j by intersecting $F(u)$'s of all such objects u. However, if the given data is not totally consistent, the intersection can be empty. If the intersection is empty, some of $F(u)$ in the given table would be wrong or some condition attribute would be missing. In this case, the union would show the possible range. Based on these ideas, $\hat{F}_P(u_j)$ is defined. Taking union when the intersection is empty intersection is similar to Dubois and Prade's combination rule [1] in Dempster-Shafer theory of evidence.

We define the following object sets for a given condition attribute set $P \subseteq C$:

$$Conf_P = \left\{u \in U \mid \bigcap \delta_P(u) = \emptyset \right\}, \tag{3}$$

$$P_*(X) = \{u \in U \mid \hat{F}_P(u) \subseteq X\}, \tag{4}$$

$$P^*(X) = \{u \in U \mid \hat{F}_P(u) \cap X \neq \emptyset\}, \tag{5}$$

$$Bn_P(X) = P^*(X) - P_*(X), \tag{6}$$

Table 2. A table of failure conjectures by users

conjecture	(case,user)	func.1	func.2	func.3	func.4	func.5	cause	$\tilde{F}_C$
u_1	(P_1,E_1)	yes	yes	no	no	yes	$\{B,C\}$	$\{B,C\}$
u_2	(P_1,E_2)	yes	yes	no	no	yes	$\{A,B,C\}$	$\{B,C\}$
u_3	(P_2,E_1)	yes	yes	no	yes	yes	$\{A,B\}$	$\{A,B\}$
u_4	(P_3,E_2)	yes	yes	yes	yes	yes	$\{A\}$	$\{A,C\}$
u_5	(P_4,E_1)	yes	yes	no	no	no	$\{C\}$	$\{C\}$
u_6	(P_3,E_1)	yes	yes	yes	yes	yes	$\{C\}$	$\{A,C\}$
u_7	(P_5,E_2)	no	yes	yes	yes	yes	$\{A,C\}$	$\{A,C\}$
u_8	(P_6,E_2)	yes	no	no	yes	no	$\{B\}$	$\{B\}$

where $X \subseteq V_d$. $Conf_P$ is a set of conflicting objects. $P_*(X)$ is a lower object set of X and $P^*(X)$ is an upper object set of X. $Bn_P(X)$ is a set of boundary objects. If $u \in P_*(X)$, the decision attribute value of u is in X with no conflict with given data. In other words, if $u \in P_*(X)$, the decision attribute value of u is surely in X as far as the given decision table is correct. On the other hand, if $u \in P^*(X)$, the decision attribute value of $u \in X$ is at least possible. Moreover, if $u \in Bn_P(X)$, the decision attribute value of $u \in X$ is only possible. If $u \notin P^*(X)$, the decision attribute value of u would never take in X. When $F(u_j)$, $j = 1, 2, \ldots, n$ are singletons, $P_*(X)$ and $P^*(X)$ are coincide with lower approximation $\underline{P}(O(X))$ and upper approximation $\overline{P}(O(X))$ of the classical rough sets [9], respectively, where $O(X) = \{u \in U \mid F(u) \subseteq X\}$. Then $P_*(X)$ and $P^*(X)$ can be seen as extensions of lower and upper approximations of the classical rough sets. Then the pair $(P_*(X), P^*(X))$ is called rough sets with respect to decision attribute value set X under imprecise decision table $\mathcal{I}$.

Example 1. Consider Table 2 showing conjectures u_1, u_2, $\ldots$, u_8 by two users E_1 and E_2 about failure causes of 6 cases P_1, P_2, $\ldots$, P_6 in a complex system from 5 functions func.1, func.2, $\ldots$, func.5. There are three possible causes A, B and C. In this table, $U = \{u_i \mid i = 1, 2, \ldots, 8\}$, $C = \{$func.1(f1), func.2(f2), func.3(f3), func.4(f4), func.5(f5)$\}$, $V_a = \{$yes, no$\}$ for $a = $ f1$, \ldots, $f5 and $V_{\text{cause}} = \{$A, B, C$\}$. Then $V = \{$yes, no, A, B, C$\}$. The second column of Table 2 shows a pair (P_i, E_i) of case P_i and user E_i. The pair shows that the failure cause of P_i is conjectured by E_i and the result is shown in the column of "cause". For each conjecture u_i, $\tilde{F}_C(u_i)$ is shown in the rightmost column of Table 2. The set of conflicting objects and some lower and upper object sets are obtained as

$$Conf_P = \{u_4, u_6\}, \qquad C_*(\{B,C\}) = \{u_1, u_2, u_5, u_8\},$$
$$C^*(\{B,C\}) = \{u_1, u_2, \ldots, u_8\}, \quad Bn_P(\{B,C\}) = \{u_3, u_4, u_6, u_7\},$$
$$C_*(\{A,C\}) = \{u_4, u_5, u_6, u_7\}, \quad C^*(\{A,C\}) = \{u_1, u_2, \ldots, u_7\},$$
$$Bn_P(\{A,C\}) = \{u_1, u_2, u_3\}, \quad C_*(\{B\}) = \{u_8\},$$
$$C^*(\{B\}) = \{u_3, u_4, u_6, u_8\}, \quad Bn_P(\{B\}) = \{u_3, u_4, u_6\}.$$

Remark 1. We may have the different definitions of lower and upper object sets:

$$\underline{P}(X) = \left\{u \in U \mid \bigcup \delta_P(u) \subseteq X\right\}, \quad \overline{P}(X) = \left\{u \in U \mid \bigcup \delta_P(u) \cap X \neq \emptyset\right\}.$$

These definitions are based on a passive approach while definitions (4) and (5) are based on an active approach. In the passive approach, we do not trust very much in the imprecise information so that we take a union of imprecise decision attribute values to estimate the range. On the other hand, in the active approach, we trust imprecise information to some extent so that we take an intersection of the imprecise decision attribute values as far as it is non-empty.

2.3 Properties of Rough Sets Under Imprecise Decision Tables

Let $X, Y \subseteq V_d$ and $P \subseteq C$. For lower and upper object sets, we have the following properties:

$$P_*(V_d) = P^*(V_d) = U, \quad P_*(\emptyset) = P^*(\emptyset) = \emptyset, \tag{7}$$

$$P_*(X \cap Y) = P_*(X) \cap P_*(Y), \quad P^*(X \cup Y) = P^*(X) \cup P^*(Y), \tag{8}$$

$$X \subseteq Y \Rightarrow P_*(X) \subseteq P_*(Y), \quad P^*(X) \subseteq P^*(Y), \tag{9}$$

$$P_*(X \cup Y) \supseteq P_*(X) \cup P_*(Y), \quad P^*(X \cap Y) \subseteq P^*(X) \cap P^*(Y), \tag{10}$$

$$P_*(V_d - X) = U - P^*(X), \quad P^*(V_d - X) = U - P_*(X), \tag{11}$$

$$P_*(X) = P_*\left(\bigcup \hat{F}_P(P_*(X))\right), \tag{12}$$

where $\hat{F}_P(O) = \{\hat{F}_P(u) \mid u \in O\}$ for an object set $O \subseteq U$. Note that we do not always have $X \subseteq \bigcup \hat{F}_P(P^*(X))$ but $\bigcup \hat{F}_P(P_*(X)) \subseteq X$. Moreover, none of $P_*(X) = P^*\left(\bigcup \hat{F}_P(P_*(X))\right)$, $P^*(X) = P^*\left(\bigcup \hat{F}_P(P^*(X))\right)$ and $P^*(X) = P_*\left(\bigcup \hat{F}_P(P^*(X))\right)$ holds. However, we have $P_*(X) \subseteq P^*\left(\bigcup \hat{F}_P(P_*(X))\right)$, $P^*(X) \supseteq P^*\left(\bigcup \hat{F}_P(P^*(X))\right)$ and $P^*(X) \subseteq P_*\left(\bigcup \hat{F}_P(P^*(X))\right)$.

We have $\bigcap \delta_P(X) \subseteq \bigcap \delta_Q(X)$ but $\bigcup \delta_P(X) \supseteq \bigcup \delta_Q(X)$ for $P \subseteq Q \subseteq C$. Therefore, neither $P_*(X) \subseteq Q_*(X)$ nor $P^*(X) \supseteq Q^*(X)$ always hold for $P \subseteq Q \subseteq C$. On the contrary, we always have $Conf_P \supseteq Conf_Q$ and $\delta_P(u) \subseteq \delta_Q(u)$ for $P \subseteq Q \subseteq C$ and $u \in U$.

3   Attribute Reduction of Imprecise Decision Tables

3.1   Structure-Preserving Reducts

By the definitions of the set of conflicting objects, lower and upper object sets and the set of boundary objects, we may induce many structures on U. Structure we stands for in this paper is a subfamily $\mathcal{F}$ of $\{Conf_C, C_*(S_d), C^*(S_d), Bn_C(S_d) \mid S_d \subseteq V_d\}$. Structure-preserving reducts have been proposed in the framework of variable precision rough sets by Inuiguchi [4]. In this subsection, we introduce structure-preserving reducts into rough sets under imprecise decision tables.

Given a structure $\mathcal{F}$, any elementary set in $\mathcal{F}$ is a set of objects depending on attribute set C since it is one of the set of conflicting objects, lower object set, upper object set and the set of boundary objects. Then let us denote a generic elementary set in $\mathcal{F}$ by $F(C)$. A $\mathcal{F}$-preserving reduct is defined as follows.

Definition 1. *Attribute set P is called an $\mathcal{F}$-preserving reduct if and only if*

($\mathcal{F}$1) *$F(P) = F(C)$ for all $F(C) \in \mathcal{F}$, and*
($\mathcal{F}$2) *There is no $Q \subset P$ such that $F(Q) = F(C)$ for all $F(C) \in \mathcal{F}$.*

We consider the following structures in this paper: $\mathcal{C} = \{Conf_C\}$, $\mathcal{L} = \{C_*(S_d) \mid S_d \subseteq V_d\}$, $\mathcal{U} = \{C^*(S_d) \mid S_d \subseteq V_d\}$, $\mathcal{B} = \{Bn_C(S_d) \mid S_d \subseteq V_d\}$, $\mathcal{L}_k = \{C_*(S_d) \mid S_d \subseteq V_d, |S_d| = k\}$, $\mathcal{U}_k = \{C^*(S_d) \mid S_d \subseteq V_d, |S_d| = k\}$ and $\mathcal{B}_k = \{Bn_C(S_d) \mid S_d \subseteq V_d, |S_d| = k\}$, where $k \in [1, q-1]$ is an integer and q is a number of decision attribute values, i.e., $q = |V_d|$, where $|A|$ shows a cardinality of set A.
    We have the following theorem.

Theorem 1. *The following assertions are valid:*

(a) *The concepts of $\mathcal{U}$-preserving reduct, $\mathcal{L}$-preserving reduct, $\mathcal{B}$-preserving reduct, $\mathcal{U}_1$-preserving reduct, $\mathcal{L}_{q-1}$-preserving reduct and $\mathcal{B}_1$-preserving reduct are equivalent.*

(b) *The concept of $\mathcal{U}_k$-preserving reduct is equivalent to that of $\mathcal{L}_{q-k}$-preserving reduct for any $k \in [1, q-1]$.*

(c) *A $\mathcal{U}_k$-preserving reduct satisfies ($\mathcal{U}_l$1) and ($\mathcal{L}_{q-l}$1) for all $l \geq k$.*

(d) *If $k \leq q/2$, the concept of $\mathcal{B}_k$-preserving reduct is equivalent to that of $\mathcal{U}_k$-preserving reduct. Otherwise, the concept of $\mathcal{B}_k$-preserving reduct is equivalent to that of $\mathcal{L}_k$-preserving reduct.*

Proof. We show (a) and (d) because (b) and (c) can be proven in the same way as (a). From (11), the equivalence between $\mathcal{U}_1$- and $\mathcal{L}_{q-1}$-preserving reducts can be easily shown. For (a) and (d), it suffices to prove that (i) a $\mathcal{U}_1$-preserving reduct P satisfies ($\mathcal{U}$1) and ($\mathcal{L}$1) and (ii) a $\mathcal{B}_k$-preserving reduct Q satisfies ($\mathcal{U}_k$1) and ($\mathcal{L}_k$1). First let us prove (i). Since P is a $\mathcal{U}_1$-preserving reduct, P satisfies ($\mathcal{U}_1$1), i.e., $P^*(\{v_d\}) = C^*(\{v_d\})$ for all $v_d \in V_d$. From (8), $P^*(S_d) = \bigcup_{v_d \in S_d} P^*(\{v_d\}) = \bigcup_{v_d \in S_d} C^*(\{v_d\}) = C^*(S_d)$ for any $S_d \subseteq V_d$. Then P satisfies ($\mathcal{U}$1). Moreover, from (11) and ($\mathcal{U}$1), we have $P_*(S_d) = U - P^*(V_d - S_d) = U - C^*(V_d - S_d) = C_*(S_d)$ for any $S_d \subseteq V_d$. Then P satisfies ($\mathcal{L}$1). Thus (i) is shown. Now, let us prove (ii). Suppose $Q^*(V_d) \neq C^*(V_d)$ for some V_d such that $|V_d| = k$. Then there exists $u \in U$ such that $u \in C^*(V_d) - Q^*(V_d)$ or $u \in Q^*(V_d) - C^*(V_d)$. When $\bigcap \delta_Q(u) \neq \emptyset$, $Q_*(V_d) \supseteq C_*(V_d)$ because $\hat{F}_Q(u) \subseteq \hat{F}_C(u)$. This implies $u \in C^*(V_d) - Q^*(V_d)$. Let $Z = C^*(V_d) - Q^*(V_d)$. Z is not empty because $u \in Z$. We have $C^*(V_d) = Q^*(V_d) \cup Z$ and $Z \cap Q^*(V_d) = \emptyset$. Since Q is a $\mathcal{B}_k$-preserving reduct, we have $Q^*(V_d) - Q_*(V_d) = C^*(V_d) - C_*(V_d)$. From this, we should have $C_*(V_d) = Q_*(V_d) \cup Z$. On the contrary, we have $C_*(V_d) \subseteq Q_*(V_d)$ due to $\hat{F}_Q(u) \subseteq \hat{F}_C(u)$. A contradiction. The case when $\bigcap \delta_Q(u) = \emptyset$, we have $Q_*(V_d) \subseteq C_*(V_d)$ because $\hat{F}_Q(u) \supseteq \hat{F}_C(u)$. We obtain a contradiction in this case, too, in the same way as we did in the other case but with replacements of $Q^*(V_d)$ and $Q_*(V_d)$ by $C^*(V_d)$ and $C_*(V_d)$, respectively. Therefore, we have ($\mathcal{U}_k$1), i.e., $Q^*(V_d) = C^*(V_d)$ for any V_d such that $|V_d| = k$. The proof for ($\mathcal{L}_k$1) can be done in the same way. □

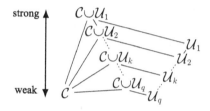

Fig. 1. Strong-weak relations among three kinds of reducts

From Theorem 1(c), we know that the smaller k is, the stronger the required preservation in the $\mathcal{U}_k$-preserving reduct is. From Theorem 1, we need to consider only $\mathcal{C}$-preserving reducts and $\mathcal{U}_k$-preserving reducts for $k \in [1, q-2]$. $\mathcal{C}$-preserving reducts are minimal sets of condition attributes which do not expand the inconsistency. $\mathcal{U}_k$-preserving reducts are minimal sets of condition attributes which preserve all certain objects of $S_d \subset V_d$ for all $|S_d| \le q - k$.

As demonstrated in Example 2 shown later, for any $k \in [1, q-1]$, no $\mathcal{U}_k$-preserving reduct can always preserve ($\mathcal{C}1$). Conversely, no $\mathcal{C}$-preserving reduct can always preserve ($\mathcal{U}_k 1$) for any $k \in [1, q-1]$. Then, we may define $\mathcal{C} \cup \mathcal{U}_k$-preserving reduct for any $k \in [1, q-1]$. The strong-weak relations among those reducts are depicted in Fig. 1. $\mathcal{C} \cup \mathcal{U}_1$-preserving reducts are the strongest so that the number of condition attributes composing a $\mathcal{C} \cup \mathcal{U}_1$-preserving reduct is minimal. The larger k is, the weaker $\mathcal{C} \cup \mathcal{U}_k$-preserving reducts and $\mathcal{U}_k$-preserving reducts are.

3.2   Indiscernibility Matrices

To enumerate all reducts, the indiscernibility matrix method [10] is useful. In this paper, we extend the conventional indiscernibility matrix to cases of $\mathcal{C}$-, $\mathcal{U}_k$- and $\mathcal{C} \cup \mathcal{U}_k$-preserving reducts for $k \in [1, q-1]$.

To apply the indiscernibility matrix method, the condition preserving the structure should be decomposable to conditions obtained from the pair-wise comparisons. Fortunately, all reducts we are discussing can satisfy this requirement. By the pair-wise comparisons, we should first consider whether the unification of two objects is allowed or not, where "unification of two objects" stands for the identification of two objects by reducing condition attributes. If it is not allowed, we collect condition attributes which differ between two objects.

At the beginning, let us discuss $\mathcal{C}$-preserving reducts. Let P be a condition attribute set. We express condition that P satisfies ($\mathcal{C}1$) as a Boolean function. If two objects u_i and u_j are both in $Conf_{\mathcal{C}}$, they can be unified. If one object is in $Conf_{\mathcal{C}}$ and the other is not, they cannot be unified and the following set of statements, M_{ij} is calculated:

$$M_{ij} = \{a \in C \mid a(u_i) \ne a(u_j), a \in C\}. \tag{13}$$

the condition corresponding to M_{ij} means that at least one of the statements in M_{ij} should be satisfied. If none of two objects u_i and u_j are in $Conf_{\mathcal{C}}$, we should check the intersection of $\delta_C(u_i)$ and $\delta_C(u_j)$. If $\bigcap(\delta_C(u_i) \cup \delta_C(u_j)) = \emptyset$, they cannot be unified and M_{ij} of (13) is calculated. If $\bigcap(\delta_C(u_i) \cup \delta_C(u_j)) \neq \emptyset$, the following set of statements, $\hat{M}_{ij}$ is calculated:

$$\hat{M}_{ij} = \left\{ d(u_i) \in \bigcap(\delta_C(u_i) \cup \delta_C(u_j)) \wedge d(u_j) \in \bigcap(\delta_C(u_i) \cup \delta_C(u_j)) \right\} \cup M_{ij}, \tag{14}$$

where $d(u)$ shows the decision value of u and $\wedge$ is a conjunction operation. $\hat{M}_{ij}$ is an extension of M_{ij} and the first term of the union in the right-hand side of (14) is extended term. This term is required because the unification of u_i and u_j implies that $\hat{F}_P(u_i) = \hat{F}_P(u_j) \subseteq \bigcap(\delta_C(u_i) \cup \delta_C(u_j))$. Then the condition corresponding to $\hat{M}_{ij}$ can be described as if none of statements in M_{ij} is satisfied, both of $d(u_i) \in \bigcap(\delta_C(u_i) \cup \delta_C(u_j))$ and $d(u_j) \in \bigcap(\delta_C(u_i) \cup \delta_C(u_j))$ should be satisfied.

From the discussion above, the (i,j)-component of the discernibility matrix M^C for C-preserving reducts is defined as follows:

$$M_{ij}^{\mathcal{C}} = \begin{cases} C, & \text{if } \{u_i, u_j\} \subseteq Conf_C, \\ M_{ij}, & \text{if } \{u_i, u_j\} \not\subseteq Conf_C \text{ and } \bigcap(\delta_C(u_i) \cup \delta_C(u_j)) = \emptyset, \\ \hat{M}_{ij}, & \text{if } \{u_i, u_j\} \not\subseteq Conf_C \text{ and } \bigcap(\delta_C(u_i) \cup \delta_C(u_j)) \neq \emptyset. \end{cases} \tag{15}$$

All C-preserving reducts are obtained as prime implicants of a Boolean function,

$$f^{\mathcal{C}} = \bigwedge_{i,j: u_i, u_j \in U} \bigvee M_{ij}^{\mathcal{C}}. \tag{16}$$

If a prime implicant is represented as $a_1 \in P \wedge a_2 \in P \wedge a_3 \in P$, the corresponding C-preserving reduct candidate is $\{a_1, a_2, a_3\}$. If a prime implicant is represented as $d(u_4) \in Z_4 \neq \emptyset \wedge d(u_5) \in Z_5 \neq \emptyset \wedge a_6 \in P \wedge a_7 \in P$, the corresponding C-preserving reduct candidate is $\{a_6, a_7\}$. In such a way, we obtain C-preserving reduct candidates corresponding to all prime implicants. The C-preserving reducts are all minimal set among all C-preserving reduct candidates.

Now let us describe a discernibility matrix with respect to $\mathcal{U}_k$-preserving reducts for $k \in [1, q-1]$. By a similar discussion, taking care of objects $u \in U$ such that $|\hat{F}_C(u)| \leq q - k$, we define a discernibility matrix $M^{\mathcal{U}_k}$ whose (i,j)-component $M_{ij}^{\mathcal{U}_k}$ is defined by

$$M_{ij}^{\mathcal{U}_k} = \begin{cases} C, & \text{if } \{u_i, u_j\} \cap Conf_C \neq \emptyset, |\hat{F}_C(u_i)| \leq q - k \text{ and} \\ & \hat{F}_C(u_i) = \hat{F}_C(u_j) = \bigcup(\delta_C(u_i) \cup \delta_C(u_j)), \\ C, & \text{if } \{u_i, u_j\} \cap Conf_C \neq \emptyset, |\hat{F}_C(u_i)| > q - k \text{ and} \\ & |\hat{F}_C(u_j)| > q - k, \\ C, & \text{if } \{u_i, u_j\} \cap Conf_C = \emptyset, |\hat{F}_C(u_i)| \leq q - k \text{ and} \\ & \hat{F}_C(u_i) = \hat{F}_C(u_j), \\ C, & \text{if } \{u_i, u_j\} \cap Conf_C = \emptyset, |\hat{F}_C(u_i)| > q - k, |\hat{F}_C(u_j)| > q - k \\ & \text{and } |\bigcap(\delta_C(u_i) \cup \delta_C(u_j))| > q - k, \\ M_{ij}, & \text{otherwise.} \end{cases} \tag{17}$$

All $\mathcal{U}_k$-preserving reducts are obtained as prime implicants of a Boolean function,

$$f^{\mathcal{U}_k} = \bigwedge_{i,j:u_i,u_j \in U} \bigvee M_{ij}^{\mathcal{U}_k}. \tag{18}$$

Finally, we describe the discernibility matrix with respect to $\mathcal{C} \cup \mathcal{U}_k$-preserving reducts for $k \in [1, q-1]$. By a similar discussion, we define a discernibility matrix $M^{\mathcal{CU}_k}$ whose (i,j)-component $M_{ij}^{\mathcal{CU}_k}$ is defined by

$$M_{ij}^{\mathcal{CU}_k} = \begin{cases} C, & \text{if } \{u_i, u_j\} \subseteq Conf_C \text{ and } \hat{F}_C(u_i) = \hat{F}_C(u_j), \\ C, & \text{if } \{u_i, u_j\} \subseteq Conf_C, \hat{F}_C(u_i) \neq \hat{F}_C(u_j), |\hat{F}_C(u_i)| > q - k \\ & \text{and } |\hat{F}_C(u_j)| > q - k, \\ C, & \text{if } \{u_i, u_j\} \cap Conf_C = \emptyset \text{ and } \hat{F}_C(u_i) = \hat{F}_C(u_j), \\ C, & \text{if } \{u_i, u_j\} \cap Conf_C = \emptyset, \hat{F}_C(u_i) \neq \hat{F}_C(u_j), \\ & \text{and } \left| \hat{F}_C(u_i) \cap \hat{F}_C(u_j) \right| > q - k, \\ M_{ij}, & \text{otherwise.} \end{cases} \tag{19}$$

$\mathcal{C} \cup \mathcal{U}_k$-preserving reducts are obtained as prime implicants of a Boolean function,

$$f^{\mathcal{CU}_k} = \bigwedge_{i,j:u_i,u_j \in U} \bigvee M_{ij}^{\mathcal{CU}_k}. \tag{20}$$

Example 2. Consider the imprecise decision table given in Table 2. Due to limited space, we cannot write all components of discernibility matrix M^C. Then we show some of them, $M_{13}^C = \{d(u_1) \in \{B\} \wedge d(u_3) \in \{B\}, f4 \in P\}$, $M_{34}^C = \{f3 \in P\}$ and $M_{47}^C = \{f1 \in P, f2 \in P\}$. Calculating all other components, we obtain the following Boolean function F^C:

$$F^{\mathcal{U}} = (d(u_1) \in \{B\} \wedge d(u_2) \in \{B\} \wedge d(u_3) \in \{B\} \wedge f1 \in P \wedge f3 \in P \wedge f5 \in P)$$
$$\vee (d(u_1) \in \{C\} \wedge d(u_2) \in \{C\} \wedge d(u_5) \in \{C\} \wedge f1 \in P \wedge f3 \in P \wedge f4 \in P)$$
$$\vee (d(u_1) \in \{B, C\} \wedge d(u_2) \in \{B, C\} \wedge f1 \in P \wedge f3 \in P \wedge f4 \in P \wedge f5 \in P).$$

Then $\{f1, f3, f5\}$, $\{f1, f3, f4\}$ and $\{f1, f3, f4, f5\}$ are candidates. Taking minimal sets, we obtain two $\mathcal{C}$-preserving reducts, $\{f1, f3, f5\}$ and $\{f1, f3, f4\}$. Then we understand condition attributes f1 and f3 are very important not to expand the inconsistency. However they are not sufficient to avoid the expansion of the inconsistency and at least one of condition attributes f5 and f4 should be added.

Similarly, based on discernibility matrices $M^{\mathcal{U}_k}$ and $M^{\mathcal{CU}_k}$ for $k = 1, 2$, we obtain $\{f3, f4, f5\}$ as a unique $\mathcal{U}_1$-preserving reduct, $\{f2, f5\}$ and $\{f4, f5\}$ as $\mathcal{U}_2$-preserving reducts, $\{f1, f3, f4, f5\}$ as a unique $\mathcal{C} \cup \mathcal{U}_1$-preserving reduct and $\{f1, f2, f5\}$ and $\{f1, f4, f5\}$ as $\mathcal{U}_2$-preserving reducts. Note that, for $\mathcal{U}_1$-preserving reduct $\{f3, f4, f5\}$, $\mathcal{U}_2$-preserving reducts $\{f2, f5\}$ and $\{f4, f5\}$, u_7 becomes inconsistent as well as u_4 and u_6. Thus those reducts do not preserve (C1).

4   Conclusions

In this paper, we have proposed a rough set approach to imprecise decision tables. Based on an active approach, rough sets have been defined under imprecise decision tables. Three kinds of reducts have been proposed and investigated. Discernibility matrices corresponding to those kinds of reducts have been proposed to compute all reducts. The computational complexity is same as the conventional discernibility matrix method [10]. By the proposed approach, we can analyze attribute importance even by imprecise data.

Many other approaches would be conceivable. For example, we may introduce structure enhancing reducts [5] and information source-wise approaches. Moreover, we may discuss rule induction based on the proposed rough sets. Those are future topics in rough set approach to imprecise decision tables.

Acknowledgments

This work has been supported by the Grant-in-Aid for Scientific Research (B) No.17310098 and the Grant-in-Aid for Exploratory Research No.18651078.

References

1. Dubois, D., Prade, H.: Representation and combination of uncertainty with belief functions and possibility measures. Computational Intelligence 4, 244–264 (1988)
2. Greco, S., Matarazzo, B., Slowinski, R.: Rough set theory for multicriteria decision analysis. European Journal of Operational Research 129, 1–47 (2001)
3. Inuiguchi, M.: Generalization of rough sets and rule extraction. In: Peters, J.F., Skowron, A., Grzymała-Busse, J.W., Kostek, B.z., Świniarski, R.W., Szczuka, M.S. (eds.) Transactions on Rough Sets I. LNCS, vol. 3100, pp. 96–119. Springer, Heidelberg (2004)
4. Inuiguchi, M.: Attribute reduction in variable precision rough set model. International Journal of Uncertainty, Fuzziness and Knowledge-Based Systems 14, 461–479 (2006)
5. Inuiguchi, M.: Structure-based attribute reduction in variable precision rough set models. Journal of Advanced Computational Intelligence and Intelligent Informatics 10, 657–665 (2006)
6. Kryszkiewicz, M.: Rough set approach to incomplete information systems. Information Sciences 112, 39–49 (1998)
7. Lipski, W.: On semantic issues connected with incomplete information databases. ACM Transactions on Database Systems 4(3), 262–296 (1979)
8. Orłowska, E., Pawlak, Z.: Representation of nondeterministic information. Theoretical Computer Science 29, 27–39 (1984)
9. Pawlak, Z.: Rough sets. International Journal of Information and Computer Sciences 11, 341–356 (1982)
10. Skowron, A., Rauser, C.M.: The discernibility matrix and functions in information systems. In: Słowinski, R. (ed.) Intelligent Decision Support: Handbook of Applications and Advances of the Rough Sets Theory, pp. 331–362. Kluwer Academic Publishers, Dordrecht (1992)
11. Słowiński, R., Vanderpooten, D.: A Generalized Definition of Rough Approximations Based on Similarity. IEEE Transactions on Data and Knowledge Engineering 12(2), 331–336 (2000)
12. Yao, Y.Y., Lin, T.Y.: Generalization of Rough Sets Using Modal Logics. Intelligent Automation and Soft Computing 2(2), 103–120 (1996)
13. Ziarko, W.: Variable precision rough set model. J. Comput. Syst. Sci. 46(1), 39–59 (1993)

Computing Approximations of Dominance-Based Rough Sets by Bit-Vector Encodings

Chien-Chung Chan[1,2] and Gwo-Hshiung Tzeng[3,4]

[1] Department of Computer Science, University of Akron, Akron, OH, 44325-4003, USA
chan@uakron.edu
[2] Department of Information Science and Applications, Asia University, Wufeng, Taichung 41354, Taiwan
[3] Department of Business and Entrepreneurial Administration, Kainan University, No. 1 Kainan Road, Luchu, Taoyuan County 338, Taiwan
ghtzeng@mail.knu.edu.tw
[4] Institute of Management of Technology, National Chiao Tung University, 1001 Ta-Hsueh Road, Hsinchu 300, Taiwan
ghtzeng@cc.nctu.edu.tw

Abstract. This paper introduces a mechanism for computing approximations of Dominance-Based Rough Sets (DBRS) by bit-vector encodings. DBRS was introduced by Greco et al. as an extension of Pawlak's classical rough sets theory by using dominance relations in place of equivalence relations for approximating sets of preference ordered decision classes. Our formulation of dominance-based approximation spaces is based on the concept of indexed blocks introduced by Chan and Tzeng. Indexed blocks are sets of objects indexed by pairs of decision values where approximations of sets of decision classes are defined in terms of exclusive neighborhoods of indexed blocks. In this work, we introduced an algorithm for updating indexed blocks incrementally, and we show that the computing of dominance-based approximations can be accomplished more intuitively and efficiently by encoding indexed blocks as bit-vectors. In addition, bit-vector encodings can simplify the definitions of lower and upper approximations greatly. Examples are given to illustrate presented concepts.

Keywords: Rough sets, Dominance-based rough sets, Multiple criteria decision analysis (MCDA), Approximate reasoning.

1 Introduction

Dominance-based rough sets (DBRS) introduced by Greco et al. [1, 2, 3] extend Pawlak's classical rough sets (CRS) [8, 9, 10] by considering attributes, called criteria, with preference-ordered domains and by substituting the indiscernibility relation in CRS with a dominance relation that is reflexive and transitive. It is also assumed that decision classes are ordered by some preference ordering. A consistent preference model is taken to be one that respects the dominance

C.-C. Chan et al. (Eds.): RSCTC 2008, LNAI 5306, pp. 131–141, 2008.

principle when assigning actions (objects) to the preference ordered decision classes. The dominance principle requires that if action x dominates action y, then x should be assigned to a class not worse than y. Given a total ordering on decision classes, in DBRS, the sets to be approximated are called the upward union and downward union of decision classes [5].

In [13], a dominance-based approximation space for quantitative and totally ordered multiple criteria decision tables are represented by a family of indexed blocks, which are sets of objects indexed by pairs of decision values. The basic idea is to use a binary relation on decision values as indices for grouping objects based on dominance principle. Inconsistency is defined as a result of violating the dominance principle. A set of ordered pairs is derived from objects violating dominance principle involving a pair of decision values, which is used as the index for the set of ordered pairs. For example, objects that are consistently assigned to decision class i form a set of ordered pairs with decision index (i, i). A set of ordered pairs with decision index (i, j), $i \neq j$, corresponds to objects that violate dominance principle involving decision values i and j. Each indexed set of ordered pairs induces a set of objects called a block that is indexed by the same pair of decision values. These blocks are called indexed blocks, which are granules of a dominance-based approximation space. Rules were introduced for computing the reduction of inconsistency when criteria are aggregated one by one incrementally. Approximations of any union of sets of decision classes in a dominance-based approximation space represented by indexed blocks are based on the concept of neighborhoods of indexed blocks [13].

Following the indexed blocks approach, this paper introduces a mechanism for computing indexed blocks from indexed ordered sets when criteria are combined incrementally. In addition, we introduce a bit-vector representation for indexed block granules to facilitate the computing of approximations. New definitions of approximations are given in terms of bit-vector encodings.

The remainder of this paper is organized as follows. In Section 2, related concepts of rough sets, dominance-based rough sets, and indexed blocks are reviewed. In Section 3, we introduce an algorithm for updating indexed blocks when criteria are combined incrementally. In Section 4, we introduce bit-vector encoding of indexed blocks. Definitions of approximations based on bit-vectors are given here. Then, we show how to compute those approximations. Finally, conclusion is given in Section 5.

2   Related Concepts

2.1   Information Systems and Rough Sets

In rough sets theory [8, 9, 10], information of objects in a domain is represented by an information system $IS = (U, A, V, f)$, where U is a finite set of objects, A is a finite set of attributes, $V = \cup_{q \in A} V_q$ and V_q is the domain of attribute q, and $f : U \times A \to V$ is a total information function such that $f(x, q) \in V_q$ for every $q \in A$ and $x \in U$. In many applications, we use a special case of information systems called *decision tables* to represent data sets. In a decision table $(U, C \cup D = \{d\})$, there is a designated attribute $\{d\}$ called *decision attribute* and attributes in C are called *condition attributes*. Each attribute q in $C \cup D$ is associated with an equivalence relation R_q on the set of objects of U such that for each x and $y \in U$, xR_qy means $f(x,q) = f(y,q)$. For each x and $y \in U$, we say that x and y are *indiscernible* on attributes $P \subseteq C$ if and only if xR_qy for all $q \in P$.

2.2   Dominance-Based Rough Sets

In dominance-based rough sets attributes with totally ordered domains are called *criteria*. More precisely, each criterion q in C is associated with an outranking relation [11] S_q on the set of objects of U such that for each x and $y \in U, xS_qy$ means $f(x,q) \geq f(y,q)$. For each x and $y \in U$, we say that x *dominates* y on criteria $P \subseteq C$ if and only if xS_qy for all $q \in P$. The dominance relations are taken to be total pre-ordered, i.e., strongly complete and transitive binary relations [5].

Dominance-based rough sets approach is capable of dealing with inconsistencies in MCDA problems based on the *principle of dominance*, namely: given two objects x and y, if x dominates y, then x should be assigned to a class not worse than y. Assignments of objects to decision classes are inconsistent if the dominance principle is violated. The sets of decision classes to be approximated are considered to have upward union and downward union properties. More precisely, let $Cl = \{Cl_t | t \in T\}, T = \{1, 2, ..., n\}$, be a set of decision classes such that for each $x \in U$, x belongs to one and only one $Cl_t \in \mathbf{Cl}$ and for all r, s in T, if $r > s$, the decision from Cl_r is preferred to the decision from Cl_s. Based on this total ordering of decision classes, the upward union and downward union of decision classes are defined respectively as:

$$Cl_t^{\geq} = \cup_{s \geq t} Cl_s, \quad Cl_t^{\leq} = \cup_{s \leq t} Cl_s, \quad t = 1, 2, ..., n.$$

An object x is in $Cl_t^{\geq}$ means that x at least belongs to class Cl_t, and x is in $Cl_t^{\leq}$ means that x at most belongs to class Cl_t.

2.3   Indexed Blocks as Granules

In [13], the concept of indexed blocks was used to represent approximation spaces derived from decision tables with multiple criteria. Indexed blocks are sets of objects indexed by pairs of decision values.

Let $(U, C \cup D = \{d\})$ be a multi-criteria decision table where condition attributes in C are criteria and decision attribute d is associated with a total preference ordering. For each condition criterion q and a decision value d_i of d let $\min_q(d_i)$ denote the minimum value of q among objects with decision value d_i, and $\max_q(d_i)$ denote the maximum value.

For each condition criterion q, the mapping $I_q(i,j) : D \times D \to \wp(V_q)$ is defined as

$$I_q(i,j) = \{f(x,q) = v | v \geq \min_q(d_j) \ and \ v \leq \max_q(d_i), \ for \ i < j; i, j = 1, ..., V_D\},$$

$I_q(i,j) = I_q(j,i)$ if $i > j$, and $I_q(i,i) = \{f(x,q)|f(x,d) = i$ and $f(x,q) \notin \cup_{i<j} I_q(i,j)\}$, where $\wp(V_q)$ denotes the power set of V_q For simplicity, the set of values $I_q(i,j)$ is denoted as $[\min_q(j), \max_q(i)]$ or simply as $[\min_j, \max_i]$ for a decision value pair i and j with $i < j$.

Intuitively, the set $I_q(i,i)$ denotes the values of criterion q where objects can be consistently labeled with decision value i. For $i < j$, values in $I_q(i,j)$ are conflicting or inconsistent in the sense that objects with higher values of criterion q are assigned to a lower decision class or vice versa, namely, the dominance principle is violated.

For each $I_q(i,j)$ and $i \neq j$, the corresponding set of *ordered pairs* $[I_q(i,j)]$: $D \times D \rightarrow \wp(U \times U)$ is defined as $[I_q(i,j)] = \{(x,y) \in U \times U | f(x,d) = i, f(y,d) = j$ *such that* $f(x,q) \geq f(y,q)$ *for* $f(x,q), f(y,q) \in I_q(i,j)\}$. For each set $[I_q(i,j)]$ of ordered pairs, the restrictions of $[I_q(i,j)]$ to i and j are defined as:

$[I_q(i,j)]_i = \{x \in U|$ there exists $y \in U$ such that $(x, y) \in [I_q(i,j)]\}$ and
$[I_q(i,j)]_j = \{y \in U|$ there exists $x \in U$ such that $(x, y) \in [I_q(i,j)]\}$
The corresponding indexed block $B_q(i,j) \subseteq U$ of $[I_q(i,j)]$ is defined as

$$B_q(i,j) = [I_q(i,j)]_i \cup [I_q(i,j)]_j.$$

Example 1. For convenience, we will use the multiple criteria decision table taken from [6], which is shown in Table 1. The inconsistent intervals for each criterion q_1, q_2, and q_3 are shown in Tables 2, 3, and 4.

Table 1. A multi-criteria decision table

U	1	2	3	4	5	6	7	8	9	10	11	12	13	14	15	16	17
q_1	1.5	1.7	0.5	0.7	3	1	1	2.3	1	1.7	2.5	0.5	1.2	2	1.9	2.3	2.7
q_2	3	5	2	0.5	4.3	2	1.2	3.3	3	2.8	4	3	1	2.4	4.3	4	5.5
q_3	12	9.5	2.5	1.5	9	4.5	8	9	5	3.5	11	6	7	6	14	13	15
d	2	2	1	1	3	2	1	3	1	2	2	2	1	2	3	3	3

Table 2. Inconsistent intervals $I_{q1}(i, j)$

$D \times D$	1	2	3
1	[ ]	[0.5, 2]	[ ]
2		[ ]	[2.3, 2.5]
3			[2.7, 3.0]

Table 3. Inconsistent intervals $I_{q2}(i, j)$

$D \times D$	1	2	3
1	[0.5, 0.5]	[1, 3]	[ ]
2		[ ]	[3.3, 5]
3			[5.5, 5.5]

Table 4. Inconsistent intervals $I_{q3}(i, j)$

$D \times D$	1	2	3
1	[1.5, 2.5]	[3.5, 8]	[ ]
2		[ ]	[9, 14]
3			[15, 15]

From the intervals in the above tables, we can derive the following sets of ordered pairs.

The sets of ordered pairs for criterion q_1 are:

$[I_{q1}(1, 1)] = [I_{q1}(2, 2)] = \varnothing$,

$[I_{q1}(1, 2)] = \{(4, 12), (3, 12), (7, 12), (7, 6), (9, 12), (9, 6), (14, 12), (14, 6),$
    $(14, 13), (14, 1), (14, 2), (14, 10), (14, 15)\}$,

$[I_{q1}(1, 3)] = \varnothing$,

$[I_{q1}(2, 3)] = \{(11, 8), (11, 16)\}$,

$[I_{q1}(3, 3)] = \{(17, 17), (5, 5)\}$.

The sets of ordered pairs for criterion q_2 are:

$[I_{q2}(1, 1)] = \{(4, 4)\}$,

$[I_{q2}(2, 2)] = \varnothing$,

$[I_{q2}(1, 2)] = \{(7, 13), (3, 13), (3, 6), (14, 13), (14, 6), (9, 13), (9, 6), (9, 10),$
    $(9, 1), (9, 12)\}$,

$[I_{q2}(1, 3)] = \varnothing$,

$[I_{q2}(2, 3)] = \{(11, 8), (11, 16), (15, 8), (15, 16), (15, 5), (2, 8), (2, 16), (2, 5)\}$,

$[I_{q2}(3, 3)] = \{(17, 17)\}$.

The sets of ordered pairs for criterion q_3 are:

$[I_{q3}(1, 1)] = \{(4, 4), (3, 3)\}$,

$[I_{q3}(2, 2)] = \varnothing$,

$[I_{q3}(1, 2)] = \{(9, 10), (9, 6), (14, 10), (14, 6), (14, 12), (7, 10), (7, 6), (7, 12),$
    $(7, 13)\}$,

$[I_{q3}(1, 3)] = \varnothing$,

$[I_{q3}(2, 3)] = \{(2, 5), (2, 8), (11, 5), (11, 8), (1, 5), (1, 8), (15, 5), (15, 8),$
    $(15, 16)\}$,

$[I_{q3}(3, 3)] = \{(17, 17)\}$.

From the above sets of ordered pairs, we can derive their corresponding indexed blocks for criteria q_1, q_2, and q_3 as shown in Tables 5, 6, and 7. Three rules for updating indexed blocks by combining criteria incrementally were introduced in [13]. In Section 3, we will provide an algorithm for updating indexed blocks. The set of indexed blocks after combining q_1, q_2, and q_3 is shown in Table 8. It represents the dominance-based approximation space with each indexed block as a granule generated by the criteria $\{q_1, q_2 q_3\}$ from the multiple criteria decision table shown in Table 1.

2.4   Approximating Sets of Decision Classes

Typical approximated decision classes considered are sets of decision classes satisfying upward and downward union properties [1, 2, 3, 6]. Using indexed

Table 5. Indexed blocks $B_{q1}(i, j)$

$D \times D$	1	2	3
1	$\emptyset$	{1, 2, 3, 4, 6, 7, 9, 10, 12, 13, 14, 15}	$\emptyset$
2		$\emptyset$	{8, 11, 16}
3			{5, 17}

Table 6. Indexed blocks $B_{q2}(i, j)$

$D \times D$	1	2	3
1	{4}	{1, 3, 6, 7, 9, 10, 12, 13, 14, }	$\emptyset$
2		$\emptyset$	{2, 5, 8, 11, 15, 16}
3			{17}

Table 7. Indexed blocks $B_{q3}(i,j)$.

$D \times D$	1	2	3
1	{3, 4}	{6, 7, 9, 10, 12, 13,14}	$\emptyset$
2		$\emptyset$	{1, 2, 5, 8, 11, 15, 16}
3			{17}

Table 8. Indexed blocks $B_{\{q1,q2,q3\}}(i, j)$

$D \times D$	1	2	3
1	{3, 4, 7}	{6, 9, 14}	$\emptyset$
2		{1, 2, 10, 12, 13, 15}	{8, 11}
3			{5, 16, 17}

blocks as granules, approximations to any combination of decision classes can be computed. This kind of approximations is formulated by using the concept of neighborhoods of indexed blocks introduced in [13]. For a criterion q and for each decision value $i \in V_d$, and for each indexed block $B_q(i,i)$ of $[I_q(i,i)]$, the *exclusive neighborhood of* $B_q(i,i)$ is defined as

$$ENB(B_q(i,i)) = \{B_q(k,i)|k \geq 1 \ and \ k < i\} \cup \{B_q(i,k)|k \geq 1 \ and \ k > i\}.$$

The exclusive neighborhood of $B_q(i,i)$ corresponds to sets of objects which have inconsistent decision class assignments associated with decision i. It is clear that neighborhood of $B_q(i,i)$ can be defined as the union of $\{B_q(i,i)\}$ and its exclusive neighborhood.

Let $(U, C \cup D = \{d\})$ be a multiple criteria decision table. The lower approximation of a decision class in a dominance-based approximation space generated by C is the indexed block $B_C(i,i)$, and the boundary set of the decision class is the exclusive neighborhood of $B_C(i,i)$, $ENB(B_C(i,i))$. The upper approximation is the union of lower approximation and boundary set, i.e., the neighborhood of $B_C(i,i)$ The above definitions are applicable to approximations of sets of decision classes. However, the complication involved is how to update the exclusive neighborhood for a set of decision classes. One formulation was given in [13]. In this paper, we will show in Section 4 that it is easier to use bit-vector encodings of the neighborhoods of indexed blocks.

3   Updating Indexed Blocks by Combination of Criteria

In the following, we will give a computational procedure for realizing the three rules introduced in [13] when combining two criteria to update indexed blocks incrementally.

Procedure for updating indexed blocks by two criteria q_1 and q_2:

Step 1. For each decision class i, update $B_{\{q_1,q_2\}}(i,i) = [I_{q_1}(i,i)]_i \cup [I_{q_2}(i,i)]_i$.

Step 2. For each decision pairs (i,j) and $i < j$:

Step 2.1 Compute $[I_{\{q_1,q_2\}}(i,j)] = [I_{q_1}(i,j)] \cap [I_{q_2}(i,j)]$;

Step 2.2 Compute the indexed block:
$B_{\{q_1,q_2\}}(i,j) = [I_{\{q_1,q_2\}}(i,j)]_i \cup [I_{\{q_1,q_2\}}(i,j)]_j$;

Step 2.3 Compute $[[I_{q_1}(i,j)] - [I_{\{q_1,q_2\}}(i,j)]]_i$ and $[[I_{q_1}(i,j)] - [I_{\{q_1,q_2\}}(i,j)]]_j$;

Step 2.4 Compute $[[I_{q_2}(i,j)] - [I_{\{q_1,q_2\}}(i,j)]]_i$ and $[[I_{q_2}(i,j)] - [I_{\{q_1,q_2\}}(i,j)]]_j$;

Step 2.5 Compute
$B_i = [[I_{q_1}(i,j)] - [I_{\{q_1,q_2\}}(i,j)]]_i \cup [[I_{q_2}(i,j)] - [I_{\{q_1,q_2\}}(i,j)]]_i$,
then update $B_{\{q_1,q_2\}}(i,i) = B_{\{q_1,q_2\}}(i,i) \cup B_i - B_{\{q_1,q_2\}}(i,j)$;

Step 2.6 Compute
$B_j = [[I_{q_1}(i,j)] - [I_{\{q_1,q_2\}}(i,j)]]_j \cup [[I_{q_2}(i,j)] - [I_{\{q_1,q_2\}}(i,j)]]_j$,
then update $B_{\{q_1,q_2\}}(j,j) = B_{\{q_1,q_2\}}(j,j) \cup B_j - B_{\{q_1,q_2\}}(i,j)$.

Example 2. Consider combining criteria q_1 and q_2 in the multiple criteria decision table given in Table 1 for updating decision pair $(1,2)$. The inconsistent intervals of criteria q_1 and q_2 are shown in Tables 2 and 3.

Step 1. For each decision class i, update $B_{\{q_1,q_2\}}(i,i) = [I_{q_1}(i,i)]_i \cup [I_{q_2}(i,i)]_i$:
$[I_{q1}(1,1)] = \varnothing$ and $[I_{q2}(1,1)] = \{(4,4)\}$ $[I_{q2}(1,1)]_1 = \{4\}$,
so $B_{\{q_1,q_2\}}(1,1) = \varnothing \cup \{4\} = \{4\}$.
$[I_{q1}(2,2)] = \varnothing$ and $[I_{q2}(2,2)] = \varnothing$, so $B_{\{q_1,q_2\}}(2,2) = \varnothing$.

Step 2. For each decision pairs (ij) and $i < j$:

Step 2.1 Compute $[I_{\{q_1,q_2\}}(i,j)] = [I_{q_1}(i,j)] \cap [I_{q_2}(i,j)]$:
$[I_{q1}(1,2)] = \{(4,12), (3,12), (7,12), (7,6), (9,12), (9,6), (14,12), (14,6),$
$(14,13), (14,1), (14,2), (14,10), (14,15)\}$ and
$[I_{q2}(1,2)] = \{(7,13), (3,13), (3,6), (9,13), (9,10), (9,1)\}$.
$[I_{\{q_1,q_2\}}(1,2)] = [I_{q1}(1,2)] \cap [I_{q2}(1,2)] = \{(9,12), (9,6), (14,6), (14,13)\}$.

Step 2.2 Compute the indexed block:
$B_{\{q_1,q_2\}}(i,j) = [I_{\{q_1,q_2\}}(i,j)]_i \cup [I_{\{q_1,q_2\}}(i,j)]_j$:
$[I_{\{q_1,q_2\}}(1,2)]_1 = \{9, 14\}$ and $[I_{\{q_1,q_2\}}(1,2)]_2 = \{6, 12, 13\}$,
so $B_{\{q_1,q_2\}}(1,2) = \{6, 9, 12, 13, 14\}$.

Step 2.3 Compute $[[I_{q_1}(i,j)] - [I_{\{q_1,q_2\}}(i,j)]]_i$ and $[[I_{q_1}(i,j)] - [I_{\{q_1,q_2\}}(i,j)]]_j$:
$[I_{q_1}(1,2)] - [I_{\{q_1,q_2\}}(1,2)] = \{(4,12), (3,12), (7,12), (7,6), (14,12), (14,1),$
$(14,2), (14,10), (14,15)\}$
$[[I_{q_1}(1,2)] - [I_{\{q_1,q_2\}}(1,2)]]_1 = \{3, 4, 7, 14\}$ and
$[[I_{q_1}(1,2)] - [I_{\{q_1,q_2\}}(1,2)]]_2 = \{1, 2, 6, 10, 12, 15\}$.
Step 2.4 Compute $[[I_{q_2}(i,j)] - [I_{\{q_1,q_2\}}(i,j)]]_i$ and $[[I_{q_2}(i,j)] - [I_{\{q_1,q_2\}}(i,j)]]_j$:
$[I_{q_2}(1,2)] - [I_{\{q_1,q_2\}}(1,2)] = \{(7,13), (3,13), (3,6), (9,13), (9,10), (9,1)\}$.
$[[I_{q_2}(1,2)] - [I_{\{q_1,q_2\}}(1,2)]]_1 = \{3, 7, 9\}$ and
$[[I_{q_2}(1,2)] - [I_{\{q_1,q_2\}}(1,2)]]_2 = \{1, 6, 10, 13\}$.
Step 2.5 Compute
$B_i = [[I_{q_1}(i,j)] - [I_{\{q_1,q_2\}}(i,j)]]_i \cup [[I_{q_2}(i,j)] - [I_{\{q_1,q_2\}}(i,j)]]_i$,
then update $B_{\{q_1,q_2\}}(i,i) = B_{\{q_1,q_2\}}(i,i) \cup B_i - B_{\{q_1,q_2\}}(i,j)$:
$B_1 = [[I_{q_1}(1,2)] - [I_{\{q_1,q_2\}}(1,2)]]_1 \cup [[I_{q_2}(1,2)] - [I_{\{q_1,q_2\}}(1,2)]]_1$
$= \{3, 4, 7, 14\} \cup \{3, 7, 9\} = \{3, 4, 7, 9, 14\}$.
$B_{\{q_1,q_2\}}(1,1) = B_{\{q_1,q_2\}}(1,1) \cup B_1 - B_{\{q_1,q_2\}}(1,2)$
$= \{4\} \cup \{3, 4, 7, 9, 14\} - \{6, 9, 12, 13, 14\} = \{3, 4, 7\}$.
Step 2.6 Compute
$B_j = [[I_{q_1}(i,j)] - [I_{\{q_1,q_2\}}(i,j)]]_j \cup [[I_{q_2}(i,j)] - [I_{\{q_1,q_2\}}(i,j)]]_j$,
then update $B_{\{q_1,q_2\}}(j,j) = B_{\{q_1,q_2\}}(j,j) \cup B_j - B_{\{q_1,q_2\}}(i,j)$:
$B_2 = [[I_{q_1}(1,2)] - [I_{\{q_1,q_2\}}(1,2)]]_2 \cup [[I_{q_2}(1,2)] - [I_{\{q_1,q_2\}}(1,2)]]_2$
$= \{1, 2, 6, 10, 12, 15\} \cup \{1, 6, 10, 13\} = \{1, 2, 6, 10, 12, 13, 15\}$.
$B_{\{q_1,q_2\}}(2,2) = B_{\{q_1,q_2\}}(2,2) \cup B_2 - B_{\{q_1,q_2\}}(1,2)$
$= \emptyset \cup \{1, 2, 6, 10, 12, 13, 15\} - \{6, 9, 12, 13, 14\} = \{1, 2, 10, 15\}$.

4   Bit-Vector Encodings of Indexed Blocks

In the following, we consider encoding indexed blocks as bit-vectors. For a multiple criteria decision table with N objects and K decision values, each object is encoded as a bit-vector of K bits. The encoding of a decision table can be represented by an $N \times K$ Boolean matrix generated from the table of indexed blocks as follows. Each indexed block is encoded by a vector of N bits where 1 means an object is in the block. The encoding for decision class i is computed by taking a logical OR of indexed blocks in the i-th row and i-th column of the indexed block table.

Let each decision class be encoded as a vector $\mathbf{v}$ of K bits where the i-th bit is set to 1 for a decision class with decision value i. A vector of all zero bits is called a $NULL$ vector, denoted by $\mathbf{0}$. Two bit-vectors $\mathbf{v}_1$ and $\mathbf{v}_2$ are *compatible*, if the logical AND of $\mathbf{v}_1$ and $\mathbf{v}_2$ is not an $NULL$ vector, i.e., $\mathbf{v}_1 \wedge \mathbf{v}_2 \neq \mathbf{0}$. Two compatible vectors $\mathbf{v}_1$ and $\mathbf{v}_2$ are *equal* if they have exactly the same bit patterns, i.e., $\mathbf{v}_1 \wedge \mathbf{v}_2 = \mathbf{v}_1 = \mathbf{v}_2$. A bit-vector $\mathbf{v}_1$ is a *subvector* of $\mathbf{v}_2$, $\mathbf{v}_1 \subseteq \mathbf{v}_2$, if $\mathbf{v}_1[i] = 1$ then $\mathbf{v}_2[i] = 1$ for all bit $i = 1, \ldots, K$.

Let $(U, C \cup D = \{d\})$ be a multiple criteria decision table. Let $E(x)$ denote the decision bit-vector encoding of object $x \in U$. Then, the lower approximation $\underline{C}\mathbf{v}$ and upper approximation $\overline{C}\mathbf{v}$ for an encoded decision class $\mathbf{v}$ are defined as

Table 9. Bit-vector encoding for objects in indexed block $(1, 1)$

1	2	3	4	5	6	7	8	9	10	11	12	13	14	15	16	17
		1	1			1										

Table 10. Bit-vector encoding for objects in indexed block $(1, 2)$

1	2	3	4	5	6	7	8	9	10	11	12	13	14	15	16	17
					1			1					1			

Table 11. Logical OR of objects in indexed blocks $(1, 1)$ and $(1, 2)$

1	2	3	4	5	6	7	8	9	10	11	12	13	14	15	16	17
		1	1		1	1		1					1			

Table 12. Bit-vector encoding of indexed blocks of Table 8

U	1	2	3	4	5	6	7	8	9	10	11	12	13	14	15	16	17
d_1	0	0	1	1	0	1	1	0	1	0	0	0	0	1	0	0	0
d_2	1	1	0	0	0	1	0	1	1	1	1	1	1	1	1	0	0
d_3	0	0	0	0	1	0	0	1	0	0	1	0	0	0	0	1	1

$\underline{C}\mathbf{v} = \{x|\ E(x) \subseteq \mathbf{v}\}$ and
$\overline{C}\mathbf{v} = \{x|E(x) \wedge \mathbf{v} \neq \mathbf{0}\}$.
The boundary set of $\mathbf{v}$ is $BN(\mathbf{v}) = \overline{C}\mathbf{v} - \underline{C}\mathbf{v}$.

Example 3. Consider the indexed blocks in Table 8. The set of objects in the indexed block $(1, 1)$ is $\{3, 4, 7\}$, and the objects of indexed block $(1, 2)$ is $\{6, 9, 14\}$. Their bit-vector encodings are shown in Table 9 and Table 10, respectively. The result of taking a logical OR of the above two vectors is shown in Table 11.

Applying the procedure to all indexed blocks of Table 8, the resulted encoding for all three decision values $\{1, 2, 3\}$ is shown in Table 12.

Example 4. Consider the encodings in Table 12.
The bit-vector of decision class 1 is the vector $\mathbf{v}_{\{1\}} = (1, 0, 0)$. Thus,
$\underline{C}\mathbf{v}_{\{1\}} = \{3, 4, 7\}$,
$\overline{C}\mathbf{v}_{\{1\}} = \{3, 4, 6, 7, 9, 14\}$, and
$BN(\mathbf{v}_{\{1\}}) = \{6, 9, 14\}$.
The bit-vector of decision class 2 is the vector $\mathbf{v}_{\{2\}} = (0, 1, 0)$. Thus,
$\underline{C}\mathbf{v}_{\{2\}} = \{1, 2, 10, 12, 13, 15\}$,
$\overline{C}\mathbf{v}_{\{2\}} = \{1, 2, 6, 8, 9, 10, 11, 12, 13, 14, 15\}$, and
$BN(\mathbf{v}_{\{2\}}) = \{6, 8, 9, 11, 14\}$.
The bit-vector of decision classes $\{1, 2\}$ is the vector $\mathbf{v}_{\{1,2\}} = (1, 1, 0)$. Thus,
$\underline{C}\mathbf{v}_{\{1,2\}} = \{1, 2, 3, 4, 6, 7, 9, 10, 12, 13, 14, 15\}$,
$\overline{C}\mathbf{v}_{\{1,2\}} = \{1, 2, 3, 4, 6, 7, 8, 9, 10, 11, 12, 13, 14, 15\}$, and
$BN(\mathbf{v}_{\{1,2\}}) = \{8, 11\}$.

5 Conclusion

In this paper we introduced an algorithm for updating indexed blocks by combining criteria incrementally. The algorithm can be used to generate indexed blocks representation of dominance-based approximation spaces. Then, a bit-vector encodings of indexed blocks as a representation of dominance- based approximation spaces was considered. It simplifies the definitions of lower and upper approximations of sets of decision classes greatly. We believe that it provides a solid foundation for designing efficient computations of approximations, and it provides new ways for understanding and studying dominance-based rough sets. One of our future works is to develop rule induction algorithms based on the bit-vector encodings.

References

1. Greco, S., Matarazzo, B., Slowinski, R.: Rough approximation of a preference relation by dominance relations, ICS Research Report 16/96, Warsaw University of Technology, Warsaw. European Journal of Operational Research 117(1), 63–83 (1999)
2. Greco, S., Matarazzo, B., Slowinski, R.: A new rough set approach to evaluation of bankruptcy risk. In: Zopounidis, C. (ed.) Operational Tools in the Management of Financial Risks, pp. 121–136. Kluwer Academic Publishers, Dordrecht (1998)
3. Greco, S., Matarazzo, B., Slowinski, R.: The use of rough sets and fuzzy sets in MCDM. In: Gal, T., Stewart, T., Hanne, T. (eds.) Advances in Multiple Criteria Decisions Making, ch. 14, pp. 14.1–14.59. Kluwer Academic Publishers, Dordrecht (1999)
4. Greco, S., Matarazzo, B., Slowinski, R.: Rough sets theory for multicriteria decision analysis. European Journal of Operational Research 129(1), 1–47 (2001)
5. Greco, S., Matarazzo, B., Slowinski, R.: Rough sets methodology for sorting problems in presence of multiple attributes and criteria. European Journal of Operational Research 138(2), 247–259 (2002)
6. Greco, S., Matarazzo, B., Slowinski, R., Stefanowski, J.: An algorithm for induction of decision rules consistent with the dominance principle. In: Ziarko, W., Yao, Y. (eds.) RSCTC 2000. LNCS (LNAI), vol. 2005, pp. 304–313. Springer, Heidelberg (2001)
7. Greco, S., Matarazzo, B., Slowinski, R., Stefanowski, J., Zurawski, M.: Incremental versus non-incremental rule induction for multicriteria classification. In: Peters, J.F., Skowron, A., Dubois, D., Grzymała-Busse, J.W., Inuiguchi, M., Polkowski, L. (eds.) Transactions on Rough Sets II. LNCS, vol. 3135, pp. 33–53. Springer, Heidelberg (2004)
8. Pawlak, Z.: Rough sets: basic notion. International Journal of Computer and Information Science 11(15), 344–356 (1982)
9. Pawlak, Z.: Rough sets and decision tables. In: Skowron, A. (ed.) SCT 1984. LNCS, vol. 208, pp. 186–196. Springer, Heidelberg (1985)
10. Pawlak, Z., Grzymala-Busse, J., Slowinski, R., Ziarko, W.: Rough sets. Communication of ACM 38(11), 89–95 (1995)

11. Roy, B.: Methodologie Multicritere d'Aide a la Decision. Economica, Paris (1985)
12. Fan, T.F., Liu, D.R., Tzeng, G.H.: Rough set-based logics for multicriteria decision analysis. European Journal of Operational Research 182(1), 340–355 (2007)
13. Chan, C.-C., Tzeng, G.-H.: Dominance-based rough sets using indexed blocks as granules. In: Wang, G., Li, T., Grzymała-Busse, J.W., Miao, D., Skowron, A., Yao, Y. (eds.) RSKT 2008. LNCS (LNAI), vol. 5009, pp. 244–251. Springer, Heidelberg (2008)

A Framework for Multiagent Mobile Robotics: Spatial Reasoning Based on Rough Mereology in Player/Stage System

Lech Polkowski and Paweł Ośmiałowski

Polish–Japanese Institute of Information Technology
Koszykowa 86, 02008 Warsaw, Poland
polkow@pjwstk.edu.pl, newchief@king.net.pl

Abstract. Problems of multiagent mobile robotics concern teams of mobile robots organized to perform an ordained task. Dynamic problems of navigation in multiagent environment require a theory of spatial reasoning. We propose here a spatial theory based on rough mereology along with an implementation in the software system Player/Stage. The proposed theoretical–software system provides a platform for analysis of tasks of multiagent mobile robotics.

Keywords: mobile robotics, spatial reasoning, rough mereology, player/stage software system.

1   Introduction

Qualitative Spatial Reasoning is a basic ingredient in a variety of problems in mobile robotics, see, e.g., [1], [4]. Spatial reasoning which deals with objects like solids, regions etc., by necessity refers to and relies on mereological theories of concepts based on the opposition part–whole [3]. Mereological ideas have been early applied toward axiomatization of geometry of solids, see [6], [11].

Mereological theories rely either on the notion of a part [7] or on the notion of objects being connected [3]. Our approach to spatial reasoning is developed within the paradigm of rough mereology, see sect. 2. Rough mereology is based on the predicate of being a part to a degree and thus it is a natural extension of mereology based on part relation.

2   Rough Mereology

Rough mereology, see, e.g., [9] begins with the notion of a *rough inclusion* which is a parameterized relation μ_r such that for any pair of objects u, v the formula u is μ_r v means that u is a *part of v to a degree r* where $r \in [0,1]$.

The following is the list of basic postulates for rough inclusions; el is the element (ingredient) relation of a mereology system based on a part relation, see [7]; informally, it is a partial ordering on the given class of objects defined from the strict order set by part relation.

C.-C. Chan et al. (Eds.): RSCTC 2008, LNAI 5306, pp. 142–149, 2008.

RM1. u is μ_1 v $\iff$ u is el v (a part in degree 1 is equivalent to an element).
RM2. u is μ_1 $v \implies$ for all w (w is μ_r u $\implies$ w is μ_r v) (monotonicity of μ).
RM3. u is μ_r $v \wedge s \leq r \implies u$ is μ_s v (assuring the meaning "a part to degree *at least r*").

In our applications to spatial reasoning, objects will be regions in Euclidean spaces, notably 2D space, like rectangles, squares, discs, and the rough inclusion applied will predominantly be the one defined by the equation,

$$u \text{ is } \mu_r^0 \, v \text{ iff } \frac{|u \cap v|}{|u|} \geq r, \tag{1}$$

where $|u|$ is the area of the region u. A variant in which area is replaced with cardinality is also used in some contexts, notably grid objects.

3   Mereogeometry

We are interested in introducing into the mereological world defined by μ^0 a geometry in whose terms it will be possible to express spatial relations among objects; a usage for this geometry will be found in navigation and control tasks of multiagent mobile robotics, cf. [1], [4].

We first introduce a notion of distance κ in our rough mereological universe by letting,

$$\kappa(u, v) = min\{max \, u, max \, w : \, u \text{ is } \mu_u \, v, v \text{ is } \mu_w \, u\}. \tag{2}$$

Observe that mereological distance differs essentially from the standard distance: the closer are objects, the greater is the value of κ.

We now introduce the notion of betweenness $T(u, v)$,

$$z \text{ is } T(u, v) \iff \text{ for all w } \kappa(z, w) \, \in [\kappa(u, w), \kappa(v, w)]. \tag{3}$$

Here, $[,]$ means the non–oriented interval. We check that T satisfies the axioms of Tarski [12] for *betweenness*.

Proposition 1. *1. z is $T(u, u) \implies z = u$ (identity).*
   2. v is $T(u, w)$ and z is $T(v, w) \implies v$ is $T(u, z)$ (transitivity).
   3. v is $T(u, z)$ and v is $T(u, w)$ and $u \neq v \implies z$ is $T(u, w)$ or w is $T(u, z)$ (connectivity).

Proof. By means of κ, the properties of betweenness in our context are translated into properties of betweenness in the real line which hold by the Tarski theorem [12], Theorem 1.

3.1   Nearness

We apply κ to define in our context the functor N of nearness proposed in van Benthem [2],

$$w \text{ is } N(u, v) \iff (\kappa(w, u) = r, \kappa(u, v) = s \implies s < r). \tag{4}$$

Here, nearness means that w is closer to u than v is to u.
   Then the following hold, i.e., N does satisfy all axioms for nearness in [2],

Proposition 2. *1. z is $N(u,v)$ and v is $N(u,w) \implies z$ is $N(u,w)$ (transitivity).*
2. z is $N(u,v)$ and u is $N(v,z) \implies u$ is $N(z,v)$ (triangle inequality).
3. $non(z$ is $N(u,z))$ (irreflexivity).
4. $z = u$ or z is $N(z,u)$ (selfishness).
5. z is $N(u,v) \implies z$ is $N(u,w)$ or w is $N(u,v)$ (connectedness).

We now may introduce the notion of equidistance as a functor $Eq(X,Y)$ defined as follows,

$$z \text{ is } Eq(u,v) \iff \kappa(z,u) = \kappa(z,v). \tag{5}$$

3.2 Betweenness

In addition to betweenness T, we make use of a betweenness functor T_B [2],

$$z \text{ is } T_B(u,v) \iff [\text{for all } w \ (Z \text{ is } w \text{ or } z \text{ is } N(u,w) \text{ or } z \text{ is } N(v,w))]. \tag{6}$$

Proposition 3. *The functor T_B of betweenness does satisfy the Tarski axioms.*

3.3 Examples

We give some examples of specific contexts in which functors defined above can be useful.

Example 1. We adopt as objects topologically connected unions of finitely many cubes in the unit grid on the space R^d (topological connectedness will be defined recursively: (1). a single cube is connected; (2) given a connected union C and a cube c, the union $C \cup c$ is connected if c is adjacent by the edge or a vertex to a cube in C). We adopt as the rough inclusion the function $\mu^1(C,D,r)$ iff $\frac{n(C \cap D)}{n(C)} \geq r$, where $n(C)$ is the number of cubes in C. One checks that: *the connected union E is between (in the sense of T_B) disjoint cubes c,d whenever E contains C and D and E consists of a minimal number of cubes for E being connected.*

Example 2. We consider a context in which objects are rectangles positioned regularly, i.e., having edges parallel to axes in R^2. The measure μ is μ^0 of (1). In this setting, given two disjoint rectangles C, D, the only object between C and D is the *extent* $ext(C,D)$ of C,D, , i.e., the minimal rectangle containing the union $C \cup D$. To see this, one can consider two squares C,D of identical size centered on the axis $x = 0$, and solve the problem analytically by showing that there is no other rectangle nearer to C and D than $ext(C,D)$ (this requires solving a set of linear inequalities); then, the general case follows by observing that linear shrinking or stretching of an edge does not change the area relations.

A line segment may be defined via the auxiliary notion of a pattern; we introduce this notion as a functor Pt.

We let

$$Pt(u,v,z) \iff z \text{ is } T_B(u,v) \text{ or } u \text{ is } T_B(z,v) \text{ or } v \text{ is } T_B(u,z).$$

We will say that a finite sequence $u_1, u_2, ..., u_n$ of objects *belong in a line segment* whenever $Pt(u_i, u_{i+1}, u_{i+2})$ for $i = 1, , ..., n-2$; formally, we introduce the functor *Line* of finite arity defined via

$$Line(u_1, u_2, ..., u_n) \Longleftrightarrow \text{for all } i < n - 1.Pt(u_i, u_{i+1}, u_{i+2}).$$

Example 3. With reference to Example 2, rectangles C, D and their extent $ext(C, D)$ form a line segment.

3.4   Extensions in the Robotic Context

We can model robots in a team by means of their extents (understood as safety regions about robots). Then for robots a, b, c, we say that b is $T_B(a, c)$ (robot b is between robots a, c) in case the rectangle $ext(b)$ is contained in the extent of rectangles $ext(a), ext(c)$. This allows for a partial betweenness which models in a more precise manner the relations between a, b, c: b is $T_B(a, c, r)$ in case $ext(b)$ is $\mu_r^0(ext[ext(a), ext(c)])$.

4   Implementation in Player/Stage Software System

Player/Stage is an Open-Source software system designed for many UNIX-compatible platforms, widely used in robotics laboratories [8]. Main two parts are Player – message passing server (with bunch of drivers for many robotics devices, extendable by plug–ins) and Stage – a plug–in for Player's bunch of drivers which simulates existence of real robotics devices that operate in the simulated 2D world. Player/Stage offers client–server architecture. Many clients can connect to one Player server, where clients are programs (robot controllers) written by a roboticist who can use Player client-side API. Player itself uses drivers to communicate with devices, in this activity it does not make distinction between real and simulated hardware. It gives roboticist means for testing programmed robot controller in both real and simulated world.

Among all Player drivers that communicate with devices (real or simulated), there are drivers not intended for controlling hardware, instead those drivers offer many facilities for sensor data manipulation, for example, camera image compression, retro–reflective detection of cylindrical markers in laser scans, path planning. One of the new features added to Player version 2.1 is the PostGIS driver: it connects to PostgreSQL database in order to obtain and/or update stored vector map layers.

PostGIS itself is an extension to the PostgreSQL object–relational database system which allows GIS (Geographics Information Systems) objects to be stored in the database [10]. It also offers new SQL functions for spatial reasoning. Maps which to be stored in SQL database can be created and edited by graphical tools like uDig or by C/C++ programs written using GEOS library of GIS functions. PostGIS, uDig and GEOS library are projects maintained by Refractions Research.

A map can have many named layers, and for each layer a table in SQL database is created. We can assume that layer named *obstacles* consists of objects which a robot cannot walk through. Other layers can be created in which we

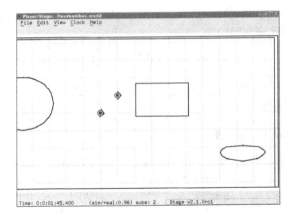

Fig. 1. Stage simulator in use - two iRobot Roomba robots inside simulated world

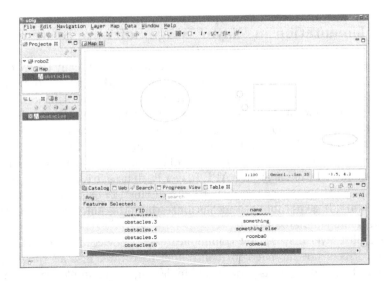

Fig. 2. uDig application in use - modification of *obstacles* layer

can divide robot's workspace into areas with an assigned attribute which for example tells whether a given area is occupied by an obstacle or not. During our experimentations, we have created a plug-in for Players bunch of drivers which constantly tracks changes of position of every robot and updates *obstacles* layer so robots are marked as obstacles. As a result, the map stored in SQL database is kept always up to date. This feature is also useful in multi-agent environments: at any time a robot controller can send a query to SQL database server regarding every other robot position.

A roboticist can write a robot controller using Player client-side API which obtains information about current situation through the *vectormap* interface.

Additionally, to write such a program, PostgreSQL client–side API can be used in order to open direct connection to the database server on which our mereogeometry SQL functions are stored together with map database. These functions can be called using this connection, results are sent back to the calling program. This gives robot controller program ability to perform spatial reasoning based on rough mereology.

Using PostGIS SQL extensions we have created our mereogeometry SQL functions [5]. Rough mereological distance is defined as such:

```
CREATE FUNCTION meredist(object1 geometry, object2 geometry)
RETURNS DOUBLE PRECISION AS
$$
    SELECT min(degrees.degree) FROM
        ((SELECT
                ST_Area(ST_Intersection(extent($1), extent($2)))
                / ST_Area(extent($1))
                AS degree)
        UNION (SELECT
                ST_Area(ST_Intersection(extent($1), extent($2)))
                / ST_Area(extent($2))
                AS degree))
        AS degrees;

$$ LANGUAGE SQL STABLE;
```

Having mereological distance function we can derive nearness predicate:

```
CREATE FUNCTION merenear(obj geometry, o1 geometry, o2 geometry)
RETURNS BOOLEAN AS
$$
    SELECT meredist($1, $2) > meredist($3, $2)
$$ LANGUAGE SQL STABLE;
The equi-distance can be derived as such:
CREATE FUNCTION mereequ(obj geometry, o1 geometry, o2 geometry)
RETURNS BOOLEAN AS
$$
    SELECT (NOT merenear($1, $2, $3))
            AND (NOT merenear($1, $3, $2));

$$ LANGUAGE SQL STABLE;
```

Our implementation of the betweenness predicate makes use of a function that produces an object which is an extent of given two objects:

```
CREATE FUNCTION mereextent(object1 geometry, object2 geometry)
RETURNS geometry AS
$$
    SELECT GeomFromWKB(AsBinary(extent(objects.geom))) FROM
        ((SELECT $1 AS geom)
        UNION (SELECT $2 AS geom))
        AS objects;

$$ LANGUAGE SQL STABLE;
```

The betweenness predicate is defined as follows:

```
CREATE FUNCTION merebetb(obj geometry, o1 geometry, o2 geometry)
RETURNS BOOLEAN AS
$$
    SELECT
        meredist($1, $2) = 1
        OR meredist($1, $3) = 1
        OR
            (meredist($1, $2) > 0
            AND meredist($1, $3) > 0
            AND meredist(mereextent($2, $3),
                    mereextent(mereextent($1, $2), $3)) = 1);

$$ LANGUAGE SQL STABLE;
```

Using the betweenness predicate we can check if three objects form a pattern:

```
CREATE FUNCTION merepattern
    (object1 geometry, object2 geometry, object3 geometry)
RETURNS BOOLEAN AS
$$
    SELECT merebetb($3, $2, $1)
        OR merebetb($1, $3, $2)
        OR merebetb($2, $1, $3);

$$ LANGUAGE SQL STABLE;
```

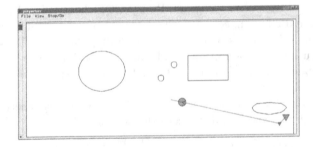

Fig. 3. Playernav - a Player client-side application used to set a goal points for server-side planner driver

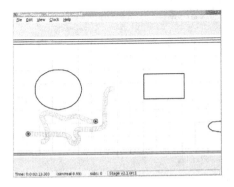

Fig. 4. *Show trails* is a nice option in Stage which can be used to track robot trajectory. Here we see two Roomba robots avoiding obstacles and each other. The robot controller program was using *meredist* function in order to choose free space area as a local target. This method of navigation suffers from local optima problem: a robot can start to spin around one place between obstacles.

Also having pattern predicate we can check if four objects form a line:

```
CREATE FUNCTION mereisline4
    (obj1 geometry, obj2 geometry, obj3 geometry, obj4 geometry)
RETURNS BOOLEAN AS
$$
    SELECT merepattern($1, $2, $3) AND merepattern($2, $3, $4);

$$ LANGUAGE SQL STABLE;
```

Those predicates can be used in global navigation tasks. We can create an additional map layer for navigational markers. Whenever the target is set, a robot planner should form a path across markers. The path itself can be a group of objects representing areas free of obstacles and forming a mereological line. A robot should follow this path by going from one area centroid to another until the goal is reached. If the changes in the world are expected (e.g. in multi–robot environments) a planner should update the path.

References

1. AISB 1997: Spatial Reasoning in Mobile Robots and Animals. In: Proceedings AISB 1997 Workshop. Manchester Univ., Manchester (1997)
2. van Benthem, J.: The Logic of Time. Reidel, Dordrecht (1983)
3. Gotts, N.M., Gooday, J.M., Cohn, A.G.: A connection based approach to commonsense topological description and reasoning. The Monist 79(1), 51–75 (1996)
4. Kuipers, B.J., Byun, Y.T.: A qualitative approach to robot exploration and map learning. In: Proceedings of the IEEE Workshop on Spatial Reasoning and Multi-Sensor Fusion, pp. 390–404. Morgan Kaufmann, San Mateo (1987)
5. Ladanyi, H.: SQL Unleashed. Sams Publishing (1997)
6. De Laguna, T.: Point, line, surface as sets of solids. J. Philosophy 19, 449–461 (1922)
7. Leśniewski, S.: On the foundations of mathematics. Topoi 2, 7–52 (1982)
8. Osmialowski, P.: Player and Stage at PJIIT Robotics Laboratory. Journal of Automation, Mobile Robotics and Intelligent Systems 2, 21–28 (2007)
9. Polkowski, L.: On connection synthesis via rough mereology. Fundamenta Informaticae 46, 83–96 (2001)
10. Ramsey, P.: PostGIS Manual. Postgis.pdf file downloaded from Refractions Research home page
11. Tarski, A.: Les fondements de la géométrie des corps. In: Supplement to Annales de la Sociéte Polonaise de Mathématique, Cracow, pp. 29–33 (1929)
12. Tarski, A.: What is elementary geometry? In: The Axiomatic Method with Special Reference to Geometry and Physics, pp. 16–29. North-Holland, Amsterdam (1959)

Natural versus Granular Computing: Classifiers from Granular Structures

Piotr Artiemjew

University of Warmia and Mazury
Olsztyn, Poland
artem@matman.uwm.edu.pl

Abstract. In data sets/decision systems, written down as pairs $(U, A \cup \{d\})$ with objects U, attributes A, and a decision d, objects are described in terms of attribute–value formulas. This representation gives rise to a calculus in terms of descriptors which we call a *natural computing*. In some recent papers, the idea of L. Polkowski of computing with granules induced from similarity measures called rough inclusions have been tested. In this work, we pursue this topic and we study granular structures resulting from rough inclusions with classification problem in focus. Our results show that classifiers obtained from granular structures give better quality of classification than natural exhaustive classifiers.

Keywords: rough inclusions, granular computing, classification of data.

1 Introduction

Heuristics for data classification augmenting standard rough set based classifiers have been studied by this author recently in a series of papers, see, e.g., [1], [2]. Those heuristics have been based on idea of L. Polkowski to apply granulation of knowledge by means of rough inclusions to preprocessing of training sets in order to reduce noise and uncertainty in data, see, e.g., [6], [7], [8]. Granulation proceeds according to a formal model proposed in Polkowski, opera.cit. Former papers on this subject explored granulation based on rough inclusions induced from the Łukasiewicz t–norm and variants of rough inclusions based on residual implications of t–norms, see, e.g., [1], [2], [6]. Rough inclusions are similarity measures introduced in [10] in the framework of rough mereology, see, e.g., [9]. The impact which granulation of knowledge has on information content of a given data set has been evaluated by the quality of the exhaustive rough set classifier expressed by accuracy and coverage, see, e.g., [11] for these parameters definitions.

Granular structures have been introduced into data sets by means of the process consisting in: computing granules of a given radius by means of chosen rough inclusion μ [7,8] as sets of the form $\{v \in U : \mu(v, u, r)\}$, where $u \in U$ runs over objects in the given training data set. Randomly chosen from those granules covering of the training set was subject to some strategy of attribute factoring, see, e.g., [7], [8]. The decision system obtained in this way was used as the new training set to induce exhaustive classification rules which were then applied to the

C.-C. Chan et al. (Eds.): RSCTC 2008, LNAI 5306, pp. 150–159, 2008.

test part of data and compared to results of classification by the exhaustive classifier induced from the original, non–granulated training set. That strategy was exploited in former papers and it is pursued in this paper. In this paper, we study a new version of granulation by means of a variant of weak rough inclusion, see, e.g., [5], [6], see the formula (2) in what follows, which relies on a parameter called a catch radius. To give a comparison with former approaches, we include results of tests with formerly studied granulation of training objects, and granulation of rules from training objects. All experiments presented in this paper, have been carried out with Heart Disease and Pima Indians Diabetes data sets from UCI Repository [12]. Voting by granules goes on lines described e.g., in [3].

2   Rough Inclusions Applied in the Tests. Granulation of Data

Data sets are formally described as decision systems, see [4]. Rough inclusions which are used in this paper belong in a class of rough inclusions defined by metrics see [5], [6], [7], [8]. The basic rough inclusion is μ_L, induced from the Łukasiewicz t–norm [5]

$$\mu_L(u, v, r) \text{ if and only if } \frac{|IND(u,v)|}{|A|} \geq r, \tag{1}$$

where $IND(u,v) = \{a \in A : a(u) = a(v)\}$.

A modification of this rough inclusion takes into account the value distribution of attributes, [5], and for a given value of ε, we let,

$$\mu_L^{\varepsilon}(v, u, r) \text{ iff } |\{a \in A : ||a(v) - a(u)|| < \varepsilon\}| \geq r \cdot |A|, \tag{2}$$

where $||x - y||$ is the metric in the real line.

Given a rough inclusion, or its variant, μ, a granule about the object u and of the radius r, is the set $\{v : \mu(v, u, r)\}$.

The idea of a granulated data set was proposed in [8]: given a granulation radius r, the set $G(r, \mu)$ of all granules of the radius r is formed. From this set, a covering $Cov(r, \mu, \mathcal{G})$ of the set of objects U is chosen by means of a strategy $\mathcal{G}$, which is usually a random choice of granules with irreducibility checking.

Given the covering $Cov(r, \mu, \mathcal{G})$, attributes in the set A are factored through granules to make a new attribute set. For an attribute $a \in A$, and a granule g, the new attribute a_G is defined as

$$a_G(g) = \mathcal{S}(\{a(v) : v \in g\}), \tag{3}$$

where the strategy $\mathcal{S}$ applied here is majority voting with random tie resolution. A new information system: $(U^G = C(r, \mu, \mathcal{G}), A_G = \{a_G : a \in A\})$ is called a *granular reflection* of the given information system [5], [8]. A number of tests, see [1], [2], have witnessed effectiveness of granular approach in classification problems. Results of tests with granulated structures are compared to the results given by the exhaustive classifier (see, e.g., [11] for a public domain exhaustive classifier): for Heart Disease data, accuracy=0.804, coverage=1.0, for Pima Indians Diabetes data set, accuracy=0.6528, coverage=0.9972.

3   Tests with Granules of Training Objects

Granules were computed here according to (2) with $r = 1$, i.e., a granule $g^\varepsilon(u) = \{v : ||a(u) - a(v)|| < \varepsilon$ for each a $\in A\}$. Objects in the granule $g = g^\varepsilon(u)$ vote for decision class assignment at u (see [3] for voting scheme discussion) by computing for each class c of the factor $p(c) = \frac{|\{training\ objects\ in\ g\ in\ c\}|}{size\ c\ in\ training\ set}$, and the class assigned to u is the one with the largest value of p. Fig. 1 shows results for Heart Disease data set, and Fig. 2 for Pima Indians Diabetes data set [12] by means of CV-5 cross–validation.

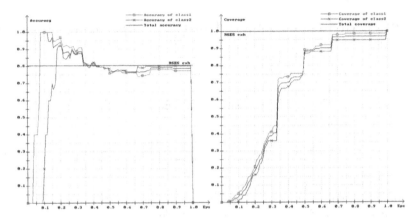

Fig. 1. CV-5; Heart Disease; Algorithm 1_v1; Best result for $\varepsilon = 0.88$, accuracy=0.790172, coverage $= 0.97037$

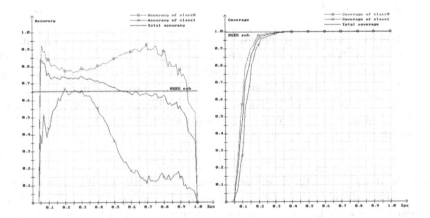

Fig. 2. CV-5; Pima Indians Diabetes; Algorithm 1_v1; Best result for $\varepsilon = 0.3$, accuracy=0.742736, coverage $= 0.99085$

4   Tests with Granules of Exhaustive Set of Training Rules

A parallel test have been carried out with rules obtained from the training set by exhaustive algorithm. In case of a rule r, the symbol $a(r)$ stands for the value of a in the premise of the rule r. Thus, rules are treated similarly to objects, except for voting procedure: rules in the granule $g = g^\varepsilon(u)$, vote by computing for each class c of the factor $q(c) = \dfrac{\text{sum of supports of rules in g pointing to c}}{\text{size c in training set}}$ and assigned is a class with the largest q.

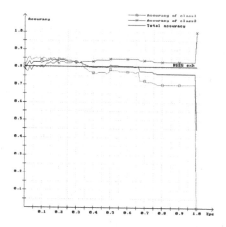

Fig. 3. CV-5; Heart Disease; Algorithm 3_v1; Best result for $\varepsilon = 0.15$, accuracy=0.844444, coverage = 1.0

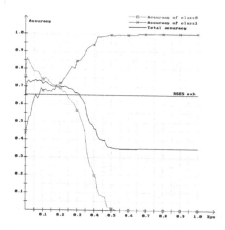

Fig. 4. CV-5; Pima Indians Diabetes; Algorithm 3_v1; Best result for $\varepsilon = 0.06$, accuracy=0.738562, coverage = 1.0

Figs. 3,4 show results obtained by means of CV-5 cross–validation with Heart Disease and Pima Indians Diabetes data sets.

5   Tests with Granules of Granular Reflections

In this approach, the training data set is granulated, and for each granule g in the chosen covering $Cov(r, \mu, \mathcal{G})$ of U, its *granular reflection* i.e., the granular object $o(g) = \{(a_G, a_G(g)) : a \in A\}$ is formed. The set $\{o(g) : g \in Cov(r, \mu, \mathcal{G})\}$ of granular reflection is treated as a training set for classification exactly on lines of sect.3. Parameters here are a granulation radius r and ε.

Table 1. CV-5; Heart Disease; Algorithm 2_v1; r_gran=granulation radius, optimal_eps=best optimal epsilon, acc=accuracy, cov=coverage, trn=training set. Best results for $r = 0.769231$, $\varepsilon = 0.88$: accuracy=0.790172, coverage=0.97037.

r_gran	optimal eps	acc	cov	trn
nil	nil	0.804	1.0	216
0.0	1.0	0.551852	1.0	1.0
0.0769231	1.0	0.551852	1.0	1.8
0.153846	1.0	0.551852	1.0	2.2
0.230769	1.0	0.551852	1.0	3.2
0.307692	1.0	0.551852	1.0	6.2
0.384615	1.0	0.529629	1.0	12.2
0.461538	0.99	0.789828	0.692593	24.8
0.538462	0.78	0.776477	0.944444	113.6
0.615385	0.99	0.758454	0.951852	116.0
0.692308	0.88	0.778198	0.966666	178.2
0.769231	0.88	0.790172	0.97037	209.8
0.846154	0.88	0.790172	0.97037	216.0
0.923077	0.88	0.790172	0.97037	216.0
1.0	0.88	0.790172	0.97037	216.0

Table 2. CV-5; Pima Indians Diabetes; Algorithm 2_v1; r_gran=granulation radius, optimal_eps=best optimal epsilon, acc=accuracy, cov=coverage, trn=training set. Best results for $r = 0.625$, $\varepsilon = 0.29$: accuracy=0.744864, coverage=0.989542.

r_gran	optimal eps	acc	cov	trn
nil	nil	0.6528	0.9972	615
0.0	1.0	0.649673	1.0	1.0
0.125	1.0	0.649673	1.0	19.4
0.25	0.47	0.663576	0.994771	114.4
0.375	0.31	0.711504	0.983006	328.2
0.5	0.3	0.729724	0.986928	520.4
0.625	0.29	0.744864	0.989542	607.6
0.75	0.3	0.742736	0.99085	614.8
0.875	0.3	0.742736	0.99085	614.8
1.0	0.3	0.742736	0.99085	614.8

Tables 1,2 show results obtained with CV-5 cross–validation for Heart Disease and Pima Indians Diabetes, respectively.

6 Tests with Graded Granules of Training Objects

In this case granulation goes according to (2) with the general radius r. Thus, a granule $g(u, r, \mu_\delta^\varepsilon)$ consists of objects v such that at least $r \cdot 100$ percent of attributes a satisfy $||a(u) - a(v)|| \leq \varepsilon$. The parameter r is called here the *catch radius*. Voting goes like in sect. 3.

Tables 3,4 show results obtained with CV-5 cross–validation for Heart Disease and Pima Indians Diabetes, respectively.

Table 3. CV-5; Heart Disease; Algorithm 1_v2. r_catch=catch radius, ε_{opt}=optimal epsilon, acc=accuracy, cov=coverage. Best result for $r_{catch} = 0.307692$, $\varepsilon_{opt} = 0.04$: accuracy=0.859259, coverage=1.0.

r_catch	ε_{opt}	acc	cov
nil	nil	0.804	1.0
0.153846	0.0	0.777778	1.0
0.230769	0.06	0.82963	1.0
0.307692	0.04	0.859259	1.0
0.384615	0.07	0.855555	1.0
0.461538	0.09	0.844445	1.0
0.538462	0.1	0.848148	1.0
0.615385	0.11	0.840741	1.0
0.692308	0.19	0.844445	1.0
0.769231	0.33	0.837037	1.0
0.846154	0.36	0.839483	0.992592
0.923077	0.53	0.829141	0.996296
1.0	0.88	0.790172	0.97037

Table 4. CV-5; Pima Indians Diabetes; Algorithm 1_v2. r_catch=catch radius, ε_{opt}=optimal epsilon, acc=accuracy, cov=coverage. Best result for $r_{catch} = 0.5$, $\varepsilon_{opt} = 0.09$: accuracy=0.747712, coverage=1.0.

r_catch	ε_{opt}	acc	cov
nil	nil	0.6528	0.9972
0.25	0.09	0.730719	1.0
0.375	0.11	0.745098	1.0
0.5	0.09	0.747712	1.0
0.625	0.15	0.745098	1.0
0.75	0.31	0.747712	1.0
0.875	0.3	0.73464	1.0
1.0	0.3	0.742736	0.99085

7   Tests with Graded Granules of Exhaustive Set of Rules from Training Set

In this case, granulation proceeds as in sect. 6 and voting according to sect. 4. Tables 5,6 show results obtained with CV-5 cross–validation for Heart Disease and Pima Indians Diabetes, respectively.

Table 5. CV-5; Heart Disease; Algorithm 3_v2. r_catch=catch radius, ε_{opt}=optimal epsilon, acc=accuracy, cov=coverage. Best result for $r_{catch} = 0.153846$, $\varepsilon_{opt} = 0$: accuracy=0.840741, coverage=1.0.

r_catch	optimal eps	acc	cov
nil	nil	0.804	1.0
0.0	0.0	0.448148	1.0
0.0769231	0.0	0.833333	1.0
0.153846	0.0	0.840741	1.0
0.230769	0.06	0.837037	1.0
0.307692	0.14	0.798672	0.992592
0.384615	0.74	0.724598	0.996296
0.461538	1.0	0.470371	1.0

Table 6. CV-5; Pima Indians Diabetes; Algorithm 3_v2. r_catch=catch radius, ε_{opt}=optimal epsilon, acc=accuracy, cov=coverage. Best result for $r_{catch} = 0.25$, $\varepsilon_{opt} = 0.14$: accuracy=0.637909, coverage=1.0.

r_catch	optimal eps	acc	cov
nil	nil	0.6528	0.9972
0.0	0.0	0.350327	1.0
0.125	0.06	0.601307	1.0
0.25	0.14	0.637909	1.0
0.375	0.09	0.431988	0.971242
0.5	0.96	0.201307	0.6

8   Tests with Graded Granules of Granular Reflections

In this last case, parallel to sect. 5, granular reflections are formed in accordance with 1 and then $\{o(g) : g \in Cov(r, \mu, \mathcal{G})\}$ is granulated according to (2). Voting goes as in sect. 3. In this case there are two granulation radii: the radius of the starting granulation procedure by means of (1) and the radius of the second granulation procedure on granular reflections according to general (2); the second radius is called the *catch radius*. Results are given in Tables 7, 8 for Heart Disease and Pima Indians Diabetes, respectively, obtained with CV-5 cross–validation. Results are given against granulation radii in terms of the optimal catch radii at which the best accuracy of classification has been obtained.

Table 7. CV-5; Heart Disease; Algorithm 2_v2. r_gran=granulation radius, optimal_r_catch=optimal catch radius, ε_{opt}=optimal epsilon, acc=accuracy, cov=coverage, trn=training set. Best result for r_{gran} = 0.846154, optimal r_{catch} = 0.307692, ε_{opt} = 0.04: accuracy=0.859259, coverage=1.0.

r_gran	optimal_r_catch	optimal eps	acc	cov	trn
nil	nil	nil	0.804	1.0	216
0.0769231	0.538462	0.3	0.592104	0.933333	1.4
0.153846	0.538462	0.29	0.583084	0.948148	2.6
0.230769	0.307692	0.01	0.618408	0.981481	3.2
0.307692	0.615385	0.17	0.769323	0.914815	5.4
0.384615	0.384615	0.01	0.785059	0.981481	10.8
0.461538	0.384615	0.02	0.835755	0.992593	26.2
0.538462	0.461538	0.04	0.848148	1.0	115.0
0.615385	0.461538	0.04	0.855556	1.0	114.6
0.692308	0.692308	0.19	0.844444	1.0	177.8
0.769231	0.384615	0.08	0.855556	1.0	210.0
0.846154	0.307692	0.04	0.859259	1.0	216.0
0.923077	0.307692	0.04	0.859259	1.0	216.0
1.0	0.307692	0.04	0.859259	1.0	216.0

Table 8. CV-5; Pima Indians Diabetes; Algorithm 2_v2. r_gran=granulation radius, optimal_r_catch=optimal catch radius, ε_{opt}=optimal epsilon, acc=accuracy, cov=coverage, trn=training set. Best result for r_{gran} = 0.625, optimal r_{catch} = 0.375, ε_{opt} = 0.1: accuracy=0.74902, coverage=1.0.

r_gran	optimal_r_catch	optimal eps	acc	cov	trn
nil	nil	nill	0.6528	0.9972	615
0.0	0.625	0.44	0.650559	0.998693	1.0
0.125	0.5	0.14	0.667585	0.983006	20.2
0.25	0.75	0.16	0.698365	0.992157	118.0
0.375	0.625	0.1	0.734641	1.0	329.0
0.5	0.375	0.05	0.741176	1.0	521.6
0.625	0.375	0.1	0.74902	1.0	607.6
0.75	0.5	0.09	0.747712	1.0	614.8
0.875	0.5	0.09	0.747712	1.0	614.8
1.0	0.5	0.09	0.747712	1.0	614.8

9   Conclusions

In this paper we have presented results of our study on inducing classifiers from granulated data sets. Procedures of granulation are to a large extent randomized, e.g., in choice of a granular covering as well as in choice of values of factored attributes on granules. Therefore, a strict analysis of this heuristic is difficult

and the main estimate of the quality of this method is in results of classification on test data sets. We have chosen here as a similarity measure applied in
granulation process, the rough inclusion μ_L obtained from the Lukasiewicz t–
norm, and its graded variant μ_L^ε in which the Euclidean standard distance in the
real line is applied. Granulation mechanism is applied here in few cases: with μ_L
to training objects (sect. 3), to rules induced from the training set by an exhaustive classifier (sect.4); with μ_L^ε (r=1) (see formula (2)) to granules of granulated
objects (granular reflections) (sect. 5); with general μ_L^ε of formula (2) to granules
of training objects (sect. 6); finally with general μ_L^ε of formula (2) to granules
of rules induced from granulated data set by an exhaustive classifier (sect. 7).
As exemplary data sets Heart Disease and Pima Indians Diabetes from UCI
Repository were chosen. Results show a substantial improvement in quality of
classification in comparison with the classical exhaustive classifier, e.g., for Pima
Indians Diabetes, the exhaustive classifier accuracy was 0.6528, and in sect. 2
accuracy was 0. 7427 (Fig.2), similar results were obtained in other cases. The
results show that the method promises to give good results and further study is
oriented toward additional heuristics which may further improve the classifiers
based on granulation.

References

1. Artiemjew, P.: On classification of data by means of rough mereological granules of
   objects and rules. In: Wang, G., Li, T., Grzymała-Busse, J.W., Miao, D., Skowron,
   A., Yao, Y. (eds.) RSKT 2008. LNCS (LNAI), vol. 5009, pp. 221–228. Springer,
   Heidelberg (in print, 2008)
2. Artiemjew, P.: Rough mereological classifiers obtained from weak variants of rough
   inclusions. In: Wang, G., Li, T., Grzymała-Busse, J.W., Miao, D., Skowron, A., Yao,
   Y. (eds.) RSKT 2008. LNCS (LNAI), vol. 5009, pp. 229–236. Springer, Heidelberg
   (in print, 2008)
3. Bazan, J.G.: A comparison of dynamic and non–dynamic rough set methods for
   extracting laws from decision tables. In: Polkowski, L., Skowron, A. (eds.) Rough
   Sets in Knowledge Discovery, vol. 1, pp. 321–365. Physica Verlag, Heidelberg (1998)
4. Pawlak, Z.: Rough Sets: Theoretical Aspects of Reasoning about Data. Kluwer,
   Dordrecht (1991)
5. Polkowski, L.: On the idea of using granular rough mereological structures in classification of data. In: Wang, G., Li, T., Grzymała-Busse, J.W., Miao, D., Skowron,
   A., Yao, Y. (eds.) RSKT 2008. LNCS (LNAI), vol. 5009, pp. 213–220. Springer,
   Heidelberg (in print, 2008)
6. Polkowski, L.: The paradigm of granular rough computing. In: Zhang, D., Wang, Y.,
   Kinsner, W. (eds.) ICCI 2007, pp. 145–163. IEEE Computer Society, Los Alamitos
   (2007)
7. Polkowski, L.: Formal granular calculi based on rough inclusions (a feature talk). In:
   Zhang, Y.-Q., Lin, T.Y. (eds.) IEEE GrC 2006, pp. 9–18. IEEE Press, Piscataway
   (2006)
8. Polkowski, L.: Formal granular calculi based on rough inclusions (a feature talk).
   In: Hu, X., Liu, Q., Skowron, A., Lin, T.Y., Yager, R.R., Zhang, B. (eds.) IEEE
   GrC 2005, pp. 57–62. IEEE Press, Piscataway (2005)

9. Polkowski, L.: Toward rough set foundations. Mereological approach (a plenary lecture). In: Tsumoto, S., Słowiński, R., Komorowski, J., Grzymała-Busse, J.W. (eds.) RSCTC 2004. LNCS (LNAI), vol. 3066, pp. 8–25. Springer, Heidelberg (2004)
10. Polkowski, L., Skowron, A.: Rough mereology: a new paradigm for approximate reasoning. International Journal of Approximate Reasoning 15(4), 333–365 (1997)
11. RSES, http://logic.mimuw.edu.pl/rses
12. UCI Repository, http://www.ics.uci.edu/~mlearn/databases/

Inducing Better Rule Sets by Adding Missing Attribute Values

Jerzy W. Grzymala-Busse[1,2] and Witold J. Grzymala-Busse[3]

[1] Department of Electrical Engineering and Computer Science University of Kansas,
Lawrence, KS 66045, USA
[2] Institute of Computer Science, Polish Academy of Sciences,
01–237 Warsaw, Poland
[3] Touchnet Information Systems, Inc.,
Lenexa, KS 66219, USA

Abstract. Our main objective was to verify the following hypothesis: for some complete (i.e., without missing attribute vales) data sets it is possible to induce better rule sets (in terms of an error rate) by increasing incompleteness (i.e., removing some existing attribute values) of the original data sets. In this paper we present detailed results of experiments on one data set, showing that some rule sets induced from incomplete data sets are significantly better than the rule set induced from the original data set, with the significance level of 5%, two-tailed test. Additionally, we discuss criteria for inducing better rules by increasing incompleteness and present graphs for some well-known data sets.

1 Introduction

In this paper we show that by increasing incompleteness of data sets (i.e., by removing attribute values in a data set) we may improve quality of the rule sets induced from such modified data sets. In our experiments we replaced randomly existing attribute values in the original data sets by symbols that were recognized by the rule induction module as missing attribute values. In other words, the rule sets were induced from data sets in which some values were erased using a Monte Carlo method. The process of such replacements was done incrementally, with an increment equal to 5% of the total number of attribute values of a given data set.

We distinguish three different kinds of missing attribute values: *lost values* (the values that were recorded but currently are unavailable) [1,2,3,4], *attribute-concept values* (these missing attribute values may be replaced by any attribute value limited to the same concept) [5], and *"do not care" conditions* (the original values were irrelevant) [4,6,7,8]. A *concept* (class) is a set of all cases classified (or diagnosed) the same way.

We assumed that for each case at least one attribute value was specified, i.e., they are not missing. Such an assumption limits the percentage of missing attribute values used for experiments; for example, for the *wine* data set, starting

C.-C. Chan et al. (Eds.): RSCTC 2008, LNAI 5306, pp. 160–169, 2008.

Table 1. Data sets used for experiments

Data set		Number of	
	cases	attributes	concepts
Bankruptcy	66	5	2
Breast cancer - Slovenia	277	9	2
Hepatitis	155	19	2
Image segmentation	210	19	7
Iris	150	4	3
Lymphography	148	18	4
Wine	178	12	3

from 70% of randomly assigned missing attribute values, this assumption was violated. Additionally, we assumed that all decision values were specified.

For rule induction from incomplete data we used the MLEM2 data mining algorithm, for details see [9]. We used rough set methodology [10,11], i.e., for a given interpretation of missing attribute vales, *lower* and *upper approximations* were computed for all concepts and then rule sets were induced, *certain* rules from lower approximations and *possible* rules from upper approximations. Note that for incomplete data there is a few possible ways to define approximations, we used *concept* approximations [4,5].

As follows from our experiments, some of the rule sets induced from such incomplete data are better than the rule sets induced form original, complete data sets. More precisely, the error rate, a result of ten-fold cross validation, is significantly lower, with the significance level of 5%, than the error rate for rule sets induced from the original data.

2   Experiments

In our experiments seven typical data sets were used, see Table 1. All of these data sets are available from the UCI ML Repository, with the exception of the *bankruptcy* data set. These data sets were completely specified (all attribute values were completely specified), with the exception of *breast cancer - Slovenia* data set, which originally contained 11 cases (out of 286) with missing attribute values. These 11 cases were removed.

In two data sets: *bankruptcy* and *iris* all attributes were numerical. These data sets were processed as numerical (i.e., discretization was done during rule induction by MLEM2). The *image segmentation* data set was converted into symbolic using a discretization method based on agglomerative cluster analysis (this method was described, e.g., in [12]).

Preliminary results [13] show that, for some data sets by increasing incompleteness we may improve rule sets. Therefore we decided to conduct extensive

Table 2. Wine data set. Certain rule sets.

Percentage of lost values	Average error rate	Standard deviation	Z score
0	7.66	1.32	
5	7.17	1.74	1.22
10	7.13	2.00	1.20
15	8.76	1.85	−2.66
20	7.06	1.38	1.72
25	7.27	1.55	1.06
30	6.20	1.39	**4.17**
35	6.55	1.16	**3.43**
40	6.8	1.28	**2.56**
45	7.73	1.48	−0.21
50	7.21	0.82	1.58
55	8.01	1.29	−1.05
60	7.30	1.00	1.00
65	8.41	0.98	0.98

Table 3. Wine data set. Possible rule sets.

Percentage of lost values	Average error rate	Standard deviation	Z score
0	7.66	1.32	
5	7.21	1.92	1.06
10	7.32	1.34	0.98
15	8.46	1.75	−2.01
20	7.17	1.72	1.23
25	7.64	1.63	0.05
30	6.33	1.15	**4.15**
35	6.57	1.12	**3.44**
40	6.22	1.29	**4.27**
45	7.79	1.30	−0.39
50	7.12	0.68	**2.00**
55	7.68	0.98	−0.06
60	6.89	0.78	**2.74**
65	8.31	1.17	−2.04

Table 4. Wine data set. Size of rule sets.

Percentage of lost values lost values	Certain rule set Number of		Possible rule set Number of	
	rules	conditions	rules	conditions
0	20	65	20	65
10	25	89	21	73
20	34	108	28	90
30	38	117	46	149
40	47	140	54	166
50	62	246	70	204
60	59	156	61	148

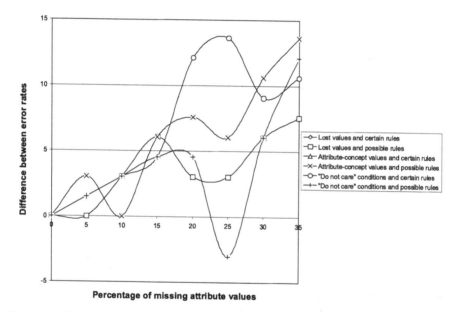

Fig. 1. Bankruptcy data set. Difference between error rates for testing with complete data sets and data sets with missing attribute values.

experiments, repeating 30 times the ten-fold cross validation experiment (changing the random case ordering in data sets) for every percentage of lost values and then computing the Z score using the well-known formula

$$Z = \frac{\overline{X_1} - \overline{X_2}}{\sqrt{\frac{s_1^2 + s_2^2}{30}}},$$

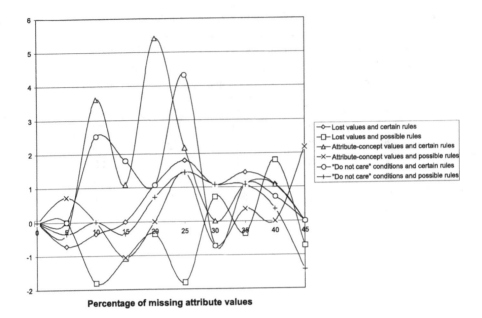

Fig. 2. Breast cancer - Slovenia data set. Difference between error rates for testing with complete data sets and data sets with missing attribute values.

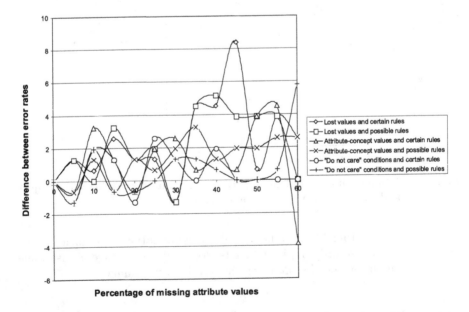

Fig. 3. Hepatitis data set. Difference between error rates for testing with complete data sets and data sets with missing attribute values.

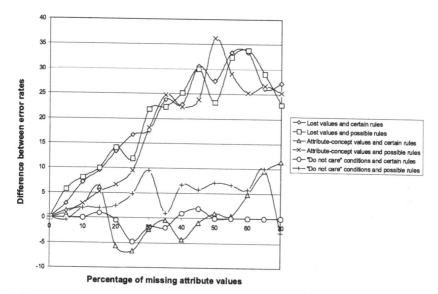

Fig. 4. Image segmentation data set. Difference between error rates for testing with complete data sets and data sets with missing attribute values.

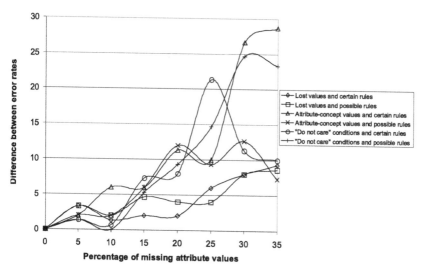

Fig. 5. Iris data set. Difference between error rates for testing with complete data sets and data sets with missing attribute values.

where $\overline{X_1}$ is the mean of 30 ten-fold cross validation experiments for the original data set, $\overline{X_2}$ is the mean of 30 ten-fold cross validation experiments for the data set with given percentage of lost values, s_1 and s_2 are sample standard deviations for original and incomplete data sets, respectively.

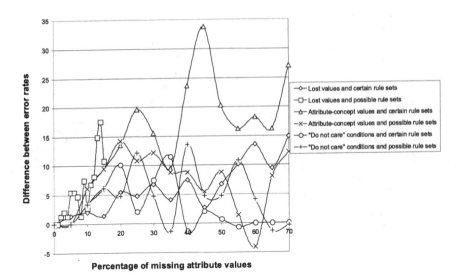

Fig. 6. Lymphography data set. Difference between error rates for testing with complete data sets and data sets with missing attribute values.

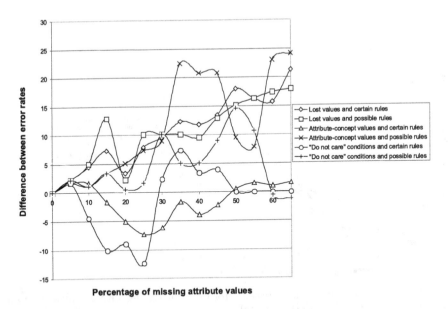

Fig. 7. Wine data set. Difference between error rates for testing with complete data sets and data sets with missing attribute values.

Note that though rule sets were induced from incomplete data, for testing such rule sets the original, complete data were used so that the results for incomplete data are fully comparable with results for the original data sets. Obviously, if

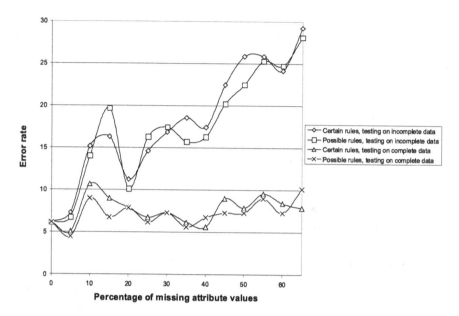

Fig. 8. Wine data set. Testing on complete and incomplete data sets, all missing attribute values are interpreted as *lost*.

the Z score is larger than 1.96, the rule set induced from the data set with given percentage of lost values is significantly better than the corresponding rules set induced from the original data set, with the significance level of 5%, two-tailed test. As follows from Tables 2 and 3, there are three and five rules sets better than the rule sets induced from the original data sets, for certain and possible rule sets, respectively. In Tables 2 and 3, the corresponding Z scores are presented in bold font. Additionally, in only one case for certain rule sets and for two cases for possible rule sets the rule sets induced from incomplete data are worse than the rule sets induced from the original data.

The problem is how to recognize a data set that is a good candidate for improving rule sets by increasing incompleteness. One possible criterion is a large difference between two error rates: one induced from incomplete data and tested on incomplete data and the other induced from incomplete data and tested on the original data set. The corresponding differences of these error rates are presented on Figures 1–7.

Another criterion of potential usefulness of inducing rules from incomplete data is the graph of an error rate for rule sets induced from incomplete data and tested on original, complete data. Such graphs were presented in [13]. In this paper we present these graphs, restricted to the *wine* data set and to *lost values* on Figure 8. A good candidate is characterized by the flat graph, roughly speaking, parallel to the *percentage of missing attribute values* axis. It is clear that the *wine* data set satisfies both criteria. Note that all graphs, presented in Figures 1–8, were plotted for single experiments of ten-fold cross validation.

Because of the space limitation, we cannot present more experimental results in this paper, but it is clear that the main objective of this paper is proven: for some data sets it is possible to improve the quality of rule sets by increasing incompleteness of data sets (or replacing existing attribute values by symbols of missing attribute values).

The question is why sometimes we may improve the quality of rule sets by increasing incompleteness of the original data set. As follows from Table 4, the size of the induced rule sets form incomplete data, both in terms of the number of rules and the total number of conditions, is larger for incomplete data. This fact follows from the MLEM2 algorithm: MLEM2 is less likely to induce simpler rules if the search space is smaller. A possible explanation for occasional improvement of the quality of rule sets is redundancy of information in some data sets, such as *wine* data set, so that it is still possible to induce not only good but sometimes even better rule sets than the rule set induced from the original data set.

3   Conclusions

As follows form our experiments, there are some cases of the rule sets, induced from incomplete data sets, with an error rate (result of ten-fold cross validation) significantly smaller (with a significance level of 5%, two-tailed test) than the error rate for the rule set induced from the original data set. Thus, we proved that there exists an additional technique for improving rule sets, based on increasing incompleteness of the original data set (by replacing some existing attribute values by symbols of missing attribute values). Note that this technique is not always successful. A possible criterion for success are based on large difference between the error rate for rule sets induced from incomplete data and tested on original data and on incomplete data. As follows from Figures 1–7, *image segmentation, iris* and *lymphography* data sets are also, potentially, good candidates for improving rule sets based on increasing incompleteness of the original data sets. Another criterion is a flat graph for an error rate versus percentage of missing attribute vales for rule sets induced from incomplete data and tested on original, complete data.

References

1. Grzymala-Busse, J.W., Wang, A.Y.: Modified algorithms LEM1 and LEM2 for rule induction from data with missing attribute values. In: Proceedings of the Fifth International Workshop on Rough Sets and Soft Computing (RSSC 1997) at the Third Joint Conference on Information Sciences (JCIS 1997), pp. 69–72 (1997)
2. Stefanowski, J., Tsoukias, A.: On the extension of rough sets under incomplete information. In: Zhong, N., Skowron, A., Ohsuga, S. (eds.) RSFDGrC 1999. LNCS (LNAI), vol. 1711, pp. 73–82. Springer, Heidelberg (1999)
3. Stefanowski, J., Tsoukias, A.: Incomplete information tables and rough classification. Computational Intelligence 17, 545–566 (2001)
4. Grzymala-Busse, J.W.: Rough set strategies to data with missing attribute values. In: Workshop Notes, Foundations and New Directions of Data Mining, in conjunction with the 3rd International Conference on Data Mining, pp. 56–63 (2003)

5. Grzymala-Busse, J.W.: Three approaches to missing attribute values—a rough set perspective. In: Proceedings of the Workshop on Foundation of Data Mining, in conunction with the Fourth IEEE International Conference on Data Mining, pp. 55–62 (2004)
6. Grzymala-Busse, J.W.: On the unknown attribute values in learning from examples. In: Raś, Z.W., Zemankova, M. (eds.) ISMIS 1991. LNCS, vol. 542, pp. 368–377. Springer, Heidelberg (1991)
7. Kryszkiewicz, M.: Rough set approach to incomplete information systems. In: Proceedings of the Second Annual Joint Conference on Information Sciences, pp. 194–197 (1995)
8. Kryszkiewicz, M.: Rules in incomplete information systems. Information Sciences 113, 271–292 (1999)
9. Grzymala-Busse, J.W.: MLEM2: A new algorithm for rule induction from imperfect data. In: Proceedings of the 9th International Conference on Information Processing and Management of Uncertainty in Knowledge-Based Systems, pp. 243–250 (2002)
10. Pawlak, Z.: Rough sets. International Journal of Computer and Information Sciences 11, 341–356 (1982)
11. Pawlak, Z.: Rough Sets. Theoretical Aspects of Reasoning about Data. Kluwer Academic Publishers, Dordrecht (1991)
12. Chmielewski, M.R., Grzymala-Busse, J.W.: Global discretization of continuous attributes as preprocessing for machine learning. International Journal of Approximate Reasoning 15, 319–331 (1996)
13. Grzymala-Busse, J.W., Grzymala-Busse, W.J.: Improving quality of rule sets by increasing incompleteness of data sets. In: Proceedings of the Third International Conference on Software and Data Technologies, pp. 241–248 (2008)

Rule Induction: Combining Rough Set and Statistical Approaches

Wojciech Jaworski

Faculty of Mathematics, Computer Science and Mechanics
Warsaw University, Banacha 2, 02-097 Warsaw, Poland
wjaworski@mimuw.edu.pl

Abstract. In this paper we propose the hybridisation of the rough set concepts and statistical learning theory. We introduce new estimators for rule accuracy and coverage, which base on the assumptions of the statistical learning theory. Then we construct classifier which uses these estimators for rule induction. These estimators allow us to select rules describing statistically significant dependencies in data. We test our classifier on benchmark datasets and show its applications for KDD.

Keywords: Rough sets, quality measures, accuracy, coverage, significance, rule induction, rule selection.

1 Introduction

Rough set theory [7] and statistical learning theory [11] provide two different methodologies for reasoning from data.

The rough set concept theory is a theoretical framework for describing and inferring knowledge. Examined knowledge is imperfect. It is imprecise due to vague concepts involved in knowledge representation and it is based on incomplete data. The central point of the theory is the idea of concept approximation by the sets of objects that certainly belongs to the concept and the set of those which may belong to the concept on the basis of possessed data.

The main goal of statistical learning theory is to provide a framework for studying the problem of inference. For this purpose, there are introduced statistical assumptions about the way the data is generated. A probabilistic model of data generation process, which is the core of the theory, establishes the formalisation of relationships between past and future observations.

While rough set theory provides an intuitive description of relationships in data, stereotypes that express general yet imprecise truths, statistical learning theory measures the significance and correctness of discovered dependencies.

The combination of both approaches provides us tools for building simple, human understandable classifiers, whose quality will be guaranteed by the statistical assumptions.

In this paper we propose the hybridisation of the rough set approach and statistical learning theory. We define the probabilistic model of data generation process. We recall rough set concepts in this new setting. Then we show how to

C.-C. Chan et al. (Eds.): RSCTC 2008, LNAI 5306, pp. 170–180, 2008.

extend set approximations from a sample to the set of all objects. Our attitude is similar to the idea of inductive extensions of approximation spaces presented, for example, in [8,9].

We introduce measures of approximation quality: accuracy and coverage. Taking advantage of the underlying probabilistic model we estimate values of the above indices on the set of all objects using a sample. We propose two estimators: one based on Hoeffding inequality [6], and second based on the optimal probability bound presented in [4,5].

The statistical nature of estimators leads us to the index, the measure called a significance. The significance measures how often sample-based accuracy and coverage estimations are correct. The trade-off relation between these three measures allow us to balance the approximation between fitting to the sample and generalisation.

The properties of accuracy and coverage were thoroughly studied in [10]. The author proposed the probabilistic definition of the indices, yet he neither defined an underlying probability model nor showed the trade-off between accuracy or coverage and significance. Quality measures are also examined from the statistical point of view in [3], but without placing them in the rough set context.

Gediga and Düntsch propose in [2] an application of statistical techniques in rough set data analysis, yet they did not incorporate the assumptions on the data generating process required by these techniques into the presented model.

In order to show how the estimators behave in practice we developed a simple rule based classifier. Estimated indices guarantee the quality of each rule, decide how accurate rules are acceptable and how many objects have to match the rule in order to make it significant. We test the classifier on benchmark datasets obtained from [1].

Test results revealed that the obtained classifier generates highly relevant rules. Each rule is assigned with its accuracy and coverage estimations. Rules cover only that part of universe for which it is possible to predict decision with high accuracy. As a consequence the classifier is able to judge whether it has enough knowledge to classify a certain object.

2 Probabilistic Model

We propose the following definition of the problem of induction. We are given a domain organised in terms of *objects* possessing *attributes*. Depending on the nature of domain, objects are interpreted as, e.g. cases, states, processes, patients, observations. Attributes are interpreted as features, variables, characteristics, conditions, etc.

Let $\mathbb{U}$ be a finite set of objects for a given domain. We denote $\mathbb{U}$ as *universe*. Let A be a non-empty finite set of attributes such that $a : \mathbb{U} \to V_a$ for every $a \in A$ and let $d \notin A$ such that $d : \mathbb{U} \to V_d$ be a decision attribute. We introduce a probability measure $P_\#$ on $2^{\mathbb{U}}$ according to the following formula:

$$\forall_{X \subseteq \mathbb{U}} \, P_\#(X) = \frac{|X|}{|\mathbb{U}|},$$

where $|\cdot|$ denotes the number of elements in a set.

Statistical learning theory [11] assumes that the phenomena underlying generated data have statistical nature and the observed objects are independent, identically distributed random variables.

Formally we introduce a probability space $(\Omega, 2^\Omega, P)$. Observed objects $u_1, u_2, \ldots, u_i, \ldots$ are values of independent random variables $U_1, U_2, \ldots, U_i, \ldots$. Each U_i is a function $U_i : \Omega \to \mathbb{U}$. The distribution of U_i is identical to $P_\#$, i.e.:

$$\forall_i \forall_{X \subseteq \mathbb{U}} P_\#(X) = P(\{\omega \in \Omega \mid U_i(\omega) \in X\}) = P(U_i^{-1}(X))$$

Let $U \subseteq \mathbb{U}$ be a non-empty, finite set of observed objects called a *sample*. U, together with the values of attributes for elements of U, is our knowledge about the domain. We denote elements of U by $u_1, \ldots, u_n$, where u_i is a realisation (or value) of the random variable U_i. We represent this knowledge in terms of the triple $\mathcal{A} = (U, A, d)$, usually denoted as a *decision system*.

3   Set Approximations

Classification is the task of finding the dependence between the attribute values and the value of decision. The rough set theory [7] provides tools and methodology for performing classification.

The basic concept of rough set theory is the indiscernibility relation. Let $\mathcal{A} = (U, A, d)$ be a decision system and $B \subseteq A$.

$$IND_\mathcal{A}(B) = \{(u, u') \in U^2 \mid \forall a \in B \; a(u) = a(u')\}$$

is called the *B-indiscernibility relation*. The B-indiscernibility is an equivalence relation. We will denote its equivalence class generated by object u as $[u]_B$.

The notion of indiscernibility is used to define set approximations. A given set $X \subseteq U$ may be approximated using only the information contained in $B \subset A$ by constructing the *B-lower* and *B-upper approximations of* X, denoted $\underline{B}X$ and $\overline{B}X$ respectively, where $\underline{B}X = \bigcup\{[u]_B \mid [u]_B \subseteq X\}$ and $\overline{B}X = \bigcup\{[u]_B \mid [u]_B \cap X \neq \emptyset\}$.

In the case of classification, we approximate sets of objects that possess a given decision. Let $X_v = \{u \in U \mid d(u) = v\}$. The objects in $\underline{B}X_v$ can be with certainty classified as members of decision class v on the basis of knowledge represented by B, while the objects in $U \setminus \overline{B}X_v$ definitely are not members of decision class v on the basis of knowledge represented by B.

For a given set of attributes B, formulae of the form $a = v$, where $a \in B$ and $v \in V_a$ are called *descriptors* over B. The set of *conditional formulae* over B is defined as the least set containing all descriptors over B and closed with respect to the propositional connectives $\wedge$ (conjunction), $\vee$ (disjunction) and $\neg$ (negation).

Let φ be a conditional formula over B. $\|\varphi\|_\mathcal{A}$ denotes the meaning of φ in the decision system $\mathcal{A}$, which is the set of all objects in U with the property φ. These sets are defined as follows:

1. if φ is of the form $a = v$, then $\|\varphi\|_\mathcal{A} = \{x \in U \mid a(x) = v\}$;
2. $\|\varphi \wedge \varphi'\|_\mathcal{A} = \|\varphi\|_\mathcal{A} \cap \|\varphi'\|_\mathcal{A}$; $\|\varphi \vee \varphi'\|_\mathcal{A} = \|\varphi\|_\mathcal{A} \cup \|\varphi'\|_\mathcal{A}$; $\|\neg\varphi\|_\mathcal{A} = U \setminus \|\varphi\|_\mathcal{A}$.

Every indiscernibility class can be represented by means of conditional formulae composed out of conjunction of descriptors. Let $B = \{a_1, \ldots, a_n\}$, $u \in U$ and let $v_1, \ldots, v_n$ be such that $a_i(u) = v_i$. In such a case

$$[u]_B = \{u' \in U | \forall a \in B\ a(u) = a(u')\} = ||a_1 = v_1 \wedge \cdots \wedge a_n = v_n||_{\mathcal{A}}.$$

We express the lower approximation by means of a conditional formula $\underline{B}X = ||\varphi_1 \vee \cdots \vee \varphi_k||_{\mathcal{A}}$, such that $\varphi_1, \ldots, \varphi_k$ are formulae representing indiscernibility classes that compose $\underline{B}X$. Similarly, there exist $\psi_1, \ldots, \psi_l$ such that $\overline{B}X = ||\psi_1 \vee \cdots \vee \psi_l||_{\mathcal{A}}$.

A *decision rule* for $\mathcal{A}$ is any expression of the form $\varphi \rightarrow d = v$, where φ is a conditional formula, $v \in V_d$ and $||\varphi||_{\mathcal{A}} \neq \emptyset$. A decision rule $\varphi \rightarrow d = v$ is *true* in $\mathcal{A}$ if, and only if, $||\varphi||_{\mathcal{A}} \subseteq ||d = v||_{\mathcal{A}}$. A decision rule describes the dependence between a decision class and its approximation.

4 Extended Approximations

In the above section we considered set approximations that described the dependence between the attribute values and the value of decision for objects in U. Now, we extend set approximations on the whole universe $\mathbb{U}$. The extended approximations of all decision classes will compose a classifier.

The assumption that past and future observations are both sampled independently from the same distribution provide us with tools for extending the approximations. However, the extension will be correct only with some probability.

We represented approximations by means of conditional formulae which are interpreted in the decision system. For a given set of attributes B, extended approximations are represented by means of conditional formulae over B interpreted in the universe $\mathbb{U}$. Let φ be a conditional formula over B and let $||\varphi||_{\mathbb{U}}$ denote its meaning in the universe of all objects. The meaning is defined as follows:

1. if φ is of the form $a = v$ then $||\varphi||_{\mathbb{U}} = \{u \in \mathbb{U} | a(u) = v\}$;
2. $||\varphi \wedge \varphi'||_{\mathbb{U}} = ||\varphi||_{\mathbb{U}} \cap ||\varphi'||_{\mathbb{U}}$; $||\varphi \vee \varphi'||_{\mathbb{U}} = ||\varphi||_{\mathbb{U}} \cup ||\varphi||_{\mathbb{U}}$; $||\neg\varphi||_{\mathbb{U}} = \mathbb{U} \setminus ||\varphi||_{\mathbb{U}}$;

For every U_i we obtain from its definition[1]

$$P_{\#}(||a = v||_{\mathbb{U}}) = P_{\#}(\{u \in \mathbb{U} | a(u) = v\}) =$$

$$= P(\{\omega \in \Omega | a(U_i(\omega)) = v\}) = P(a(U_i) = v).$$

This correspondence may be easily extended on all conditional formulae.

Now, we define extended approximations using conditional formulae interpreted in the universe $\mathbb{U}$:

---

[1] The latter equality introduces a standard probabilistic notation in which 'ω', '{' and '}' are omitted in expressions with random variables.

Definition 1. *Let $X \subseteq \mathbb{U}$ and B be a set of attributes and let $Y \subseteq \mathbb{U}$ be such that*

$$Y = \|\varphi\|_{\mathbb{U}},$$

where φ is a conditional formula over B. The set $Y \subseteq \mathbb{U}$ is called B-α-κ-approximation of X when

$$P_{\#}(X \mid Y) \geq \alpha \text{ and } P_{\#}(Y \mid X) \geq \kappa.$$

We denote α as the approximation accuracy *and we denote κ as the* approximation coverage.

On the contrary to the standard approximations defined in a decision system this definition does not construct a set Y, it only states whether a given set possesses a property of being an α-κ-approximation.

Accuracy and coverage are indices of the approximation quality. Accuracy measures the probability that an object belonging to the approximation belongs also to the approximated set. Coverage measures the percent of objects in a set that are included in its approximation. When the approximation accuracy is equal to 1 and the coverage is maximised the approximation may be considered as *lower* and when the approximation coverage is equal to 1 and the accuracy is maximised the approximation may be considered as *upper*.

Accuracy and coverage are defined by means of the underlying probability distribution, according to which the sample is drawn. Since we are given only a sample and we do not know the probability distribution, we must estimate values of the indices using the sample and probabilistic inequalities of the form

$$P\big(P_{\#}(X \mid Y) \geq f_n(U_1, \ldots, U_n)\big) \geq \gamma_n.$$

The above inequality may be interpreted in the following way: if we draw $\{(u_1^i, u_2^i, \ldots, u_n^i)\}_{i=1}^{\infty}$, an infinite sequence of n-element samples, then according to the law of large numbers

$$P\big(P_{\#}(X \mid Y) \geq f_n(U_1, \ldots, U_n)\big) =$$

$$= \lim_{k \to \infty} \frac{1}{k} \cdot |\{i \leq k \mid P_{\#}(X \mid Y) \geq f_n(u_1^i, \ldots, u_n^i)\}|.$$

Hence γ_n describes how frequent it is true that $P_{\#}(X \mid Y) \geq f_n(u_1^i, \ldots, u_n^i)$ or, in other words how likely $P_{\#}(X \mid Y) \geq f_n(u_1^i, \ldots, u_n^i)$ is to happen in one occurrence. γ_n is a measure called *significance*.

We propose two methods of deriving estimators of the accuracy and the coverage on the basis of sample. The first bases on the Hoeffding inequality [6]:

Theorem 1. *Let $Z_1, \ldots, Z_n$ be identically distributed independent random variables. Assume that each $Z_i \in [0, 1]$. Then, for every $\varepsilon > 0$, the following inequality takes place:*

$$P(EZ_1 \leq \frac{1}{n} \sum_{i=1}^{n} Z_i + \varepsilon) \geq 1 - e^{-2n\varepsilon^2}. \tag{1}$$

Assume that Y is an α-κ-approximation for the set X. Let U be a sample and let $\{U_1, \ldots, U_n\} = U \cap Y$. For the purpose of accuracy estimation we declare that

$$Z_i = \begin{cases} 0, & \text{when } U_i \in X \\ 1, & \text{when } U_i \notin X \end{cases} .$$

Since

$$EZ_1 = P(Z_1 = 1) = P(U_1 \notin X \mid U_1 \in Y) = 1 - P_\#(X \mid Y),$$

we obtain the following inequality

$$P\left((1 - P_\#(X \mid Y)) \le \frac{1}{n} \sum_{i=1}^{n} Z_i + \varepsilon\right) \ge 1 - e^{-2n\varepsilon^2}$$

Now, we take the advantage of the law of large numbers and the fact that we know the realisation of the sample U. We calculate a realisation for each Z_i in the following way

$$z_i = \begin{cases} 0, & \text{when } u_i \in X \\ 1, & \text{when } u_i \notin X \end{cases} ,$$

where u_i is i-th u_k such that $u_k \in Y$. The statement

$$(1 - P_\#(X \mid Y)) - \frac{1}{n} \sum_{i=1}^{n} z_i \le \varepsilon$$

is likely to happen with significance $1 - e^{-2n\varepsilon^2}$.

n denotes the number of variables Z_i. It is equal from the definition to the number of elements in the sample that belong to Y. On the other hand $Z_i = 1$ if and only if the corresponding U_i does not belong to X. Since U_i have to belong to U and Y we obtain

$$n = |U \cap Y| \quad \text{and} \quad \frac{1}{n} \sum_{i=1}^{n} z_i = \frac{|(U \cap Y) \setminus X|}{|U \cap Y|} = 1 - \frac{|U \cap Y \cap X|}{|U \cap Y|}.$$

If we assume that significance is equal to γ we obtain

$$\varepsilon = \sqrt{\frac{\ln(1 - \gamma)}{-2|U \cap Y|}}$$

and the approximation accuracy is estimated from (1) with the significance γ according to the formula

$$P_\#(X \mid Y) \ge \frac{|U \cap Y \cap X|}{|U \cap Y|} - \sqrt{\frac{\ln(1 - \gamma)}{-2|U \cap Y|}}.$$

The coverage estimator is developed in the analogous way from (1), and the following estimator is obtained

$$P_\#(Y \mid X) \ge \frac{|U \cap Y \cap X|}{|U \cap X|} - \sqrt{\frac{\ln(1 - \gamma)}{-2|U \cap X|}}.$$

Table 1. Exemplary decision system

	a	d
u_0	1	1
u_1	0	0
u_2	0	0
$\vdots$	$\vdots$	$\vdots$
u_{100}	0	0

We illustrate the trade-off between these three numerical factors using the following example. Consider decision system presented in Table 1. We obtain the following lower approximation for the objects in the system:

$$\underline{\{a\}}||d = 0||_{\mathcal{A}} = ||a = 0||_{\mathcal{A}}, \quad \underline{\{a\}}||d = 1||_{\mathcal{A}} = ||a = 1||_{\mathcal{A}}.$$

Yet we cannot state that $||a = 0||_U$ is an approximation of $||d = 0||_U$ with a 100% accuracy, since there may exist an object u_{101} in $\mathbb{U} \setminus U$ such that $a(u_{101}) = 0$ and $d(u_{101}) = 1$. The given decision system suggests that such an occurrence is unlikely, yet still it is possible.

We estimate the approximation accuracy with significance 95%:

$$P_{\#}(||d = 0||_U \mid ||a = 0||_U) \geq \frac{|\ ||d = 0 \wedge a = 0||_{\mathcal{A}}|}{|\ ||a = 0||_{\mathcal{A}}|} - \sqrt{\frac{\ln(1 - 0.95)}{-2|\ ||a = 0||_{\mathcal{A}}|}} =$$

$$= \frac{100}{100} - \sqrt{\frac{\ln(0.05)}{-200}} = 0.88.$$

Hence, the accuracy of the approximation of the set $||d = 0||_U$ by means of $||a = 0||_U$ is greater than 88% with significance 95%. On the other hand, for the approximation $\underline{\{a\}}||d = 1||_U = ||a = 1||_U$, we do not obtain any significant accuracy estimation.

Hoeffding inequality provides us with a simple analytic formula for the approximation accuracy, yet the obtained estimator is not optimal. That is why we propose the second estimator based on the bound proposed in [4]. It results in an optimal estimator.

Theorem 2. *Let $Z_1, \ldots, Z_n$ be identically distributed independent random variables such that $Z_i \in \{0,1\}$, $i = 1, \ldots, n$. Then, the following inequality takes place:*

$$P\left(EZ_1 > g_{n,\gamma}(\frac{1}{n}\sum_{i=1}^{n} Z_i)\right) < \gamma,$$

where, for a given $k < n$, $g_{n,\gamma}$ satisfies the equation

$$\sum_{i=0}^{k} \binom{n}{i} g_{n,\gamma}(\frac{k}{n})^i (1 - g_{n,\gamma}(\frac{k}{n}))^{n-i} = \gamma$$

and $g_{n,\gamma}(1) = 1$. $g_{n,\gamma}$ provides the optimal (most sharp) bound of EZ_1.

The second estimator does not provide any analytic formula for an estimator value, yet $g_{n,\gamma}(\frac{k}{m})$ may be calculated using an algorithm proposed in [4].

According to the second estimator the accuracy of the approximation of the set $||d = 0||_U$ by means of $||a = 0||_U$ is greater than 97% with significance 95%.

5   Rule Induction Algorithm

Extended approximations of all decision classes compose a classifier. Unfortunately an extended approximation for a given set is not uniquely defined. Many algorithms for calculating approximations were developed. Often the approximations are represented by means of decision rules.

In order to illustrate the link of theory with practical results we propose a simple algorithm for rule induction. The algorithm generates a classifier calculating extended approximations for all decision classes. Each approximation is represented as a set of decision rules whose predecessors are conjunctions of descriptors. For each rule, the accuracy, the coverage and the significance are calculated. The algorithm is parametrised by minimal levels of significance and accuracy and it induces all the rules that satisfy these minimal levels of indices. As a consequence induced rules do not cover all objects, and the classifier has not enough knowledge to recognise some objects. On the other hand all the classified objects are certified to be classified correctly with a very high probability.

The algorithm works as follows: In the k-th step the algorithm tries to induce rules whose predecessors possess k descriptors In the 0th step it checks using the estimator whether there is a decision value v such that the rule with empty predecessor and decision value v would have the desired accuracy and significance. If the answer is positive, then the rule is generated and the rule induction process ends. Otherwise, all the possible rule predecessors with one selector are generated and checked using the estimator. Then the second selector is added, and so on.

The algorithm uses two heuristics that speed it up: it does not try to generate a rule that is more specific than any existing rule and it checks whether there is enough objects matching to the rule predecessor to make it significant.

The algorithm ends when no more rules may be created.

In the case when during classification several rules may be applied to a given object, we choose the rule with the greatest accuracy.

Many more effective algorithms for rule generation that the one described above were developed (for example, in RSES system). However, our objective was to illustrate the theory with a practical application and to show the link between set approximations and induced rules only.

6   Tests

To evaluate the performance of the algorithm, 3 benchmark data sets were selected: *chess, nursery, census94*. The data sets are obtained from the repository of University of California at Irvine [1].

Each data set is split into a training and a test set. For *census94* data sets the original partition available in the repository was used in the experiments. The remaining data sets (*chess* and *nursery*) ware randomly split into a training and a test part with the split ratio 2 to 1.

All the selected sets are the data sets from UCI repository that have data objects represented as vectors of attributes values and have the size between a few thousand and several tens thousand of objects.

Chess and *nursery* have only nominal attributes. *Census94* possess both nominal and numeric attributes. The numeric attributes were discretised.

Table 2 presents test results obtained using the estimator based on Thm. 1. Table 3 presents test results obtained using the estimator based on Thm. 2. In both cases rules were induced with significance 95%.

The tests results show that the algorithm generates a small number of highly relevant rules which makes it useful for knowledge discovery. The fact that it estimates accuracy and coverage for each rule provide us with an insight into

Table 2. Test results obtained using the estimator based on Thm. 1

dataset	min accuracy	number of rules	classifier accuracy	classifier coverage
nursery	0.900000	42	0.985617	0.778395
chess	0.900000	80	0.952963	0.954944
census94	0.950000	32	0.951100	0.502610
census94	0.900000	83	0.899346	0.758307
census94	0.800000	107	0.812987	0.998894

Table 3. Test results obtained using the estimator based on Thm. 2

dataset	min accuracy	number of rules	classifier accuracy	classifier coverage
nursery	0.900000	112	0.989269	0.884722
chess	0.900000	310	0.957419	0.968085
census94	0.950000	92	0.951274	0.590873

Table 4. Part of 53 rules induced from *census94* dataset with significance 0.95 and minimal accuracy 0.85

Accuracy	Coverage	Rule
0.874541	0.388500	sex=Female → class=<=50K
0.938863	0.417531	marital-status=Never-married → class=<=50K
0.883111	0.310253	relationship=Not-in-family → class=<=50K
0.958077	0.198552	relationship=Own-child → class=<=50K
0.967552	0.195818	age=17-23 → class=<=50K
0.893863	0.143788	age=24-28 → class=<=50K
0.899943	0.064978	hours-per-week=18-24 → class=<=50K
0.940021	0.168487	capital-gain=7000-99999 → class=>50K
0.843932	0.050181	occupation=Machine-op-inspct, hours-per-week=40 → class=<=50K
0.879734	0.052272	occupation=Handlers-cleaners → class=<=50K
0.827732	0.040772	occupation=Adm-clerical, education=Some-college → class=<=50K
0.901273	0.048653	education=11th → class=<=50K

the internal structure of data. Table 4 illustrates the above statements presenting a part of rules induced from *census94* dataset.

7   Conclusions

The hybridisation of roughs sets and statistical learning theory resulted in the concept of extended approximation and statistical estimators for rule accuracy and coverage.

These estimators may be used with any rule induction algorithm. They guarantee the relevance of induced rules.

Extended approximations create a theoretical background for the classification. They indicate the connection between lower and upper approximations and rules induced from sample.

The theory and algorithms may be further developed to make them suitable for handling missing values, numerical attributes and other types of data.

Acknowledgment. The research has been partially supported by grants N N516 368334 and N N206 400234 from Ministry of Science and Higher Education of the Republic of Poland and by the grant Innovative Economy Operational Programme 2008-2012 (Priority Axis 1. Research and development of new technologies) managed by Ministry of Regional Development of the Republic of Poland.

References

1. Asuncion, A., Newman, D.J.: UCI Machine Learning Repository. University of California, School of Information and Computer Science, Irvine (2007), http://www.ics.uci.edu/~mlearn/MLRepository.html
2. Gediga, G., Düntsch, I.: Statistical techniques for rough set data analysis. In: Polkowski, L., et al. (eds.) Rough set methods and applications: New developments in knowledge discovery in information systems, pp. 545–565. Physica Verlag, Heidelberg (2000)
3. Guillet, F., Hamilton, H.J. (eds.): Quality Measures in Data Mining. Studies in Computational Intelligence, vol. 43. Springer, Heidelberg (2007)
4. Jaworski, W.: Model Selection and Assessment for Classification Using Validation. In: Ślęzak, D., et al. (eds.) RSFDGrC 2005. LNCS (LNAI), vol. 3641, pp. 481–490. Springer, Heidelberg (2005)
5. Jaworski, W.: Bounds for Validation. Fundamenta Informaticae 70(3), 261–275 (2006)
6. Hoeffding, W.: Probability Inequalities for Sums of Bounded Random Variables. Journal of the American Statistical Association 58, 13–30 (1963)
7. Pawlak, Z.: Rough Sets. Theoretical Aspects of Reasoning about Data. Kluwer Academic Publishers, Dordrecht (1991)
8. Pawlak, Z., Skowron, A.: Rough sets: Some extensions. Information Sciences 177(1), 28–40 (2007)

9. Skowron, A., Swiniarski, R., Synak, P.: Approximation spaces and information granulation. In: Peters, J.F., Skowron, A. (eds.) Transactions on Rough Sets III. LNCS, vol. 3400, pp. 175–189. Springer, Heidelberg (2005)
10. Tsumoto, S.: Accuracy and Coverage in Rough Set Rule Induction. In: Alpigini, J.J., Peters, J.F., Skowron, A., Zhong, N. (eds.) RSCTC 2002. LNCS (LNAI), vol. 2475, pp. 373–380. Springer, Heidelberg (2002)
11. Vapnik, V.N.: Statistical Learning Theory. Wiley, New York (1998)

Action Rules Discovery
without Pre-existing Classification Rules

Zbigniew W. Raś[1,2] and Agnieszka Dardzińska[1,3]

[1] Univ. of North Carolina, Dept. of Computer Science, Charlotte, NC, 28223, USA
[2] Polish Academy of Sciences, Institute of Computer Science, 01-237 Warsaw, Poland
[3] Bialystok Technical Univ., Dept. of Computer Science, 15-351 Bialystok, Poland
ras@uncc.edu, adardzin@uncc.edu

Abstract. Action rules describe possible transitions of objects from one state to another with respect to a distinguished attribute. Previous research on action rule discovery usually requires the extraction of classification rules before constructing any action rule. In this paper, we present a new algorithm that discovers action rules directly from a decision system. It is a bottom-up strategy which has some similarity to systems *ERID* and *LERS*. Finally, it is shown how to manipulate the music score using action rules.

1 Introduction

An action rule is a rule extracted from a decision system that describes a possible transition of objects from one state to another with respect to a distinguished attribute called a decision attribute [13]. We assume that attributes used to describe objects in a decision system are partitioned into stable and flexible. Values of flexible attributes can be changed. This change can be influenced and controlled by users. Action rules mining initially was based on comparing profiles of two groups of targeted objects - those that are desirable and those that are undesirable [13]. An action rule is formed as a term $[(\omega) \wedge (\alpha \rightarrow \beta)] \Rightarrow (\phi \rightarrow \psi)$, where ω is a conjunction of fixed condition features shared by both groups, $(\alpha \rightarrow \beta)$ represents proposed changes in values of flexible features, and $(\phi \rightarrow \psi)$ is a desired effect of the action. The discovered knowledge provides an insight of how relationships should be managed so the undesirable objects can be changed to desirable. For example, in society, one would like to find a way to improve his or her salary from a low-income to a high-income. Another example in business area is when an owner would like to improve his or her company's profits by going from a high-cost, low-income business to a low-cost, high-income business.

Action rules introduced in [13] has been further investigated in [15][12][14]. Paper [5] was probably the first attempt towards formally introducing the problem of mining action rules without pre-existing classification rules. Authors explicitly formulated it as a search problem in a support-confidence-cost framework. The proposed algorithm is similar to Apriori [1]. Their definition of an action rule allows changes on stable attributes. Changing the value of an attribute, either stable or flexible, is linked with a cost [16]. In order to rule out action rules with undesired changes on stable attributes, authors have assigned very high cost to such changes. However, that way, the cost of

C.-C. Chan et al. (Eds.): RSCTC 2008, LNAI 5306, pp. 181–190, 2008.

action rules discovery is getting unnecessarily increased. Also, they did not take into account the dependencies between attribute values which are naturally linked with the cost of rules used either to accept or reject a rule. Algorithm *ARED*, presented in [6], is based on Pawlak's model of an information system S [9]. The goal is to identify certain relationships between granules defined by the indiscernibility relation on its objects. Some of these relationships uniquely define action rules for S.

This paper presents a new strategy for discovering action rules directly from the decision system. Action rules are built from atomic expressions following a strategy similar to *ERID* [2].

2   Background and Objectives

In this section we introduce the notion of an information system, a decision system, stable attribute, flexible attribute, and give some examples.

By an information system [9] we mean a triple $S = (X, A, V)$, where:

1. X is a nonempty, finite set of objects
2. A is a nonempty, finite set of attributes, i.e.
   $a : U \longrightarrow V_a$ is a function for any $a \in A$, where V_a is called the domain of a
3. $V = \bigcup\{V_a : a \in A\}$.

For example, Table 1 shows an information system S with a set of objects $X = \{x_1, x_2, x_3, x_4, x_5, x_6, x_7, x_8\}$, a set of attributes $A = \{a, b, c, d\}$, and a set of their values $V = \{a_1, a_2, b_1, b_2, b_3, c_1, c_2, d_1, d_2, d_3\}$.

Table 1. Decision Table S

	a	b	c	d
x_1	a_1	b_1	c_1	d_1
x_2	a_2	b_1	c_2	d_1
x_3	a_2	b_2	c_2	d_1
x_4	a_2	b_1	c_1	d_1
x_5	a_2	b_3	c_2	d_1
x_6	a_1	b_1	c_2	d_2
x_7	a_1	b_2	c_2	d_1
x_8	a_1	b_2	c_1	d_3

An information system $S = (X, A, V)$ is called a decision system, if $A = A_{St} \cup A_{Fl} \cup \{d\}$, where d is a distinguished attribute called the decision. Attributes in A_{St} are called *stable* and attributes in A_{Fl} are called *flexible*. They jointly form the set of conditional attributes. "Date of birth" is an example of a stable attribute. "Interest rate" for each customer account is an example of a flexible attribute.

In earlier works in [13][15][12][14], action rules are constructed from classification rules. This means that we use pre-existing classification rules or generate them using a rule discovery algorithm, such as *LERS* [4] or *ERID* [2], then, construct action rules either from certain pairs of these rules or from a single classification rule. For instance, algorithm $ARAS$ [14] generates sets of terms (built from values of attributes) around classification rules and constructs action rules directly from them. In this study, we propose a different approach to achieve the following objectives:

1. Extract action rules directly from a decision system without using pre-existing classification rules.
2. Extract action rules that have minimal attribute involvement.

To meet these two goals, we introduce the notion of atomic action terms and show how to build action rules from them.

3   Action Rules

In this section we give a definition of action terms, action rules, and we propose their interpretation which we call standard.

Let $S = (X, A \cup \{d\}, V)$ be a decision system, where $V = \bigcup\{V_a : a \in A\}$. First, we introduce the notion of an action term.

By an *atomic action term* we mean an expression $(a, a_1 \rightarrow a_2)$, where a is an attribute and $a_1, a_2 \in V_a$. If $a_1 = a_2$, then a is called stable on a_1.

By a set of *action terms* we mean a smallest set such that:

1. If t is an atomic action term, then t is an action term.
2. If t_1, t_2 are action terms, then $t_1 \star t_2$ is an action term.
3. If t is an action term containing $(a, a_1 \rightarrow a_2)$, $(b, b_1 \rightarrow b_2)$ as its sub-terms, then $a \neq b$.

By the domain of an action term t, denoted by $Dom(t)$, we mean the set of all attribute names listed in t.

By an *action rule* we mean an expression $r = [t_1 \Rightarrow t_2]$, where t_1 is an action term and t_2 is an atomic action term. Additionally, we assume that $Dom(t_2) = \{d\}$ and $Dom(t_1) \subseteq A$. The domain of action rule r is defined as $Dom(t_1) \cup Dom(t_2)$.

Now, let us give an example of action rules assuming that the decision system S is represented by Table 1, a is stable and b, c are flexible attributes. Expressions $(a, a_2 \rightarrow a_2)$, $(b, b_1 \rightarrow b_3)$, $(c, c_2 \rightarrow c_2)$, $(d, d_1 \rightarrow d_2)$ are examples of atomic action terms. Expression $(b, b_1 \rightarrow b_3)$ means that the value of attribute b is changed from b_1 to b_3. Expression $(c, c_2 \rightarrow c_2)$ means that the value c_2 of attribute c remains unchanged. Expression $r = [[(a, a_2 \rightarrow a_2) \star (b, b_1 \rightarrow b_3)] \Rightarrow (d, d_1 \rightarrow d_2)]$ is an example of an action rule. The rule says that if value a_2 remains unchanged and value b will change from b_1 to b_3, then it is expected that the value d will change from d_1 to d_2. Clearly, $Dom(r) = \{a, b, d\}$.

Standard interpretation N_S of action terms in $S = (X, A, V)$ is defined as follow:

1. If $(a, a_1 \rightarrow a_2)$ is an atomic action term, then
$N_S((a, a_1 \rightarrow a_2)) = [\{x \in X : a(x) = a_1\}, \{x \in X : a(x) = a_2\}].$

2. If $t_1 = (a, a_1 \rightarrow a_2) \star t$ and $N_S(t) = [Y_1, Y_2]$, then
$N_S(t_1) = [Y_1 \cap \{x \in X : a(x) = a_1\}, Y_2 \cap \{x \in X : a(x) = a_2\}].$

Now, let us define $[Y_1, Y_2] \cap [Z_1, Z_2]$ as $[Y_1 \cap Z_1, Y_2 \cap Z_2]$ and assume that $N_S(t_1) = [Y_1, Y_2]$ and $N_S(t_2) = [Z_1, Z_2]$. Then, $N_S(t_1 \star t_2) = N_S(t_1) \cap N_S(t_2)$.

Let $r = [t_1 \rightarrow t_2]$ be an action rule, where $N_S(t_1) = [Y_1, Y_2]$, $N_S(t_2) = [Z_1, Z_2]$. Support and confidence of r are defined as follow:

1. $sup(r) = card(Y_1 \cap Z_1)$.
2. $conf(r) = [\frac{card(Y_1 \cap Z_1)}{card(Y_1)}] \cdot [\frac{card(Y_2 \cap Z_2)}{card(Y_2)}]$.

The definition of a confidence should be interpreted as an optimistic confidence. It requires that $card(Y_1) \neq 0$, $card(Y_2) \neq 0$, $card(Y_1 \cap Z_1) \neq 0$, and $card(Y_2 \cap Z_2) \neq 0$. Otherwise, the confidence of action rule is zero.

Coming back to the example of S given in Table 1, we can find many action rules associated with S. Let us take $r = [[(a, a_2 \rightarrow a_2) \star (b, b_1 \rightarrow b_2)] \Rightarrow (d, d_1 \rightarrow d_2)]$ as an example of the action rule. Then,

$N_S((a, a_2 \rightarrow a_2)) = [\{x_2, x_3, x_4, x_5\}, \{x_2, x_3, x_4, x_5\}],$
$N_S((b, b_1 \rightarrow b_2)) = [\{x_1, x_2, x_4, x_6\}, \{x_3, x_7, x_8\}],$
$N_S((d, d_1 \rightarrow d_2)) = [\{x_1, x_2, x_3, x_4, x_5, x_7\}, \{x_6\}],$
$N_S((a, a_2 \rightarrow a_2) \star (b, b_1 \rightarrow b_2)) = [\{x_2, x_4\}, \{x_3\}].$

Clearly, $sup(r) = 2$ and $conf(r) = 1 \cdot 0 = 0$.

Assume that $L([Y, Z]) = Y$ and $R([Y, Z]) = Z$. The new algorithm *ARD* for constructing action rules is similar to *ERID* [2] and *LERS* [4]. So, to present this algorithm, it is sufficient to outline the strategy for assigning marks to atomic action terms and show how terms of length greater than one are built. Only positive marks yield action rules. Action terms of length k are built from unmarked action terms of length $k-1$ and unmarked atomic action terms of length one. Marking strategy for terms of any length is the same as for action terms of length one.

Now, let us assume that $S = (X, A \cup \{d\}, V)$ is a decision system and λ_1, λ_1 denote minimum support and confidence, respectively. Each $a \in A$ uniquely defines the set $C_S(a) = \{N_S(t_a) : t_a$ is an atomic action term built from elements in $V_a\}$. By t_d we mean an atomic action term built from elements in V_d.

Marking strategy for atomic action terms

For each $N_S(t_a) \in C_S(a)$ do

if $L(N_S(t_a)) = \emptyset$ or $R(N_S(t_a)) = \emptyset$ or $L(N_S(t_a \star t_d)) = \emptyset$ or $R(N_S(t_a \star t_d)) = \emptyset$, **then** t_a is **marked negative**.

if $L(N_S(t_a)) = R(N_S(t_a))$ **then** t_a **stays unmarked**

if $card(L(N_S(t_a \star t_d))) < \lambda_1$ **then** t_a is **marked negative**

if $card(L(N_S(t_a \star t_d))) \geq \lambda_1$ and $conf(t_a \rightarrow t_d) < \lambda_2$ **then** t_a **stays unmarked**

if $card(L(N_S(t_a \star t_d))) \geq \lambda_1$ and $conf(t_a \to t_d) \geq \lambda_2$ **then** t_a is **marked positive** and the action rule $[t_a \to t_d]$ is printed.

Now, to clarify *ARD* (Action Rules Discovery) strategy for constructing action rules, we go back to our example with S defined by Table 1 and with $A_{St} = \{b\}$, $A_{Fl} = \{a, c, d\}$. We are interested in action rules which may reclassify objects from the decision class d_1 to d_2. Additionally, we assume that $\lambda_1 = 2$, $\lambda_2 = 1/4$.

All atomic action terms for S are listed below:
For Decision Attribute in S:

$$N_S(t_{12}) = [\{x_1, x_2, x_3, x_4, x_5, x_7\}, \{x_6\}]$$

For Classification Attributes in S:

$t_1 = (b, b_1 \to b_1)$, $t_2 = (b, b_2 \to b_2)$, $t_3 = (b, b_3 \to b_3)$, $t_4 = (a, a_1 \to a_2)$, $t_5 = (a, a_1 \to a_1)$, $t_6 = (a, a_2 \to a_2)$, $t_7 = (a, a_2 \to a_1)$, $t_8 = (c, c_1 \to c_2)$, $t_9 = (c, c_2 \to c_1)$, $t_{10} = (c, c_1 \to c_1)$, $t_{11} = (c, c_2 \to c_2)$, $t_{12} = (d, d_1 \to d_2)$.

Following the first loop of *ARD* algorithm we get:

$N_S(t_1) = [\{x_1, x_2, x_4, x_6\}, \{x_1, x_2, x_4, x_6\}]$ Not Marked $/Y_1 = Y_2/$
$N_S(t_2) = [\{x_3, x_7, x_8\}, \{x_3, x_7, x_8\}]$ Marked "-" $/card(Y_2 \cap Z_2) = 0/$
$N_S(t_3) = [\{x_5\}, \{x_5\}]$ Marked "-" $/card(Y_2 \cap Z_2) = 0/$
$N_S(t_4) = [\{x_1, x_6, x_7, x_8\}, \{x_2, x_3, x_4, x_5\}]$ Marked "-" $/card(Y_2 \cap Z_2) = 0/$
$N_S(t_5) = [\{x_1, x_6, x_7, x_8\}, \{x_1, x_6, x_7, x_8\}]$ Not Marked $/Y_1 = Y_2/$
$N_S(t_6) = [\{x_2, x_3, x_4, x_5\}, \{x_2, x_3, x_4, x_5\}]$ Marked "-" $/card(Y_2 \cap Z_2) = 0/$
$N_S(t_7) = [\{x_2, x_3, x_4, x_5\}, \{x_1, x_6, x_7, x_8\}]$ Marked "+"
/rule $r_1 = [t_7 \Rightarrow t_{12}]$ has $conf = 1/2 \geq \lambda_2$, $sup = 2 \geq \lambda_1/$
$N_S(t_8) = [\{x_1, x_4, x_8\}, \{x_2, x_3, x_5, x_6, x_7\}]$ Not Marked
/rule $r_1 = [t_8 \Rightarrow t_{12}]$ has $conf = [2/3] \cdot [1/5] < \lambda_2$, $sup = 2 \geq \lambda_1/$
$N_S(t_9) = [\{x_2, x_3, x_5, x_6, x_7\}, \{x_1, x_4, x_8\}]$ Marked "-" $/card(Y_2 \cap Z_2) = 0/$
$N_S(t_{10}) = [\{x_1, x_4, x_8\}, \{x_1, x_4, x_8\}]$ Marked "-" $/card(Y_2 \cap Z_2) = 0/$
$N_S(t_{11}) = [\{x_2, x_3, x_5, x_6, x_7\}, \{x_2, x_3, x_5, x_6, x_7\}]$ Not Marked $/Y_1 = Y_2/$

Now, we build action terms of length two from unmarked action terms of length one.

$N_S(t_1 \star t_5) = [\{x_1, x_6\}, \{x_1, x_6\}]$ Not Marked $/Y_1 = Y_2/$
$N_S(t_1 \star t_8) = [\{x_1, x_4\}, \{x_2, x_6\}]$ Marked "+"
/rule $r_1 = [[t_1 \star t_8] \Rightarrow t_{12}]$ has $conf = 1/2 \geq \lambda_2$, $sup = 2 \geq \lambda_1/$
$N_S(t_1 \star t_{11}) = [\{x_2, x_6\}, \{x_2, x_6\}]$ Not Marked $/Y_1 = Y_2/$
$N_S(t_5 \star t_8) = [\{x_1, x_8\}, \{x_6, x_7\}]$ Marked "-"
/rule $r_1 = [[t_5 \star t_8] \Rightarrow t_{12}]$ has $conf = 1/2 \geq \lambda_2$, $sup = 1 < \lambda_1/$
$N_S(t_5 \star t_{11}) = [\{x_6, x_7\}, \{x_6, x_7\}]$ Not Marked $/Y_1 = Y_2/$
$N_S(t_8 \star t_{11}) = [\emptyset, \{x_2, x_3, x_5, x_6, x_7\}]$ Marked "-" $/card(Y_1) = 0/$

Finally (there are only 3 classification attributes in S), we build action terms of length three from unmarked action terms of length one and length two.

Fig. 1. Example score of a Pentatonic Minor Scale played in the key of C

Scale	sma	ii	iii	iv	v	vi	vii	viii	ix	x	xi	xii	#
Pentatonic Minor	m	3	2	2	3								4

Fig. 2. Representation of a Pentatonic Minor Scale

Only, the term $t_1 \star t_5 \star t_8$ can be built. It is an extension of $t_5 \star t_8$ which is already marked as negative. So, the algorithm *ARD* stops and two action rules are constructed: $[[(b, b_1 \rightarrow b_1) \star (c, c_1 \rightarrow c_2)] \Rightarrow (d, d_1 \rightarrow d_2)]$, $[(a, a_2 \rightarrow a_1) \Rightarrow (d, d_1 \rightarrow d_2)]$. Following the notation used in previous papers on action rules mining (see [6], [14], [13], [12]), the first of the above two action rules will be presented as $[[(b, b_1) \star (c, c_1 \rightarrow c_2)] \Rightarrow (d, d_1 \rightarrow d_2)]$.

4   Application Domain and Experiment

Music Information Retrieval (MIR) is chosen as the application area for our research. In [11], authors present the system *MIRAI* for automatic indexing of music by instruments and emotions. When *MIRAI* receives a musical waveform, it divides that waveform into segments of equal size and then its classifiers identify the most dominating musical instruments and emotions associated with each segment and finally with the musical waveform. In [7], [8] authors follow another approach and present a Basic Score Classification Database (*BSCD*) which describes associations between different scales, regions, genres, and jumps. This database is used to automatically index a piece of music by emotions. In this section, we show how to use action rules extracted from *BSCD* assuming that we need to change the emotion either from the retrieved or submitted piece of music by minimally changing its score. By a score, in MIR area, we mean a written form of a musical composition.

To introduce the problem, let's start with Figure 1 showing an example of a score of a Pentatonic Minor Scale played in the key of C on a piano. As we can see, 8 notes are played: $A\sharp, G, A\sharp, F, D\sharp, G, C$ and C. The ordered sequence of the same notes without repetitions $[A\sharp, C, D\sharp, F, G]$ uniquely represents that score. Now, we explain the process of computing its numeric representation $[2, 3, 2, 2]$. The score is played in the key of $A\sharp$ which becomes the root. Its second note C is 2 tones up from $A\sharp$. The third note $D\sharp$ is three tones up from C. The fourth note F is two tones up from $D\sharp$, and finally G is two tones up from F. This is how the sequence of jumps $[2, 3, 2, 2]$ with root $A\sharp$ is generated.

Essentially any combination of notes $A\sharp, C, D\sharp, F, G$ can be played while still remaining within the constraints of a C Pentatonic Minor Scale on a piano. This scale is illustrated in Figure 2. Accordingly one plays the root, plays 3 tones up, then 2 tones up then 2 tones up, and then 3 tones up (*m* means *mode*). The first note, or in musical terms, the "Root" is a C note. It means that the remaining four notes are all in the key

Table 2. Basic Score Classification Database

J_1	J_2	J_3	J_4	J_5	Scale	Region	Genre	Emotion	sma
2	2	3	2		Pentatonic Major	Western	Blues	melancholy	s
3	2	1	1	2	Blues Major	Western	Blues	depressive	s
3	2	2	3		Pentatonic Minor	Western	Jazz	melancholy	s
3	2	1	1	3	Blues Minor	Western	Blues	dramatic	s
3	1	3	1	3	Augmented	Western	Jazz	feel-good	s
2	2	2	2	2	Whole Tone	Western	Jazz	push-pull	s
1	2	4	1		Balinese	Balinese	ethnic	neutral	s
2	2	3	2		Chinese	Chinese	ethnic	neutral	s
2	3	2	3		Egyptian	Egyptian	ethnic	neutral	s
1	4	1	4		Iwato	Iwato	ethnic	neutral	s
1	4	2	1		Japanese	Japanese	Asian	neutral	s
2	1	4	1		Hirajoshi	Hirajoshi	ethnic	neutral	s
1	4	2	1		Kumoi	Japanese	Asian	neutral	s
2	2	3	2		Mongolian	Mongolian	ethnic	neutral	s
1	2	4	3		Pelog	Western	neutral	neutral	s
2	2	3	2		Pentatonic Majeur	Western	neutral	happy	m
2	3	2	3		Pentatonic 2	Western	neutral	neutral	m
3	2	3	2		Pentatonic 3	Western	neutral	neutral	m
2	3	2	2		Pentatonic 4	Western	neutral	neutral	m
2	2	3	3		Pentatonic Dominant	Western	neutral	neutral	m
3	2	2	3		Pentatonic Minor	Western	neutral	sonorous	m
1	3	3	2		Altered Pentatonic	Western	neutral	neutral	m
3	2	1	1	2	Blues	Western	Blues	depressive	m
4	3				Major	neutral	neutral	sonorous	a
3	4				Minor	neutral	neutral	sonorous	a
4	3	4			Major 7th Major	neutral	neutral	happy	a
4	3	3			Major 7th Minor	neutral	neutral	not happy	a
3	4	4			Minor 7th Major	neutral	neutral	happy	a
3	4	3			Minor 7th Minor	neutral	neutral	not happy	a
2	2	3	3		Major 9th	neutral	neutral	happy	a
2	1	4	3		Minor 9th	neutral	neutral	not happy	a
2	2	1	2	3	Major 11th	neutral	neutral	happy	a
2	1	2	2	3	Minor 11th	neutral	neutral	not happy	a
4	4				Augmented	neutral	neutral	happy	a
3	3	3			Diminished	neutral	neutral	not happy	a

of C Pentatonic Minor Scale on a piano. However, from the score itself, we have no idea about its key or scale. We can only discern the jumps between the notes and the repeated notes.

To tackle the above problem, authors in [7] built a Basic Score Classification Database (BSCD) which describes associations between different scales, regions, genres, and jumps (see Table 2). The attribute J_i means i-th jump. When a music piece is submitted to QAS associated with BSCD, each note one by one, is drawn into the array

Table 3. Possible Representative Jump Sequences for the Input Sequence

Root	J_1	J_2	J_3	J_4
$A^\sharp$	2	3	2	2
G	3	2	3	2
F	2	3	2	3
$D^\sharp$	2	2	3	2
C	3	2	2	3

Fig. 3. Example of a Music Score

of incoming signals. Assuming that the score is represented by Figure 1, QAS will generate five optional sequences:

$[A\sharp, C, D\sharp, F, G]$, $[G, A\sharp, C, D\sharp, F]$, $[F, G, A\sharp, C, D\sharp]$, $[D\sharp, F, G, A\sharp, C]$, or $[C, D\sharp, F, G, A\sharp]$.

In the first case $A\sharp$ is the root, in the second G is the root, in the third F, in the fourth $D\sharp$, and in the fifth C is the root. Clearly, at this point, QAS has no idea which note is the root and the same which sequence out of the 5 is a representative one for the input sequence of notes $A\sharp, G, A\sharp, F, D\sharp, G, C$ and C. Table 3 gives numeric representation of these five sequences.

Paper [8] presents a heuristic strategy for identifying which sequence out of these five sequences is a representative one for the input score. The same, on the basis of associations between sequences of jumps and emotions which can be extracted from $BSCD$, we can identify the emotion which invokes in most of us the above input score.

What about changes to the input score so the scale associated with that score will change the way user wants. Action rules extracted from $BSCD$ can be used for that purpose and they guarantee the smallest number of changes needed to achieve the goal. Example of an action rule extracted from $BSCD$ is given below:

$$[(J_1, 3 \to 2) \star (J_2, 2 \to 3)] \Rightarrow (Scale, PentatonicMinor \to Egyptian).$$

For instance, this rule can be applied to a music score represented by a sequence of 25 notes (Figure 3). They are $[A\sharp, G, A\sharp, C, C, D\sharp, D, C, C, F, C, A\sharp, C, A\sharp, G, A, G, G, D\sharp, G, C, D\sharp, A\sharp, C, C]$.

The ordered sequence of the same 25 notes without repetitions $[A\sharp, C, D, D\sharp, F, G]$ uniquely represents that score. Assume now, that the score is played in the key of G. So, $[3, 2, 2, 1, 2]$ is its numeric representation.

The classifier trained on Table 1, based on Levenshtein's distance [8], identified the sequence $[3, 2, 2, 3]$ as the closest one to $[3, 2, 2, 1, 2]$. Action rule $[(J_1, 3 \rightarrow 2) \star (J_2, 2 \rightarrow 3)] \Rightarrow (Scale, PentatonicMinor \rightarrow Egyptian)]$, extracted from Table 1, converts that score to $[A\sharp, G, A, C, C, D\sharp, D, C, C, F, C, A, C, A\sharp, G, A, G, G, D\sharp, G,$ $C, D\sharp, A, C, C]$.
Please notice that $A\sharp$ is changing to A only if the note C follows it in the input score.

This example shows how to use action rules to manipulate the music score. Following the same approach, we can manipulate music emotions, genre, and region.

5   Conclusion and Future Work

We presented an algorithm that discovers action rules from a decision table. The proposed algorithm generates a complete set of shortest action rules without using pre-existing classification rules. During the experiment with several data sets, we noticed that the flexibility of attributes are not equal. For example, the social condition was most likely less flexible than the health condition in one of the data set used in our experiment, and this may have to be considered. Future work shall address this issue as well as further analysis of the algorithm with more real world data sets.

Acknowledgment

This material is based in part upon work supported by the National Science Foundation under Grant Number *IIS-0414815*, the Ministry of Science and Higher Education in Poland under Grant *N N519 404734*, and Bialystok Technical University under Grant Number *S/WI/1/08*. Any opinions, findings, and conclusions or recommendations expressed in this material are those of the author(s) and do not necessarily reflect the views of the National Science Foundation. Also, we acknowledge the help from Rory Lewis in the application part of the paper.

References

1. Agrawal, R., Srikant, R.: Fast algorithm for mining association rules. In: Proceeding of the Twentieth International Conference on VLDB, pp. 487–499 (1994)
2. Dardzińska, A., Raś, Z.: Extracting rules from incomplete decision systems. In: Foundations and Novel Approaches in Data Mining, Studies in Computational Intelligence, vol. 9, pp. 143–154. Springer, Heidelberg (2006)
3. Greco, S., Matarazzo, B., Pappalardo, N., Slowiński, R.: Measuring expected effects of interventions based on decision rules. J. Exp. Theor. Artif. Intell. 17(1-2), 103–118 (2005)
4. Grzymala-Busse, J.: A new version of the rule induction system LERS. Fundamenta Informaticae 31(1), 27–39 (1997)
5. He, Z., Xu, X., Deng, S., Ma, R.: Mining action rules from scratch. Expert Systems with Applications 29(3), 691–699 (2005)
6. Im, S., Ras, Z.W.: Action rule extraction from a decision table: ARED. In: An, A., Matwin, S., Raś, Z.W., Ślęzak, D. (eds.) Foundations of Intelligent Systems. LNCS (LNAI), vol. 4994, pp. 160–168. Springer, Heidelberg (2008)

7. Lewis, R., Ras, Z.W.: Rules for processing and manipulating scalar music theory. In: Proceedings of the International Conference on Multimedia and Ubiquitous Engineering (MUE 2007), Seoul, South Korea, April 26-28, 2007, pp. 819–824. IEEE Computer Society, Los Alamitos (2007)
8. Lewis, R., Jiang, W., Ras, Z.W.: Mining scalar representations in a non-tagged music database. In: An, A., Matwin, S., Raś, Z.W., Ślęzak, D. (eds.) Foundations of Intelligent Systems. LNCS (LNAI), vol. 4994, pp. 445–454. Springer, Heidelberg (2008)
9. Pawlak, Z.: Information systems - theoretical foundations. Information Systems Journal 6, 205–218 (1981)
10. Qiao, Y., Zhong, K., Wang, H.-A., Li, X.: Developing event-condition-action rules in real-time active database. In: Proceedings of the 2007 ACM symposium on Applied computing, pp. 511–516. ACM, New York (2007)
11. Ras, Z.W., Zhang, X., Lewis, R.: MIRAI: Multi-hierarchical, FS-tree based music information retrieval system. In: Kryszkiewicz, M., Peters, J.F., Rybinski, H., Skowron, A. (eds.) RSEISP 2007. LNCS (LNAI), vol. 4585, pp. 80–89. Springer, Heidelberg (2007)
12. Raś, Z.W., Dardzińska, A.: Action rules discovery, a new simplified strategy. In: Esposito, F., Raś, Z.W., Malerba, D., Semeraro, G. (eds.) ISMIS 2006. LNCS (LNAI), vol. 4203, pp. 445–453. Springer, Heidelberg (2006)
13. Raś, Z.W., Wieczorkowska, A.: Action-Rules: How to increase profit of a company. In: Zighed, D.A., Komorowski, J., Żytkow, J.M. (eds.) PKDD 2000. LNCS (LNAI), vol. 1910, pp. 587–592. Springer, Heidelberg (2000)
14. Raś, Z., Wyrzykowska, E., Wasyluk, H.: ARAS: Action rules discovery based on agglomerative strategy, in Mining Complex Data. In: Raś, Z.W., Tsumoto, S., Zighed, D.A. (eds.) MCD 2007. LNCS (LNAI), vol. 4944, pp. 196–208. Springer, Heidelberg (2008)
15. Tsay, L.-S., Ras, Z.W.: Discovering the concise set of actionable patterns. In: An, A., Matwin, S., Raś, Z.W., Ślęzak, D. (eds.) Foundations of Intelligent Systems. LNCS (LNAI), vol. 4994, pp. 169–178. Springer, Heidelberg (2008)
16. Tzacheva, A., Ras, Z.W.: Constraint based action rule discovery with single classification rules. In: An, A., Stefanowski, J., Ramanna, S., Butz, C.J., Pedrycz, W., Wang, G. (eds.) RSFDGrC 2007. LNCS (LNAI), vol. 4482, pp. 322–329. Springer, Heidelberg (2007)

Hierarchical Learning in Classification of Structured Objects

Tuan Trung Nguyen

Polish-Japanese Institute of Information Technology
ul. Koszykowa 86, 02-008 Warsaw, Poland
nttrung@pjwstk.edu.pl

Abstract. We discuss the hierarchical learning approach applied to the recognition of structured objects. Learning algorithms for such objects usually display high complexity and typically require a priori assumptions on the subject domain. Hierarchical learning is designed to alleviate many problems associated with structured object recognition. It helps steer searches for solutions toward more promising paths in the otherwise computationally prohibitive search spaces by breaking the original task into simpler, more manageable subtasks. It provides for an effective interactive mechanism to transfer the additional domain knowledge expressed by external human experts into low level operators. The design and the implementation of hierarchical learning and domain knowledge elicitation, based on approximate reasoning and rough mereology constitute an excellent example of Granular Computing at work.

Keywords: Rough mereology, concept approximation, machine learning, hierarchical learning, handwritten digit recognition.

1 Introduction

Machine learning can be broadly understood as a process in which a machine (a computer system) changes its structure, its programs, or its data so that its expected future performance improves. The changes may involve existing components, or *ab initio* the synthesis of entirely new ones [5].

From a bit more technical point of view, the principal task of machine learning is to reconstruct a decision function f that associates input data with their outputs, by way of assuming a hypothesis about the function f in the form of another function h, selected from a known class H of functions. Ideally, h should return the same output values as f for the same inputs. In practice, we try to approach that agreement in output values as close as possible.

The choice of a particular hypothesis class H implicitly defines a hypothesis space that a system can ever represent and therefore can ever learn. The construction of the desired h can then be viewed as a search through that space for optimal candidate hypotheses. An immediate consequence is that learning complex tasks where the hypothesis spaces are usually very large may entail prohibitive computational costs.

C.-C. Chan et al. (Eds.): RSCTC 2008, LNAI 5306, pp. 191–201, 2008.

In pattern recognition, searching through huge potential solution spaces poses challenging problems in all stages, from feature selection through classifiers construction to classification of novel samples. The most popular methods to alleviate the complexity problem is to employ heuristic search strategies as well as to approximate intermediate or partial solutions.

To make the matters worse, machine learning tasks are a good example of *inverse problems*, where an attempt is made to establish some causal factors or subsurface structures from the available observation data. Inverse problems are known to be ill-posed [13], with small deviations in data leading to amplified aberrations in solutions.

It is widely acknowledged that good feature selection would not be possible without knowledge on the domain of interest [2]. Domain (or background) knowledge can serve as additional search control tools. Usually fast and efficient greedy searches have limits in the patterns they can discover, while complex and more elaborated, more exhaustive strategies typically display high computational costs.

Pattern recognition in general benefits from a vast gamut of popular approaches such as discriminant analysis, statistical learning, decision trees, neural networks or genetic algorithms, commonly referred to as *inductive* learning methods, i.e. methods that generalize from observed training examples by finding features that empirically distinguish positive from negative training examples. Though these methods allow for highly effective learning systems, there often exist proven bounds on the performance of the classifiers they can construct, especially when the samples involved exhibit complex internal structures, such as optical characters, facial images or time series data. Such samples contain substantial structural or relational information that sometimes might prove impossible to quantify in feature vector forms, and it is believed that *analytical* learning methods based on structural analysis of training examples are more suitable in dealing with such samples. These methods, however, can only be efficiently employed using many a priori assumptions on the sample domain [5].

As an attempt to tackle all the above mentioned problems, in this paper, we present a scheme for incorporating domain knowledge about structured samples into the learning process. The knowledge is provided by a hypothetical expert that will interact with the classification system during a later phase of the learning process, providing certain 'guidance' to the difficult task of adaptive searching for correct classifiers. The main underlying assumption is that when the feature space is as large as in the case of structured samples, algorithms seeking to approximate human reasoning will perform better when equipped with domain knowledge provided by a human expert. This external knowledge will be used to steer the search process to more promising areas more quickly or to fine tune the construction of component patterns that would be difficult to find greedily (See Fig.1). Learning from external domain knowledge sources constitutes an integral part of the intensively pursued research over Knowledge-rich Data Mining, as stipulated in, e.g. [1].

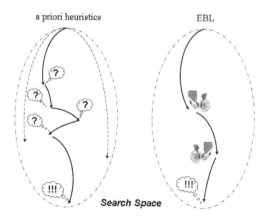

Fig. 1. A priori vs. explanation based learning

In distinction to most popular domain knowledge based approaches widely used in recognition systems, ours concentrates on specific difficult, error-prone samples encountered during the learning phase. The expert will pass the correct classification of such cases to the system along with his explanation on how he arrived at his decision on the class identity of the sample. The system then attempts to translate this knowledge into its own descriptive language and primitives, to rebuild its classifiers. The novel element here is that these explanations will not be passed as predefined, but rather will be provided by the expert in a two way dialog along with the evolution of the learning system.

In many aspects, our approach resemble the explanation-based learning *(EBL)* approach. EBL allows a solution to a sample problem to be generalized into a form that later could be used to solve conceptually similar problems [10]. The generalization process is driven by the explanation why the solution worked. In Pattern Recognition, EBL typically is used to produce more general classification rules by analyzing the classification of a sample object [5]. The main advantage of EBL is that to learn a concept, it requires a much smaller number of sample objects than other approaches. However, our method goes beyond the typical EBL schemes with the introduction of a multi layer approximate reasoning hierarchy that will help to transfer the domain knowledge to the learning process in a much more natural and efficient form.

It is noteworthy to observe that our approach, based on approximate reasoning scheme and granular computing, though developed independently, has much in common with theories and methods of Cognitive Science. For example, one of the most fundamental assumption of Unified Theory of Cognition [6] stipulates that human perception is inherently hierarchical and theories on such perception should be deliberately approximate. Most, if not all, cognitive architectures such as SOAR, ACT-R, Prodigy or recently developed ICARUS [4] are based on *knowledge and data chunking*, which follows the hierarchical structure of human perception. Chunking resembles in many ways the layered reasoning paradigm.

On the other hand, cognitive architectures seem not to incorporate the approximation of internal predicates or goal seeking strategies to a large extent, while the approximation of concepts and their binding relations is at the core of our approach.

We describe the process of transferring the expert's reasoning scheme into the recognition system, based on the rough mereology approach to concept approximation [9],[12].

2   Hierarchical Learning

The main assumption of the hierarchical learning approach posits that effective classification of complex structured samples should be conducted in subsequent steps rather than in a single all-out, wrap-up attempt. This postulate has several profound motivations.

2.1   Divide and Conquer

First, the internal relational structures of the subject samples naturally call for the need of breaking the learning process into respective simpler sub-problems and trying to compute local solutions before combining them into a larger one for the original task. This divide and conquer paradigm proved to be effective in reducing the complexity of problems involving searches through huge potential solution spaces (See Fig.2).

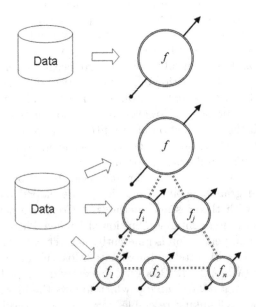

Fig. 2. All-out approach versus decomposition

2.2 Dealing with Ill-posed Problems

As pointed out in [13], inverse problems, concerned of finding a solution f to a functional equation

$$A(f) = d$$

where A can be understood as a model for a phenomenon, $f \in \mathcal{F}$ represents a function of some causal factors of the model, chosen from a class $\mathcal{F}$ of candidate functions, and d denotes some actual observation data pertaining to the phenomenon, are generally ill-posed, which means the solution f might not exist, might not be unique, and most importantly, might not be stable. Namely, with a small deviation δ in the output data d_δ, we have

$$R_\delta(f) = \|A(f) - d_\delta\|$$

not tending to zero even if δ tends to zero, where $\| \cdot \|$ is any divergence metrics appropriate for f, meaning arbitrarily small deviations in data may cause large deviations in solutions.

In particular, fundamental pattern recognition problems such as class probability density function estimation from a wide set of potential densities, or parametric estimation of optimal feature subsets, are ill-posed.

On the other hand, if the model A can be decomposed into a combination of simpler sub-models A_i, e.g. those involving search spaces with lower Vapnik-Chervonenkis (VC) dimensions, or those for which respective stable sub-solutions f_i can be found inexpensively, chances are that we will be able to assemble a solution f from sub-solutions f_i, which will be better than a solution computed in an all-out attempt for the original problem. However, the challenge in this approach is that there is no known automatic method for the computation of effective decompositions of A.

In the hierarchical learning approach, we assume that the decomposition scheme will be provided by an external human expert in an interactive process. Knowledge acquired from human expert will serve as guidance to break the original model A into simpler, more manageable sub-models A_i, organized in a lattice-like hierarchy. They would correspond to subsequent levels of abstractions in the hierarchy of perception and reasoning of the human expert.

2.3 Narrowing the Potential Search Space

As stated in [13], the problem of estimating f from a large set $\mathcal{F}$ of possible candidate solutions is ill-posed. One way to alleviate this problem is to employ the so-called Structural Risk Minimization (SRM) technique. The technique, in short, is based on a theorem on the risk bounds, which essentially states that

$$R(\alpha) \leq R_{emp}(\alpha) + CI(\alpha)$$

which means the risk functional $R(\alpha)$, expressing how far we are from the desired solution for a parameter α from a general parameter set S, is bounded by the

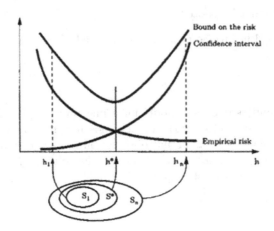

Fig. 3. Actual risk bounds across search spaces (Vapnik, *The Nature of Statistical Learning Theory*, Springer-Verlag, 1999)

sum of the empirical risk $R_{emp}(\alpha)$ and a confidence interval $CI(\alpha)$ containing the Vapnik-Chervonenkiss dimension of the function space S. This dependency is shown on Fig.3.

Instead of optimizing α over an arbitrary set of possible parameters S, we use the bounds to find a set S^* for which the risk bound is minimal, and then perform the search for the solution α^* within S^*. For more details, see [13].

The hierarchical learning approach, by reducing the complexity of the original learning problem by decomposing it into simpler ones, tries to optimize the corresponding search spaces on subsequent levels of the learning hierarchy, and is similar in function to the SRM technique.

2.4   Granular Computing

The hierarchical learning approach takes advantage of additional domain knowledge provided by human experts. In order to best employ this knowledge, it relies on the observation that human thinking and perception in general, and their reasoning while performing classification tasks in particular, can:

- inherently comprise different levels of abstraction,
- display a natural ability to switch focus from one level to another,
- operate on several levels simultaneously.

Such processes are natural subjects for the *Granular Computing* paradigm, which encompasses theories, methods, techniques and tools for such fields as problem solving, information processing, human perception evaluation, analysis of complex systems and many others. It is built around the concept of *information granule*, which can be understood as a collection of *"values that are drawn together by indistinguishability, equivalence, similarity, or proximity"* [14]. Granular Computing follows the human ability to perceive things at different levels of abstraction (*granularity*), to concentrate on a particular level of interest while preserving the ability to instantly switch to another level in case of need. This allows to obtain different levels of knowledge and, which is important, a better understanding of the inherent structure of this knowledge.

The concept of information granule is closely related to the imprecise nature of human reasoning and perception. Granular Computing therefore provides excellent tools and methodologies for problems involving flexible operations on imprecise or approximated concepts expressed in natural language.

2.5   Ontology Matching

The knowledge on training samples that comes from an expert obviously reflects his perception about the samples. The language used to describe this knowledge is a component of the expert's ontology which is an integral part of his perception. In a broad view, an ontology consists of a vocabulary, a set of concepts organized in some kind of structure, and a set of binding relations amongst those concepts [3]. We assume that the expert's ontology when reasoning about complex structured samples will have the form of a multi-layered hierarchy, or a *lattice*, of concepts. A concept on a higher level will be synthesized from its children concepts and their binding relations. The reasoning thus proceeds from the most primitive notions at the lowest levels and work bottom-up towards more complex concepts at higher levels.

Hierarchical learning, together with the transfer of knowledge expressed in natural languages from external experts to low-level computer operators, constitutes an excellent illustration of *Granular Computing* in action.

3   Implementation

We present in brief an implementation of the discussed hierarchical learning approach. For more details, see [7].

We assume an architecture that allows a learning system to consult a human expert for advices on how to analyze a particular sample or a set of samples. Typically this is done in an iterative process, with the system subsequently incorporating knowledge elicited on samples that could not be properly classified in previous attempts.

The ontology matching aims to translate the components of the expert's ontology, such as single concepts and binding relations, expressed in natural language, which may have the form of, e.g.

"A Six is a digit that has a closed belly below a slanted neck."

or, in a more formal fashion:

$[CLASS(u) = \text{`6'}] \equiv a, b$ are parts of u; "Below"(b,a); "SNeck"(a); "CBelly"(b)

into low-level operators such as pixel counts, formulas of pixel coordinates, distances, etc.

3.1   Approximation of Concepts

A foreign concept C is approximated by a domestic pattern (or a set of patterns) p in term of a rough inclusion measure $Match(p, C) \in [0, 1]$. Such measures take root in the theory of rough mereology [9], and are designed to deal with the notion of inclusion to a degree. An example of concept inclusion measures would be:

$$Match(p, C) = \frac{|\{u \in T : Found(p, u) \land Fit(C, u)\}|}{|\{u \in T : Fit(C, u)\}|}$$

where T is a common set of samples used by both the system and the expert to communicate with each other on the nature of expert's concepts, $Found(p, u)$ means a pattern p is present in u and $Fit(C, u)$ means u is regarded by the expert as fit to his concept C.

Our principal goal is, for each expert's explanation, to find sets of patterns Pat, Pat_1,...,Pat_n and a relation $\Im_d$ so as to satisfy the following *quality requirement*:

if $(\forall i : Match(Pat_i, EFeature_i) \geq p_i) \land (Pat = \Im_d(Pat_1, ..., Pat_n))$
then $Quality(Pat) > \alpha$

where $p_i : i \in \{1, .., n\}$ and α are certain cutoff thresholds, while the *Quality* measure, intended to verify if the target pattern Pat fits into the expert's concept of sample class k, can be any, or combination, of popular quality criteria such as *support*, *coverage*, or *confidence* [10].

In other words, we seek to translate the expert's knowledge into the domestic language so that to generalize the expert's reasoning to the largest possible number of training samples. More refined versions of the inclusion measures would involve additional coefficients attached to, e.g., $Found$ and Fit test functions. Adjustment of these coefficients based on feedback from actual data may help optimize the approximation quality.

For example, let us consider a handwritten digit recognition task.

When explaining his perception of a particular digit image sample, the expert may employ concepts such as *'Circle'*, *'Vertical Strokes'* or *'West Open Belly'*. The expert will explain what he means when he says, e.g. *'Circle'*, by providing a decision table (U, d) with reference samples, where d is the expert decision to which degree he considers that *'Circle'* appears in samples $u \in U$. The samples in U may be provided by the expert, or may be picked up by him among samples explicitly submitted by the system, e.g. those that had been misclassified in previous attempts.

The use of rough inclusion measures allows for a very flexible approximation of foreign concept. A stroke at 85 degree to the horizontal in a sample image can still be regarded as a vertical stroke, though obviously not a 'pure' one. Instead of just answering in a *'Yes/No'* fashion, the expert may express his degrees of belief using such natural language terms as *'Strong'*, *'Fair'*, or *'Weak'*.

Table 1. Perceived features

	Circle
u_1	Strong
u_2	Weak
...	...
u_n	Fair

Table 2. Translated features

	DPat	Circle
u_1	252	Strong
u_2	4	Weak
...	...	...
u_n	90	Fair

The expert's feedback will come in the form of a decision table (See Table 1.):

The translation process attempts to find domestic feature(s)/pattern(s) that approximate these degrees of belief (See Tab. 2). Domestic patterns satisfying the defined quality requirement can be quickly found, taking into account that sample tables submitted to experts are usually not very large. Since this is essentially a rather simple supervised learning task that involves feature selection, many strategies can be employed. In [8], genetic algorithms equipped with greedy heuristics are reported successful for a similar problem. Neural networks also prove suitable for effective implementation.

Similarly, we can approximate the expert's perception on relations between parts of a sample (See Tab. 3.). The corresponding low- level feature may be expressed by, for instance, $S_y < B_y$, which tells whether the median center of the stroke is placed closer to the upper edge of the image than the median center of the belly. (See Tab. 4)

Table 3. Perceived relations

	V Stroke	W Belly	Above
u_1	Strong	Strong	Strong
u_2	Fair	Weak	Weak
...	...	...	...
u_n	Fair	Fair	Weak

Table 4. Translated relations

	#V_S	#NES	$S_y < B_y$	Above
u_1	0.8	0.9	(Strong,1.0)	(Strong, 0.9)
u_2	0.9	1.0	(Weak, 0.1)	(Weak, 0.1)
...	...	...	...	...
u_n	0.9	0.6	(Fair, 0.3)	(Weak, 0.2)

The expert's perception "A '6' is something that has a 'vertical stroke' 'above' a 'belly open to the west'" is eventually approximated by a classifier in the form of a rule:

if $S(\#BL\_SL > 23)$ **AND** $B(\#NESW > 12\%)$ **AND** $S_y < B_y$ **then** CL='6',

where S and B are designations of pixel collections, #BL_SL and #NESW are numbers of pixels with appropriate topological features, and $S_y < B_y$ concerns the centers of gravity of the two collections.

We compared the performances gained by a standard learning approach with and without the aid of the domain knowledge. The additional knowledge, passed by a human expert on popular classes as well as some atypical samples allowed to reduce the time needed by the learning phase from 205 minutes to 168 minutes, which means an improvement of about 22 percent without loss in classification quality. In case of screening classifiers, i.e. those that decide a sample *does not* belong to given classes, the improvement is around 40 percent. The representational samples found are also slightly simpler than those computed without using the background knowledge.

4   Conclusion

A conceptual description as well as details on the implementation of the hierarchical learning approach are laid out. Hierarchical learning together with knowledge elicitation techniques, supported by experiment results, show the combined strength of approximate reasoning, granular computing and rough mereology at work. A discussion aimed at positioning the approach among well established methods in related fields is also presented.

Acknowledgment. The author would like to express gratitude to Professor Andrzej Skowron for his insightful comments and his invaluable support during the work on this paper. This work has been supported by the grant N N516 368334 from Ministry of Science and Higher Education of the Republic of Poland and by the grant Innovative Economy Operational Programme 2007-2013 (Priority Axis 1. Research and development of new technologies) managed by Ministry of Regional Development of the Republic of Poland.

References

1. Domingos, P.: Toward knowledge-rich data mining. Data Minning and Knowledge Discovery 15(1), 21–28 (2007)
2. Duda, R.O., Hart, P.E., Stork, D.G.: Pattern Classification. Wiley-Interscience Publication, Chichester (2000)
3. Fensel, D.: Ontologies: A Silver Bullet for Knowledge Management and Electronic Commerce. Springer, New York (2003)
4. Langley, P., Laird, J.E.: Cognitive architectures: Research issues and challenges. Technical Report (2002)
5. Mitchell, T.M.: Machine Learning. McGraw-Hill, New York (1997)
6. Newell, A.: Unified theories of cognition. Harvard University Press, Cambridge (1994)
7. Nguyen, T.T.: Domain knowledge assimilation by learning complex concepts. In: Greco, S., Hata, Y., Hirano, S., Inuiguchi, M., Miyamoto, S., Nguyen, H.S., Słowiński, R. (eds.) RSCTC 2006. LNCS (LNAI), vol. 4259, pp. 617–626. Springer, Heidelberg (2006)
8. Oliveira, L.S., Sabourin, R., Bortolozzi, F., Suen, C.Y.: Feature selection using multi-objective genetic algorithms for handwritten digit recognition. In: International Conference on Pattern Recognition (ICPR 2002), pp. 568–571 (2002)
9. Polkowski, L., Skowron, A.: Rough mereology: A new paradigm for approximate reasoning. Journal of Approximate Reasoning 15(4), 333–365 (1996)
10. Shavlik, J.W., Towell, G.G.: An approach to combining explanation-based and neural learning algorithms. Connection Science 1, 231–254 (1989)

11. Skowron, A.: Rough sets in perception-based computing. In: Pal, S.K., Bandy-opadhyay, S., Biswas, S. (eds.) PReMI 2005. LNCS, vol. 3776, pp. 21–29. Springer, Heidelberg (2005)
12. Skowron, A., Polkowski, L.: Rough mereological foundations for design, analysis, synthesis, and control in distributed systems. Information Sciences 104(1-2), 129–156 (1998)
13. Vapnik, V.N.: The Nature of Statistical Learning Theory (Information Science and Statistics). Springer, Heidelberg (1999)
14. Zadeh, L.A.: From imprecise to granular probabilities. Fuzzy Sets and Systems 154(3), 370–374 (2005)

A Comparison of the LERS Classification System and Rule Management in PRSM

Jerzy W. Grzymala-Busse[1] and Yiyu Yao[2]

[1] Department of Electrical Engineering and Computer Science, University of Kansas,
Lawrence, KS 66045, USA and
Institute of Computer Science Polish Academy of Sciences, 01-237 Warsaw, Poland
jerzy@ku.edu
http://lightning.eecs.ku.edu/index.html
[2] Department of Computer Science, University of Regina,
Regina, Saskatchewan, Canada S4S 0A2
yyao@cs.uregina.ca

Abstract. The LERS classification system and rule management in probabilistic rough set models (PRSM) are compared according to the interpretations of rules, quantitative measures of rules, and rule conflict resolution when applying rules to classify new cases. Based on the notions of positive and boundary regions, probabilistic rules are semantically interpreted as the positive and boundary rules, respectively. Rules are associated with different quantitative measures in LERS and PRSM, reflecting different characteristics of rules. Finally, the rule conflict resolution method used in LERS may be applied to PRSM.

1 Introduction

Rule induction is one of the most important applications of rough set theory [5,6,8,9,11,17]. In the standard rough set model, one typically interprets rules induced from the positive region (i.e., the lower approximation) of a concept (class) as *certain* rules and rules induced from the boundary region (i.e., the difference of upper approximation and lower approximation) as *uncertain* or *plausible* rules. One may associate quantitative measures to rules. For example, the precision of a rule, also called accuracy and confidence, is the conditional probability that a rule correctly indicates the concept given the set of all cases matching the rule. From the point view of precision, the interpretation of certain and uncertain rules is reasonable, as the precision of a certain rule is 1 and precision of a plausible rule is between 0 and 1.

The lack of consideration for the degree of overlap of an equivalence class and a concept had motivated many authors to consider probabilistic rough set models (PRSM). Pawlak, Wong, Ziarko [10] proposed to use 0.5 as a threshold to define probabilistic rough set approximations. Yao and Wong [13,14,15] proposed the decision-theoretic rough set model (DTRSM) in which a pair of threshold parameters for defining probabilistic approximations can be determined based on the

C.-C. Chan et al. (Eds.): RSCTC 2008, LNAI 5306, pp. 202–210, 2008.

well established Bayesian decision theory. That is, the probabilistic approximations defined by the parameters would incur minimal risk in deciding the positive, boundary and negative regions. Based on intuitive arguments, Ziarko [16] proposed variable precision rough set model (VPRSM) for probabilistic approximations. Once probabilistic approximations are introduced, one can similarly derive rules [6,11,12].

There is a semantics difficulty with interpreting probabilistic rules induced from the probabilistic positive region, since they are also uncertain (i.e., precision < 1). From the precision point of view, there is no difference between probabilistic rules induced from probabilistic positive and boundary regions, except for their levels of precision. However, this important problem has not received much attention until recently. A solution to the problem is offered by the decision-theoretic rough set model. Given a class, its positive, boundary and negative regions represent three different types of decisions. For example, consider classifying a set of patients according to a particular disease. A patient in the positive region needs "immediately treatment", a patient in the boundary requires "further investigation", and a patient in the negative region does not require any treatment. With respect to the first two cases, the notions of positive rules and boundary rules have been introduced [13]. They properly reflect the semantics interpretations of rules induced in PRSM.

Another important issue that need to be considered in PRSM is rule conflict resolution when rules are applied to classify new cases. Many studies focus more on rule induction and pay less attention to rule evaluation where rule conflict resolution must be considered. A solution for rule conflict resolution has been explored in LERS [2,3,4,5], where bucket brigade algorithm [1,7] is adopted and modified. In addition, LERS use different quantitative measure to characterize rules.

Based on the above discussion, we present a comparative study of LERS classification system and rule management in PRSM. This comparison enables us to pool together advantages of the two approaches in an attempt to obtain better rule induction algorithms within rough set theory.

2   Rule Induction

First we are going to present LEM2 (Learning from Examples Module, version 2) methodology of rule induction based on attribute-value pair blocks. LEM2 is one of rule induction modules of the LERS (Learning from Examples based on Rough Sets) data mining system.

2.1   Blocks of Attribute-Value Pairs

We assume that the input data sets are presented in the form of a *decision table*. An example of a decision table is shown in Table 1. Rows of the decision table represent *cases*, while columns are labeled by *variables*. The set of all cases will be denoted by U. In Table 1, $U = \{1, 2, ..., 19\}$. Independent variables are called *attributes* and a dependent variable is called a *decision* and is denoted by d. The

Table 1. A complete decision table

Case	Width	Attributes Gauge	Decision Quality
1	wide	heavy	good
2	wide	heavy	good
3	wide	heavy	good
4	wide	medium	good
5	wide	medium	good
6	wide	medium	bad
7	wide	light	good
8	wide	light	good
9	wide	light	bad
10	wide	light	bad
11	narrow	heavy	good
12	narrow	heavy	good
13	narrow	heavy	good
14	narrow	heavy	bad
15	narrow	medium	good
16	narrow	medium	good
17	narrow	medium	bad
18	narrow	light	bad
19	narrow	light	bad

set of all attributes will be denoted by A. In Table 1, $A = \{Width, Gauge\}$. Any decision table defines a function ρ that maps the direct product of U and A into the set of all values. For example, in Table 1, $\rho(1, Width) = wide$. A decision table with an incompletely specified function ρ will be called *incomplete*.

An important tool to analyze complete decision tables is a block of an attribute-value pair. Let a be an attribute, i.e., $a \in A$ and let v be a value of a for some case. For complete decision tables if $t = (a, v)$ is an attribute-value pair then a *block* of t, denoted $[t]$, is a set of all cases from U that for attribute a have value v. Each attribute-value pair represents one piece of knowledge about a decision table or a property of cases. These pieces of knowledge and the corresponding blocks will serve as a basis of rule induction.

For Table 1, we have,

$$[(Width, wide)] = \{1, 2, 3, 4, 5, 6, 7, 8, 9, 10\},$$
$$[(Width, narrow)] = \{11, 12, 13, 14, 15, 16, 17, 18, 19\},$$
$$[(Gauge, heavy)] = \{1, 2, 3, 11, 12, 13, 14\},$$
$$[(Gauge, medium)] = \{4, 5, 6, 15, 16, 17\},$$
$$[(Gauhe, light)] = \{7, 8, 9, 10, 18, 19\}.$$

Moreover, the two important blocks related with the decision *Quality*, called *concepts*, are:

$$[(Quality, good)] = \{1, 2, 3, 4, 5, 7, 8, 11, 12, 13, 15, 16\},$$
$$[(Quality, bad)] = \{6, 9, 10, 14, 17, 18, 19\}.$$

These blocks represent knowledge about the entire decision table. Rule induction is essential to find relationship between the blocks defined by attributes and the blocks defined by a decision.

The notion of blocks can be used to explain the basic concepts of the rough set theory [8,9]. Let B be a nonempty subset of A. The indiscernibility relation $IND(B)$ is a relation on U defined for $x, y \in U$ as follows:

$$(x, y) \in IND(B) \ \ if \ and \ only \ if \ \rho(x, a) = \rho(y, a) \ for \ all \ a \in B.$$

The indiscernibility relation $IND(B)$ is an equivalence relation. Equivalence classes of $IND(B)$ are called *elementary sets* of B and are denoted by $[x]_B$. The indiscernibility relation $IND(B)$ may be computed using the idea of blocks of attribute-value pairs. More specifically, the elementary blocks of $IND(B)$ are intersections of the corresponding blocks of attribute-value pairs, i.e., for any case $x \in U$,

$$[x]_B = \bigcap \{[(a, v)] | a \in B, \rho(x, a) = v\}.$$

In other words, the elementary block containing x is intersection all blocks defined by values of x all attributes in B.

For Table 1, the elementary sets of $IND(A)$ are given by:

$$[1]_A = [(Width, wide)] \cap [(Gauge, heavy)] = \{1, 2, 3\} = [2]_A = [3]_A,$$
$$[4]_A = [(Width, wide)] \cap [(Gauge, medium)] = \{4, 5, 6\} = [5]_A = [6]_A,$$
$$[7]_A = [(Width, wide)] \cap [(Gauge, light)] = \{7, 8, 9, 10\} = [8]_A = [9]_A = [10]_A,$$
$$[11]_A = [(Width, narrow)] \cap [(Gauge, heavy)] = \{11, 12, 13, 14\} =$$
$$[12]_A = [13]_A = [14]_A,$$
$$[15]_A = [(Width, narrow)] \cap [(Gauge, medium)] = \{15, 16, 17\} = [16]_A = [17]_A,$$
$$[18]_A = [(Width, narrow)] \cap [(Gauge, light)] = \{18, 19\} = [19]_A.$$

It follows that the elementary blocks of $IND(A)$ are $\{1, 2, 3\}$, $\{4, 5, 6\}$, $\{7, 8, 9, 10\}$, $\{11, 12, 13, 14\}$, $\{15, 16, 17\}$ and $\{18, 19\}$.

2.2   Rules in LERS

Based on the elementary blocks of the equivalence relation induced by a subset B of the attribute set A, one can define a pair of lower and upper approximations for each concept $D_i \subseteq U$. That is,

$$\underline{apr}_B(D_i) = \bigcup \{[x]_B \mid [x]_B \subseteq D_i\}$$

$$= \bigcup \{[x]_B \mid P(D_i \mid [x]_B) = 1\};$$

$$\overline{apr}_B(D_i) = \bigcup \{[x]_B \mid [x]_B \cap D_i \neq \emptyset\}$$

$$= \bigcup \{[x]_B \mid P(D_i \mid [x]_B) > 0\},$$

where $P(D_i \mid [x]_D) = |D_i \cap [x]_B|/|[x]_B|$ is the conditional probability and $|\cdot|$ is the cardinality of a set.

Thus, the lower approximations of the concepts from Table 1 are:

$$\underline{apr}_A([(Quality, good)]) = \{1, 2, 3\},$$
$$\underline{apr}_A([(Quality, bad)]) = \{18, 19\},$$

And the upper approximations of the concepts from Table 1 are:

$$\overline{apr}_A([(Quality, good)]) = \{1, 2, ..., 17\},$$
$$\overline{apr}_A([(Quality, bad)]) = \{4, 5, ..., 19\}.$$

The LERS data mining system computes lower and upper approximations for every concept and then induces rules using one of the selected modules. Rules induced from lower and upper approximations are called *certain* and *possible*, respectively [2].

The LEM2 algorithm search for rules by using a family of blocks such that their intersection is either a subset of the concept or has an overlap with the concept [3]. In the LERS format, every rule is associated with three numbers: the total number of attribute-value pairs on the left-hand side of the rule, the total number of cases correctly classified by the rule during training, and the total number of training cases matching the left-hand side of the rule, i.e., the rule domain size.

For Table 1, the LEM2 module of LERS induces the following rule sets:
the certain rule set:

2, 3, 3
(Gauge, heavy) & (Width, wide) -> (Quality, good),
2, 2, 2
(Gauge, light) & (Width, narrow) -> (Quality, bad),

and the following possible rule set:

1, 7, 10
(Width, wide) -> (Quality, good),
1, 6, 7
(Gauge, heavy) -> (Quality, good),
1, 4, 6
(Gauge, medium) -> (Quality, good),
1, 4, 9
(Width, narrow) -> (Quality, bad),
1, 4, 6
(Gauge, light) -> (Quality, bad),
1, 2, 6
(Gauge, medium) -> (Quality, bad).

2.3   Rules in PRSM

The set $\text{POS}_B(D_i) = \underline{apr}_B(D_i)$ is called the positive region of D_i, and the set $\text{BND}_B(D_i) = \overline{apr}_B(D_i) - \underline{apr}_B(D_i)$ is called the boundary region of D_i. According to the two regions, one can form two types of rules called *positive* and *boundary* rules, respectively [13].

If an elementary block is in the positive region of a decision class, one obtains a positive rule; if the elementary block is in the boundary region, one obtains one or several boundary rules. In particular, in the VPRSM format, each rule is associated with two numbers: the conditional probability and marginal probability [6].

For Table 1, we have the following positive rules:

(P1). $(Width, wide)$ & $(Gauge, heavy) \longrightarrow (Quality, good)$,   1.00, 0.158,
(P2). $(Width, narrow)$ & $(Gauge, light) \longrightarrow (Quality, bad)$,   1.00, 0.158,

and the boundary rules:

(B1). $(Width, wide)$ & $(Gauge, medium) \longrightarrow (Quality, good)$,   0.67, 0.158,
(B2). $(Width, wide)$ & $(Gauge, medium) \longrightarrow (Quality, bad)$,   0.33, 0.158,
(B3). $(Width, wide)$ & $(Gauge, light) \longrightarrow (Quality, good)$,   0.50, 0.211,
(B4). $(Width, wide)$ & $(Gauge, light) \longrightarrow (Quality, bad)$,   0.50, 0.211,
(B5). $(Width, narrow)$ & $(Gauge, heavy) \longrightarrow (Quality, good)$,   0.25, 0.211,
(B6). $(Width, narrow)$ & $(Gauge, heavy) \longrightarrow (Quality, bad)$,   0.75, 0.211,
(B7). $(Width, narrow)$ & $(Gauge, medium) \longrightarrow (Quality, good)$,   0.67, 0.158,
(B8). $(Width, narrow)$ & $(Gauge, medium) \longrightarrow (Quality, bad)$, 0.33, 0.158,

The two types of rules lead to two types of different decision. A positive rule suggests a definite and positive decision regarding the class of a case, and a boundary rule suggests a tentative and boundary decision regarding the class of a case. Semantically, these two classes are different [13].

In probabilistic approaches to rough sets, such as decision-theoretic model [13,14] and variable precision model [16], we have the parameterized approximations:

$$\underline{apr}_B(D_i) = \bigcup\{[x]_B \mid P(D_i \mid [x]_B) \geq \alpha\},$$
$$\overline{apr}_B(D_i) = \bigcup\{[x]_B \mid P(D_i \mid [x]_B) > \beta\},$$

with $\alpha > \beta$. They are referred to as the α-level lower approximation and β-level upper approximation. Similarly, the α-level positive region and the (α, β)-level boundary region can be introduced. Again, we have two types of rules corresponding the the two region.

Suppose $\alpha = 0.75$ and $\beta = 0.50$. For Table 1, rule (B6) becomes a 0.75-level positive rule, and only rules (B1) and (B7) remain to be $(0.75, 0.50)$-level boundary rules. On the other hand, for comparison, the previous rule sets, used in the VPRSM methodology, presented in the LERS format, are:

2, 3, 3
(Width, wide) & (Gauge, heavy) -> (Quality, good),
2, 2, 3
(Width, wide) & (Gauge, medium) -> (Quality, good),
2, 1, 3
(Width, wide) & (Gauge, medium) -> (Quality, bad),
2, 2, 4
(Width, wide) & (Gauge, light) -> (Quality, good),
2, 2, 4
(Width, wide) & (Gauge, light) -> (Quality, bad),
2, 3, 4
(Width, narrow) & (Gauge, heavy) -> (Quality, good),
2, 1, 4
(Width, narrow) & (Gauge, heavy) -> (Quality, bad),
2, 2, 3
(Width, narrow) & (Gauge, medium) -> (Quality, good),
2, 1, 3
(Width, narrow) & (Gauge, medium) -> (Quality, bad) and
2, 2, 2
(Width, narrow) & (Gauge, light) -> (Quality, bad).

With the additional information: $|U| = 19$, rules with the LERS format may be easily converted into VPRSM format, the converse is not true. The conditional probability is a ratio of the second LERS number to the third LERS number, the marginal probability is the ratio of the third LERS number to the cardinality of the universe. By the way, the cardinality of the universe is the same for all rules so it does not need to be recorded for a specific rule.

3   Rule Conflict Resolution

The classification system of LERS is a modification of the *bucket brigade algorithm* [1,7]. The decision to which concept a case belongs is made on the basis of three factors: specificity_factor, strength_factor, and support. They are defined as follows: *specificity_factor* is either the *specificity*, i.e., the total number of attribute-value pairs on the left-hand side of the rule or may be selected by the user to be equal to one. *Strength_factor* is either the *strength*, i.e., total number of cases correctly classified by the rule during training or *rough measure*, i.e., the ratio of the strength to the total number of training cases matching the left-hand side of the rule. For completely specified data sets the rough measure is identical with the conditional probability of the concept given the rule domain. The third factor, *support*, is defined as the sum of scores of all matching rules from the concept, where the score of the rule is the product of its strength_factor and specificity_factor. The concept C for which the support, i.e., the following expression

$$\sum_{matching\ rules\ R\ describing\ C} Strength\_factor(R) * Specificity\_factor(R)$$

is the largest is the winner and the case is classified as being a member of C. Note that the user may exclude support, i.e., the case might be classified only on the basis of its scores associated with rules.

In the classification system of LERS, if complete matching is impossible, all partially matching rules are identified. These are rules with at least one attribute-value pair matching the corresponding attribute-value pair of a case. For any partially matching rule R, the additional factor, called *matching _factor* is computed. Matching_factor (R) is defined as the ratio of the number of matched attribute-value pairs of R with a case to the total number of attribute-value pairs of R. Again, the user may choose the matching_factor to be equal to one. In partial matching, the concept C for which the following expression is the largest

$$\sum_{\substack{partially\ matching \\ rules\ R\ describing\ C}} Matching\_factor(R) * Strength\_factor(R)$$
$$* Specificity\_factor(R)$$

is the winner and the case is classified as being a member of C.

In general the LERS classification system uses four binary parameters: specificity_factor (either equal to specificity or switched to integer one), strength_factor (either the total number of well-classified training cases or the rough measure), support (either product of scores for each matching rule or each rule participates on its own), and finally matching_factor (either as defined or equal to integer one). Thus the user of the LERS classification system may apply one of 16 different strategies [5]. In the VPRSM methodology, classification is based on conditional probability, one of 16 LERS strategies (in [5] this strategy, based only on the conditional probability, is the strategy # 15). Note that the choice of the classification strategy is crucial and that the best strategy is based on specificity $= 1$, strength, support, and matching_factor [5].

4   Conclusions

The LERS system induces rules based on attribute-value pairs. Since LERS keep three important quantities of rules, namely, the total number of attributes on the left-hand side of the rule, the total number of cases correctly classified by the rule, and the total number of cases matching the left-hand side of the rule, LERS can be easily applied to discover probabilistic rules. Based on two decision-theoretic rough set model, two types of rules, known as positive rules and boundary rules, can be introduced. LERS system can easily learn the two types of rules. In addition, the rule conflict resolution strategy of LERS can be applied to rule applications and evaluation in probabilistic rough set models.

References

1. Booker, L.B., Goldberg, D.E., Holland, J.F.: Classifier systems and genetic algorithms. In: Carbonell, J.G. (ed.) Machine Learning. Paradigms and Methods, pp. 235–282. The MIT Press, Menlo Park (1990)
2. Grzymala-Busse, J.W.: Knowledge acquisition under uncertainty—A rough set approach. Journal of Intelligent & Robotic Systems 1, 3–16 (1988)

3. Grzymala-Busse, J.W.: A new version of the rule induction system LERS. Fundamenta Informaticae 31, 27–39 (1997)
4. Grzymala-Busse, J.W.: MLEM2: A new algorithm for rule induction from imperfect data. In: Proceedings of the 9th International Conference on Information Processing and Management of Uncertainty in Knowledge-Based Systems (IPMU 2002), pp. 243–250 (2002)
5. Grzymala-Busse, J.W., Wang, C.P.B.: Classification methods in rule induction. In: Proceedings of the Fifth Intelligent Information Systems Workshop, pp. 120–126 (1996)
6. Grzymala-Busse, J.W., Ziarko, W.: Data mining based on rough sets. In: Wang, J. (ed.) Data Mining: Opportunities and Challenges, pp. 142–173. Idea Group Publ. (2003)
7. Holland, J.H., Holyoak, K.J., Nisbett, R.E.: Induction. In: Processes of Inference, Learning, and Discovery. The MIT Press, Menlo Park (1986)
8. Pawlak, Z.: Rough Sets. International Journal of Computer and Information Sciences 11, 341–356 (1982)
9. Pawlak, Z.: Rough Sets. Theoretical Aspects of Reasoning about Data. Kluwer Academic Publishers, Dordrecht (1991)
10. Pawlak, Z., Wong, S.K.M., Ziarko, W.: Rough sets: probabilistic versus deterministic approach. International Journal of Man-Machine Studies 29, 81–95 (1988)
11. Stefanowski, J.: Algorithms of Decision Rule Induction in Data Mining. Poznan University of Technology Press, Poznan (2001)
12. Tsumoto, S., Tanaka, H.: PRIMEROSE: Probabilistic Rule Induction Method based on Rough Sets and Resampling Methods. Computational Intelligence 11, 389–405 (1995)
13. Yao, Y.Y.: Decision-theoretic rough set models. In: Yao, J., Lingras, P., Wu, W.-Z., Szczuka, M.S., Cercone, N.J., Ślęzak, D. (eds.) RSKT 2007. LNCS (LNAI), vol. 4481, pp. 1–12. Springer, Heidelberg (2007)
14. Yao, Y.Y., Wong, S.K.M.: A decision theoretic framework for approximating concepts. International Journal of Man-machine Studies 37, 793–809 (1992)
15. Yao, Y.Y., Wong, S.K.M., Lingras, P.: A decision-theoretic rough set model. In: Proceedings of the 5th Symposium on Methodologies for Intelligent Systems, pp. 17–24 (1990)
16. Ziarko, W.: Variable precision rough sets model. Journal of Computer and Systems Sciences 46, 39–59 (1993)
17. Ziarko, W.: Optimal decision making with data-acquired decision tables. In: Proceedings of the Intelligent Information Systems Symposium, pp. 75–85 (2000)

Similarity Relation in Classification Problems*

Andrzej Janusz

Warsaw University, Faculty of Mathematics, Informatics and Mechanics,
ul. Banacha 2, 02-097 Warszawa, Poland
janusza@mimuw.edu.pl

Abstract. This paper presents a methodology of constructing robust classifiers based on a concept called a Hierarchic Similarity Model (HSM). The hierarchic similarity is interpreted as a relation between pairs of complex objects. This relation can be derived from an information system by examining the domain related aspects of similarity. In the paper, global similarity is decomposed into many local similarities by analogy with the process of perceiving similar objects. For the purpose of estimating local relations some well-known rough sets methods are used, as well as context knowledge provided by a domain expert. Then the rules modeling interactions between local similarities are constructed and used to assess the degree of a global similarity of complex objects. The obtained relation can be used to construct classifiers which may successfully compete with other popular methods like boosted decision trees or k-NN algorithm. An implementation of the proposed models in the R script language is provided together with an empirical evaluation of the similarity based classification accuracy for some common datasets. This paper is a continuation of the research started in [1].

1 Introduction

The notion of similarity has been in the scope of interest of researchers for many years ([2], [3], [4]). Knowing how to discriminate similar cases or objects from those which are dissimilar in a context of a decision class would enable us to conduct an accurate classification and to detect unusual situations or behaviors. Although human mind is capable of learning this relation from examples, mathematicians, computer scientists, philosophers and psychologist have not come up with a single methodology of building similarity models appropriate for a wide range of complex object classes or domains.

A variety of methods were used in order to construct such models and define a relation which would combine an intuitive structure with good predictive

---

* The author would like to thank professor Andrzej Skowron for the inspiration and the useful remarks and also Aleksandra Janusz-Ochab and Marcin Szczuka for their support in writing and editing this paper. This research was supported by the grant N N516 368334 from Ministry of Science and Higher Education of the Republic of Poland and by the Innovative Economy Operational Programme 2007-2013 (Priority Axis 1. Research and development of new technologies).

C.-C. Chan et al. (Eds.): RSCTC 2008, LNAI 5306, pp. 211–222, 2008.

power. Among those a huge share was based on some distance measures. In that approach, objects are treated as points in a metric space of their attributes and the similarity is a decreasing function of the distance between them. Objects are regarded as similar if they are close enough in this space. Such models may be generalized by introducing a list of parameters to the similarity function, e.g. weights of attributes. Tuning them results in the relation better fitting to a dataset. Algorithms for computationally efficient optimization of parameters for common similarity measures in the context of information systems were described in [5].

One may argue that the relation of this kind is very intuitive because objects which have many similar values of attributes are likely to be similar. However, researchers like Amos Twersky ([2], [6]) proved empirically that in some contexts, similarity does not necessarily have features like symmetry or subadditivity implied by distance measures. This situation occurs particularly often when we compare objects of great complexity. The explanation for this may lie in the fact that complex objects can be similar in some aspects and dissimilar in others. A dependency between local and global similarities may be highly nonlinear and in order to model it we need to learn this dependency from the data relying on the domain knowledge provided by an expert.

Attempts to construct such models of a similarity have been made by researchers such as Andrzej Skowron, Hung Son Nguyen or Jan Bazan ([7], [8]). In their models, aspects of local similarity were extracted from *a similarity ontology* provided by a domain expert and *the Case-Based Reasoning* approach was used in order to find the most similar object. In this paper a slightly different approach of modeling a similarity, called *the Hierarchic Similarity Model*, is presented. It aims at encapsulating natural features of similarity argued by Amos Twersky and the ability of learning dependencies between local and global similarities from the data.

In the following section some necessary formal definitions are introduced and then, in Section 3 the proposed methodology of constructing similarity models is described. Section 4 describes experiments conducted on three well-known datasets and compares the similarity based classification accuracy with other common classification methods. Finally, the last section presents some conclusions and plans for future work.

2   Preliminaries

Construction of the Hierarchic Similarity Model (HSM) involves working on imprecise concepts described within *information systems* and as such may be well-handled in a framework provided by the rough set theory proposed by Zdzisław Pawlak in 1982 [9].

2.1   Basic Notation

In the rough set theory, an information system $I = (U, A)$ may be seen as a tabular representation of knowledge about a considered universe. Every row of the information system corresponds to a single object and is called *an instance*. The set of all instances from the information system is marked as U. Every column of I corresponds to *an attribute*. The set of all attributes is labeled by A.

A *decision table* $T = (U, A, d)$ is an information system with one distinguished attribute d called *a decision*.

Similarity may be seen as a relation defined over $U \times U$. Its features vary and depend on a domain of instances from an information system. This relation itself may very often have a subjective nature and as such, may be impossible to model directly from the data. For that reason, similarity needs to be considered in a specific context. We can formulate the following definition:

Definition 1. *Let $T = (U, A, d)$ be a decision table and τ denote a similarity relation over the set $U \times U$. We will say that τ is a similarity relation in the context of the decision d if*

$$\forall_{u_1, u_2 \in U} \ (u_1, u_2) \in \tau \Rightarrow d(u_1) = d(u_2)$$

In other words, τ is a similarity relation in the context of the decision attribute d if it is consistent with an equivalence relation determined by the decomposition of U into decision classes.

2.2   Similarity Based Classification

Having defined the concept of similarity τ in the context of a decision attribute we can construct a similarity function φ_τ which describes a degree of likeness between every pair of instances $(u_1, u_2) \in U \times U$. Such a function exists as we can always take:

$$\varphi_\tau(u_1, u_2) = \begin{cases} 1 & \text{for } (u_1, u_2) \in \tau \\ 0 & \text{otherwise} \end{cases}$$

We will say that (u_1, u_2) is in *relation of the highest similarity* $\hat{\tau}$ if the inequality $\varphi_\tau(u_1, u_2) \geq \varphi_\tau(u_1, \acute{u}_2)$ holds for all $\acute{u}_2 \in U$. Let us assume that we have a similarity function φ_τ at our disposal and u is an instance with an unknown decision value $d(u)$. Now we can construct a simple classification rule:

$$\left((u, \acute{u}) \in \hat{\tau} \wedge d(\acute{u}) = d_i \right) \Rightarrow d(u) = d_i \tag{1}$$

As we can see, the problem of classifying instances with regard to similarity in the context of a decision can be reduced to estimating the function φ_τ.

However, it is possible to define a similarity based classification rule without using concepts of the similarity function and the relation of the highest similarity. Such a rule may be derived from an intuition that two instances are more likely to be from the same decision class when there are many instances which were recognized as similar to the first instance and which had a decision value equal to the decision of the second instance from the pair. As an example of such a type of classification rule for the pair $(u, \acute{u})$ one may give:

$$\left(card(\{w \in U : d(w) = d(\acute{u}) \wedge (u, w) \in \tau\}) \text{ is maximal} \right) \Rightarrow d(u) = d(\acute{u}) \tag{2}$$

where $card(X)$ is a cardinality of the set X.

This approach however, has a drawback. When the examined instance does not have any similar instances within the decision table or has only a few, the prediction based on such a rule may be unreliable. In that case, such instances should be left unclassified or prediction should be made based on the decision values of the not-dissimilar instances. In practice, some instance-weighting techniques may also be used.

In the conducted experiments both types of presented classification rules were used. The comparison of the obtained results can be found in Section 4.

2.3   Domain Knowledge Representation

Studying human intuitive perception of similarity one may notice that people, when explaining why they consider two complex objects to be alike, frequently use rough concepts. For example, asked why we think that the presented cars are similar, we might answer that both have *a similar size* or *comparable driving parameters*. Both, size and driving parameters are different contexts for the similarity of cars. If we are interested in classifying cars regarding their type, we may want to examine those concepts and learn how they affect similarity in the context of the type of cars.

One of convenient ways of representing important concepts and relations between them is a domain ontology. In the similarity setting this ontology can be formed in a tree-like structure, with a global similarity in the context of the decision attribute placed in the root node and likeness in single non-decision attributes at leaves. We will call this type of a domain knowledge representation a *similarity ontology* (see also [10], pages 721-723). The difference between this structure and a classic tree is that in the similarity ontology a child node may have many parent nodes at different levels of the hierarchy. Branches of this structure correspond to the relation of children *having impact* on the parent. Every level of the similarity ontology may be interpreted as a different abstraction level of the considered similarity and every node may be treated as a different aspect of similarity in the context of the decision attribute. Those aspects are often called *local similarities* (compare to local relations in [10]). Concepts which are lower in the hierarchy are less complex than those above them and as such are easier to learn from the data. Figure 1 presents an exemplary similarity ontology of cars.

Domain knowledge may also be used to assess local similarities between pairs of instances. Experts may use it to define a priori local similarity functions or to reinforce learning of these functions by labeling some part of available data. This process will be detailed in the next section.

3   Hierarchic Similarity Model

The Hierarchic Similarity Model is a methodology of constructing similarities in a context of a decision attribute. Its motivation derives from works of philosophers such as Edmund Husserl [11] or Alfred Schütz [12] and psychologists such as Amos Tversky [6]. By using domain knowledge represented in a form of similarity

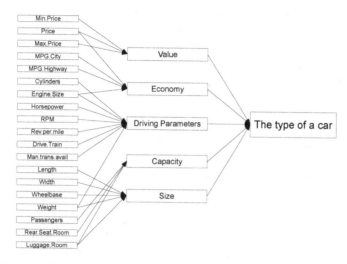

Fig. 1. A simple similarity ontology of cars in the context of their type

ontology it is possible to incorporate desired features to the created model and to avoid those unwanted. This flexibility allows to employ the HSM in solving the classification problem in a wide range of domains.

3.1   Assessing Local Similarities

The first step in the construction of the model is to decide which aspects of similarity are important in the context of the decision attribute and to create a similarity ontology. In the case when our domain knowledge is not sufficient for this purpose, an expert's supervision is needed.

After selecting concepts, assessment of local similarities is conducted. Starting from leaves and following the '*have impact*' relation coded in branches, we assign a degree of the local similarity to every pair of instances. This approach differs from the standard Case-Based Reasoning approach ([13], [8]) where single instances are checked for belonging to each of concepts from the domain ontology and the similarity assessment is done only for the decision concept in the root.

There are a few strategies to carry out the assessment of local similarities process and the choice should depend on the availability of a domain knowledge. Three of them are listed below:

1. Experts label a part of pairs and assess their degree of local similarities. This subset of data is used to build classifiers for each concept in the similarity ontology with the exception of the global similarity at the root. Unlabeled pairs need to be classified in consistency with the decision attribute. To achieve this one may use the decisions as regular attributes. In this case, new instances with an unknown decision should be treated as if their decision class was a missing value. This approach is analogical to the method used for learning hierarchic concepts in [7].

2. Experts help in defining *a priori* local similarity functions $\varphi_\tau|_{\alpha_i,d_j}$, where α_i corresponds to the i-th node of the similarity ontology and $d_j = d(u_2)$ is the decision attribute value of the instance u_2. This function measures the degree of similarity to instances from the decision class d_i in a context of the concept α_i. Function $\varphi_\tau|_{\alpha_i,d_j}$ takes as arguments values of only those attributes which are children of the concept α_i. Constructing separate functions within every decision class makes local similarities more consistent with the decision attribute and is a simple way of avoiding an unwanted symmetry among the instances with different decision values. To compute the value of any local similarity function only the information about the decision of the second instance from each pair is needed. It allows to assess directly the local similarity between the new instance and the training examples from the decision table.

3. Local similarity functions are constructed through a semi-supervised grouping method like *constraint-based clustering* [14]. First, for each local similarity concept from the ontology, pairs of instances from the decision table are clustered. Grouping is done separately for the instances from different decision classes to avoid transitivity of the local similarity relations. Then, the local similarity degree between two instances may be defined as a function of some cluster distance measure or experts may define it after examination of cluster representatives. In this case, new instances have to be assigned to a specific cluster before the computation of their local similarity degree. There are many ways of finding that cluster, e.g. the nearest cluster according to some distance measure may be chosen or the one for which adding the considered instance would have a minimal negative impact at the *silhouette information* value.

Constructing a model of similarity, one may combine any of those strategies in order to grant desired features to the global relation.

3.2 Constructing the Global Model

When local similarities are computed, the construction of the global model may begin. The idea is to select all children of the global similarity concept placed at the root and to create the second decision table $S = (U \times U, A', d')$, called a *similarity table*. As instances of this table we take all pairs of objects from the initial decision table T. A set $A' = \{\alpha_i : \text{the } i\text{-th node is a child of the root}\}$ consists of concepts from the nodes of the the children of the root, whereas the decision d' tells whether the both instances are from the same decision class.

The initial global similarity model $\Delta = (\Theta, \Theta')$ has a form of a pair of certain[1] decision rule sets:

$$\Theta = \left\{ \left(\bigwedge_{i=1}^{p} \alpha_i(u_1, u_2) = y_i \right) \Rightarrow \left(d'(u_1, u_2) = TRUE \right) : (u_1, u_2) \in U^2, \alpha_i \in A' \right\}$$

---

[1] A rule is considered to be certain if its confidence factor is equal to 1.

called *a positive rule set* and

$$\Theta'=\left\{\left(\bigwedge_{j=1}^{p}\alpha_j(u_1,u_2)=y_j\right)\Rightarrow\left(d'(u_1,u_2)=FALSE\right):(u_1,u_2)\in U^2,\alpha_i\in A'\right\}$$

called *a negative rule set*. Instances u_1, u_2 are considered to be similar if they match at least one rule from the set Θ and do not match any rule from the set Θ'. The similarity function $\varphi_\tau\big(\alpha_1(u_1,u_2),\ldots,\alpha_p(u_1,u_2),d(u_2)\big)$ should be defined as a non-decreasing function of cardinality and quality of the set $\{\theta\in\Theta:\theta$ match $(u_1,u_2)\}$, as well as non-increasing function of cardinality and quality of the set $\{\theta'\in\Theta':\theta'$ match $(u_1,u_2)\}$. As an example of such a function one may give

$$\varphi_\tau(u_1,u_2)\quad=\quad\sum_{\theta\text{ match }(u_1,u_2)}Supp(\theta)\quad-\quad\sum_{\theta'\text{match }(u_1,u_2)}Supp(\theta')\qquad(3)$$

where $Supp(\theta)$ is a *support factor* of a rule θ. In the case the pair (u_1,u_2) does not fit any rule of the model, the similarity function value may depend on the maximal number of true propositions of rules from rule sets of the model.

Those conditions result from the intuitive way of perceiving the similarity. Researchers seem to agree that people's similarity judgment depends on common and differing features of objects and is biased by the class of the objects with which they compare the examined object ([3], [4], [15]).

3.3 Examining Dependencies within the Model

Having constructed the initial global model Δ it is necessary to learn the higher abstraction level dependencies between local similarities. Rules from sets Θ and Θ' describe how individual aspects of similarity contribute to the global model. Although both sets consist of certain rules, during the evaluation of resemblance of a new, yet unseen instance to the known instance from the table T, one may find many matching but contradicting rules. This may occur because some of the local similarities can turn out to be irrelevant or their estimation is improper. Another reason for that may be the unusual nature of the examined instance.

In order to diagnose the relevance of local similarities, one needs to construct *partial models* Δ_i defined as a pair of sets (Θ_i,Θ'_i), where Θ_i and Θ'_i are correspondingly positive and negative rule sets, created without the use of information about local similarity $\alpha_i\in A'$. Depending on the algorithm used for rule selection, partial models should be computed independently of the initial global model or should become subsets of this model, created by subtracting all rules which have the similarity α_i in propositions.

The initial global model Δ and partial models Δ_i can be used to classify instances from the validation set V using one of the classification rules described in the Subsection 2.2. The decision values of those instances are known but they were not used during the learning process. Knowledge about the predicted and actual classes of instances from V can be used to identify situations in which the model Δ is likely to be unreliable. Intuitively, if the initial model classifies the instance correctly and the partial model Δ_i is incorrect, then there is a big chance that the local similarity α_i is relevant in the context of the decision class of the examined instance. Alternatively, if the model Δ classifies the instance incorrectly, whereas Δ_i predicts the true decision class of the instance, the similarity aspect α_i may not be important for the global relation. Dependencies like these may be seen as meta-rules telling us which combinations of predictions made by the partial models should be trusted and as such can be extracted using rule mining algorithms such as the *Apriori* algorithm [16]. For that purpose, a new decision table $W = (V, A'', d)$ has to be created, where V is the validation set and the set of attributes A'' consists of predicted decision values of instances $v \in V$, made by the initial model Δ and the successive partial models Δ_i. Decision attribute d is the information about the actual decision class of the instances. The final hierarchic model Ω consists of the initial global similarity, all partial models and the set of *validation rules* Φ consisting of the decision rules extracted from the table W.

Conducted experiments show that the set Φ improves the estimation of similarity relation in the context of decision class by elimination of unnecessary or false rules from the initial model Δ. As shown in Section 4, using validation rules not only improves classification accuracy, but it also makes it possible to identify unusual instances. This feature makes applications of the HSM very useful in domains like medicine or finance, where the identification of the patients who need special treatment or unusual market behavior may be crucial.

3.4   Classifying New Instances

One of the methods of measuring the quality of a similarity model is the estimation of the classification accuracy. The decision class of instances from a test set should be predicted using the decision rule described in the Subsection 2.2.

The assessment of local similarities should be conducted in consistency with the strategy chosen during the model construction. Pairs of instances with computed local similarity degrees should be tried to match the rules from the initial global model Δ and partial models Δ_i. If all models agree on the decision value of the test instance, the instance chosen by the initial model Δ should be taken as the most similar in the context of the decision class. In case of a disagreement between models from Ω, a new instance should be created analogically to the instances from the validation set W and validation rules from the set Φ should be applied. The most similar instance is chosen from the decision class pointed by the validation rules, according to the similarity degree computed using the initial global model.

Additionally, the test instance can be identified as unusual or unique if the degree of similarity to the most similar instance from the decision table T does not exceed a certain level or there are many contrary validation rules matching the instance.

4   Experimental Evaluation

Some experiments were carried out in order to assess the similarity based classification accuracy. Models were built for three datasets of different size and type. Two of those sets (*Pendigits* and *Nursery*) were taken from the UCI Machine Learning Repository[2]. The first one consists of 10992 instances from 10 decision classes, each of instances is described by 16 numeric attributes. The second dataset has 12960 instances from 5 decision classes, 8 nominal attributes. Because there were only 2 instances from the fifth decision class ('recommend'), they were removed from the set. The third dataset (*Cars93*) is a part of the *MASS* standard R library. It consists of 93 instances with a mixture of numeric and nominal attributes from 6 decision classes. As an environment of the experiment the *R system* was chosen.

The similarity ontology for datasets from the UCI repository was derived directly from the data description available with the datasets. For the set *Cars93* a domain expert was asked to build a hierarchy of concepts shown in Figure 1. Simple distance based similarity functions were used to assess local similarity degrees. The '0 − 1' distance measure was deployed to handle nominal attributes. For the *Cars93* dataset the Manhattan and the weighted Manhattan measures were tried. Weights were defined separately for each of decision classes by the domain expert. The local similarity degrees were discretized using the maximum discernibility method described in [17]. The rules for the initial global models and partial models were induced using the Apriori algorithm implemented in the *arules* library. For each dataset, the minimal support factor of induced rules was set to 10 instances and the confidence factor was set to 1.0 in order to produce only certain rules. The Apriori algorithm was also used for the computation of validation rules from the set Φ, but in this case the confidence factor was lowered to 0.80. The similarity function was defined as in (3). Both decision rules proposed in Subsection 2.2 were tried.

Pendigits dataset was provided with separate train and test sets. *Nursery* dataset was randomly divided into two subsets. 4319 instances (1/3 of the set) were used as a test set and the rest (8639 instances) served as a training set. Models were built for 5 independent splits and results were averaged. For the *Cars93* table, the classification accuracy was estimated using the *leave-one-out* cross-validation test. The HSM classification results were compared with the C4.5 decision tree, the k-NN algorithm and the AdaBoost[3] boosting method implemented in the *Weka* system. Default parameter values were used. The obtained results are shown in Table 1. As we can see, classification score of the HSM was among the best results for the examined datasets. Worth mentioning is the fact that the similarity model proved to be a robust tool for classification regardless of the data type and the structure. None of the tried decision rules turned out to be significantly better, although the models which were using the first rule achieved higher accuracy.

---

[2] http://archive.ics.uci.edu/ml/

[3] The C4.5 decision tree was used as the base classifier in boosting.

Table 1. Comparison of the classification accuracy. Θ denotes the initial model and Ω denotes the final HSM. The percentage accuracy is given.

Method\Dataset:	Cars93	Nursery	Pendigits
C4.5	84.95	95.97	92.05
k-NN	63.44	78.35	97.26
AdaBoost+C4.5	90.32	**99.49**	97.28
HSM+rule1 Θ	87.10	98.86	96.77
HSM+rule1 Ω	**91.40**	99.14	**97.63**
HSM+rule2 Θ	90.32	98.21	95.22
HSM+rule2 Ω	90.32	99.01	97.14

A big advantage of the HSM over the rest of classification methods was the ability of recognizing unusual instances. The instance from a test set was marked as *unusual* if it was labeled with different decision classes by the initial model and non-trivial[4] partial models. Otherwise, the instance was marked as *usual*. This feature helps in better understanding of the concept of similarity in the given domain and may increase the classification accuracy of the model as some dedicated classifiers can be used for instances identified as unusual. Experimental results seem to support this thesis, e.g. for the *Pendigits* dataset, 232 test instances (about 6.70%) was marked as unusual using (1) as a decision rule. A prediction of the decision class among instances marked as usual was made with 99.55% accuracy, thus using a dedicated classifier to unusual instances would almost certainly have a positive impact on the overall performance of the model.

Using the information about the similarity of pairs of instances increases the complexity of the Hierarchic Similarity Model and makes the construction time and the memory requirements much higher than in the case of other methods. The construction time of the HSM was an order of magnitude higher than the construction time of C4.5 decision tree. This drawback makes application of the HSM for large datasets very awkward and enforces usage of data sampling techniques. Fortunately, this problem can be partially overcome by the use of the parallel computing as the initial model and all the partial models can be constructed independently.

5   Conclusions

In the paper, the problem of learning a similarity relation from data for the classification purpose was brought up. A methodology of constructing classifiers which are based on such a relation, called the Hierarchic Similarity Model, has been proposed. The model aims at incorporating the natural features of similarity in a specific context of a decision attribute. In the model, a similarity between pairs of instances is examined at different levels of the hierarchy derived from the similarity ontology provided by a domain expert. Then, the initial model and the

---

[4] A model is regarded as non-trivial if it consists of at least one positive rule and is not completely biased by one decision class.

partial models consisting of the positive and the negative rule sets are induced. Finally, the dependencies between local similarities are learned with the use of the validation set.

The conducted experiments show that combining the HSM with one of the decision rules described in the Subsection 2.2 leads to the construction of robust classifiers and additionally, in some cases, makes it possible to identify unusual instances. Although the computation cost of the model is very high, the capability of using the parallel computing makes it a promising method for the classification tasks.

The HSM may also be useful in the prediction of behaviors of complex objects dynamically changing over time. Examining the similarity models fixed in a series of time points may allow the identification of rules governing the process of the change and eventually may lead to better understanding of the process. If we are able to construct a reliable global similarity model for a financial market or hospitalized patients, the ability of recognizing similar states will enable us to successfully plan our investment or a patient's treatment. All of those reasons are the motivation for the author to continue further studies on the concept of similarity and its applications in solving classification problems.

References

1. Janusz, A.: A similarity relation in machine learning. Master's thesis, Warsaw University, Faculty of Mathematics, Informatics and Mechanics (2007) (in polish)
2. Gati, I., Tversky, A.: Studies of similarity. In: Rosch, E., Lloyd, B. (eds.) Cognition and Categorization, pp. 81–99. L. Erlbaum Associates, Hillsdale (1978)
3. Hahn, U., Chater, N.: Understanding similarity: A joint project for psychology, case based reasoning, and law. Artificial Intelligence Review 12, 393–427 (1998)
4. Goldstone, R., Medin, D., Gentner, D.: Relational similarity and the nonindependence of features in similarity judgments. Cognitive Psychology 23, 222–262 (1991)
5. Nguyen, S.H.T.: Regularity analysis and its applications in data mining. Ph.D thesis, Warsaw University, Faculty of Mathematics, Informatics and Mechanics, Part II: Relational Patterns (1999)
6. Tversky, A.: Features of similarity. Psychological Review 84, 327–352 (1977)
7. Bazan, J., Nguyen, S.H., Nguyen, H.S., Skowron, A.: Rough set methods in approximation of hierarchical concepts. In: Tsumoto, S., Słowiński, R., Komorowski, J., Grzymała-Busse, J.W. (eds.) RSCTC 2004. LNCS (LNAI), vol. 3066, pp. 346–355. Springer, Heidelberg (2004)
8. Bazan, J., Kruczek, P., Bazan-Socha, S., Skowron, A., Pietrzyk, J.J.: Automatic planning of treatment of infants with respiratory failure through rough set modeling. In: Greco, S., Hata, Y., Hirano, S., Inuiguchi, M., Miyamoto, S., Nguyen, H.S., Słowiński, R. (eds.) RSCTC 2006. LNCS (LNAI), vol. 4259, pp. 418–427. Springer, Heidelberg (2006); see also the extended version in Fundamenta Informaticae 85 (2008)
9. Pawlak, Z.: Information systems, theoretical foundations. Information Systems 3(6), 205–218 (1981)
10. Skowron, A., Stepaniuk, J.: Ontological framework for approximation. In: Ślęzak, D., Wang, G., Szczuka, M.S., Düntsch, I., Yao, Y. (eds.) RSFDGrC 2005. LNCS (LNAI), vol. 3641, pp. 718–727. Springer, Heidelberg (2005)

11. Husserl, E.: The Crisis of European Sciences and Transcendental Phenomenology. Northwestern University Press, Evanston (1970); German original written in 1937
12. Schütz, A.: The Phenomenology of the Social World. Northwestern University Press, Evanston (1967)
13. Nguyen, S.H., Bazan, J., Skowron, A., Nguyen, H.S.: Layered learning for concept synthesis. In: Peters, J.F., Skowron, A., Grzymała-Busse, J.W., Kostek, B.z., Świniarski, R.W., Szczuka, M.S. (eds.) Transactions on Rough Sets I. LNCS, vol. 3100, pp. 187–208. Springer, Heidelberg (2004)
14. Basu, S.: Semi-supervised Clustering: Probabilistic Models, Algorithms and Experiments. PhD thesis, The University of Texas at Austin (2005)
15. Smyth, B., McClave, P.: Similarity vs. diversity. In: Aha, D.W., Watson, I. (eds.) ICCBR 2001. LNCS (LNAI), vol. 2080, pp. 347–361. Springer, Heidelberg (2001)
16. Agrawal, R., Imielinski, T., Swami, A.: Mining association rules between sets of items in large databases. In: Proc.of the 1993 ACM SIGMOD International Conference on Management of Data SIGMOD 1993, Washington, DC, pp. 207–216 (1993)
17. Nguyen, H.S.: On efficient handling of continuous attributes in large data bases. Fundamenta Informaticae 48(1), 61–81 (2001)

Probabilistic Granule Analysis

Ivo Düntsch* and Günther Gediga

Department of Computer Science,
Brock University,
St. Catharines, Ontario, Canada, L2S 3A1
{duentsch,gediga}@brocku.ca

Abstract. We present a semi–parametric approach to evaluate the reliability of rules obtained from a rough set information system by replacing strict determinacy by predicting a random variable which is a mixture of latent probabilities obtained from repeated measurements of the decision variable. It is demonstrated that the algorithm may be successfully used for unsupervised learning.

1 Introduction

A simple and widely used form of data operationalization is the

$$\text{OBJECT} \mapsto \text{ATTRIBUTE VALUES}$$

relationship, where each object is described by its values with respect to properties chosen from a defined set Ω of features, and which is usually represented as a data table.

Rough set data analysis (RSDA), introduced in the early 1980s [1] uses the simple observation that each occurring feature vector determines a unique sets of objects – namely, all those objects which have these features – to construct rule systems on the basis of the granularity given by observed data; furthermore, feature reduction – a major issue in data analysis – can be achieved within these systems.

Although RSDA uses a only few parameters which need simple statistical *estimation* procedures, its results should be controlled using statistical *testing* procedures, in particular, when the method is used for modeling and prediction of events. If the claim of RSDA to be a fully fledged instrument for data analysis and prediction is to hold, the following issues must be addressed:

1. Significance of rules,
2. Model selection in case of competing rules,
3. Unreliability of measurements.

In earlier work, we have developed a procedure to determine the statistical significance of rough set rules based on randomization methods, and a method of model selection which combines the principle of indifference with the maximum entropy

---

* Ivo Düntsch gratefully acknowledges support from the Natural Sciences and Engineering Research Council of Canada.

C.-C. Chan et al. (Eds.): RSCTC 2008, LNAI 5306, pp. 223–231, 2008.

principle [2,3]. The results support the view that a rule based method of data analysis does not, in principle, perform worse than traditional numerical methods, even on continuous data. Indeed, the direct comparison of linear discriminant analysis with RSDA based procedures by [4] on the Iris data [5] shows that the classification capability of non–parametric RSDA is as good as the parametric statistical method.

Traditionally, RSDA has concentrated on finding deterministic rules for the description of dependencies among attributes based on the *nominal scale assumption*: Once a deterministic rule has been found from a data set, it is tacitly assumed to hold without any error. Thus, in some sense, the theory is driven by the empirical data. However, if a measurement error is assumed to be an immeasurable part of the data, the pure RSDA approach may produce inaccurate results. On the one hand, even deterministic rules may be due to chance, and thus may not be reproducible; on the other hand, indeterministic information may be due to inaccurate measurement or the idiosyncrasies of a particular data set, thus possibly masking a theoretically deterministic situation.

In order to capture the uncertainty arising from measurement errors in a statistically sound way, we have proposed some 10 years ago the concept of *probabilistic information systems* [6], which may be viewed as an extension of the variable precision system of [7]. In the present contribution we take the opportunity to re–iterate this approach and extend it using well known procedures of classical test theory of psychometrics.

2 Definitions and Notation

We assume familiarity with the basic notions of RSDA and will just briefly recall the necessary concepts. A *decision system* is a tuple $\mathscr{I} = \langle U, y, V_y, \Omega, (V_x)_{x \in \Omega} \rangle$, where

1. $U = \{a_1, \ldots, a_N\}$ is a finite set of objects.
2. $\Omega = \{x_1, \ldots, x_T\}$ is a finite set of mappings $x : U \to V_x$. Each x_i is called an *(independent) attribute*.
3. y is a mapping from U to V_y, called the *decision attribute*.
4. The functional dependency $\Omega \Rightarrow y$ holds, i.e.

$$\text{If } x(a) = x(b) \text{ for all } x \in \Omega, \text{ then } y(a) = y(b).$$

This condition guarantees that the system is consistent.

If $\emptyset \neq X \subseteq \Omega$, we interpret X as a mapping $U \to \prod_{x \in X} V_x$ which assigns to each object $a \in U$ its feature vector $X(a) = x^X(a)$ with respect to the attributes in X; we will call $X(a)$ an *X–granule*; if $X = \Omega$, we will simply speak of a *granule*.

Each X – granule $X(a)$ can be understood as a piece of information about a set of objects in U given by the features in X, namely all those $b \in U$ for which $X(b) = X(a)$. The equivalence relation on U induced by this condition is denoted by ψ_X, i.e. for $a, b \in U$,

$$a \equiv_{\psi_X} b \iff X(a) = X(b). \tag{2.1}$$

Objects which are in the same class – and which are said to *belong to the same granule* – cannot be distinguished with the knowledge given by X.

Similarly, we define ψ_y on U by

$$a \equiv_{\psi_y} b \text{ iff } y(a) = y(b),$$

which gives us our target classification.

Suppose that $\emptyset \neq X \subseteq \Omega$. If a class M of ψ_X is contained totally within a class L of ψ_y, then $X(a)$ determines $y(b)$ for all $a, b \in M$. Such an M is called a *deterministic class of* ψ_X, and

$$\text{If } a, b \in M, \text{ then } y(a) = y(b) \tag{2.2}$$

is called a *deterministic X – rule*. Otherwise, M intersects exactly the classes $L_1, \ldots, L_k$ of ψ_y with associated values $l_1, \ldots, l_k$ in V_y, and we call

$$\text{If } a \in M, \text{ then } y(a) = l_1 \text{ or } \ldots \text{ or } y(a) = l_k \tag{2.3}$$

an *indeterministic X – rule*. The collection of all X – rules is denoted by $X \to y$, and – with some abuse of language – will sometimes be called a rule (of the information system).

The statistic

$$\gamma(X \to y) = \frac{|\bigcup\{M : M \text{ is a deterministic class of } X\}|}{|U|} \tag{2.4}$$

is called the *approximation quality of X* (with respect to y); it is the main indicator for the quality of feature reduction in RSDA [8]. It may be worthy of mention that this γ is only one of a whole family of such indicators, each of which may serve as useful approximation quality [9].

For our further discussion, we fix the following parameters:

- $U = \{a_1, \ldots, a_N\}$ is the set of objects.
- $\Omega = \{x_1, \ldots, x_T\}$ is the set of attributes.
- $G = \{g_1, \ldots, g_M\}$ is the set of granules. and T_i is the class of ψ_Ω associated with g_i, and $v(g_i) := |T_i|$.
- y is the decision attribute, $V_y = \{r_1, \ldots, r_D\}$ its set of values, and M_j is the class of ψ_y associated with r_j.
- For all $1 \leq i \leq M$, and $1 \leq j \leq D$, $\xi(i,j) := |T_i \cap M_j|$.

3   Probabilistic Decision Systems

In (deterministic) rule based systems a rule is either true or false, and a condition which holds for almost all cases will not contribute to the RSDA approximation quality. In the context of RSDA various remedies have been proposed which, instead of predicting hard decision values or intervals, regard the decision attribute as a random variable. For example, in standard rough set inclusion, deterministic rules for an indiscernibility class S and a decision class M are replaced by conditional probabilities which in the simplest case take the form

$$p(M|S) = \frac{|M \cap S|}{|M|}, \tag{3.1}$$

Table 1. A decision system

g_i	Ω		$y = r_1$	$y = r_2$	$v(g_i)$
	x_1	x_2	$\xi(i,1)$	$\xi(i,2)$	
g_1	0	1	5	1	6
g_2	1	0	2	8	10
	Σ		7	9	16

These considerations lead to probabilistic decision systems, sometimes called *Bayesian rough set models*, as structures of the from $\langle \mathscr{I}, Y \rangle$, where $\mathscr{I}$ is a classical RSDA information system, and $Y : G \times V_y \to [0,1]$ is a random variable; such structures have recently been an object of investigation, see e.g. [10,11,12]. Probabilistic rules have the form $x \to Y_j(x)$ which are pairs $\langle x, Y_j(x) \rangle$ where $x \in G$, and $Y_j(x)$ is the probability that x belongs to the decision class associated with r_j. Rough membership functions may be used to produce probabilistic decision systems such as the one shown in Table 1. There, we have $|U| = 16$, $\Omega = \{x_1, x_2\}$, and $V_y = \{r_1, r_2\}$, and both independent attributes are binary. Note that – up to indiscernibility – there are two granules, g_1, g_2. The rule system provided by the rough inclusion of (3.1) is obtained as

$$\langle 0,1 \rangle \to \{\langle 1, \tfrac{5}{6} \rangle, \langle 2, \tfrac{1}{6} \rangle\},$$
$$\langle 1,0 \rangle \to \{\langle 1, \tfrac{2}{10} \rangle, \langle 2, \tfrac{8}{10} \rangle\}.$$

Statistics such as rough inclusion are to some extent useful, however in principle they are subject to the same restrictions that the original problem poses, namely, that possible errors are not modeled within the system. In this sense, the problems persists, albeit with different, yet still "hard", boundaries for rule accuracy.

Computing the a–posteriori probability Y that a data element is assignable to a certain class requires distributional assumptions about the a priori distributions; estimation of priors is an inherent problem of Bayesian analysis. In most applications, however, it is not possible to observe the a priori distributions, and

"A statistical problem is how to accurate are 'estimations' ...with regards to the *true* regions" [12].

If the observed rules are stable, then they should be the same for a different population. However, rules obtained from a second instance of a decision system may look quite different from the original one, even if the underlying structure is unchanged.

The well known test–retest paradigm of psychometrics offers a solution to the problem by using a distributional family such as a mixture of normal distributions or a mixture of triangle distributions, and a parameter fitting procedure given a learning data set. Since the true classification variable Y is principally unknown, we suppose that it is a mixture

$$Y = \sum_{1 \le r \le R} \omega_r Y_r, \tag{3.2}$$

of i.i.d. realizations Y_r based on an index R of unknown size and with unknown weights ω_r^i, for which $\sum_r \omega_r^i = 1$. It is safe to regard the Y_r as repeated measurements of the

Table 2. Rule finding algorithm

$R := 0$, $\Delta(AIC) = 1$.
while $R \lesssim M$ and $\Delta(AIC) \gtrsim 0$ **do**
   $R := R + 1$
   Compute the best mapping $g : \{1, \ldots, M\} \to \{1, \ldots, R\}$ in terms
   of the product of the maximum likelihood of the Y replicas.
   Compute the number of parameters.
   Compute AIC_R for $L_R(\text{max})$.
   if $R = 1$ **then**
      $\Delta(AIC) := AIC_1$
   else
      $\Delta(AIC) := AIC_{R-1} - AIC_R$
   end if
end while

decision variable. In this way, the effects of an immeasurable measurement error are controlled and thus, the reliability of the rules can be tested in a statistically sound way.

The tasks now are

1. To estimate the best number R of replicas.
2. To estimate the parameters ω_r for each $1 \leq r \leq R$.

If we use the granules g_j to predict Y, the maximal number R of basic distributions is bounded by the number M of granules; equality occurs just when each granule g_j determines its own Y_j. In general, this need not to be the case, and it may happen that the same Y_j can be used to predict the class value of more than one granule; this will be indicated by a function

$$g : \{1, \ldots, M\} \twoheadrightarrow \{1, \ldots, R\},$$

which maps the (set of indices of) the granules onto a set of (indices of) mixture components of Y.

In any estimation procedure, numerous models are produced, and one needs to decide which of these offers the best description of the data. Two standard procedures for model selection based on the size of the empirical data set and the number of parameters are the *Akaike Information Criterion* AIC [13] and Schwarz's *Bayesian Information Criterion* BIC [14]

$$AIC = 2 \cdot (P - \ln(L(\text{max})))$$
$$BIC = 2 \cdot \left(\frac{\ln(K)}{2} \cdot P - \ln(L(\text{max})) \right).$$

Here, $L(\text{max})$ is the maximum likelihood of the data which may be obtained by optimizing the relevant binomial distribution by hill–climbing methods such as the EM algorithm [15]. The lower AIC (and BIC respectively), the better the model. AIC and BIC are similar, but the penalty for parameters is higher in BIC then in AIC.

An algorithm to find the most appropriate model in our context using the AIC was first described in [6]. It starts by searching for the optimal granule mapping based on a set Ω of (mutually) predicting attributes and a set Y of replicated decision attributes.

Finding the best mapping g is a combinatorial optimization problem, which can be approximated by hill-climbing methods, whereas the computation of the maximum likelihood estimators, given a fixed mapping g, is straightforward: One computes the multinomial parameters $\hat{\pi}_t(i_k)$ of the samples i defined by g for every replication y_t of Y and every value $r_k \in \{r_1, \ldots, r_Y\}$, and computes the mean value

$$\hat{\pi}(i_k) = \frac{\sum_{t=1}^{s} \hat{\pi}_t(i_k)}{s}, \tag{3.3}$$

from which the likelihood can be found (Table 2).

4   Unsupervised Learning and Semi–parametric Distribution Estimates

In [6] we have shown that the AIC search algorithm may be used as a procedure for unsupervised learning. We have exemplified the procedure with Fisher's iris data [5] resulting in a classification quality of 85% which is quite acceptable for an unsupervised learning procedure. In the analysis, we have assumed that the attributes measure the same variable up to some scaling constants and that therefore the z – transformed attributes may be used as a basis for the analysis. Upon closer inspection, it turns out the estimation of the mixture distributions is not a pure non–parametric procedure, because the standardization to z–values is, of course, a form of parametrization before the clustering procedure has started: The assumption "the attributes measure the same variables up to a some scaling constants" generates new variables which are assumed to be comparable on a standard scale.

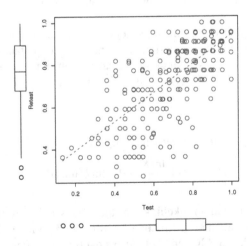

Fig. 1. Test-Retest situation

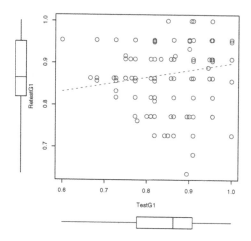

Fig. 2. Test–Retest distribution group 1

To adjust the situation we may use the measureables to transform the data as suggested by the classical test theory of psychometrics: A test X may be retested or be used in a parallel form X' to estimate the reliability of the test, and the z-transformations of test and retest should be used to computed the reliability. Our original approach shows that a two–group representation combined with a non–parametric mixture of the distribution of the test values can be performed, and that there are no extra costs in terms of additional assumptions or parameters; in other words, it's simply for free. If the test–retest-paradigm is enhanced by further retesting, or if the test can be split additionally (e.g. by summing up odd and even items within the test to form test-values), it is easy to estimate more latent classes and their distribution estimates.

We shall illustrate the procedure with a typical example. An intelligence test applied to 331 subjects was tested and retested two weeks later using a parallel form of the test items (same solving principle, but different layout). Figure 1 shows the mean item solving probabilities of the subjects.

This procedure is routine part of the standardization of a psychometric test. Furthermore, test and retest are assumed to be identical in their expectation and variance. If these assumptions hold, the assumptions for searching the best-AIC-mapping to a decision attribute with two values ("solvers" and "non–solvers") holds as well.

Applying the algorithm we observe a clear cut optimum with two groups. Group 1 consists of 52,6% of the subsects showing a joint test–retest distribution given in Figure 2. This group of subjects shows a high probability to solve the test items ("solvers"). The group is rather homogeneous, because the correlation of test and retest value is very low.

Group 2 consists of 47,4% of the subjects showing a joint test–retest distribution given in figure 3. This group has a much lower probability to solve the test items than the subjects in group 1 ("non–solvers"). Because the test–retest correlation is substantial, we have to argue that this group is not the final representation; owing to the restriction of

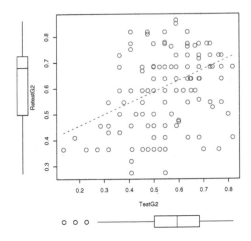

Fig. 3. Test–Retest distribution group 2

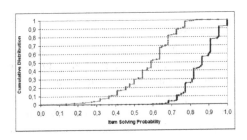

Fig. 4. Cumulative distributions of test values

only two measurements, the best-AIC-mapping cannot squeeze out more groups from the data.

The cumulative distributions of the test values in the groups can now be used to classify the subjects (Figure 4).

One can see, for example, that a subject showing a score of 0.6 is very likely to be a member of group 2, whereas a subject showing a score of 0.9 is member of group 1.

5   Conclusion and Outlook

We have proposed a mixture model which enables traditional RSDA to handle possible measurement errors in the decision variable. The method makes only mild distributional assumptions which makes it well suited for the non–invasive approach of RSDA. In future work, we will extend the approach to predict unseen cases from partially known information and investigate estimations of semi–parametric mixture distributions and re–classification of latent groups in the context of RSDA. We will also apply our approach to estimate the reliability of data discretization procedures.

References

1. Pawlak, Z.: Rough sets. Internat. J. Comput. Inform. Sci. 11, 341–356 (1982)
2. Düntsch, I., Gediga, G.: Statistical evaluation of rough set dependency analysis. International Journal of Human–Computer Studies 46, 589–604 (1997)
3. Düntsch, I., Gediga, G.: Uncertainty measures of rough set prediction. Artificial Intelligence 106, 77–107 (1998)
4. Browne, C., Düntsch, I., Gediga, G.: IRIS revisited: A comparison of discriminant and enhanced rough set data analysis. In: Polkowski, L., Skowron, A. (eds.) Rough sets in knowledge discovery, vol. 2, pp. 345–368. Physica–Verlag (1998)
5. Fisher, R.A.: The use of multiple measurements in taxonomic problems. Ann. Eugen. 7, 179–188 (1936)
6. Gediga, G., Düntsch, I.: Statistical tools for rule based data analysis. In: Komorowski, J., Düntsch, I., Skowron, A. (eds.) Workshop on Synthesis of Intelligent Agent Systems from Experimental Data, ECAI 1998 (1998)
7. Ziarko, W.: Variable precision rough set model. Journal of Computer and System Sciences 46 (1993)
8. Pawlak, Z.: Rough sets: Theoretical aspects of reasoning about data. System Theory, Knowledge Engineering and Problem Solving, vol. 9. Kluwer, Dordrecht (1991)
9. Gediga, G., Düntsch, I.: Rough approximation quality revisited. Artificial Intelligence 132, 219–234 (2001)
10. Pawlak, Z., Skowron, A.: Rough sets: Some extensions. Information Sciences 177, 28–40 (2007)
11. Pawlak, Z.: A rough set view on Bayes' theorem. International Journal of Intelligent Systems 18, 487 (2003)
12. Slezak, D., Ziarko, W.: The investigation of the Bayesian rough set model. International Journal of Approximate Reasoning 40, 81–91 (2005)
13. Akaike, H.: Information theory and an extension of the maximum likelihood principle. In: Petrov, B.N., Cáski, F. (eds.) Second International Symposium on Information Theory, Budapest, Akademiai Kaidó, pp. 267–281 (1973); Reprinted in: Kotz, S., Johnson, N.L. (eds.) Breakthroughs in Statistics, vol. I, pp. 599–624. Springer, New York (1992)
14. Schwarz, G.: Estimating the dimension of a model. Annals of Statistics 6, 461–464 (1978)
15. Redner, R.A., Walker, H.F.: Mixture densities, maximum likelihood and the EM algorithm. SIAM Review 26, 195–236 (1984)

Paraconsistent Case-Based Reasoning Applied to a Restoration of Electrical Power Substations

Helga Gonzaga Martins, Germano Lambert-Torres, Luiz Eduardo Borges da Silva, Claudio Inácio de Almeida Costa, and Maurilio Pereira Coutinho

Federal University at Itajuba, Av. BPS, 1303, 37.500-503 Itajuba, MG, Brazil
{helgagonzaga,germanoltorres,leborgess,
maurilio.coutinho}@gmail.com

Abstract. This paper presents a connection of two techniques applied in Artificial Intelligence to solve problems of restoration of electrical power substations. The techniques are: Case-based Reasoning – CBR and the Four-Valued Annotated Paraconsistent Logic – 4vAPL. This linking process happens in the manipulation of the functions of belief, disbelief, expertise and temporality of the 4vAPL for the recovery of cases to determine process diagnostics of a CBR. The domain of CBR is applied in the restoration of an electrical power substation. The 4vAPL is the support applied to the problems that present inconsistent, partial, undefined information. Thus, it approaches the system under study to real situations.

Keywords: Intelligent System, Decision Support System, Paraconsistent Logic, CBR, Case-based Reasoning, Restoration of Electrical Power Substations.

1 Introduction

In the restoration of an electrical substation, the aim is to reintegrate it to the electrical power system, readjusting it effectively and quickly, as closely as possible, to the configuration prior to the failure [1]. Along the recovery of the substation, each restoration procedure is evaluated as well as the influence of the operation conditions in the system, so as to validate what has been accomplished and released for the next execution. The operation of a substation is complex due to the number of variables that must be manipulated. The operator has to be able to manipulate several kinds of data and information in order to respond to a variety of requirements concerning the supervision and control.

Digital technology introduced in substations and the application of Artificial Intelligence (AI) techniques, have made the automation process possible, as well as the enhancement of the operation quality [2, 3]. An automatic substation restoration system aims to normalize the operations in a substation after its components being switched off partially or totally, reintegrating it to the system in a stable fashion. AI may also be feasible in the switching automation or in the components restoration following manual or forced outages [1]. The restoration of the substation normal operation configuration after an incident is structured in pre-established criteria of engineering studies which includes the reasoning to identify the actions for restoration, to

C.-C. Chan et al. (Eds.): RSCTC 2008, LNAI 5306, pp. 232–241, 2008.

validate the measurements and diagnostics, and to structure the switching plan. All these decision-based functions may be automatized through the use of Artificial Intelligence techniques [1- 4].

The AI techniques applied to automation are varied due to the increasing technological advances and the large number of research which have been accomplished. This work presents the link of the two techniques applied in Artificial Intelligence to solve the electrical substation restoration problems. These techniques are the Four-Valued Annotated Paraconsistent Logic – 4vAPL and Case-based Reasoning– CBR. In the following section the functions used in 4vAPL will be presented, these will later be incorporated by a CBR system. This linking procedure will be applied to the restoration of a substation, whose configuration is used in the Minas Gerais State Power Company – CEMIG, in Brazil.

2 Considerations of Four-Valued Annotated Paraconsistent Logic – 4vAPL

In [5-8], the interpretations of 3vAPL, starting from the Unitary Square on Cartesian Plan – USCP are presented. The values of the *Degree of Expertise* vary in the closed real interval [0,1], as the *Degrees of Belief μ_1 and Disbelief μ_2*, in this way, one may interpret a point obtained from a triple (μ_1,μ_2,e) that is located in the Analyzer Unit Cube, shown in Fig 1. The lattice regions in Fig 1 represent well defined regions, since they interpret the expert opinion deciding for a diagnostics referring to axis x, Dx, or for a diagnostics referring to axis y, Dy, or still, opting for Inconsistent (I) or Paracomplete ($\perp$) Regions, which represent problematic regions, once they have points that allow the interpretation of inconsistencies or contradictions.

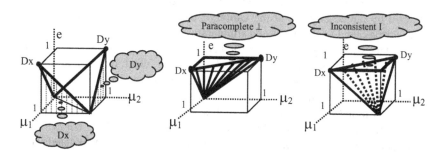

Fig. 1. Representation of Diagnostic Regions and Inconsistent – Paracomplete Regions

The regions that are located inside the Unitary Cube and involved by the regions mentioned above are named unstable regions and they behave in different ways according to the Degree of Expertise. These regions are named [5]:

- $\perp{\to}q{\neg}Dx$, $\perp{\to}q{\neg}Dy$ – Paracomplete tending to almost not Dx (Dy)
- $I{\to}qDx$, $I{\to}qDy$ – Inconsistent tending to almost Dx (Dy)

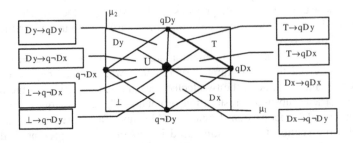

Fig. 2. Definition of Regions for the Para-Expert Algorithm

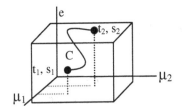

Fig. 3. Temporality on the Unitary Cube of 4vAPL

- Dx→qDx – Dx tending to almost Dx, and Dy→qDy – Dy tending to almost Dy
- Dx→q¬Dy – Dx tending to almost not Dy, and Dy→q¬Dx – Dy tending to almost not Dx.

Observing the behaviors in Expert Systems, based in 3vAPL, the algorithm constructed by the description of the Unitary Cube enables the design of a computer program for practical applications and simulations of different situations.

The Unitary Cube Regions for a particular Degree of Expertise e = 0.5 are shown in Fig 2, according to [5], and expanded to all the other degrees in the closed interval [0,1]. A point may be analyzed by moving along the Unitary Analyzer Cube, as shown in Fig 3, according to [7]. At time t_1 the point is found on position s_1, at time t_2 the point is found on position s_2, in such a way that as time flows, the point describes a curve C in the interior of the Unitary Analyzer Cube. This behavior allows the introduction of one more variable, time t, to the Three-valued Annotated Paraconsistent Logic- 3vAPL; thus extending it to Four-valued Annotated Paraconsistent Logic- 4vAPL.

In 4vAPL, the point in the Unitary Cube is represented by a quadruple (μ_1,μ_2,e,t). The intention of introducing one more annotated variable to represent the point is, to able to analyze the behavioral evolution of the Experts.

Hence, a Neophyte (Expert of degree e=0), facing its inexperience, will acquire experience as the time variable flows. The Degree of Expertise is expected to increase in order to define between two diagnostics Dx or Dy. The Degree of Expertise has the behavior of a *classical case* when found at the top of the Unitary Cube, roughly speaking. This analysis may be done for any level of expertise. The essence of the fourth dimension time is to visualize the behavior of the Experts in decision making of a specific system.

3 Implementation of Algorithm "CBR Para-Expert" from 3vAPL

To implement the "Para-Expert" algorithm in a CBR, whose domain is restoration of electrical power substations, a model, that suggests control actions restricted only to the cases of the knowledge base, is proposed. The New Matching Degree is calculated according to the equation (1):

$$
\text{NMD}^{k} = \frac{\sum\limits_{i=1}^{m} \omega_i^n \omega_i^{P_k} \left(1 - \left(\frac{(x_i - y_i)}{R_i} \right) \right)}{\sqrt{\sum\limits_{i=1}^{m} \left(\omega_i^n \right)^2 \sum\limits_{i=1}^{m} \left(\omega_i^{P_k} \right)^2}} \times e \tag{1}
$$

For: $i = 1,...m$ (descriptions); $k = 1,...r$ (previous cases);
$e \equiv$ Degree of Expertise of 3vAPL which describes the degree of pertinence in relation to Dx;
ω_i^n, $\omega_i^{P_k} \equiv$ weight of ith description on weight vector from new and previous cases;
x_i, y_i = value of description in new and previous cases;
R_i = extension value of the description scale.
$1 - \left(\frac{(x_i - y_i)}{R_i} \right) \equiv$ denotes the similarity in the ith description between of new case and previous case.

The denominator terms of equation above normalize the vectors weight by the determinations of its Euclidian lengths. The similarity function is based in the pertinence (weight) of description values for the diagnostic. The similarity between the value of the present description from the new case and the value of same present description in the previous case of memory is taken as being the difference between the unit and a rate between the weights that each one of these values have for the diagnostic of the case in memory, with the extension value of the description scale.

The architecture projected to implement the model is seen in Fig 4. In the first module, the inputs of a new case and the memory cases are accomplished through similarities. In the second module, the New Matching Degree is calculated. *The New Matching Degree - NMD is calculated from the 3vAPL Degree of Expertise of each case. For the cases from the knowledge base, the value of NMD is maximum, equal to 1, and for the new cases its value is lower than 1, except for the new case which is equal to any case from the knowledge base.* In the third module, the NMD is ranked in decreasing order. With this procedure, all the degrees from the knowledge base cases will be ordered first and then comes the ordering of the new cases. In this way, the search is restricted to only the cases from the initial knowledge base, and not the whole memory.

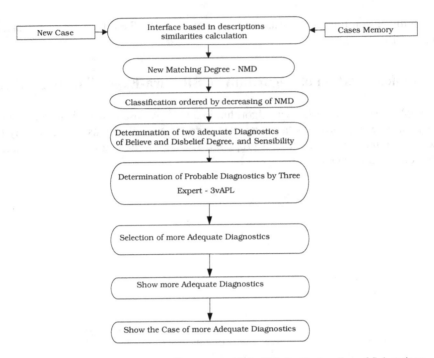

Fig. 4. Architecture of a Recovery Prototype with 3vAPL for Restoration of Substations

4 Operation Procedures and Data for Automatic Restoration of Substations

This section describes some characteristics of a typical electrical substation of an electrical power system, Minas Gerais State Power Company – CEMIG, Brazil. The substation in study is of Main and Transfer Buses type, one of the most frequent models used by CEMIG, Fig 5 [1].

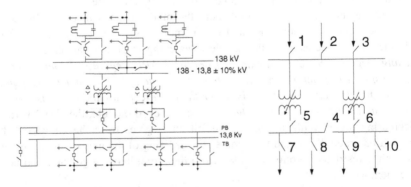

Fig. 5. Electrical substation in study and its simplification

Considering that the three transmission lines (1, 2, 3) that supply energy to the substation have the same amount, 35 MW. The nominal power of each transformer is 50 MVA. The system has three kinds of loads during the day: light, moderate, and peak (heavy), in such a way that the demands are altered in the period. Supposing that the four loads (7, 8, 9, 10) are balanced, then their demand description is as follows: (a) light load - 10 MW; (b) moderate load - 15 MW; (c) peak load - 25 MW.

After a contingency, the restoration will include reasoning for the identification of the need for actions and for a switching plan. The configuration of the substation presents restrictions and priorities, such as: switch 4 may only be turned on when only one of the transformers is energized; switch 7 must be turned on under any circumstance after a contingency. The desirable diagnostics for the restoration of the normal configuration after the occurrence of a contingency are:

- *Optimum Diagnostics (O)* – Diagnostics that is *desirable* to reestablish the normal condition of the substation, without violating any technical condition neither failing to supply any expected need.
- *Correct Diagnostics (C)* – Diagnostics that *is correct* to reestablish the normal condition of the substation, that is, it does not violate any technical condition, but fails to supply some expected need.
- *Incorrect Diagnostics (I)* – Diagnostics that *is incorrect* to reestablish the normal condition of the substation, in such a way that it violates some technical condition and/or fails to supply some expected need.
- *Minimum Diagnostics (MIN)* – Diagnostics that does not change any operational condition of the substation.

The substation operator controls the position of five switches: 4, 7, 8, 9 and 10, according to Fig 5. For this, he must observe the position of the ten switches and the substation loading level. Switches 1, 2, 3, 5 and 6 are operated automatically by the protection system. The operator's actions may be described as in Table 1.

Table 1. Specification of Descriptions and their Extensions

Attributes	Descriptions	Extension of Importance Scale	Attributes	Descriptions	Extension of Importance Scale
Condition	1 - Switch 1	0 - off / 1 – on	Decision	4 - Switch 4	0 - off / 2 – on
Condition	2 - Switch 2	0 - off / 1 – on	Decision	7 - Switch 7	0 - off / 4 – on
Condition	3 - Switch 3	0 - off / 1 – on	Decision	8 - Switch 8	0 - off / 1 – on
Condition	4 - Switch 4	0 - off / 2 – on	Decision	9 - Switch 9	0 - off / 1 – on
Condition	5 - Switch 5	0 - off / 3 – on	Decision	10 - Switch 10	0 - off / 1 – on
Condition	6 - Switch 6	0 - off / 2 – on			
Condition	7 - Switch 7	0 - off / 4 – on			
Condition	8 - Switch 8	0 - off / 1 – on			
Condition	9 - Switch 9	0 - off / 1 – on	(*)	1 – light	
Condition	10 - Switch 10	0 - off / 1 – on		2 – moderate	
Condition	L – Load	(*)		3 – peak	

Table 2. Knowledge Base

C	1	2	3	4	5	6	7	8	9	10	L	D	C	1	2	3	4	5	6	7	8	9	10	L	D
1	1	0	0	0	0	2	0	0	1	1	1	D1	17	0	0	1	0	0	2	0	0	1	1	1	D7
2	1	0	1	0	0	2	0	0	1	1	2	D1	18	0	1	0	0	0	2	0	0	1	1	2	D8
3	1	0	0	0	3	0	4	1	0	0	1	D1	19	1	1	0	0	0	2	0	0	1	1	3	D8
4	0	0	1	0	3	0	4	1	0	0	1	D2	20	1	0	0	0	3	2	4	1	0	0	2	D9
5	1	1	0	0	3	0	4	1	0	0	2	D2	21	1	1	1	0	3	0	4	1	0	0	3	D9
6	0	1	0	0	3	0	4	1	0	0	1	D2	22	1	1	0	0	3	2	4	1	1	1	2	D10
7	0	1	0	0	0	2	0	0	1	1	1	D3	23	1	1	1	0	3	2	4	1	1	1	3	D10
8	0	1	1	0	0	2	0	0	1	1	2	D3	24	0	1	0	0	3	2	4	0	1	1	1	D11
9	1	0	0	0	0	2	0	0	0	1	3	D4	25	1	0	0	0	3	2	4	1	0	1	1	D12
10	0	1	0	0	0	2	0	0	0	1	3	D4	26	0	1	1	0	3	2	4	0	1	0	3	D13
11	0	0	1	0	0	2	0	0	1	0	3	D4	27	1	0	0	0	3	2	4	0	1	0	2	D13
12	1	0	0	0	0	2	0	0	1	1	2	D5	28	1	0	1	0	3	2	4	0	0	1	3	D14
13	1	1	1	0	0	2	0	0	1	1	3	D5	29	1	0	0	0	3	2	4	0	0	1	2	D14
14	1	1	0	0	3	0	4	1	0	0	1	D6	30	0	0	1	0	3	0	4	0	0	0	3	D15
15	1	1	1	0	3	0	4	1	0	0	1	D6	31	0	0	1	0	3	2	4	0	0	0	3	D15
16	0	0	1	0	0	2	0	0	1	1	2	D7	32	0	1	0	0	3	2	1	1	1	0	1	D16

The knowledge base consists of 32 cases, describing the initial estates of the switches and their diagnostics as shown in Table 2.

Table 3 depicts the possible diagnostics and the identification referring to their descriptions.

Table 3. Diagnostics and their characteristics

Diagnostics	4	7	8	9	10	Diagnostics	4	7	8	9	10
D1	2	4	1	1	0	D9	0	4	1	0	0
D2	2	4	1	0	1	D10	0	4	1	1	1
D3	2	4	0	1	1	D11	0	4	0	1	1
D4	2	4	0	0	0	D12	0	4	1	0	1
D5	2	4	0	0	1	D13	0	4	0	1	0
D6	2	4	1	1	1	D14	0	4	0	0	1
D7	2	4	0	1	0	D15	0	4	0	0	0
D8	2	4	1	0	0	D16	0	4	1	1	0

The knowledge base is the result of the expert's experience, thus, it is a very particular situation of each substation since the occurrences are experiences located in their areas and the restoration to normal configuration is specific of each situation. Despite this fact, tables 2 and 3 may be obtained from the substation past switching or from theoretical studies carried out at the substation. Once the diagnostics and the restoration characteristics for the knowledge base are defined, the objective now is to apply the CBR Para-Expert program to solve the restoration problem.

When a new configuration occurs, it will be presented to the algorithm which will execute the CBR Para-Expert program resulting in the following diagnostics: O, C, I,

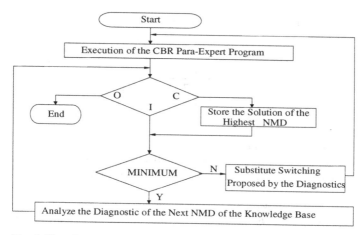

Fig. 6. Flowchart of the Strategies to Determine the Optimum Diagnostic

and MIN. The CBR Para-Expert algorithm will be repeated several times until it reaches its aim of restoration, that is, until it reaches the Optimum Diagnostics. The rules of the *Strategies to Determine the Optimum* Diagnostic are shown in the flowchart of Fig 6. When translating the analysis of the 4vAPL into the scenario of automatic restoration one may visualize the behavior of the diagnostics during the execution of the CBR Para-Expert program until it finds the *Optimum Diagnostics* for the configuration of the substation after the contingency.

5   Application of the "CBR Para-Expert" in Automatic Restoration of Substations

The configuration of an electrical power substation analyzed after a contingency is shown in Fig 7. It presents an "incorrect" diagnostics according to the rules defined previously.

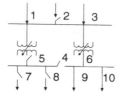

Fig. 7. Substation arrangement with incorrect diagnostics

The input case is of the type **[10100200111]**. According to this description, the system configuration is such that the substation loading level is light, value **1**. The switches present the following characteristics: 1-on; 2-off; 3-on; 4-off; 5-off; 6-on; 7-off; 8-off; 9-on; and 10-on. By applying the algorithm of Fig. 6 proposed for the restoration:

   a)   Execution of CBR Para-Expert for the new case [10100200111].

The three first solutions with the highest Matching Degree are shown in Table 4.

Table 4. Three first solutions for the case [10100200111]

N	NMD	Case	Diag.	4	7	8	9	10
1	0.7454	1	D1	2	4	1	1	0
2	0.7454	17	D7	2	4	0	1	0
3	0.7037	2	D1	2	4	1	1	0

b) Determination of the Diagnostics type
The diagnostics for the configuration is "CORRECT", but not "OPTIMUM", because none of the technical conditions are violated; however not all the possible needs are supplied.
c) Stores the Highest New Matching Degree.   NMD = 0.7454
d)Determination of the characteristics of the diagnostics - Minimum or Non-Minimum.

The diagnostics is "NON-MINIMUM" because the final position of the switches is not the same as the initial position.
e) Substitution of the Solution
In this case, the proposed switching plan is substituted by **Diagnostics 1**, such that the new analyzed case will be: *[10120241101]*.
f) Program Execution for the new case [10120241101]
The three first solutions with the highest Matching Degree are shown in Table 5.

Table 5. Three first solutions for the case [10120241101]

N	NMD	Case	Diag.	4	7	8	9	10
1	0.3509	15	D6	2	4	1	1	1
2	0.3509	25	D12	0	1	1	0	1
3	0.3509	32	D16	0	1	1	1	0

g) Determination of the Diagnostics type
The diagnostics for the configuration is "OPTIMUM", because none of the technical conditions are violated, and all the possible needs are supplied. Thus, Fig 8 represents the diagnostics behavior.

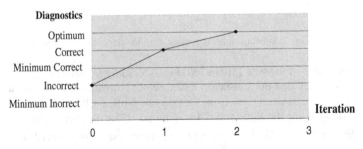

Fig. 8. Diagnostics behavior in the restoration of a substation

6 Conclusions

This paper presents an alternative approach using Artificial Intelligence techniques in the search for the optimization of actions and procedures in restoration of electrical power substations. The proposed approach joins the Annotated Paraconsistent Logic and Case-Based Reasoning to establish the rules for the *Strategies to Determine the Optimum Diagnostics*. Thus, the consistency between the recovered cases and their diagnostics is guaranteed. These strategies describe the reasoning for the identification of actions and a switching plan necessary for the automation of the restoration of electrical power substations.

The 4vAPL and CBR techniques applied to the restoration of substations present advantages once they take into account the situations closer to reality, treating the inconsistencies and the different kinds of diagnostics in a non-trivial fashion. This makes the systems more reliable and consistent. This study case confirms the consistency of the method.

The proposed methodology could be equally applied to other substation arrangements such as duple bus, ring, one and half circuit breakers, among others.

Acknowledgments. The authors would like to thank CNPq, CAPES, and FAPEMIG, Brazilian research funding agencies, for the research scholarships, which supported this work.

References

1. Lambert-Torres, G., Ribeiro, G.M., Costa, C.I.A., Alves da Silva, A.P., Quintana, V.H.: Knowledge Engineering Tool for Training Power-Substation Operators. IEEE Transactions on Power Systems 12(2), 694–699 (1997)
2. CIGRÉ: Practical use of expert systems in planning and operation of power systems. TF 38.06.03, Électra 146, 30–67 (1993)
3. CIGRÉ: Exploring user requirements of expert systems in power system operation and control. TF 39.03, Électra 146, 68–84 (1993)
4. Lambert-Torres, G., Rossi, R., Ribeiro, G.M., Valiquette, B., Mukhedkar, D.: Computer program package for power system protection and control. In: 34th CigrÉ Session Biennal Metting, Paris, France, August 31 - September 5, 1992, pp. 39–304 (1992)
5. Martins, H.G., Lambert-Torres, G., Pontin, L.F.: Annotated Paraconsistent Logic. In: COMMUNICAR (ed.) (2007) (in portuguese)
6. Costa, D., Newton, C.A.: Inconsistent Formal Systems, presented originally as thesis in 1963 - College of Philosophy, Sciences and Letters of the University of the Parana, Curitiba, Parana, Brazil (1993)
7. Lambert-Torres, G., Costa, C.I.A., Martins, H.G.: Decision- Making System based on Fuzzy and Paraconsistent Logics. In: Abe, J.M., da Silva Filho, J.I. (eds.) Logic, Artificial Intelligence and Robotics – Frontiers in Artificial Intelligence and Applications, LAPTEC 2001, pp. 135–146. IOS Press, Amsterdam (2001)
8. Martins, H.G., Lambert-Torres, G., Pontin, L.F.: Extension from NPL2v to NPL3v. In: Lambert-Torres, p.G., Abe, J.M., Mucheroni, M.L., Cruvinel, P.E. (eds.) Advances in Intelligent Systems and Robotics, vol. I, pp. 9–17. IOS Press, Amsterdam (2003)

Solving the Attribute Reduction Problem with Ant Colony Optimization

Hong Yu[1,2], Guoyin Wang[1], and Fakuan Lan[1]

[1] Institute of Computer Science and Technology, Chongqing University of Posts and
Telecommunications, Chongqing, 400065, P.R. China
[2] Department of Computer Science, University of Regina, Regina, Saskatchewan
Canada S4S 0A2
{yuhong,wanggy}@cqupt.edu.cn

Abstract. Attribute reduction is an important process in rough set theory. More minimal attribute reductions are expected to help clients make
decisions in some cases, though the minimal attribute reduction problem
(MARP) is proved to be a NP-hard problem. In this paper, we propose
a new heuristic approach for solving the MARP based on the ant colony
optimization (ACO) metaheuristic. We first model the MARP as finding an assignment which minimizes the cost in a graph. Afterward, we
introduce a preprocessing step that removes the redundant data in a
discernibility matrix through the absorbtion operator, the goal of which
is to favor a smaller exploration of the search space at a lower cost. We
then develop a new algorithm R-ACO for solving the MARP. Finally,
the simulation results show that our approach can find more minimal
attribute reductions more efficiently in most cases.

1 Introduction

Rough set theory, proposed by Pawlak [5] in 1982, is a valid mathematical tool
to deal with imprecise, uncertain, and vague information. It has been developed
and applied to many fields such as decision analysis, machine learning, data
mining, pattern recognition, and knowledge discovery successfully.

In these applications, it is typically assumed that the values of objects are
represented by an information table. The notion of a reduction plays an essential
role in analyzing an information table [5]. In many cases, the minimal (optimal)
attribute reduction is expected. Unfortunately, it is proven to be a NP-hard
problem [13] to compute the minimal attribute reduction problem(MARP) of
an information table. Thus, many heuristic methods have been proposed and
examined for finding the set of all reductions or a single reduction [9,10,11,12,14].

Ant colony optimization (ACO) [1], is a stochastic approach to solve different
hard combination optimization problems [1,7] such as traveling salesman problems, vehicle routing problem, constraint satisfaction problem, machine learning,
etc. The main idea of ACO is to model the problem as a search for a minimum
cost path in a graph. Artificial ants walk through this graph, looking for good
paths. Each ant has a rather simple behavior so that it will typically only find

C.-C. Chan et al. (Eds.): RSCTC 2008, LNAI 5306, pp. 242–251, 2008.
© Springer-Verlag Berlin Heidelberg 2008

rather poor-quality paths on its own. Better paths are found as the emergent result of the global cooperation among ants in the colony.

Since the MARP is a NP-hard problem, inspired by the character of ant colony optimization, some researchers [3,4] have focused on solving the problem with ACO. To combat the efficiency and gain more minimal attribute reductions, we will propose a new heuristic approach for solving the MARP based on ACO in this paper. We transfer the MARP to a constraint satisfaction problem. The goal is to find an assignment which satisfies the minimum cost in a graph.

As we know, the notion of the discernibility matrix introduced by Skowron and Rauszer [6] is important in computing cores and attribute reductions. In fact, there usually exists redundant data in the matrix. We can remove the redundant data from the matrix in order to reduce the exploration of the search space. In other words, the methods can make improvement in time and space. Therefore, a preprocessing step is needed before reduction, which is described in Section 4.

This paper is organized as follows. First, we introduce some definitions and terminologies about the minimal attribute reduction problem. In Section 3, a model of solving the MARP with ACO is proposed. Section 4 develops a preprocessing step by removing the redundant data in a discernibility matrix through the absorbtion operator. A new algorithm R-ACO for solving the MARP is given in Section 5. The experiment results in Section 6 show that the approach to solve the MARP with ACO can find more minimal reductions more efficiently in most cases. Some conclusions will be given in Section 7.

2   The Attribute Reduction Problem

Let us first review the relevant definitions and terminologies [6,11].

Definition 1 (Information Table). *An information table is the following tuple: $I = (U, Atr = C \cup D, V, f)$ where $U = \{x_1, x_2, \ldots, x_n\}$ is a finite non-empty set of objects, $C = \{a_1, a_2, \ldots, a_m\}$ is a finite non-empty set of attributes and also called the conditional attribute set, $D = \{d\}$ is the decision attribute set, V is the set of possible feature values, f is the information function, given an object and a feature, f maps it to a value $f : U \times Atr \to V$.*

Definition 2 (Discernibility Matrix). *Given a consistent information table $I = (U, C \cup D, V, f)$, its discernibility matrix $M = (M_{i,j})$ is a $|U| \times |U|$ matrix, in which the element $M_{i,j}$ for an object pair (x_i, x_j) is defined by:*

$$M_{i,j} = \begin{cases} \{a \mid a \in C \land a(x_i) \neq a(x_j)\} & if\ d(x_i) \neq d(x_j) \\ \emptyset & else \end{cases}$$

Definition 3 (Discernibility Function). *The discernibility function of a discernibility matrix is defined by:*

$$f(M) = \bigwedge_{\substack{1 \leq i \leq |U|-1 \\ i+1 \leq j \leq |U|}} \left(\bigvee_{\substack{a_k \in M_{i,j} \\ M_{i,j} \neq \emptyset}} a_k \right).$$

Table 1 is the discernibility matrix of an information table. The discernibility function can be denoted by $f(M) = \{a \vee b \vee c \vee d \vee e\} \wedge \{a \vee c\} \wedge \{a \vee e\} \wedge \{c\} \wedge \{a \vee b \vee d\} \wedge \{b \vee c \vee d \vee e\}$.

Table 1. A Discernibility Matrix

$\emptyset$	$\{a,b,c,d,e\}$	$\{a,c\}$	$\emptyset$	$\emptyset$
$\{a,b,c,d,e\}$	$\emptyset$	$\emptyset$	$\{a,e\}$	$\{c\}$
$\{a,c\}$	$\emptyset$	$\emptyset$	$\{a,b,d\}$	$\{b,c,d,e\}$
$\emptyset$	$\{a,e\}$	$\{a,b,d\}$	$\emptyset$	$\emptyset$
$\emptyset$	$\{c\}$	$\{b,c,d,e\}$	$\emptyset$	$\emptyset$

The discernibility function can be transformed to a disjunctive form as $f(M) = \bigvee R_q$, where R_q is a conjunction of some attributes. Each conjunctor $R_p = a_1 \wedge a_2 \wedge \cdots \wedge a_q$ is a reduction, denoted by $R_p = \{a_1, a_2, \ldots, a_q\}$. We can acquire the minimal attribute reductions based on this Boolean calculation. However, the computation is very complex when considering the scale of the problem since it is a NP-hard problem.

3 Model of Solving MARP with ACO

Previously, [3,4] have used the ACO approach to solve the MARP. The model they used is a complete graph whose nodes represent conditional attributes, with the edges between them denoting the choice of the next conditional attribute. The search for the minimal attribute reduction is then an ant traversal through the graph where a minimum number of nodes are visited that satisfies the traversal stopping criterion. The ant terminates its traversal and outputs the attribute subset as a candidate of attribute reductions.

A suitable heuristic desirability of traversing between attributes could be any subset evaluation function - for example, the rough set dependency measure [3] or an mutual information entropy based measure [4]. However, the relevant operations cost too much time because the operations are all in the space $U \times C$. On the other hand, the approaches are also heuristic.

To combat this problem, with more reductions especially minimal reductions expected, we propose a new model R-Graph to solve the MARP with ACO.

Firstly, let us review the constraint satisfaction problem(CSP) [8].

Definition 4 (Constraint Satisfaction Problem). *A constraint satisfaction problem is defined by a triple (B, Dom, Con) such that $B = \{B_1, B_2, \ldots, B_k\}$ is a finite set of k variables, Dom is a function which maps every variable to its domain $Dom(B_p)$, and Con is a set of constraints.*

A solution of a CSP(B, Dom, Con) is an assignment $A = \{< B_1, v_1 >, \cdots, < B_k, v_k >\}$, which is a set of variable-value pairs, where $v_i \in Dom(B_i)$. The cost of an assignment A is denoted by $cost(A)$. An optimal solution of a CSP$(B,$

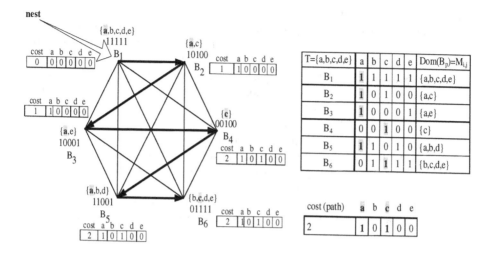

Fig. 1. R-graph of Table 1

$Dom, Con)$ is a complete assignment for all the variables in B, which satisfies all the constraints in Con with the minimal cost.

Let us come back to the discernibility matrix M. If we take a unique $M_{i,j}$ as a variable B_p, then $Dom(B_p) = M_{i,j}$. We can transfer an attribute reduction problem to a CSP and define the following model R-graph to describe the attribute reduction problem.

Definition 5 (R-graph Associated with CSP). *A R-graph associates a vertex with each value of the tuple* $T = < v_1, \ldots, v_m >$ *to be permuted, and* T *is a tuple of* $|C| = m$ *values, where* C *is the conditional attribute set of an information table* I. *There is an extra vertex corresponding to the nest, from which ants will start their paths. Hence, the R-graph associated with a CSP(B, Dom, Con) is a complete oriented graph* $G = (V, E)$ *such that:* $V = \{< B_i, v >| B_i \in B$ *and* $v \in Dom(B_i)\}, E = \{(< B_i, v >, < B_j, w >) \in V^2 \mid B_i \neq B_j\}$.

Definition 6 (Path in a R-graph). *A path in a R-graph* $G = (V, E)$ *is a sequence of vertices of* V. *We only consider elementary paths, which do not contain any cycles.*

Definition 7 (Cost of a Path). *The goal of R-graph is to find a complete assignment which minimizes the cost. Because the values in every* B_i *is in* $T = < v_1, \ldots, v_m >$, *ants going over the same values in the path will have a cost of* 0. *Therefore, the more duplicate values selected in a path, the less cost a path is. Hence, the cost of a path* π, *denoted by* $cost(\pi)$, *is the number of the different values that appeared in the path* π.

A solution of a R-graph G is an assignment $A = \{B_1 \leftarrow r_1, \cdots, B_i \leftarrow r_i, \cdots, B_k \leftarrow r_k\}$, where $r_i \in Dom(B_i)$. Ants deposit pheromone on edges of the R-graph; the amount of pheromone laying on an edge (v_i, v_j) is denoted as $\tau_{v_i v_j}$.

Ants start from the nest, and successively visit each of the other k vertices. Here, all ants have a constraint condition such that they must visit every vertex.

According to [6], if we choose one attribute from every element of the discernibility matrix, then the set of the selected attributes can compose a super attribute reduction. Therefore, the path(assignment) of a R-graph is also a super reduction of the information table. For Table 1, we can gain a reduction $f(M) = \{a\} \wedge \{a\} \wedge \{c\} \wedge \{a\} \wedge \{c\} = \{a\} \wedge \{c\} = \{a, c\}$. The path is described as thicker lines(or shadow letters) in Fig.1, where 0-1 representation is used to encode the variables. For example, B_2 can be encoded as $\{10100\}$. The path is denoted by 10100 with cost 2 corresponding to an ant from nest to food passes the values a and c. Recall that, $\{a, c\}$ is the minimal attribute reduction of Table 1. In summary, if ants can find the minimal cost paths in a R-graph, then ants find the minimal attribute reductions of the information table. In fact, we prefer which values in the tuple T are selected in a path to how to build a path.

4   Discernibility Matrix Simplification

Paths in a R-graph are in the reductions of the corresponding discernibility matrix. However, since there are many redundant data in the matrix, we may need to reduce the search space through removing the redundant elements of the matrix.

Consider the reduction space S, which is the subset of discernibility matrix M with no uniform elements, denoted by $S = \{B_k \mid B_k = M_{i,j} \wedge (\forall s \forall t B_s \neq B_t)\}$. Obviously, the reduction computation based on a discernibility matrix is the computation based on Boolean calculation [6]. Let $B_i \in S$, if $B_j \in S$, and $B_j \subseteq B_i$, B_i is called an absorbed discernibility attribute element by B_j. We then have $B_j \cap B_i = B_j, (\vee(B_i)) \wedge (\vee(B_i)) = \vee(B_j), (\wedge(B_j)) \vee (\wedge(B_i)) = \wedge(B_j)$. The property pertaining to Boolean logic can be used to reduce the reduction space, in other words, a minimal reduction space(MRS) can be acquired by removing all absorbed discernibility attribute items. Algorithm AMRS(Acquire Minimal Reduction Space) is given in Fig.2.

In addition, we know that the reduction based on the simplified matrix MRS is equal to the reduction on the original discernibility matrix M[15].

In order to judge if a discernibility element is absorbed by another, we define the absorbtion operator &&, which is an and-operator one by one bit. Considering Table 1, $B_1 = \{a, b, c, d, e\} = \{11111\}$ and $M_{i,j} = M_{1,3} = \{a, c\} = \{10100\}$, then $B_k \&\& M_{i,j} = \{11111\} \&\& \{10100\} = \{10100\}$. Obviously, this leads to $B_k \&\& M_{i,j} = M_{i,j}$. We then have $M_{i,j} \subseteq B_k$, which means B_k should be removed.

The elements with $|B_i| = 1$ are the core attributes. We can remove the core from MRS and add it to the reduction later. The size of the $MRS[P][Q]$ will be extended to $(P+1) \times (Q+1)$, where each cell of the 0-th row stores the number of the 1's in the corresponding column. Each cell of the 0-th column stores the number of the 1's in the corresponding row. These values can be acquired easily through Algorithm AMRS, and they can be used as heuristic information in the next section.

Algorithm AMRS: Acquire Minimal Reduction Space
Input: $I = (U, Atr = C \cup D, V, f)$, $|U| = n, |C| = m$
Output: the Minimal Reduction Space MRS
Begin
    $MRS = \Phi$; $absorded = 0$;
    for $i= 1$ to n do
        for $j=i+1$ to n do
            if $(d(x_i) \neq d(x_j))$ then
                Compute $M_{i,j}$;
            end if
            for each B_k in MRS and $B_k \neq \Phi$ do
                if $(B_k \subseteq M_{i,j})$ then   //if $M_{i,j}$ be absorbed, i.e $B_k \&\& M_{i,j} = B_k$
                    $absorded = 1$;
                elseif $(M_{i,j} \subseteq B_k)$ //if B_k be absorbed, i.e. $B_k \&\& M_{i,j} = M_{i,j}$
                    $MRS = MRS - B_k$;
                end if
            end for
            if $(absorded = 0)$ then // if $M_{i,j}$ cannot be absorbed or B_k be absorbed
                $MRS = MRS + M_{i,j}$;   // add $M_{i,j}$ to the space MRS
            end if
            $absorded = 0$;   //reset
        end for
    end for
    Output: MRS
End

Fig. 2. The Algorithm of Acquire Minimal Reduction Space

Table 2 is the extended $MRS[P][Q]$ of Table 1, where the number in the parentheses means the number of 1's in the corresponding row or column. The R-graph can be built from Table 2 instead of Table 1, the search space is greatly reduced.

As already mentioned in Section 3, the decision method, to add an attribute to the ant traversing, used in [3,4] are all on the space $U \times C$. To compute the entropy or the dependency function, the compute time of each decision is $O(|U|^2 + |U|)/2$. Therefore, the time complexity is $O(|C| \cdot (|U|^2 + |U|)/2)$ in [3,4]. By contraries, the searching space used in R-graph is no better than $|C|$ at each decision. Considering the worst case, the items of the discernibility matrix are different from each other, then the times of decisions will be $(|U|^2 + |U|)/2$. Actually, there are usually many redundant items in the discernibility matrix, the compute times is far less than $O((|U|^2 + |U|)/2) \cdot |C|)$ in most cases.

Table 2. The $MRS[P+1][Q+1]$ of Table 1

$T = \{a,b,d,e\}$	$a(2)$	$b(1)$	$d(1)$	$e(1)$	$Dom(B_p)$
$B_1(3)$	1	1	1	0	$\{a,b,d\}$
$B_2(2)$	1	0	0	1	$\{a,e\}$

In addition, the following property holds. It can be used to cut a travel path which can not be a minimum cost path. The cut operating can make improvement in time and space.

Property 1. If $Q_{max} = maxMRS[0][q], 1 \leq q \leq Q$, the minimum attributes of the minimal reduction denoted by min_Redu, C_0 is the core, then
$min\_Redu - |C_0| \leq P - Q_{max} + 1$.

5   Algorithm to Solve MARP with ACO

After the preprocessing step as described in the last section, a simplified discernibility matrix will be obtained. A new algorithm R-ACO is proposed to find most of the minimal attribute reductions using the ant colony optimization. Let us explain some ideas used in the algorithm.

There are different variable orderings [8] for selecting a variable. We use the random ordering, which comes from MRS (the simplified discernibility matrix) directly with no other computation. The pheromone values are all initialized to a constant value e, every node (value) in B_1 is assigned an ant. That is, the initialized number of ants is $ant = MRS[1][0]$.

The termination conditions used here as same as the basic ACO [1]. That is, when the solutions are stable or the max cycles are reached, the iterations are ended. In fact, the iterations perform no more than 5 times when the solutions are convergent in our experiments, even if the max cycles is initialized to 10.

The basic ingredient of any ACO algorithm is a constructive heuristic for probabilistically constructing solutions. The transition probabilities are defined as follows:

$$P_{uv}^k = \begin{cases} \frac{\tau_{uv}^\alpha (\eta_v + \Delta\eta_v)^\beta}{\sum_{r \in B_i} \tau_{ur}^\alpha (\eta_r + \Delta\eta_r)^\beta} & \text{if } v \in B_i \\ 0 & \text{otherwise} \end{cases} \tag{1}$$

where B_i means the next selected variable, u is the current value, v is the next value in B_i, and η is the heuristic information. As we have discussed in Section 4, $MRS[0][j]$ is used as an initial η. In order to avoid convergence quickly, the heuristic information is adjusted by $\Delta\eta$ automatically. The values of parameters α and β determine the relative importance value and heuristic information. There, $1 \leq \alpha \leq 5$ and $1 \leq \beta \leq 5$ are derived from experience. To acquire more optimal solutions, the random probability is also used in the R-ACO algorithm.

Pheromone (τ) updating uses the following rule:

$$\begin{aligned} \tau &= (1 - \rho)\tau_{uv} + \Delta\tau_{uv}, & u \in B_{i-1}, v \in B_i \\ \Delta\tau_{uv} &= \sum_{k=1}^m \Delta\tau_{uv}^k & \\ \Delta\tau_{uv}^k &= \begin{cases} \frac{Q}{cost(SP_k)} & u, v \in SP_k \\ 0 & \text{else} \end{cases} \end{aligned} \tag{2}$$

where $\rho \in (0, 1]$ is a parameter for pheromone evaporation, Q is a constant, and SP is the partial solution.

The algorithm R-ACO (finding the set of attribute Reduction based on ACO) is represented in Fig.3.

Algorithm R-ACO: finding the set of attribute Reduction based on ACO
Input: MRS[$P+1$][$Q+1$] //$2 \leq Q \leq |C|$
Output: the set SG whose element is an attribute reduction
Begin
    // Initialization:
    $temp\_mincost = P - Q_{max} + 1$;   //come from Algorithm AMRS and Property 1
    Artifical ants $ant = MRS[1][0]$;   // the number of ants is ant
    $E_{uv}.pheromone = e$;   // initialize the pheromone
    $EdgeCovered_{uv} = 0$;   // the nubmer of ant that selected E_{uv}
    $Constringency$ = False;   // the termination conditions
    Initial q_0, max_cycles; // They are constants, $0 < q_0 < 10$. eg: $q_0 = 5$, $max\_cycles = 10$
    while (!$constringency$) do
        if ($ant = MRS[1][0]$) then
            set the ant ants to each of the node in B_1;
        else
            set ant ants on nodes with larger pheromone values;
        end if
        // construct solution:
        for $k = 1$ to ant do   // local search
            $SP_k = B_{1,k}$;   // initialize the partial solution is the node b_{1k} the ant k associate
            for $i \in B_2$ to B_P do
                Produce a random value q; // $0 < q < 10$
                if ($q > q_0$) then
                    Select a node v random from the set whose nodes have the maximal
                    heuristic information.
                else
                    $maxNode = \max(P_{uv}^{k})$; // Equation (1),
                    node $v \leftarrow v \in maxNode \wedge v$ is min($EdgeCovered$).
                end if
                $SP_k' = SP_k \cup \{v\}$; $EdgeCovered_{uv}$ ++;
                if ((SP_k' && SP_k) != SP_k) then
                    compute new SP_k and the $cost(SP_k)$;
                end if
                if ($cost(SP_k) > temp\_mincost$) then
                    delete SP_k; continue;
                end if
                $\Delta\eta_v$ --; // the node v has been selected by the current ant
            end for
        end for
        $SP_{min} = \min\{SP_1, SP_2..., SP_k,..., SP_{ant}\}$;
        // updating:
        if ($|SP_{minK}| < |SG|$) then
            $SG = SP_{min}$; $temp\_mincost = \min(cost(SP_k[0]))$; // $1 \leq k \leq m$
        else
            $SG = SG \cup (SPmin - SG)$;
        end if
        update the pheromone of all nodes according Equation (2); max_cycles --;
        if (SG is stable or $max\_cycles = 0$) then
            $constringency$ = true;
        end if
    end while
    Output the set SG.
End

Fig. 3. The Algorithm R-ACO: Finding the Set of Attribute Reduction Based on ACO

6   Experimental Results

In our experiments, eight databases [2] are used. We test each database on a Pentium4 PC with Algorithm R-ACO, RSACO [4] and Algorithm 3 [9]. The algorithm in [9] can be used to compute all the reductions of an information table and give us the minimal reductions to reference. The results of our experiments are shown in Table 3. $|U|$ and $|C|$ are the cardinality of the universe and the conditional attribute set, respectively. $|SG|$ is the cardinality of the set of reductions, $|att\_R|$ is the number of conditional attributes in a attribute reduction, $|att\_minR|$ is the number of conditional attributes in a minimal attribute reduction, and CPU(s) is the CPU time (by second) of the process.

Table 3. Comparison of the CPU time and Results of the Algorithms

Database	$|U|$	$|C|$	Algorithm R-ACO			Algorithm RSACO			Algorithm 3
			$|SG|$	$|att\_R|$	CPU(s)	$|SG|$	$|att\_R|$	CPU(s)	$|att\_minR|$
ZOO	101	17	6	11	0.003	3	11	6.570	11
Car	1728	6	1	6	0.000	1	6	53.589	6
Soybean-large	307	35	4	11	0.857	1	11	207.809	9
LED24	1000	24	2	18	6.361	2	16	1144.455	16
Australian	690	14	2	3	0.076	2	3	38.017	3
Tic-tac-toe	958	9	6	8	0.016	2	8	39.804	8
statlog(germa)	1000	20	3	10	1.240	1	10	641.651	7
Mushroom	4062	22	1	1	0.045	1	1	829.314	1

From Table 3, we can observe that Algorithm R-ACO is more efficient than Algorithm RSACO, and Algorithm R-ACO finds more minimal reductions. Furthermore, we can see that Algorithm R-ACO developed in this paper is a feasible solution to the MARP and the approach can acquire the minimal attribute reductions in most cases.

7   Conclusion

Attribute reduction is an important process in data mining based on rough set theory. In this paper, the minimal attribute reduction problem is studied based on the ant colony optimization metaheuristic. A model R-graph is first constructed to find an assignment which minimizes the cost in the graph. To simplify the search space, a preprocessing step that removes the redundant data in a discernibility matrix is introduced. A new algorithm R-ACO for solving minimal attribute reduction problem is proposed. Finally, the simulation results show that our approach can find more minimal attribute reductions more efficiently in most cases.

Acknowledgments

We would like to thank Professor Y.Y. Yao for his support when writing this paper. In particular, he suggested the idea for modeling the question. The research is partially funded by the China NNSF grant 60573068 & 60773113, and the Chongqing of China grant 2008BA2017 & KJ080510.

References

1. Dorigo, M., Sttzle, T.: Ant Colony Optimization. MIT Press, Cambridge (2004)
2. http://archive.ics.uci.edu/ml/
3. Jensen, R., Shen, Q.: Finding rough set reducts with ant colony optimization. In: Proc. 2003 UKWorkshop on Computational Intelligence, pp. 15–22 (2003)
4. Jiang, Y.C., Liu, Y.Z.: An Attribute Reduction Method Based on Ant Colony Optimization. In: Intelligent Control and Automation, 2006. WCICA 2006, vol. 1, pp. 3542–3546 (2006)
5. Pawlak, Z.: Rough Sets. International Journal of Computer and Information Sciences 11, 341–356 (1982)
6. Skowron, A., Rauszer, C.: The discernibility matrices and functions in information systems. In: Slowinski, R. (ed.) Intelligent Decision Support, Handbook of Applications and Advances of the Rough Sets Theory. Kluwer, Dordrecht (1992)
7. Solnon, C.: Ants can solve constraint satisfaction problems. IEEE Transactions on Evolutionary Computation 6(4), 347–357 (2002)
8. Tsang, E.P.K.: Foundations of Constraint Satisfaction. Academic Press, London (1993)
9. Wang, G.Y., Wu, Y., Fisher, P.S.: Rule Generation Based on Rough Set Theory. In: Dasarathy, B.V. (ed.) Proceedings of SPIE Data Mining and Knowledge Discovery: Theory, Tools, and Technology II, vol. 4057, pp. 181–189 (2000)
10. Wang, G.Y., Yu, H., Yang, D.C.: Decision table reduction based on conditional information entropy. Chinese Journal of Computers 25, 759–766 (2002)
11. Wang, G.Y.: Calculation Methods for Core Attributes of Decision Table. Chinese Journal of Computers 26, 611–615 (2003)
12. Wang, J., Wang, J.: Reduction algorithms based on discernibility matrix: the ordered attributes method. Journal of Computer Science and Technology 16(6), 489–504 (2001)
13. Wong, S.K.M., Ziarko, W.: On optimal decision rules in decision tables. Bulletin of Polish Academy of Sciences 33, 693–696 (1985)
14. Wu, W.Z., Zhang, M., Li, H.Z., Mi, J.S.: Knowledge reduction in random information systems via Dempster-Shafer theory of evidence. Information Sciences 174, 143–164 (2005)
15. Yao, Y.Y., Zhao, Y.: Discernibility Matrix Simplification for Constructing Attribute Reducts. Information Science (to appear)

Actor Critic Learning: A Near Set Approach

Shamama Anwar and K. Sridhar Patnaik

Dept. of Computer Science and Engineering,
Birla Institute of Technology, Mesra,
Ranchi - 835215, India
shamama3@gmail.com, ktosri@rediffmail.com

Abstract. This paper introduces an approach to reinforcement learning by cooperating agents using a near set-based variation of the Peters-Henry-Lockery rough set-based actor critic adaptive learning method. Near sets were introduced by James Peters in 2006 and formally defined in 2007. Near sets result from a generalization of rough set theory. One set X is near another set Y to the extent that the description of at least one of the objects in X matches the description of at least one of the objects in Y. The hallmark of near set theory is object description and the classification of objects by means of features. Rough sets were introduced by Zdzisław Pawlak during the early 1980s and provide a basis for perception of objects viewed on the level of classes rather than the level of individual objects. A fundamental basis for near set as well as rough set theory is the approximation of one set by another set considered in the context of approximation spaces. It was observed by Ewa Orłowska in 1982 that approximation spaces serve as a formal counterpart of perception, or observation. This article extends earlier work on an ethology-based Peters-Henry-Lockery actor critic method that is episodic and is defined in the context of an approximation space. The contribution of this article is a framework for actor-critic learning defined in the context of near sets. This paper also reports the results of experiments with three different forms of the actor critic method.

Keywords: Adaptive learning, approximation space, ethogram, ethology, actor critic, near sets, rough sets.

1 Introduction

The problem considered in this paper is how to refine and extend Peters-Henry-Lockery rough set-based actor critic learning method (see, *e.g.*, [1,4,3,15]). The Peters-Henry-Lockery approach provides an ethology-based form of the Sutton-Barto actor-critic method [2], an on-policy method that predefines the policy to select an action. Actor critic methods are temporal difference [2] methods that have a separate memory structure that explicitly represents a policy independent of a value function. The policy structure is known as *actor* because it is used to select actions and the estimated value function is known as *critic*, because it criticizes the actions made by the actor. Learning is always on-policy: the critic must learn about and critique whatever policy is currently being followed

C.-C. Chan et al. (Eds.): RSCTC 2008, LNAI 5306, pp. 252–261, 2008.

by an actor. The critique takes the form of a TD error (*i.e.*, difference between value-of-state estimates obtained at different times). The solution to the problem considered in this paper results from a near set-based actor critic method.

Near sets were introduced by James Peters in 2006 [16], formally defined in 2007 [17] and elaborated in [18]. Near sets result from a generalization of rough set theory. Briefly, one set X is near another set Y to the extent that the description of at least one of the objects in X matches the description of at least one of the objects in Y. The hallmark of near set theory is object description and the classification of objects by means of features [14]. Rough sets were introduced by Zdzisław Pawlak during the early 1980s [10,11] and provide a basis for perception of objects viewed on the level of classes rather than the level of individual objects. A fundamental basis for near set as well as rough set theory is the approximation of one set by another set considered in the context of approximation spaces. It was observed by Ewa Orłowska in 1982 that approximation spaces serve as a formal counterpart of perception, or observation [13].

The contribution of this article is a framework for actor-critic learning defined in the context of near sets. This paper also reports the results of experiments with three different forms of the actor critic method, traditional Sutton-Barto actor critic method [2], Peters-Henry-Lockery actor critic method [1,4,3,15], and a new near set-based actor critic method. A brief introduction to near sets is given in Sect. 5. Then a new near set-based actor critic method is presented in Sect. 6 (this section also includes the results of experiments with all three forms of actor critic learning).

This paper has the following organization. The traditional actor critic method is briefly presented in Sect. 2. An approach to describing organism behaviour is given in Sect. 3. The basic framework for the Peters-Henry-Lockery actor critic method is presented in Sect. 4.

2   Sutton-Barto Actor Critic

Actor critic methods are a natural extension of the idea of reinforcement comparison [2] methods to TD learning (*i.e.*, combination of Monte Carlo and dynamic programming methods [2]) and to the full reinforcement learning problem. Alg. 1 provides a representation of the basic Sutton-Barto actor-critic method described in [2].

Let S be a set of possible states, let s denote a current state and for each $s \in S$, and let $A(s)$ denote the set of actions available in state s. Put $A = \cup_{s \in S} A(s)$. the collection of all possible actions. Let a denote a possible action in the current state; let s·denote the subsequent state after action a; let $p(s, a)$ denote an action-preference and let r denote the reward for an action while in state s.

The method begins by fixing a number $\gamma \in [0, 1]$, called a *discount rate* that diminishes the estimated value of the next state; in a sense, γ captures the confidence in the expected value of the next state. Let $C(s)$ denote the number of times the actor has observed state s. The estimated value function $V(s)$ is

Algorithm 1. The actor critic Method

Input   : States $s \in S$, Actions $a \in A(s)$, Initialize γ, β
Output: Policy $\pi(s,a)$
for *(all $s \in S$, $a \in A(s)$)* **do**
  $p(s,a) \longleftarrow 0; \pi(s,a) \longleftarrow \frac{e^{p(s,a)}}{\sum_{b=1}^{|A(s)|} e^{p(s,b)}}; C(s) \longleftarrow 0;$
end
while *True* **do**
  Initialize s;
  for *($i = 0; i \leq \#$ of episodes; $i++$)* **do**
    Choose a from s using policy π;
    Take action a; observe reward r, and next state s';
    C(s)$\longleftarrow$ C(s) + 1;
    V(s) $\longleftarrow$ V(s)+ $\frac{1}{(s)}$[r - V(s)];
    δ = r + γ V(s') - V(s);
    $p(s,a) \longleftarrow p(s,a) + \beta.\delta$;
    $\pi(s,a) \longleftarrow \frac{e^{p(s,a)}}{\sum_{b=1}^{|A(s)|} e^{p(s,b)}}$;
    $s \longleftarrow s'$;
  end
end

defined as the average of the rewards received while in state s. This average may be calculated as

$$V(s) = \frac{C(s)-1}{C(s)} \cdot V_{C(s)-1}(s) + \frac{1}{C(s)} \cdot r \tag{1}$$

where $V_{C(s)-1}(s)$ denotes $V(s)$ for the previous occurrence of state s. After each action selection, the critic evaluates the quality of selected action using

$$\delta \leftarrow r + \gamma V(s') - V(s),$$

which is the error between successive estimates of expected value of a state. If $\delta > 0$, then it can be said that the expected return received from taking action a at time t is larger than the expected return in state s resulting in an increase in action preference $p(s,a)$. Conversely, if $\delta < 0$, then the action a produced a return that is worse than expected and $p(s,a)$ is decreased. The preferred action a in state s is calculated using

$$p(s,a) \leftarrow p(s,a) + \beta\delta,$$

where β is the actor's learning rate. The policy $\pi(s,a)$ is employed by an actor to choose actions stochastically using Gibbs softmax method [9] in

$$\pi(s,a) \leftarrow \frac{e^{p(s,a)}}{\sum_{b=1}^{|A(s)|} e^{p(s,b)}}.$$

3   Behaviour Description

This section briefly presents the approach to describing learning behaviour, starting first with object description useful in classifying perceived objects. Objects are known by their descriptions. An *object description* is defined by means of a tuple of function values $\phi(x)$ associated with an object $x \in X$ (see (2)). The important thing to notice is the choice of functions $\phi_i \in B$ used to describe an object of interest.

Object Description : $\phi(x) = (\phi_1(x), \phi_2(x), \ldots, \phi_i(x), \ldots, \phi_L(x)).$ (2)

The intuition underlying a description $\phi(x)$ is a recording of measurements from sensors, where each sensor is modelled by a probe function ϕ_i. Assume that $B \subseteq \mathcal{F}$ is a given set of functions representing features of sample objects $X \subseteq \mathcal{O}$. Let $\phi_i \in B$, where $\phi_i : \mathcal{O} \longrightarrow \Re$. The value of $\phi_i(x)$ is a measurement associated with a feature of an object $x \in X$. The function ϕ_i is called a *probe*. In combination, the functions representing object features provide a basis for an *object description* $\phi : \mathcal{O} \to \Re^L$, a vector containing measurements (returned values) associated with each functional value $\phi_i(x)$ in (2), where the description length $|\phi| = L$.

4   Peters-Henry-Lockery Actor Critic Method

This section briefly introduces Peters-Henry-Lockery actor critic method that is rough set-based. The basics of rough set theory are presented in [12] and are omitted, here. To set up an ethological approach to adaptive learning, put $\mathcal{B} = \{[x]_B \mid x \in \mathcal{O}\}$, a set of classes that "represent" behaviours of an organism that learns adaptively. Let D denote a decision class, e.g., $D = \{x \mid d(x) = 1\}$, a set of objects having acceptable behaviours. Let ν denote traditional rough coverage computed relative to B_*D (lower approximation of D) as shown in (3). Define $\bar{\nu}$ (average rough coverage)[1] in (3).

$$\bar{\nu} = \frac{1}{|\mathcal{B}|} \sum_{[x]_B \in \mathcal{B}} \nu\left([x]_B, B_*D\right), \text{ where, } \nu = \frac{\left|\,[x]_B, B_*D\right|}{|B_*D|}. \qquad (3)$$

where $\nu = 1$, if $B_*D = \emptyset$. From (3), it is possible to design various families of adaptive learning algorithms (see, e.g., [1,4,3,15]). For example, $\bar{\nu}$ is used to compute preference $p(s,a)$ as shown in (4).

$$p(s,a) \leftarrow p(s,a) + \beta[\delta - \bar{v}] \qquad (4)$$

where $\bar{\nu}$ is reminiscent of the idea of a reference reward used during reinforcement comparison. To complete the picture, it is assumed that learning is episodic. In

---

[1] $\bar{\nu}$ is computed at the end of each episode using an ethogram that is part of the adaptive learning cycle.

Algorithm 2. Peters-Henry-Lockery Actor Critic Method

Input : States, $s \in S$, Actions $a \in A(s)$, Initialize γ, β, $\bar{v}$.

Output: Policy $\pi(s,a)$

for *(all $s \in S$, $a \in A(s)$)* **do**

$\quad\mid\quad$ $p(s,a) \longleftarrow 0$; π $(s,a) \longleftarrow \frac{e^{p(s,a)}}{\sum_{b=1}^{|A(s)|} e^{p(s,b)}}$; $C(s) \longleftarrow 0$;

end

while *True* **do**

$\quad$ Initialize s;

$\quad$ **for** $(i = 0; i \leq \#$ *of episodes*$; i + +)$ **do**

$\quad\quad$ Choose a from s, using policy;

$\quad\quad$ Take action a; observe reward, r, and next state, s';

$\quad\quad$ $C(s) \longleftarrow C(s) + 1$;

$\quad\quad$ $V(s) \longleftarrow V(s) + \frac{1}{(s)}[r - V(s)]$;

$\quad\quad$ $\delta = r + \gamma V(s') - V(s)$;

$\quad\quad$ $p(s, a) \longleftarrow p(s, a) + \beta[\delta - \bar{v}]$;

$\quad\quad$ $\pi(s, a) \longleftarrow \frac{e^{p(s,a)}}{\sum_{b=1}^{|A(s)|} e^{p(s,b)}}$;

$\quad\quad$ $s \longleftarrow s'$;

$\quad$ **end**

$\quad$ Extract ethogram table $IS = (U_{beh}, A, d)$;

$\quad$ Discretize feature values in IS;

$\quad$ Compute $\bar{v}$ using IS;

end

keeping with the analogy of learning by a biological organism, organism behaviour observed an episode is stored in a table called an ethogram. An ethogram is a tabular representation of observed behaviours. In this case, observation observation of each behaviour is limited to a recording of a tuple $(s, a, r, V(s), d)$, i.e., state, action, reward, value of state and decision d, respectively. In simulating organism behaviour, we assume that $d = 1$ (action a accepted) and $d = 0$ (action is rejected). This is a special case of the model for object description in (2). For more details about ethograms used in ethology-based adaptive learning, see [4,7].

5   Near Sets

The basic idea in the near set approach to adaptive learning is to compare behaviour descriptions. In general, sets X, X' are considered near each other if the sets contain objects with at least partial matching descriptions. Let $\sim_B$ denote $\{(x, x') \mid f(x) = f(x') \ \forall f \in B\}$ (called the indiscernibility relation [11,12]).

Definition 1. Near Sets [17]

Let $X, Y \subseteq \mathcal{O}$, $B \subseteq \mathcal{F}$. Set X is near Y if, and only if there exists $x \in X, y \in Y, \phi_i \in B$ such that $x \sim_{\{\phi_i\}} y$.

Object recognition problems, especially in adaptive learning and images [1] and the problem of the nearness of objects have motivated the introduction of near sets (see, *e.g.*, [5,6,17,18]).

5.1 Nearness Approximation Spaces

The original generalized approximation space (GAS) model [16] has been extended as a result of recent work on nearness of objects [5]. A nearness approximation space (NAS) is a tuple

$$NAS = (\mathcal{O}, \mathcal{F}, \sim_{B_r}, N_r, \nu_{N_r}),$$

defined using set of perceived objects $\mathcal{O}$, set of probe functions $\mathcal{F}$ representing object features, indiscernibility relation $\sim_{B_r}$ defined relative to $B_r \subseteq B \subseteq \mathcal{F}$, family of neighbourhoods N_r, and neighbourhood overlap function ν_{N_r}. The relation $\sim_{B_r}$ is the usual indiscernibility relation from rough set theory restricted to a subset $B_r \subseteq B$. The subscript r denotes the cardinality of the restricted subset B_r, where we consider $\binom{|B|}{r}$, *i.e.*, $|B|$ functions $\phi_i \in \mathcal{F}$ taken r at a time to define the relation $\sim_{B_r}$. This relation defines a partition of $\mathcal{O}$ into non-empty, pairwise disjoint subsets that are equivalence classes denoted by $[x]_{B_r}$, where

$$[x]_{B_r} = \{x' \in \mathcal{O} \mid x \sim_{B_r} x'\}.$$

These classes form a new set called the quotient set $\mathcal{O}/\sim_{B_r}$, where

$$\mathcal{O}/\sim_{B_r} = \{[x]_{B_r} \mid x \in \mathcal{O}\}.$$

In effect, each choice of probe functions B_r defines a partition ξ_{B_r} on a set of objects $\mathcal{O}$, namely,

$$\xi_{B_r} = \mathcal{O}/\sim_{B_r}.$$

Table 1. Nearness Approximation Space Symbols

Symbol	Interpretation
B_r	non-empty, countable set of probe functions in B, $r \leq |B|$,
$\sim_{B_r}$	Indiscernibility relation defined using B_r,
$[x]_{B_r}$	$[x]_{B_r} = \{x' \in \mathcal{O} \mid x \sim_{B_r} x'\}$, equivalence class,
$\mathcal{O}/\sim_{B_r}$	$\mathcal{O}/\sim_{B_r} = \{[x]_{B_r} \mid x \in \mathcal{O}\}$, quotient set,
ξ_{B_r}	Partition $\xi_{\mathcal{O},B_r} = \mathcal{O}/\sim_{B_r}$,
ϕ_i	Probe function $\phi_i \in \mathcal{F}$,
r	$\binom{|B|}{r}$, *i.e.*, $|B|$ functions $\phi_i \in \mathcal{F}$ taken r at a time,
$N_r(B)$	$N_r(B) = \{\xi_{B_r} \mid B_r \subseteq B\}$, set of partitions,
ν_{N_r}	$\nu_{N_r} : \mathcal{P}(\mathcal{O}) \times \mathcal{P}(\mathcal{O}) \longrightarrow [0, 1]$, overlap function,
$N_r(B)_*X$	$N_r(B)_*X = \bigcup_{x:[x]_{B_r} \subseteq X} [x]_{B_r}$, lower approximation,
$N_r(B)^*X$	$N_r(B)^*X = \bigcup_{x:[x]_{B_r} \cap X} [x]_{B_r} \neq \emptyset$, upper approximation,
$Bnd_{N_r(B)}(X)$	$N_r(B)^*X \backslash N_r(B)_*X = \{x \in N_r(B)^*X \mid x \notin N_r(B)_*X\}$.

Algorithm 3. The Near Actor Critic Method

Input : States, s $\in$ S, Actions a $\in$ A(s), feature n.
Output: Ethogram resulting form Policy π(s,a).
for *(all s $\in$ S, a $\in$ A(s))* **do**
$\quad \mid \quad$ p(s,a) $\longleftarrow$ 0; π (s,a) $\longleftarrow \frac{e^{p(s,a)}}{\sum_{b=1}^{|A(s)|} e^{p(s,b)}}$; C(s) $\longleftarrow$ 0;
end
while *True* **do**
$\quad \mid \quad$ Initialize $\bar{v}_{a_v}$ wrt n feature nbds, s;
$\quad \mid \quad$ **for** *(i = 0; i $\le$ # of episodes; i + +)* **do**
$\quad \mid \quad \mid \quad$ Choose a from s, using policy;
$\quad \mid \quad \mid \quad$ Take action a; observe reward, r, and next state, s';
$\quad \mid \quad \mid \quad$ C(s) $\longleftarrow$ C(s) + 1;
$\quad \mid \quad \mid \quad$ V(s) $\longleftarrow$ V(s)+ $\frac{1}{(s)}[r - V(s)]$;
$\quad \mid \quad \mid \quad$ δ = r + γ V(s') - V(s);
$\quad \mid \quad \mid \quad$ p(s,a) $\longleftarrow$ p(s,a) + $\beta[\delta - \bar{v}_a]$;
$\quad \mid \quad \mid \quad$ π (s,a) $\longleftarrow \frac{e^{p(s,a)}}{\sum_{b=1}^{|A(s)|} e^{p(s,b)}}$;
$\quad \mid \quad \mid \quad$ s $\longleftarrow$ s';
$\quad \mid \quad$ **end**
$\quad \mid \quad$ Extract ethogram table $IS = (U_{beh}, A, d)$;
$\quad \mid \quad$ Discritize feature values in IS;
$\quad \mid \quad$ Compute $\bar{v}_{a_v}$ using IS;
end

Every choice of the set B_r leads to a new partition of $\mathcal{O}$. The overlap function ν_{N_r} is defined by

$$\nu_{N_r} : \mathcal{P}(\mathcal{O}) \times \mathcal{P}(\mathcal{O}) \longrightarrow [0, 1],$$

where $\mathcal{P}(\mathcal{O})$ is the powerset of $\mathcal{O}$. The overlap function ν_{N_r} maps a pair of sets to a number in $[0, 1]$ representing the degree of overlap between sets of objects with features defined by probe functions $B_r \subseteq B$. For each subset $B_r \subseteq B$ of probe functions, define the binary relation $\sim_{B_r} = \{(x, x') \in \mathcal{O} \times \mathcal{O} : \forall \phi_i \in B_r, \phi_i(x) = \phi_i(x')\}$. Since each $\sim_{B_r}$ is, in fact, the usual indiscernibility relation [12], let $[x]_{B_r}$ denote the equivalence class containing x, *i.e.*,

$$[x]_{B_r} = \{x' \in \mathcal{O} \mid \forall f \in B_r, f(x') = f(x)\}.$$

If $(x, x') \in \sim_{B_r}$ (also written $x \sim_{B_r} x'$), then x and x' are said to be *B-indiscernible* with respect to all feature probe functions in B_r. Then define a collection of partitions $N_r(B)$ (families of neighbourhoods), where

$$N_r(B) = \{\xi_{B_r} \mid B_r \subseteq B\}.$$

Families of neighbourhoods are constructed for each combination of probe functions in B using $\binom{|B|}{r}$, *i.e.*, $|B|$ probe functions taken r at a time. The family of neighbourhoods $N_r(B)$ contains a set of percepts. A *percept* is a byproduct of perception, *i.e.*, something that has been observed. For example, a class in $N_r(B)$ represents *what has been perceived about objects belonging to a neighbourhood, i.e.*, observed objects with matching probe function values.

It is now possible to introduce a near set-based form of coverage that extends the basic rough coverage model in (3). That is, we can formulate a basis for measuring the degree of overlap between each class in $N_r(B)$ and the lower approximation $N_r(B)_*X$ of a set X for each choice of r. The lower approximation $N_r(B)_*X$ defines a standard for classifying perceived objects. The notation $B_j(x)$ denotes a class in the family of neighbourhoods in $N_r(B)$, where $a \in B_r$. Put

$$v_a([x]_{B_r}, N_r(B)_*X) = \frac{|[x]_{B_r} \cap N_r(B)_*X|}{|N_r(B)_*X|}$$

where v_j is defined to be 1, if $N_r(B)_*X = \phi$. Put $B = \{[x]_{B_r} : a(x) = j, x \in \mathcal{O}\}$, a set of equivalence classes that represent action $a(x) = j$. Let D denote a decision class, *e.g.*, $D = \{x|d(x) = 1\}$, a set of object having acceptable behaviours. Define $\bar{v}_a(t)$ (near lower average coverage) with respect to an action $a(x) = j$ at time t in (5).

$$\bar{v}_a(t) = \frac{1}{|B|} \sum_{[x]_{B_r} \in B} v([x]_{B_r}, N_r(B)_*D). \tag{5}$$

6   Near Actor Critic Method

This section briefly introduces a near set-based actor critic method. Using (5), action preference is now calculated as

$$p(s, a) \leftarrow p(s, a) + \beta \cdot [\delta - \bar{v}_a],$$

where $\bar{v}_a$ is computed at the end of each episode for each feature value using an ethogram. In experimenting with near actor critic learning, different values of r are considered, starting with $r = 1$ (single feature case based on the selection of a single feature from $\{a, r, V(s)\}$) and concluding with $r = 3$ (three feature case). In a manner similar to [4], the behaviour of groups of interacting organisms that learn has been simulated using Alg. 1 and Alg. 2 as well as Alg. 3 near actor critic method.

From the plots in Fig. 1, we observe that the Peters-Henry-Lockery actor critic method fares better than the traditional actor critic method, *i.e.*, average $V(s)$ values are comparatively much higher. The plots in Fig. 1 also show that single feature near actor critic yields $V(s)$ values consistently higher than all of the other forms of the actor critic method. It can also be observed that learning during the initial episodes is much better for the near set actor critic than for the other forms of actor critic. The results in Fig. 1 are promising but are considered preliminary.

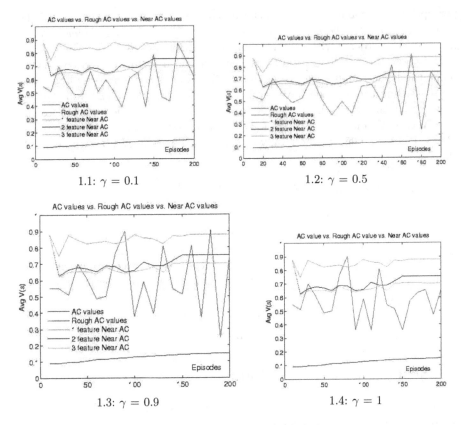

Fig. 1. Sample Experimental Results

7   Conclusion

This paper considers several approaches to actor critic learning. The proposed near set-based form of actor critic is compared with the Sutton-Barto and Peters-Henry-Lockers forms of actor critic. From the experimental results already obtained for these three methods appear to confirm the near set-based approach to biologically-inspired adaptive learning reported in [15]. Future work will include the implementation of the near actor critic method in engineering systems that learn. In addition, there is some interest in considering limiting cases and the convergence of value-of-state to an optimal value during a typical sequence of learning episodes.

Acknowledgement

The authors gratefully acknowledge the suggestions and constant support by Prof. James F. Peters , Department of Electrical and Computer Engg., University of Manitoba, Winnipeg, Manitoba R3T 5V6, Canada.

References

1. Lockery, D., Peters, J.F.: Adaptive learning by a target tracking system. Int. J. of Intelligent Computing and Cybernetics 1, 1–28 (2008)
2. Sutton, R.S., Barto, A.G.: Reinforcement Learning: An Introduction. The MIT Press, Cambridge (1998)
3. Peters, J.F.: Rough Ethology: Towards a biologically inspired study of collective behavior in Intelligent System with Approximation Spaces. In: Peters, J.F., Skowron, A. (eds.) Transactions on Rough Sets III. LNCS, vol. 3400, pp. 153–174. Springer, Heidelberg (2005)
4. Peters, J.F., Henry, C., Gunderson, D.S.: Biologically-inspired approximate adaptive learning control strategies: A rough set approach. International Journal of Hybrid Intelligent Systems 4(4), 203–216 (2007)
5. Peters, J.F., Skowron, A., Stepaniuk, J.: Nearness of Objects: Extension of Approximation Space Model. Fundamenta Informaticae 79(3-4), 497–512 (2007)
6. Peters, J.F., Skowron, A., Stepaniuk, J.: Nearness in approximation spaces. In: Lindemann, G., Schlilngloff, H., et al. (eds.) Proc. Concurrency, Specification & Programming. Informatik-Berichte 2006, Humboldt-Universität zu, Berlin, pp. 434–445 (2006)
7. Peters, J.F., Henry, C.: Reinforcement Learning with approximation Spaces. Fundamental Informaticae 71(2-3), 323–349 (2006)
8. Peters, J.F., Ramanna, S.: Feature Selection: Near Set Approach. In: Ras, Z.W., Tsumoto, S., Zighed, D.A. (eds.) 3rd Int. Workshop on Mining Complex Data (MCD 2007), ECML/PKDD-2007, vol. 4484, pp. 57–71. Springer, Heidelberg (2007)
9. Gibbs, J.W.: Elementary Principles in Statistical Mechanics. Dover, NY (1960)
10. Pawlak, Z.: Rough Sets, Institute for Computer Science, Polish Academy of Sciences, Report 431 (March 1981)
11. Pawlak, Z.: Classification of Objects by Means of Attributes, Institute for Computer Science, Polish Academy of Sciences, Report 429 (March 1981)
12. Pawlak, Z., Skowron, A.: Rudiments of rough sets. Information Sciences. An International Journal 177(1), 3–27 (2007)
13. Orlowska, E.: Applications of Rough Sets, Semantics of Vague Concepts, Institute for Computer Science, Polish Academy of Sciences, Report 469 (March 1982); Dorn, G., Weingartner, P.: Foundations of Logic and Linguistics. Problems and Solutions, March 1982, pp. 465–482. Plenum Press, London (1985)
14. Peters, J.F.: Classification of Perceptual Objects by Means of Features. Int. J. of Information Technology and Intelligent Computing 3(2), 1–35 (2007)
15. Peters, J.F., Shahfar, S., Ramanna, S., Szturm, T.: Biologically-inspired adaptive learning: A near set approach. In: Proc. Frontiers in the Convergence of Bioscience and Information Technologies,10.1109/FBIT.2007.39, pp. 403–408 (2007)
16. Peters, J.F.: Near Sets. Special Theory about Nearness of Objects. Fundamenta Informaticae 76, 1–27 (2006)
17. Peters, J.F.: Near sets. General theory about nearness of objects. Applied Mathematical Sciences 1(53), 2609–2029 (2007), http://wren.ece.umanitoba.ca/
18. Peters, J.F., Wasilewski, P.: Foundations of near sets. Information Sciences (submitted, 2008)

Compact Rule Learner on Weighted Fuzzy Approximation Spaces for Class Imbalanced and Hybrid Data

Yang Liu[1,2], Boqin Feng[1], and Guohua Bai[2]

[1] Department of Computer Science and Technology,
Xi'an Jiaotong University,
Xi'an, 710049, P.R.China
liuyang2006@gmail.com, bqfeng@mail.xjtu.edu.cn
[2] School of Engineering,
Blekinge Institute of Technology,
Ronneby, 372 25, Sweden
gba@bth.se

Abstract. Rough set theory is an efficient tool for machine learning and knowledge acquisition. By introducing weightiness into a fuzzy approximation space, a new rule induction algorithm is proposed, which combines three types of uncertainty: weightiness, fuzziness and roughness. We first define the key concepts of block, minimal complex and local covering in a weighted fuzzy approximation space, then a weighted fuzzy approximation space based rule learner, and finally a weighted certainty factor for evaluating fuzzy classification rules. The time complexity of proposed rule learner is theoretically analyzed. Furthermore, in order to estimate the performance of the proposed method on class imbalanced and hybrid datasets, we compare our method with classical methods by conducting experiments on fifteen datasets. Comparative studies indicate that rule sets extracted by this method get a better performance on minority class than other approaches. It is therefore concluded that the proposed rule learner is an effective method for class imbalanced and hybrid data learning.

Keywords: Rule induction, fuzzy rough set, weighted rough set, hybrid attributes, class imbalanced data sets.

1 Introduction

Continuous attributes and class imbalanced data sets exist in a large number of real-world domains and are recognized as two crucial problems in rough set based machine learning and knowledge acquisition [1,2]. LERS system (Learning from examples based on rough sets) is one of the most widely used rule induction systems for real-world applications [3,4]. In LERS system, algorithm LEM2 cannot cope with continuous attributes [5]. Algorithm MLEM2 is proposed as an extension of LEM2, which computes cut points by averaging any two consecutive

C.-C. Chan et al. (Eds.): RSCTC 2008, LNAI 5306, pp. 262–271, 2008.

values of a continuous attribute [6]. However, the selection of crisp cut points is crucial to the performance of a rule learner. The cut points should reflex the structure of the data and patterns. If the boundary of patterns were fuzzy and indistinguishable, the crisp cut point would not capture the actual semantics [7].

As we know, in Pawlak's rough set model, fuzziness and weightiness are not taken into account [8,9]. Some generalized rough set models were proposed to deal with hybrid attributes. Dubois et al. [10] proposed fuzzy rough set, and the properties of fuzzy rough set were analyzed in details by [11]. Currently, fuzzy rough set models were extensively applied to data reduction [12], fuzzy rough classification tree [13]. As to fuzzy attributes or numerical attributes, fuzzy set and fuzzy equivalence relations are capable of modeling the uncertainty in hybrid datasets [14].

However, these models just consider all examples equally important. Using this default hypothesis may increase overall accuracy of the generated rules, especially for highly skewed data [15]. In some cases, it may even lead to generation of only majority class label as "artificial" rules [16]. While such a solution produces accurate classification rules, there is no practical use for the user since no data model for any of the minority class can be obtained. In order to balance the accuracy on majority and minority class, weighted rough set models were proposed, where samples are associated with probabilities [17]. Based on weighted factors, attribute reduction and rule induction algorithms were proposed [18].

It can be seen from the reviews above that a weighted fuzzy approximation space is a good solution to deal with classification rule learning from imbalanced data, but it has not been discussed in this framework so far. In this study, we focus on introducing a weighted fuzzy approximation space into rule learner so that it can balance classification accuracy on minority class in order to imply useful information from minority class. By conducting systematic comparative experiments on fifteen datasets, we find that weighted fuzzy rough set based rule learner is effective for class imbalanced and hybrid data learning.

2   WFLEM2: Rule Learner on Weighted Fuzzy Approximation Spaces

In fuzzy approximation spaces, all samples are considered equally important. In order to introduce probabilities of samples into fuzzy rough set model, the concept of weightiness to represent a prior knowledge can be employed [18].

A five-tuple WFS $= \langle U, A, w, V, f \rangle$ is denoted as a *weighted fuzzy information system* or a *weighted fuzzy approximation space*, where $U = \{x^i\}|_{i=1 \sim n}$ is a nonempty and finite universe, $A = C \cap D$ is an attribute set, $C = \{c_i\}|_{i=1 \sim m}$ is a set of conditional attributes, and $D = \{d\}$ is a set of decision attributes, $w : U \to \mathbb{R}$ is a weighted distribution function on U, $V = \cup_{a \in A} V_a$ is the domain of all attributes, $f : U \times A \to V$ is an information function. For simplicity, we use x_j^i denotes the value of $f(x^i, c_j)$. Each attribute c_j is represented by a family of fuzzy sets $Term(c_j) = \{F_j^k\}|_{k=1 \sim t_j}$. F_j^k refers to kth fuzzy set of an attribute

c_j, and t_j is equal to the number of fuzzy sets on c_j. Therefore, $\mu_{F_j^k}(x_j^i)$ is the membership degree of the ith object's value of attribute c_j on the fuzzy set F_j^k

Decision attribute d classifies each object to a single class out of q classes. Each class $l|_{l=1\sim q}$ has been modeled as a fuzzy singleton with membership degree of x^i in lth class defined as follows:

$$\mu_l(x_d^i) = \begin{cases} 1, \ if x_d^i = l \\ 0, \ otherwise \end{cases}. \tag{1}$$

A fuzzy classification rule defined in space $Term(c_1) \times \ldots \times Term(c_m) \times Term_d$ can be represented in the form of :

$$\text{IF } c_1 \text{ is } F_1^{k_1} \text{ AND } \ldots \text{ AND } c_m \text{ is } F_m^{k_m} \text{ THEN } d \text{ is } l. \tag{2}$$

A fuzzy decision rule can be put into simpler form $F_1^{k_1} \times \ldots F_m^{k_m} \Rightarrow l$. It is interpreted as a fuzzy relation from m-dimension space of conditional attributes to the space of decision attribute, where $F_V = \{F_j^{k_j}\}|_{j=1\sim m}$ is called as fuzzy evidence, which represents the fuzzy values taken by one or more fuzzy conditional attributes presented in the premise part of a rule [19].

For a given weighted fuzzy approximation space WFS, weight function w is used to present a prior knowledge about samples. While a crisp equivalence relation will generate a crisp block, a fuzzy rough set model can induce a fuzzy block. Therefore, the definition of fuzzy block can be naturally derived from fuzzy rough set model.

Definition 1 Fuzzy Block. Let WFS be a weighted fuzzy approximation space, $t = (c_j, F_j^k)$ be an attribute-value pair. The *fuzzy block* of t is defined as

$$[t] = \int_{x^i \in U} \mu_{F_j^k}(x_j^i)/x^i. \tag{3}$$

Due to the fuzzy properties, the fuzzy block is a fuzzy set. This is a main difference of fuzzy block with crisp block. It is easy to find that the definition of fuzzy block is a natural extension of crisp one. If an attribute is nominal, the relation matrix and equivalence relation with respect to it will be degraded to the classical concept of block.

Some problems encountered by fuzzy lower and upper approximation set were identified [20]. The crisp lower and upper approximation set was consequently used to solve the computation problems in real-world applications [14]. We employ a weaker definition of the inclusion operator, which is defined as that a fuzzy set $A \subseteq_\alpha B$ if and only if $\forall x \in U, \max(1 - \mu_A(x), \mu_B(x)) \geqslant \alpha$.

Definition 2 α-Fuzzily Depends On. Let X be a crisp set, T be a set of attribute-value pairs. Set X *α-fuzzily depends on* T if and only if

$$\emptyset \neq [T] = \cap\{[t] \mid t \in T\} \subseteq_\alpha X, \tag{4}$$

where $A \cap B = \int_{x \in U} \min(\mu_A(x), \mu_B(x))/x$.

Let B be a nonempty lower or upper approximation set of a concept, T be a set of attribute-value pairs. Set T is an α-*minimal complex* of B if and only if B α-fuzzily depends on T and no proper subset T' of T exists such that B α-fuzzily depends on T'.

Definition 3 (α, β)-**Weighted Local Covering.** Let WFS be a weighted fuzzy information system, $\mathbb{T}$ be a nonempty collection of nonempty sets of attribute-value pairs and B be a nonempty crisp set. $\mathbb{T}$ is an (α, β)- *weighted local covering* of B if and only if the following conditions are satisfied:

1. each member T of $\mathbb{T}$ is an α-minimal complex of B,
2. $I^w(\cup_{T \in \mathbb{T}}[T], B) \leqslant \beta$, and
3. $\mathbb{T}$ is minimal, i.e., $\mathbb{T}$ has the smallest possible number of members,

where the weighted similarity function of fuzzy sets is $I^w(A, B) = (|\overline{A} \cap B|^w + |A \cap \overline{B}|^w)/|A \cup B|^w$, $|\bullet|^w = \sum_{x \in U} \mu_\bullet(x) \times w(x)$, $\overline{A} = \int_{x \in U}(1 - \mu_A(x))/x$.

It is obvious that α and β provide a new stopping condition for rule learner. When $\alpha = 0, \beta = 0, w(x) = w, \forall x \in U$ and attributes are nominal, the stopping condition is degraded into the same condition used in algorithm LEM2. Due to the strong termination criteria of algorithm LEM2, it may extract over-specified rules that would overfit training data, and the computational complexity of the algorithm increases dramatically. In contrast with LEM2, we introduce a flexible stopping condition that can be easily tuned, by which rule learner can allow an input set partially depends on a minimal complex and tolerate a local covering that does not cover a small amount of training data. These parameters are used to relax the termination requirement of algorithms in LERS system. Such a mechanism is especially valuable in case of data containing overlapping classes and having inconsistent examples.

In LERS system, algorithms explore the search space of attribute-value pairs from the lower or upper approximation set of each concept. The searching strategy can be hardly effective because the minimal complex with a large number of attribute-value pairs is apt to be obtained. In contrast to LEM2, our rule learner tends to select the attribute-value pair with the maximum value based on a new score function on attribute-value pairs.

Definition 4 Score function of an attribute-value pair. Let t be an attribute-value pair and G be a fuzzy set. The *score* of t related to G is defined as follows:

$$Score^w(t, G) = |[t] \cap G|^w, \tag{5}$$

where $| \bullet |^w = \sum_{x \in U} \mu_\bullet(x) \times w(x)$.

Let us denote that WFS is a weighted fuzzy approximation space, B is a nonempty lower or upper approximation set of a concept set. Normal operations of fuzzy set are based on Zadeh's definition [21]. Algorithm 1 presents the pseudocode of proposed rule learner.

Algorithm 1. WFLEM2 Rule Learner

Data: A weighted information system WFS $= \langle U, A, w, V, f \rangle$, a non-empty lower or upper approximation set of a concept B and thresholds α and β.

Result: A (α, β)-weighted local covering $\mathbb{T}$.

1 $G \Leftarrow B$; // G is a fuzzy set of objects
2 $\mathbb{T} \Leftarrow \emptyset$; //$\mathbb{T}$ is a crisp set of attribute-value pairs
3 **while** $I^w(\cup_{T \in \mathbb{T}}[T], B) > \beta$ **do**
4 $\quad$ $T \Leftarrow \emptyset$;
5 $\quad$ **while** $(T = \emptyset)$ **or** (**not** ($[T] \subseteq_\alpha B$)) **do**
6 $\quad\quad$ $t \Leftarrow \arg\max_{\forall t' \in T, \text{head}(t) \neq \text{head}(t')} Score^w(t, G)$;
7 $\quad\quad$ $T \Leftarrow T \cup \{t\}$;
8 $\quad\quad$ $G \Leftarrow [t] \cap G$;
9 $\quad$ **end**
10 $\quad$ **for** $\forall t \in T$ **do**
11 $\quad\quad$ **if** $[T \setminus \{t\}] \subseteq_\alpha B$ **and** $T \setminus \{t\} \neq \emptyset$ **then**
12 $\quad\quad\quad$ $T \Leftarrow T \setminus \{t\}$;
13 $\quad\quad$ **end**
14 $\quad$ **end**
15 $\quad$ $\mathbb{T} \Leftarrow \mathbb{T} \cup \{T\}$;
16 $\quad$ $G \Leftarrow B \setminus \cup_{T \in \mathbb{T}}[T]$;
17 **end**
18 **for** $\forall T \in \mathbb{T}$ **do**
19 $\quad$ **if** $I^w(\cup_{S \in \mathbb{T} \setminus \{T\}}[S], B) \leqslant \beta$ **then**
20 $\quad\quad$ $\mathbb{T} \Leftarrow \mathbb{T} \setminus \{T\}$;
21 $\quad$ **end**
22 **end**
23 **return** $\mathbb{T}$;

Now we theoretically analyze the time complexity of WFLEM2 rule learner. Assumptions are as follows: r is the number of generated rules, and q is the number of decision classes. The complexity of operations on sets is asymptotic to the total size of sets. The cardinality of a minimal complex is $O(\log n)$, and it is not longer than m. r and q are small constants in the analysis so that we can provide general complexity estimation. The finally complexity thus is a function of n. We break the analysis process into determination of the time complexity for particular steps of *WFLEM2* procedure:

Line 1: $O(1)$
Line 2: $O(1)$
Line 3-17: Iterates at most $O(r)$ times, $O(n)$ evaluates condition
Line 4: $O(1)$
Line 5-9: Iterates at most $O(\log n)$ times, $O(n)$ evaluates condition
Line 6: $O(mn)$ one sweep through G to find pair with max score
Line 7: $O(1)$
Line 8: $O(n)$ fuzzy set operation
Line 10-14: Iterates at most $O(\log n)$ times

Line 11-13: $O(n)$ fuzzy set operation
Line 15: $O(\log n)$ set operation
Line 16: $O(n)$ fuzzy set operation
Line 18-22: Iterates at most $O(r)$ times
Line 19-21: $O(n)$
Line 23: $O(1)$

Thus, the total estimated time complexity of the *WFLEM2* procedure is: $O(r) \cdot \{O(n) + O(\log n) \cdot [O(n) + O(mn) + O(n)] + O(\log n) \cdot [O(n) + O(\log n) + O(n)]\} + O(r) \cdot O(n) = O(rmn \log n)$.

Our rule learner for the problems with q classes will call the *WFLEM2* procedure q times. Therefore, the overall complexity is $qO(rmn \log n)$. As from the assumption in the beginning of this section, r and q are usually small constants. We note that for some applications the number of attributes m can be $O(1)$, or m can be $O(\log n)$. In the worst case, time complexity of the WFLEM2 learner is $O(n \log^2 n)$. We also note that the above estimation holds for decision table that supports any hybrid attributes such as nominal, numerical or fuzzy attributes.

Definition 5 Weighted Certainty Factor of a Rule. Let WFS be a weighted fuzzy information system, $F_V \Rightarrow l$ be a fuzzy decision rule. The *weighted certainty factor* of the rule can be measured by

$$\beta^w(F_V \Rightarrow l) = \frac{\sum\limits_{x^i \in U} w(x^i) \times \{[\wedge_{F \in F_V} \mu_F(x_j^i)] \wedge \mu_l(x_d^i)\}}{\sum\limits_{x^i \in U} w(x^i) \times [\wedge_{F \in F_V} \mu_F(x_j^i)]}, \tag{6}$$

where $\wedge$ is denoted as min operator.

In LERS system, the prediction of unseen cases is done using a modification of the bucket brigade algorithm, on the basis of the rule strength, specificity, support and partial matching factor. However, in algorithm WFLEM2, a basic fuzzy inference engine can take either fuzzy measurements or crisp measurements as inputs from real world to suggest a classification or decision. In proposed rule learner, we have considered fuzzy classifiers with crisp measurements from real world as inputs. The outputs that the fuzzy classifiers produce can be either a fuzzy singleton representing a class label of the pattern or the prediction certainty with which the pattern can be classified to each class. In experimental section, WFLEM2 employs a *matching factor* based procedure to classify new objects. For a given test case, the sum of weighted certainty factor of rules that match this case within each concept is computed. The concept with the biggest value wins as target for classification of the test case.

3   Experiments and Discussion

We implemented WFLEM2 in Java and tested on some benchmark problems with the same parameters and conditions. Here, threshold parameters α and β

are set to be 0.02 and 0.05. Continuous attributes are fuzzified by fuzzy c-mean method into three fuzzy clusters and a triangle membership function is employed to approximate each fuzzy set [22]. We calculate the inverse class probability as the weight of each object, i.e., the weight of x is computed by $n/|[x]_D|$, where $[x]_D$ denotes the decision equivalence class containing object x [18].

In order to evaluate the performance of WFLEM2, comparative experiments are conducted. The methods employed for the comparison comprise a weighted decision tree based method and a classical rough set based rule learner.

1. WC4.5: A weighted decision tree method proposed by [23]. The inverse class probability weight is assigned to each sample for class imbalance learning.
2. LEM2: A rough set based rule induction algorithm proposed by [4]. The continuous attributes need to be discretized first. LEM2 uses fuzzy c-means clustering algorithm as a front-end discretization technique with pre-defined three clusters.

Table 1. Description of fifteen datasets in comparison tests. C indicates the number of continuous attributes, and N indicates the number of nominal attributes.

Abbr.	Dataset	Size	Class	Attr.(C/N)	Test data	Class distribution
hea	StatLog heart disease	270	2	7/6	10F-CV	120/150
cle	Cleve database	303	2	6/8	10F-CV	138/165
bup	BUPA liver disorders	345	2	6/0	10F-CV	145/200
ion	Ionosphere database	351	2	34/0	10F-CV	126/225
hor	Horse Colic database	368	2	7/20	10F-CV	124/244
cov	Congressional voting	435	2	0/16	10F-CV	168/267
aca	Australian credit approval	690	2	6/8	10F-CV	307/383
wib	Wisconsin breast cancer	699	2	9/0	10F-CV	241/458
pim	PIMA indian diabetes	768	2	9/0	10F-CV	268/500
ann	Annealing data	898	6	9/29	10F-CV	8/40/67/99/684
gec	StatLog German credit	1000	2	7/13	10F-CV	300/700
dna	Statlog DNA	2000	3	0/180	10F-CV	464/485/1051
hyp	Hypothyroid disease	3163	2	7/18	10F-CV	151/3012
sat	Statlog satellite image	6435	6	36/0	2000	415/470/479/961/1038/1072
adu	Adult	48842	2	6/8	16281	7841/24720

A detailed description of fifteen benchmarking datasets is presented in Table 1. The datasets were obtained from the UCI ML repository [24]. The missing values in each data set are filled with mean values for continuous attributes and majority values for nominal attributes. We focus on reporting accuracy and rule complexity of three methods on fifteen datasets.

Table 2 reports accuracy of WFLEM2 and other two learners achieved on fifteen datasets. We report the average accuracy on minority class and the average accuracy on majority class. Results for all ten-fold cross-validation experiments are mean values include standard deviations. On average, WFLEM2 obtains the highest average accuracy on minority class, with WC4.5 being the second best, followed by LEM2. LEM2 obtains the highest average overall accuracy, with WC4.5 being the second best, followed by WFLEM2. Closer observation reveals that there is no universally best learner in this comparison group.

The t-statistics test was used to compare learners for thirteen ten-fold cross-validation experiments. The 5% confidence level test was performed. Table 3

Table 2. Comparison of average accuracy between three methods on fifteen datasets

Set	Accuracy of minimum class			Accuracy of maximum class			Overall accuracy		
	WC4.5	LEM2	WFLEM2	WC4.5	LEM2	WFLEM2	WC4.5	LEM2	WFLEM2
hea	82 ±1.6	73 ±3.3	74 ±3.9	68 ±2.5	80 ±2.1	82 ±2.3	75 ±1.8	77 ±1.2	78 ±2.6
cle	55 ±5.8	50 ±2.7	57 ±5.2	57 ±2.0	59 ±1.9	56 ±2.4	56 ±3.5	58 ±2.3	55 ±3.4
bup	46 ±8.2	24 ±13.5	43 ±10.3	71 ±7.3	75 ±6.2	72 ±9.4	67 ±8.3	68 ±13.0	68 ±7.5
ion	80 ±5.9	73 ±3.1	85 ±3.8	93 ±2.7	93 ±1.5	92 ±3.0	92 ±2.3	89 ±3.6	90 ±2.4
hor	95 ±1.5	95 ±2.8	96 ±2.6	98 ±1.3	97 ±1.1	96 ±2.8	96 ±1.7	97 ±1.3	95 ±1.9
cov	98 ±2.2	97 ±3.9	97 ±2.8	97 ±3.7	96 ±3.5	96 ±2.1	97 ±2.0	97 ±1.9	97 ±1.6
aca	85 ±2.4	80 ±2.0	84 ±1.7	87 ±3.8	84 ±1.5	82 ±4.7	84 ±2.2	86 ±1.8	83 ±3.6
wib	51 ±4.6	24 ±7.5	92 ±5.2	95 ±1.7	97 ±2.5	95 ±0.8	94 ±2.8	93 ±3.6	94 ±2.3
pim	47 ±5.3	41 ±5.7	52 ±5.4	87 ±2.5	89 ±3.7	86 ±2.6	73 ±5.0	75 ±5.7	74 ±4.6
ann	98 ±1.3	100 ±0.0	100 ±0.0	100 ±0.0	100 ±0.0	100 ±0.0	99 ±1.0	100 ±0.0	100 ±0.0
gec	53 ±4.1	23 ±7.2	80 ±3.8	75 ±2.3	95 ±4.5	56 ±2.8	70 ±3.2	71 ±5.7	60 ±3.0
dna	85 ±1.5	80 ±3.4	84 ±3.8	92 ±3.2	96 ±2.6	93 ±3.9	88 ±2.2	94 ±1.9	90 ±2.9
hyp	75 ±5.6	46 ±7.8	71 ±7.3	92 ±4.2	94 ±3.7	91 ±4.9	88 ±4.3	90 ±5.2	89 ±6.8
sat	69	57	66	78	85	84	75	84	81
adu	75	74	82	80	87	80	78	81	80
Mean	73	62	78	85	88	84	82	84	82
Stdev	18.5	26.8	17.0	12.7	10.8	13.6	12.7	12.2	13.5

Table 3. Results of t-test between WFLEM2 and a second learner with respect to accuracy of minority class on thirteen ten-fold cross-validation experiments. "++" denotes WFLEM2 is significantly better than a second learner, "+−" denotes WFLEM2 is no different with a second learner and "−−" denotes WFLEM2 is significantly worse than a second learner.

Set	hea	cle	bup	ion	hor	cov	aca	wib	pim	ann	gec	dna	hyp	++	+−	−−
WC4.5	−−	+−	−−	++	+−	−−	+−	++	++	++	++	+−	+−	5	5	2
LEM2	++	++	++	++	+−	+−	++	++	++	+−	++	++	++	10	3	0

Table 4. Comparison of rule set complexity on fifteen datasets

Set	Number of rules			Length per rule		
	WC4.5	LEM2	WFLEM2	WC4.5	LEM2	WFLEM2
hea	17	32	22	3.8	2.8	2.7
cle	23	43	28	2.6	3.1	2.3
bup	18	29	16	3.7	2.7	2.8
ion	5	14	3	2.3	2.0	2.1
hor	23	76	28	3.7	3.2	3.9
cov	8	35	12	1.6	2.5	1.8
aca	23	67	21	3.8	3.2	3.7
wib	18	43	16	2.4	2.1	2.7
pim	15	76	12	2.1	2.4	2.5
ann	39	103	42	4.3	3.7	4.8
gec	41	74	38	3.4	5.8	4.3
dna	48	96	53	3.4	3.6	3.2
hyp	16	46	18	2.8	2.5	3.1
sat	95	224	103	5.6	4.8	5.2
adu	59	179	62	3.7	3.3	4.4
Mean	30	76	32	3.3	3.2	3.3
Stdev	23.5	57.6	25.5	1.0	1.0	1.0

shows the results, where each entry describes t-test outcome considering difference in accuracy obtained by WFLEM2 and some second learner given in the left column. The summary columns show that the proposed algorithm is significantly better on five datasets, significantly worse on two datasets, and no difference on five datasets, when compared with WC4.5 learner.

In Table 4, we present measurements of rule complexity. The average number of rules generated by WFLEM2 is close to the average number of rules generated by WC4.5. We note that WFLEM2 generates rule sets that are on average half the size of LEM2 series, which shows a significant improvement for LEM2. Although both the number of rules and the average length per rules are reported, the latter measure gives better indication of the average complexity of an individual rule.

To summarize, WFLEM2 is characterized by accuracy that is comparable with accuracy of other two methods. The WFLEM2 is statistically better in comparison with WC4.5 and LEM2, although for majority of datasets they characterized by results of similar quality. It generates rule sets that are comparable in size to those generated by WC4.5, albeit two times smaller than those generated by LEM2. The good accuracies and compact rule sets are observed in comparison experiments.

4   Conclusion

Classical rough set based rule learner just works in nominal domain and treats each sample as equal important weightiness. In this paper, we propose a new rule induction algorithm that can exhibits on class imbalanced and hybrid datasets. This method overcomes the poor termination condition and the limitation of classical algorithms on highly skewed or hybrid datasets. The time complexity of proposed rule learner is theoretically analyzed. Experiments show that proposed rule learner gets the same results as that of other classical methods on fifteen UCI datasets. However, the performance of the proposed method on minority class is better than the classical methods with respect to class imbalanced and hybrid datasets. Therefore, the proposed rule learner can gain better performance in case of skewed and hybrid data learning.

References

1. Wu, W.Z., Mi, J.S., Zhang, W.X.: Generalized fuzzy rough sets. Inf. Sci. 151, 263–282 (2003)
2. Liu, J., Hu, Q., Yu, D.: Weighted rough set learning: Towards a subjective approach. In: Zhou, Z.H., Li, H., Yang, Q. (eds.) PAKDD 2007. LNCS (LNAI), vol. 4426, pp. 696–703. Springer, Heidelberg (2007)
3. Grzymala-Busse, J.W.: Knowledge acquisition under uncertainty - A rough set approach. J. Intel. Rob. Syst. 1, 3–16 (1988)
4. Grzymala-Busse, J.W.: LERS - a system for learning from examples based on rough sets. In: Slowinski, R. (ed.) Intelligent Decision Support: Handbook of Applications and Advances of the Rough Sets Theory, pp. 3–18. Kluwer Academic Publishers, Dordrecht (1992)
5. Grzymala-Busse, J.W.: A new version of the rule induction system LERS. Fundamenta Informaticae 31, 27–39 (1997)

6. Grzymala-Busse, J.W.: MLEM2: A new algorithm for rule induction from imperfect data. In: Proceedings of the 9th International Conference on Information Processing and Management of Uncertainty in Knowledge-Based Systems, IPMU 2002, Annecy, France, pp. 243–250 (2002)
7. Hu, Q.: Hybrid data reduction with fuzzy rough set theory for classification. In: Jonker, W., Petković, M. (eds.) SDM 2005. LNCS, vol. 3674. Springer, Heidelberg (2005)
8. Pawlak, Z.: Rough sets. International Journal of Computer and Information Sciences, 341–356 (1982)
9. Pawlak, Z., Skowron, A.: Rudiments of rough sets. Inf. Sci. 177(1), 3–27 (2007)
10. Dubois, D., Prade, H.: Rough fuzzy sets and fuzzy rough sets. Int. J. of General Systems 17(2-3), 191–209 (1990)
11. Wu, W.Z., Mi, J.S., Zhang, W.X.: Generalized fuzzy rough sets. Inf. Sci. 151, 263–282 (2003)
12. Jensen, R., Shen, Q.: Fuzzy-rough sets assisted attribute selection. IEEE T. Fuzzy Systems 15(1), 73–89 (2007)
13. Bhatt, R.B., Gopal, M.: FRCT: fuzzy-rough classification trees. Pattern Anal. Appl. 11(1), 73–88 (2007)
14. Hu, Q., Xie, Z., Yu, D.: Hybrid attribute reduction based on a novel fuzzy-rough model and information granulation. Pattern Recognition 40(12), 3509–3521 (2007)
15. Japkowicz, N., Stephen, S.: The class imbalance problem: A systematic study. Intell. Data Anal. 6(5), 429–449 (2002)
16. Chawla, N.V., Japkowicz, N., Kotcz, A.: Editorial: special issue on learning from imbalanced data sets. SIGKDD Explorations 6(1), 1–6 (2004)
17. Hu, Q., Yu, D., Xie, Z., Liu, J.: Fuzzy probabilistic approximation spaces and their information measures. IEEE T. Fuzzy Systems 14(2), 191–201 (2006)
18. Liu, J., Hu, Q., Yu, D.: A weighted rough set based method developed for class imbalance learning. Inf. Sci. 178(4), 1235–1256 (2008)
19. Yuan, Y., Shaw, M.J.: Induction of fuzzy decision trees. Fuzzy Sets and Systems 69(2), 125–139 (1995)
20. Bhatt, R.B., Gopal, M.: On the compact computational domain of fuzzy-rough sets. Pattern Recognition Letters 26(11), 1632–1640 (2005)
21. Zadeh, L.A.: Fuzzy sets. Inf. Control 8, 338–353 (1965)
22. Pal, N.R., Bezdek, J.C.: On cluster validity for the fuzzy c-means model. IEEE-FS 3, 370–379 (1995)
23. Ting, K.M.: An instance-weighting method to induce cost-sensitive trees. IEEE Trans. Knowl. Data Eng. 14(3), 659–665 (2002)
24. Blake, C.L., Merz, C.J.: UCI Repository of ML Databases (1998), http://archive.ics.uci.edu/ml/

Feature Selection Based on the Rough Set Theory and Expectation-Maximization Clustering Algorithm

Farideh Fazayeli[1], Lipo Wang[1], and Jacek Mandziuk[2]

[1] School of Electrical and Electronic Engineering
Nanyang Technological University, Singapore 639798
{fari0004,elpwang}@ntu.edu.sg
[2] Faculty of Mathematics and Information Science Warsaw University of Technology
Plac Politechniki 1,
00-661 Warsaw, Poland
mandziuk@mini.pw.edu.pl

Abstract. We study the Rough Set theory as a method of feature selection based on tolerant classes that extends the existing equivalent classes. The determination of initial tolerant classes is a challenging and important task for accurate feature selection and classification. In this paper the Expectation-Maximization clustering algorithm is applied to determine similar objects. This method generates fewer features with either a higher or the same accuracy compared with two existing methods, i.e., Fuzzy Rough Feature Selection and Tolerance-based Feature Selection, on a number of benchmarks from the UCI repository.

1 Introduction

The problem of reducing dimensionality has been investigated for a long time in a wide range of fields, e.g., statistics, pattern recognition, machine learning, and knowledge discovery. In order to reduce the input dimensionality, there exist two main approaches, i.e., feature extraction and feature selection (FS). Feature extraction maps the primitive feature space into a new space with a lower dimensionality. Two of the most popular feature extraction approaches include Principal Components Analysis [13], and Partial Least Squares [2]. There are numerous applications of feature extraction in the literature, such as image processing [9], visualization[29], and signal processing [21]. In contrast, the FS approach chooses the most informative features from the original features according to a selection method, e.g., t -statistic [17], f -statistic [15], correlation [34], separability correlation measure [7], or information gain [32]. The irrelevant and redundant features in the dataset lead to slow learning and low accuracy. Finding the subset of features that are enough informative is NP complete. Some heuristic algorithms are proposed to search through the feature space. The selected subset can be evaluated from some issues, such as the complexity of the learning algorithm and the accuracy.

C.-C. Chan et al. (Eds.): RSCTC 2008, LNAI 5306, pp. 272–282, 2008.

The Rough Set (RS) theory can be used as a tool to reduce the input dimensionality and to deal with vagueness and uncertainty in datasets. The reduction of attributes is based on data dependencies. The RS theory partitions a dataset into some equivalent (indiscernibility) classes, and approximates uncertain and vague concepts based on the partitions. The measure of dependency is calculated by a function of the approximations. The dependency measure is employed as a heuristic to guide the FS process. In order to obtain a significant measure, proper approximations of the concepts are required. Hence, the initial partitions play an important rule. Given a discrete dataset, it is possible to find the indiscernibility classes; however, in case of datasets with real-valued attributes, it is impossible to say whether two objects are the same, or to what extent they are the same, using the indiscernibility relation. A number of research groups [6, 20, 26, 27, 28, 30] extended the RS theory using the tolerant or similarity relation (termed tolerance-based Rough Set). The similarity measure between two objects is delineated by a distance function of all attributes. Two objects are considered to be similar when their similarity measure exceeds a similarity threshold value. Finding the best threshold boundary is both important and challenging. [14] used genetic algorithms to find the best similarity threshold. [8, 10, 22, 23, 25] used fuzzy similarity to cope with real-valued attributes. In this paper we use Expectation-Maximization (EM) [3, 5, 16, 24, 33, 35] clustering algorithm to determine the tolerance classes. The EM algorithm is a general statistical method for finding the maximum likelihood estimations of parameters in probabilistic models. In particular it can be applied in clustering problems. The EM algorithm allows for overlapping clusters and it is robust to noise and to highly skewed data.

The paper is organized as follows. Section 2 summarizes basics of the RS theory. A brief overview of the mixture model and EM algorithm is represented in section 3. In Section 4, the proposed method of feature selection using the RS theory and EM clustering algorithm is outlined. Section 5 shows the potential of the proposed method on some real datasets. We discuss our results and draw some conclusions in the final section.

2   Basics of the Rough Set Theory

Let $T(U, A, C, D)$ be a decision table, where U is a universe of objects, A is a set of primitive features, C is a set of conditional attribute, D is a decision attribute or class label, and $C, D \subseteq A$. For an arbitrary set $P \subseteq A$, an indiscernibility relation is defined as follows,

$$IND(P) = \{(x, y) \in U \times U : \forall a \in P, a(x) = a(y)\} \tag{1}$$

If $P \subseteq C$ and $X \subseteq U$ then the lower and upper approximations of X, with respect to P, are respectively defined as follow,

$$\underline{P}X = \{x \in U : [x]_{IND(P)} \subseteq X\} \tag{2}$$

$$\overline{P}X = \{x \in U : [x]_{IND(P)} \cap X \neq \phi\} \tag{3}$$

where

$$[x]_{IND(P)} = \{y \in U : a(y) = a(x), \forall a \in P\} \tag{4}$$

is the equivalence class of x in $U/IND(P)$.

A P-positive region of D is a set of all objects from the universe U which can be classified with certainty to one class of $U/IND(D)$ employing attributes from P,

$$POS_P(D) = \bigcup_{x \in U/IND(D)} \underline{P}X \tag{5}$$

A dependency of D on P is defined as,

$$\gamma_p(D) = \frac{|POS_P(D)|}{|U|}. \tag{6}$$

where $|A|$ is the cardinality of a set A.

A feature $a \in C$ is dispensable in P, if $\gamma_p(D) = \gamma_{p-a}(D)$; otherwise a is an indispensable attribute in P with respect to D. An arbitrary set $B \subseteq C$ is called independent if all its attributes are indispensable.

From these definitions a reduct set of features can be defined as follows, a set of features $R \subseteq C$ is called the reduct of C, if R is independent and $POS_R(D) = POS_C(D)$. In other words, the reduct is a set of attributes that conserves the partitions generated by C.

In [4] the QUICKREDUCT algorithm for determining the reduct set is proposed. It is a heuristic algorithm that avoids exhaustively generating all possible subsets. The greedy algorithm starts with an empty set and in each iteration adds the attribute that results in the greatest increase in the rough set dependency metric to the reduct set.

3   Mixture Model and EM Algorithm

The mixture model is an effective representation of the probability density function and consists of k component density functions. The objective of a mixture model is to fit the density functions to a given dataset to approximate the data distribution. The EM alogrithm can be used in solving the problem of the mixture models where Θ is the model parameters, and unknown-random variable $Y = \{y_i\}^N$ presents each object belongs to which model. That means $y_i = k$ if the $i - th$ object belongs to the component k. The EM algorithm allows for overlapping clusters hence each object can belong to more than one component. The EM algorithm is outlined in the Appendix.

Let D be a dataset with m objects and d attributes and $\mathbf{x} \in D$ be an object in the dataset. The mixture model probability density function, evaluated at $\mathbf{x}$, is defined as follows,

$$p(\mathbf{x}|\Theta) = \sum_{l=1}^{k} W_l.p(\mathbf{x}|\theta_l) \tag{7}$$

where

- W_l is the fraction of data points belonging to the cluster l, and $\sum_{l=1}^{k} W_l = 1$.
- $p(\mathbf{x}|\theta_l)$ is the cluster or component distribution models the records of the l-th cluster.
- θ_l is the model parameters of density function of cluster l. In case of Gaussian distribution, θ_l is the mean (μ_l) and covariance matrix (Σ_l).

The complete-data log-likelihood expression for this density from the data X and Y is given by:

$$\log(L(\Theta|X,Y)) = \log(P(X,Y|\Theta)) =$$

$$\sum_{i=1}^{N} \log(P(x_i|y_i)P(y)) = \sum_{i=1}^{N} \log W_{y_i} p(x_i|\theta_{y_i}) \tag{8}$$

In this work a Gaussian distribution is used. The EM algorithm is used to determine the value of mean (μ_l), covariance matrix (Σ_l), and sampling probability (W_l) for each cluster. The attribute set will affect the distribution of data and lead to the different model parameters.

The algorithm is as follows,

1. **E Step.** For each object $\mathbf{x} \in D$, compute the membership probability of $\mathbf{x}$ in each cluster $l = 1 \cdots k$ at iteration j:

$$p(y_i|x_i, \boldsymbol{\mu}^j, \boldsymbol{\Sigma}^j) = \frac{W_{y_i}^j \cdot p(x_i|\mu_{y_i}^j, \Sigma_{y_i}^j)}{p(x_i|\boldsymbol{\mu}^j, \boldsymbol{\Sigma}^j)} \tag{9}$$

2. **M Step.** Update mixture model parameters for each cluster $l = 1, 2, \cdots, k$ that maximize the value of $Q(\Theta, \Theta^{(j)})$:

$$W_l^{j+1} = \frac{1}{N} \sum_{\mathbf{x} \in D} pr(l|\mathbf{x}) \tag{10}$$

$$\mu^{j+1,l} = \frac{\sum_{\mathbf{x} \in D} \mathbf{x} \cdot pr(l|\mathbf{x})}{\sum_{\mathbf{x} \in D} pr(l|\mathbf{x})} \tag{11}$$

$$\Sigma^{j+1,l} = \frac{\sum_{\mathbf{x} \in D} pr(l|\mathbf{x})(\mathbf{x} - \mu_{j+1,l})(\mathbf{x} - \mu_{j+1,l})^T}{\sum_{\mathbf{x} \in D} pr(l|\mathbf{x})} \tag{12}$$

3. If $|L^j - L^{j+1}| \le \epsilon$, stop. Else set $j = j+1$ and go to 1. L^j is the log likelihood of the mixture model at iteration j,

$$L^j = \sum_{\mathbf{x} \in D} \log(pr^j(\mathbf{x})) = \sum_{\mathbf{x} \in D} \log(\sum_{l=1}^{k} W_l^j \cdot pr^j(\mathbf{x}|\mu_l^j, \Sigma_l^j)) \tag{13}$$

4 Proposed Method

In the proposed method, each cluster represents a tolerance class. The tolerance classes that are generated by the EM clustering algorithm for an object x are defined as:

$$Clus_P(x) = \{Y \in U| \ x, \ and \ Y \ belongs \ to \ the \ same \ cluster\} \tag{14}$$

4.1 Approximations and Dependency

In a similar way to the original RS theory, the lower and upper approximations are then delineated as follow,

$$\underline{P}X = \{x \in U : Clus_P(x) \subseteq X\} \tag{15}$$

$$\overline{P}X = \{x \in U : Clus_P(x) \cap X \neq \phi\} \tag{16}$$

Based on this, the positive region and dependency functions can respectively be defined as follow,

$$POS_P(D) = \bigcup_{x \in U/IND(D)} \underline{P}X, \tag{17}$$

$$\acute{\gamma}_P(D) = \frac{|POS_P(D)|}{|U|} \tag{18}$$

Following the above definitions, a feature selection algorithm can be constructed that uses the tolerance-based degree of dependency, $\gamma_P(D)$, to evaluate the significance of feature subsets. The proposed FS algorithm are presented in Figure 1.

EM-CLUSTERING-REDUCT(C, D).
Inputs :
C, the set of all conditional attributes;
D, the set of decision attributes;
Output :
R, the Reduct Set

(1) $R = \phi$
(2) $\acute{\gamma}_{best} = 0$
(3) **do**
(4) $\acute{\gamma}_{tmp} = \acute{\gamma}_{best}$
(5) $T = R$
(6)   **for** x **in** $(C - R)$
(7)   **if** $\acute{\gamma}_{R \cup \{x\}}(D) > \acute{\gamma}_T(D)$
(8) $T = R \cup \{x\}$
(9) $\acute{\gamma}_{best} = \acute{\gamma}_T(D)$
(10) $R = T$
(11) **until** $\acute{\gamma}_{best} == \acute{\gamma}_{tmp}$
(12) **return** R

Fig. 1. EM Clustering QuickReduct

4.2 An Illustrative Example

In this section, a simple example is used to demonstrate the procedure of the proposed method (see Table 1). There are three continuous conditional attributes

Table 1. Example Table

Object	a	b	c	q
1	-0.4	-0.3	-0.5	0
2	-0.4	0.2	-0.1	1
3	-0.3	-0.4	-0.3	0
4	0.3	-0.3	0	1
5	0.3	-0.3	0	1
6	0.2	0	0	0

and a crisp-valued class attribute in the dataset. In this example, the number of clusters is set to 3.

The greedy algorithm starts with an empty reduct set. It checks each attribute separately and chooses the attribute that has the highest dependency degree. In this example the attribute c is chosen with the dependency degree of 0.33. Then the attribute c is added to the reduct set.

$$U/clust_{\{q\}} = \{\{1, 3, 6\}, \{2, 4, 5\}\}$$

$$U/clus_{\{a\}} = \{\{3\}, \{4, 5, 6\}, \{1, 2\}\}$$

$$\acute{\gamma}_a = \frac{|\{3\}|}{|\{1, 2, 3, 4, 5, 6\}|} = \frac{1}{6} = 0.17$$

$$U/clust_{\{b\}} = \{\{1, 3, 4, 5\}, \{6\}, \{2\}\}$$

$$\acute{\gamma}_b = \frac{|\{2, 6\}|}{|\{1, 2, 3, 4, 5, 6\}|} = \frac{2}{6} = 0.33$$

$$U/clust_{\{c\}} = \{\{3\}, \{2, 4, 5, 6\}, \{1\}\}$$

$$\acute{\gamma}_c = \frac{|\{1, 3\}|}{|\{1, 2, 3, 4, 5, 6\}|} = \frac{2}{6} = 0.33$$

$$R \leftarrow \{c\}$$

The hill climbing forward selection algorithm chooses other attributes in the reduct set as follow,

$$U/clust_{\{a,c\}} = \{\{1, 3\}, \{4, 5, 6\}, \{2\}\}$$

$$\acute{\gamma}_{a,c} = \frac{|\{1, 2, 3\}|}{|\{1, 2, 3, 4, 5, 6\}|} = \frac{3}{6} = 0.5$$

$$U/clust_{\{b,c\}} = \{\{1, 3\}, \{4, 5\}, \{2, 6\}\}$$

$$\acute{\gamma}_{b,c} = \frac{|\{1,2,4,5\}|}{|\{1,2,3,4,5,6\}|} = \frac{4}{6} = 0.67$$

$$R \leftarrow \{b,c\}$$

$$U/clust_{\{a,b,c\}} : \{\{1,3\},\{4,5,6\},\{2\}\}$$

$$\acute{\gamma}_{a,b,c} = \frac{|\{1,2,3\}|}{|\{1,2,3,4,5,6\}|} = \frac{3}{6} = 0.5$$

Finally, it returns $\{b,c\}$ as the reduct set which has the same size as the reduct set provided by the Fuzzy Rough Feature Selection (FRFS) and tolerance based FS methods in [11].

5   Simulation Result

In order to evaluate the proposed method, we applied it to a number of real datasets from the UCI repository [1] in Table 2. The EM clustering algorithm from the Weka software [31] was chosen where the number of clusters was selected empirically. The obtained reducts are evaluated via the accuracy of classification. J48, JRIP, and PART classifier in the Weka [31] are chosen as the classifier algorithms.

The obtained accuracies are compared with the accuracy of the FRFS and Tolerance-based FS in [11]. In [12] the FRFS method is compared with other FS methods (such as Relief-F, PCA, and entropy-based approaches) and has been shown that the FRFS method outperformed them. Hence in this paper, the proposed method is compared with only the FRFS and Tolerance-based FS. Table 3 shows the average classification accuracy of 10-fold cross validation as a percentage. The classification algorithms are performed on the original dataset and reduced datasets were obtained by the feature selection algorithms, i.e., the FRFS [11], the Tolerance-based FS [11], and the proposed method.

Table 2. Reduct Size For FRFS, Tolerance, and EM Clustering Methods

Dataset	Objects	Features	Reduct Size		
			FRFS[a]	Tol.[b]	EMRS[c]
Glass	214	10	9	7	5
Heart	270	14	11	10	3
Ionosphere	230	35	11	10	5
Iris	150	5	5	4	4
Water2	**390**	**39**	11	8	3
Wine	178	14	10	8	8

[a] FRFS : Fuzzy Rough Set Feature Selection [11].
[b] Tol. : Tolerance-based Feature Selection [11].
[c] EMRS : The proposed method, i.e., Feature Selection using the RS theory and EM algorithm

Table 3. Classification Accuracies(%) For Unreduced, FRFS, Tolerance, and Clustering Methods

CA[a]		J48				JRIP				PART			
Dataset	FS[b]	Original[c]	FRFS[d]	Tol.[e]	EMRS[f]	Original[c]	FRFS[d]	Tol.[e]	EMRS[f]	Original[c]	FRFS[d]	Tol.[e]	EMRS[f]
Glass		67.29	69.63	69.16	69.16	69.16	67.76	67.76	69.16	67.76	68.22	69.62	69.16
Heart		76.67	78.89	80.37	79.59	79.63	81.85	82.59	79.59	73.33	78.52	80.37	79.59
Ionosphere		87.83	91.30	87.39	88.32	86.96	86.52	86.96	86.61	88.26	91.30	86.52	90.03
Iris		96.00	96.00	96.00	96.00	95.33	95.33	94.67	95.33	94.00	94.00	95.33	95.33
Water2		83.33	80.26	81.79	81.77	81.03	80.51	82.31	81.57	85.64	82.56	81.28	82.34
Wine		94.38	92.14	94.94	94.94	91.57	90.45	94.38	92.7	93.82	93.82	94.38	94.38

[a] CA : Classification Algorithm.
[b] FS : Feature Selection Algorithm used for each Classification Algorithm.
[c] Original : Original dataset.
[d] FRFS : Fuzzy Rough Set Feature Selection [11].
[e] Tol. : Tolerance-based Feature Selection [11].
[f] EMRS : The proposed method, i.e., Feature Selection using the RS theory and EM algorithm

It is evident from Table 2 that the proposed method generated fewer features compared with the two other FS methods. For the J48 classifier, the clustering based FS improved the average accuracy of the unreduced datasets except for the water2 dataset. The proposed method either unchanged or improved upon the performance of the reduced datasets with the other two FS algorithms in all but in the Ionosphere dataset. For the JRip classifier, the proposed method maintained the average accuracy of the unreduced datasets in all. It either improved or maintained the performance of the reduced dataset with the other two FS algorithms in all but two cases. For PART, the proposed method improved the average accuracy of unreduced datasets in all except the water2 dataset. It has the same behavior as the other two FS methods.

Overall, the proposed algorithm produced a smaller number of attributes compared to the other two FS algorithms and the average accuracy of classifiers is improved or in a few instances remains unchanged. For example, in the water2 dataset the proposed method chose 3 features among 39 features whereas the FRFS chose 10 and the Tolerance-based FS method chose 8 features. In addition, the proposed method has a similar average accuracy compared with the other two approaches.

6 Conclusion

In this work the EM clustering algorithm was applied to deal with the problem of determining initial tolerant classes to obtain a significant classification accuracy. Through some experiments, it was concluded that the proposed method generated a smaller size of feature sets in all datasets compared with the FRFS [11] and tolerance-based FS methods [11]. Beside that, the proposed method either improved or unchanged the average accuracy in all except a few datasets. For future work, an improvement of searching algorithm for finding the reduct set with the new definition of approximations is required. In Addition, an evaluation of the proposed method through experimental comparisons with the other methods in the literature is recommended.

References

[1] Asuncion, A., Newman, D.J.: UCI Machine Learning Repository. University of California, School of Information and Computer Science, Irvine (2007), http://www.ics.uci.edu/~mlearn/MLRepository.html

[2] Boulesteix, A.L.: PLS Dimension Reduction for Classification with Microarray Data. Statistical Applications in Genetics and Molecular Biology 3(1), Article 33 (2004)

[3] Bradley, P., Fayyad, U., Reina, C.: Scaling EM clustering to large databases. Technical report, Microsoft Research (1999)

[4] Chouchoulas, A., Shen, Q.: Rough Set-Aided Keyword Reduction for Text Categorisation. Applied Artificial Intelligence 15, 843–873 (2001)

[5] Dempster, A.P., Laird, N.M., Rubin, D.: Maximum likelihood estimation from incomplete data via the EM algorithm. Journal of The Royal Statistical Society 39, 1–38 (1977)

[6] Doherty, P., Szalas, A.: On the Correspondence between Approximations and Similarity. In: Tsumoto, S., Słowiński, R., Komorowski, J., Grzymała-Busse, J.W. (eds.) RSCTC 2004. LNCS (LNAI), vol. 3066, pp. 143–152. Springer, Heidelberg (2004)

[7] Fu, X., Wang, L.: Data dimensionality reduction with application to simplifying rbf network structure and improving classification performance. IEEE Transactions on Systems, Man, and Cybernetics, Part B 33, 399–409 (2003)

[8] Greco, S., Inuiguchi, M., Slowinski, R.: Fuzzy rough sets and multiple-premise gradual decision rules. International Journal of Approximate Reasoning 41, 179–211 (2006)

[9] Hancock, P., Burton, A., Bruce, V.: Face processing: Human perception and principal components analysis (1996)

[10] Jensen, R., Shen, Q.: Semantics-Preserving Dimensionality Reduction: Rough and Fuzzy-Rough-Based Approaches. IEEE Transactions on Knowledge and Data Engineering 16, 1457–1471 (2004)

[11] Jensen, R., Shen, Q.: Tolerance-based and Fuzzy-Rough Feature Selection. In: Proceedings of the 16th International Conference on Fuzzy Systems (FUZZ-IEEE 2007), pp. 877–882 (2007)

[12] Jensen, R., Shen, Q.: Fuzzy-Rough Sets Assisted Attribute Selection. IEEE Transactions on Fuzzy Systems 15, 73–89 (2007)

[13] Kambhatla, N., Leen, T.K.: Dimension Reduction by Local Principal Component Analysis. Neural Comp. 9, 1493–1516 (1997)

[14] Kim, D.: Data classification based on tolerant rough set. Pattern Recognition 34, 1613–1624 (2001)

[15] Lai, Y., Wu, B., Chen, L., Zhao, H.: Statistical method for identifying differential gene-gene co-expression patterns. Bioinformatics, 3146–3155 (2004)

[16] Ordonez, C., Cereghini, P.: SQLEM: Fast clustering in SQL using the EM algorithm. In: ACM SIGMOD Conference (2000)

[17] Pan, W.: A comparative review of statistical methods for discovering differentially expressed genes in replicated microarray experiments. Bioinformatics 18, 546–554 (2002)

[18] Pawlak, Z.: Rough Sets. International Journal of Computer and Information Sciences 11, 341–356 (1982)

[19] Pawlak, Z.: Rough Sets, Theoretical Aspects of Reasoning about Data. Kluwer Academic Publishers, Dordrecht (1991)

[20] Polkowski, L., Semeniuk-Polkowska, M.: On Rough Set Logics Based on Similarity Relations. Fundam. Inf. 64, 379–390 (2005)

[21] Porrill, J., Stone, J.: Independent components analysis for signal separation and dimension reduction (1997)

[22] Radzikowska, A., Kerre, E.E.: Fuzzy rough sets based on residuated lattices. In: Peters, et al. (eds.), vol. 228, pp. 278–296.

[23] Radzikowska, A., Kerre, E.E.: A comparative study of fuzzy rough sets. Fuzzy Sets and Systems 126, 137–155 (2002)

[24] Roweis, S., Ghahramani, Z.: A unifying review of Linear Gaussian Models. Neural Computation (1999)

[25] Roy, A., Pal, S.K.: Fuzzy discretization of feature space for a rough set classifier. Pattern Recogn. Lett. 24, 895–902 (2003)

[26] Skowron, A., Stepaniuk, J.: Tolerance approximation spaces. Fundam. Inf. 27, 245–253 (1996)

[27] Slowinski, R., Vanderpooten, D.: Similarity relation as a basis for rough approximations. In: Wang, P. (ed.) Advances in Machine Intelligence and Soft Computing, vol. 4, pp. 17–33. Duke University Press, Duke (1997)

[28] Slowinski, R., Vanderpooten, D.: A Generalized Definition of Rough Approximations Based on Similarity. IEEE Trans. on Knowl. and Data Eng. 12, 331–336 (2000)

[29] Torkkola, K.: Feature extraction by non-parametric mutual information maximization. Journal of Machine Learning Research 3, 1415–1438 (2003)

[30] Vakarelov, D.: A modal characterization of indiscernibility and similarity relations in Pawlaks information systems. In: Slezak, et al. (eds.), vol. 300, pp. 12–22 (plenary talk)

[31] Witten, I.H., Frank, E.: Data Mining: Practical machine learning tools and techniques, 2nd edn. Morgan Kaufmann, San Francisco (2005)

[32] Wu, Y., Zhang, A.: Feature selection for classifying high-dimensional numerical data. CVPR 2, 251–258 (2004)

[33] Xu, L., Jordan, M.: On convergence properties of the EM algorithm for Gaussian mixtures. Neural Computation 7 (1995)

[34] Yu, L., Liu, H.: Feature selection for high-dimensional data: A fast correlation-based filter solution. In: Fawcett, T., Mishra, N. (eds.) ICML, pp. 856–863. AAAI Press, Menlo Park (2003)

[35] Yuille, A.L., Stolorz, P., Utans, J.: Statistical physics, mixtures of distributions and the EM algorithm. Neural Computation 6, 334–340 (1994)

Appendix: EM Algorithm

Maximum Likelihood (ML) is a famous method for finding the model parameters for complete data. In case of incomplete data, the EM algorithm can be used for determining the parameters. Assume X is some observation data which is incomplete, and $Z = (X, Y)$ be a complete data with the density function,

$$p(\mathbf{z}|\Theta) = p(\mathbf{x}, \mathbf{y}|\Theta) = p(\mathbf{y}|\mathbf{x}, \Theta)p(\mathbf{x}|\Theta) \tag{19}$$

The complete data likelihood is defined as,

$$L(\Theta|Z) = L(\Theta|X, Y) = p(X, Y|\Theta). \tag{20}$$

The problem is to find Θ which makes the maximume likelihood for the complete data. In this case the unknown-random variable Y leads to have a variable likelihood. The EM algorithm is used for finding the parameters in 2 steps namely Expectaion Step (E-step) and Maximization Step (M-step).

In the E-step, the expected value of log-likelihood of the complete data is determined as follow,

$$Q(\Theta, \Theta^{(i-1)}) = E[\log(p(X, Y|\Theta)|X, \Theta^{(i-1)})] =$$
$$\int_{\mathbf{y} \in \Upsilon} \log p(X, \mathbf{y}|\Theta) f(\mathbf{y}|X, \Theta^{(i-1)}) d\mathbf{y} \tag{21}$$

where the notations are as follow,

X Observed-incomplete data and is constant.

$\Theta^{(i-1)}$ Current estimation of the parameter Θ and is constant.

Y Unlnown-Random variable with a presumably governed by an underlying distribution $f(\mathbf{y}|X, \Theta^{(i-1)})$.

Θ Normal variable. The objective is to adjust Θ to obtain the maximum likelihood for the complete data Z.

Then, the M-step is applied to determine the value of the Θ in the iteration i that maximizes the expected value of log-likelihood of the complete data,

$$\Theta^{(i)} = \arg\max_{\Theta} Q(\Theta, \Theta^{(i-1)}). \tag{22}$$

The EM algorithm iterates both steps alternatively till converge.

Outlier Detection Based on Granular Computing

Yuming Chen, Duoqian Miao, and Ruizhi Wang

Department of Computer Science and Technology,
The Key Laboratory of Embedded System and Service Computing, Tongji University
Shanghai, 201804, P.R. China
cym0620@163.com

Abstract. As an emerging conceptual and computing paradigm of information processing, granular computing has received much attention recently. Many models and methods of granular computing have been proposed and studied. Among them was the granular computing model using information tables. In this paper, we shall demonstrate the application of this granular computing model for the study of a specific data mining problem - outlier detection. Within the granular computing model using information tables, this paper proposes a novel definition of outliers - GrC (granular computing)-based outliers. An algorithm to find such outliers is also given. And the effectiveness of GrC-based method for outlier detection is demonstrated on three publicly available databases.

Keywords: Granular computing, outlier detection, rough sets, data mining.

1 Introduction

L. A. Zadeh introduced the concept of granular computing in 1979 under the name of information granularity [2]. And the term "granular computing" came to life with a suggestion from T. Y. Lin in the discussion of BISC Special Interest Group on Granular Computing [3]. Basic ingredients of granular computing are granules such as subsets, classes, and clusters of a universe. Furthermore, Andrzej Skowron, et al. introduced the discovery of information granules and information granules in distributed environment [4-5]. D. Q. Miao, et al. proposed an approach to web mining based on granular computing [6-9]. Specially, Y. Y. Yao and N. Zhong proposed a granular computing model using information tables [1, 10]. In an information table, each object of a finite nonempty universe is described by a finite set of attributes. Based on attribute values of objects, one may decompose the universe into parts called granules. Objects in each granule share the same or similar description in terms of their attribute values. Within this model, various methods for the construction, interpretation, and representation of granules were examined. Although the model is simple, it is powerful for the study of fundamental issues in granular computing, and has many potential applications in data mining.

Data mining is an important issue in the development of data- and knowledge-base systems. Usually, the tasks of data mining can be classified into four general

C.-C. Chan et al. (Eds.): RSCTC 2008, LNAI 5306, pp. 283–292, 2008.
© Springer-Verlag Berlin Heidelberg 2008

categories: (a) dependency detection, (b) class identification, (c) class description, and (d) outlier/exception detection [11]. In contrast to most tasks of data mining, outlier detection aims to find small groups of data objects that are exceptional when compared with the rest large amount of data, in terms of certain sets of properties. For many applications, such as fraud detection in E-commerce, it is more interesting to find the rare events than to find the common ones, from a data mining standpoint.

Outliers exist extensively in the real world. While there is no single, generally accepted, formal definition of an outlier, Hawkins' definition captures the spirit: an outlier is an observation that deviates so much from other observations as to arouse suspicions that it was generated by a different mechanism [11-12].

With increasing awareness on outlier detection in literatures, more concrete meanings of outliers are defined for solving problems in specific domains. But to our best knowledge, there are few works about outlier detection in granular computing community [14]. In this paper, we aim to exploit the granular computing model using information tables proposed by Yao for outlier detection. The basic idea is as follows. Given an information table $S = (U, A, V, f)$, where U is a non-empty finite set of objects, A a set of attributes, V the union of attribute domains, and $f : U \times A \to V$ a function such that for any $x \in U$ and $a \in A$, $f(x, a) \in V_a$. In S, each attribute subset $B \subseteq A$ determines an indiscernibility relation $IND(B)$ on U. $IND(B)$ induces a partition of U, which is denoted by $U/IND(B)$, where each element from $U/IND(B)$ is a granule (equivalence class), and the element containing $x \in U$ is called the granule containing x under relation $IND(B)$. For a given object $x \in U$ and a set of indiscernibility relations (available information/knowledge) on U, we can obtain a granule containing x under each of these indiscernibility relations. Then through calculating the degree of outlierness for each of these granules containing x, we can decide whether object x behaves normally according to the given knowledge at hand. That is, if the degrees of outlierness of the granules containing x under these indiscernibility relations are always very high, then we may consider object x as a GrC (granular computing)-based outlier in U wrt S. A GrC-based outlier in U wrt S is an element such that the granules containing it always have a high degree of outlierness in view of the given knowledge.

The paper is organized as follows. In the next section, we introduce some preliminaries that are relevant to this paper. In section 3, based on the granular computing model using information tables, we give the definition of GrC-based outliers. An algorithm to find GrC-based outliers is also given. In section 4 we give the experimental results. Section 5 concludes the paper.

2   Preliminaries

An information table is a quadruple $S = (U, A, V, f)$, where:

1. U is a non-empty finite set of objects;
2. A is a non-empty finite set of attributes;

3. V is the union of attribute domains, i.e., $V = \bigcup_{a \in A} V_a$, where V_a denotes the domain of attribute a;

4. $f : U \times A \rightarrow V$ is an information function such that for any $a \in A$ and $x \in U$, $f(x, a) \in V_a$.

In an information table $S = (U, A, V, f)$, each subset $B \subseteq A$ of attributes determines a binary relation $IND(B)$, called indiscernibility relation, defined as $IND(B) = \{(x, y) \in U \times U : \forall a \in B(f(x, a) = f(y, a))\}$.

For any two objects $u_1, u_2 \in U$, if $(u_1, u_2) \in IND(B)$ then one cannot differentiate u_1 from u_2 based solely on their values on attributes of B. We say that u_1 and u_2 are indistinguishable. Since each indiscernibility class may be viewed as a granule consisting of indistinguishable elements, u_1 and u_2 may be put into the same granule.

Given any $B \subseteq A$, relation $IND(B)$ induces a partition of U, which is denoted by $U/IND(B)$, where an element from $U/IND(B)$ is called an equivalence class or elementary set. Each equivalence class of relation $IND(B)$ is a granule. The equivalence class of $IND(B)$ that contains object $x \in U$, written $[x]_B$, is defined by collecting all objects whose value on each attribute $a \in B$ is the same as x's value:

$$[x]_B = \{y \in U : \forall a \in B(f(y, a) = f(x, a))\} \tag{1}$$

For every object $x \in U$, $[x]_B$ is called the granule containing x under relation $IND(B)$. When B is a singleton subset of A, the elements in $U/IND(B)$ are called *elementary granules*, as they are the smallest granules derivable. From the elementary granules, large granules may be built by taking a union of a family of elementary granules. One can build a hierarchy of granules [1].

3   GrC-Based Outlier Detection

3.1   Definitions

Given an information table S, we first define a granular outlier factor (GOF), which can indicate the degree of outlierness for every granule in the granular computing model using S [13]. Then we define an object outlier factor (OOF) by virtue of GOF, which can indicate the degree of outlierness for every object.

Definition 1 (Distance Between Granules). *Let $S = (U, A, V, f)$ be an information table. Given any $B \subseteq A$, relation $IND(B)$ induces a partition of U, which is denoted by $G = U/IND(B)$, where each equivalence class of relation $IND(B)$ is a granule. For any two granules $g_1, g_2 \in G$, the distance between granules g_1 and g_2 in S is defined as follows:*

$$M(g_1, g_2) = \frac{\sum_{p \in g_1, q \in g_2} \delta(p, q)}{|g_1| \times |g_2|} \tag{2}$$

where $M : G \times G \rightarrow [0, \infty]$ is a distance function such that for any $g_1, g_2 \in G$, $M(g_1, g_2)$ denotes the distance between sets g_1 and g_2. And δ is a given distance metric on U for nominal attributes.

In the above definition, to calculate the distance between any two granules, we consider the average distance between the objects in the analyzed two granules, which is adopted in the average linkage algorithm of hierarchical clustering [15].

Definition 2 (Granular Outlier Factor). Let $S = (U, A, V, f)$ be an information table. Given any $B \subseteq A$, relation $IND(B)$ induces a partition of U, which is denoted by $G = U/IND(B)$, where each equivalence class of relation $IND(B)$ is a granule. For any granule $g \in G$, the granular outlier factor of g in S is defined as follows:

$$GOF(g) = \frac{|\{g' \in G : M(g, g') > d\}|}{|G|} \qquad (3)$$

where $M(g, g')$ denotes the distance between granules g_1 and g_2, d is a given parameter, and $|K|$ denotes the cardinality of set K.

Definition 3 (Object Outlier Factor). Let $S = (U, A, V, f)$ be an information table. For any $x \in U$, the object outlier factor of x in S is defined as

$$OOF(x) = \frac{\sum\limits_{a \in A} (GOF([x]_{\{a\}}) \times W_{\{a\}}(x))}{|A|} \qquad (4)$$

where for every singleton subset $\{a\}$ of A, $W_{\{a\}} : U \to (0, 1]$ is a weight function such that for any $x \in U$, $W_{\{a\}}(x) = 1 - \frac{|[x]_{\{a\}}|}{|U|}$. $[x]_{\{a\}} = \{u \in U : f(u, a) = f(x, a)\}$ denotes the indiscernibility class of relation $IND(\{a\})$ that contains element x, i.e. the elementary granule containing x under relation $IND(\{a\})$. $GOF([x]_{\{a\}})$ denotes the granular outlier factor of granule $[x]_{\{a\}}$ and $|K|$ denotes the cardinality of set K.

In the above definition, we can see that in the granular computing model using information table S, only those elementary granules are used to calculate the object outlier factor. We do not consider other granules in the model.

Furthermore, the weight function W in the above definition expresses such an idea that outlier detection always concerns the minority of objects in the data set and the minority of objects are more likely to be outliers than the majority of objects. Since from the above definition, we can see that the more the weight, the more the object outlier factor, the minority of objects should have more weight than the majority of objects. Therefore for every $a \in A$, if the elementary granule containing x under relation $IND(\{a\})$ is small with respect to other elementary granules under relation $IND(\{a\})$, then we consider x belonging to the minority of objects in U, and assign a high weight to x.

Definition 4 (Granular Computing-based Outliers). Let $S = (U, A, V, f)$ be an information table. Let μ be a given threshold value, for any $x \in U$, if $OOF(x) > \mu$ then x is called a granular computing (GrC)-based outlier in S, where $OOF(x)$ is the object outlier factor of x in S.

3.2 Algorithm

In the worst case, the time complexity of algorithm 3.1 is $O(m \times n^2)$, and its space complexity is $O(m \times n^2)$, where m and n are the cardinalities of A and U respectively.

Algorithm 3.1

Input: information table $S = (U, A, V, f)$, where $|U| = n$ and $|A| = m$; threshold value μ, d

Output: a set O of GrC-based outliers

---

(1) For any two objects $u_1, u_2 \in U$, calculate the distance between them under a given distance metric on U, that is, $\delta(u_1, u_2)$;

(2) For every $a \in A$

(3) {

(4)   Sort all objects from U according to a given order (e.g. the lexicographical order) on domain V_a of attribute a [16];

(5)   Determine the partition $U/IND(\{a\})$;

(6)   For any $g_1, g_2 \in U/IND(\{a\})$, calculate the distance

$$M(g_1, g_2) = \frac{\sum_{p \in g_1, q \in g_2} \delta(p, q)}{|g_1| \times |g_2|}$$

(7) }

(8) For every $x \in U$

(9) {

(10)   For every $a \in A$

(11) {

(12)   Calculate the granular outlier factor of $[x]_{\{a\}}$ in S, i.e.

$$GOF([x]_{\{a\}}) = \frac{|\{g' \in U/IND(\{a\}): M([x]_{\{a\}}, g') > d\}|}{|U/IND(\{a\})|};$$

(13)   Assign weight $W_{\{a\}}(x) = 1 - \frac{|[x]_{\{a\}}|}{|U|}$ to x

(14) }

(15)   Calculate $OOF(x)$, the object outlier factor of object x in S;

(16)   If $OOF(x) > \mu$ then $O = O \cup \{x\}$

(17) }

(18) Return O.

---

4 Experimental Results

4.1 Experiment Design

In this section, following the experimental setup in [17], we use three real life data sets (*lymphography, annealing* and *cancer*) to demonstrate the performance of our algorithm against traditional distance-based method [11], FindCBLOF algorithm [18] and KNN algorithm [19]. In addition, on the *cancer* data set, we add the results of RNN-based outlier detection method for comparison, these results can be found in the work of Harkins et al. [20, 21].

For algorithm 3.1, in order to calculate the distance between any two granules, we should first calculate the distances between objects contained in these granules under a given distance metric on U. In our experiment, we adopt the *overlap metric in rough set theory*, which is defined as follows:

Definition 5. *Given an information table $S = (U, A, V, f)$, let $x, y \in U$ be any two objects between which we shall calculate the distance. The overlap metric in rough set theory is defined as*

$$\Delta(x, y) = |\{a \in A : a(x) \neq a(y)\}| \tag{5}$$

where $\Delta : U \times U \to [0, \infty]$ is a function from $U \times U$ to the non-negative real number, and $|M|$ denotes the cardinality of set M.

And in algorithm 3.1, in order to calculate the granular outlier factor for a given granule, we should specify a value for parameter d, we set $d = |A| /2$ in our experiment, where $|A|$ denotes the cardinality of attribute set A.

Furthermore, in our experiment, the two parameters needed by FindCBLOF algorithm are set to 90% and 5 separately as done in [18]. And for the KNN algorithm, the results were obtained by using the 4th nearest neighbor [19].

4.2   Lymphography Data

The first is the Lymphography data set, which can be found in the UCI machine learning repository [22]. It contains 148 instances with 19 attributes (including the class attribute). The 148 instances are partitioned into 4 classes: "normal find" (1.35%), "metastases" (54.73%), "malign lymph" (41.22%) and "fibrosis" (2.7%). Classes 1 and 4 are regarded as rare classes.

Aggarwal et. al. proposed a practicable way to test the effectiveness of an outlier detection method [17, 23]. That is, we can run the outlier detection method on a given data set and test the percentage of points which belonged to one of the rare classes (Aggarwal considered those kinds of class labels which occurred in less than 5% of the data set as rare labels [23]). Points belonged to the rare class are considered as outliers. If the method works well, we expect that such abnormal classes would be over-represented in the set of points found.

The experimental results are summarized in table 1.

Table 1. Experimental Results in Lymphography Data Set

Top Ratio (Number of Objects)	Number of Rare Class Included (Coverage)			
	GrC	DIS	FindCBLOF	KNN
5%(7)	6(100%)	5(83%)	4(67%)	5(83%)
6%(9)	6(100%)	6(100%)	4(67%)	5(83%)
8%(12)	6(100%)	6(100%)	4(67%)	6(100%)
20%(30)	6(100%)	6(100%)	6(100%)	6(100%)

In table 1, "GrC", "DIS", "FindCBLOF", "KNN" denote GrC-based, traditional distance-based, FindCBLOF and KNN-based outlier detection methods, respectively. For every objects in U, the degree of outlierness is calculated by using the four outlier detection methods, respectively. For each outlier detection method, the "Top Ratio (Number of Objects)" denotes the percentage (number) of the objects selected from U whose degrees of outlierness calculated by the method are higher than those of other objects in U. And if we use $X \subseteq U$ to contain all those objects selected from U, then the "Number of Rare Class Included" is the number of objects in X that belong to one of the rare classes. The "Coverage" is the ratio of the "Number of Rare Class Included" to the number of objects in U that belong to one of the rare classes [17].

From table 1, we can see that for the lymphography data set, GrC-based method performs best, since it can find all outliers in U when the *Top Ratio* reaches 5%. The next one is distance-based method, which can find all outliers in U when the *Top Ratio* reaches 6%. And the worst is FindCBLOF method, since it can not achieve that goal until the *Top Ratio* reaches 20%.

Furthermore, for the lymphography data set, the *false alarm rates* (i.e., the percentage of objects in set X that are actually not outliers, where X is the set of the top-n objects with highest degrees of outlierness calculated by the given method, n is the number of outliers in U) of GrC-based, distance-based, FindCBLOF and KNN-based method are 17%, 17%, 33% and 33%, respectively.

4.3   Annealing Data

The Annealing data set is found in the UCI machine learning repository [22]. The data set contains 798 instances with 38 attributes. The data set contains a total of 5 (non-empty) classes : class 1, 2, 3, 5 and U, where class 3 has 608 instances, and the remained classes have 190 instance. Classes 1, 2, 5 and U are regarded as rare classes since they are small in size. Since Annealing data set contains 6 continuous attributes, we respectively transform these continuous attributes into categorical attributes by using the automatic discretization functionality provided by the CBA software [24].

The experimental results are summarized in table 2.

Table 2. Experimental Results in Annealing Data Set

Top Ratio (Number of Objects)	Number of Rare Class Included (Coverage)			
	GrC	DIS	FindCBLOF	KNN
10%(80)	75(39%)	73(38%)	45(24%)	21(11%)
15%(105)	96(51%)	92(48%)	55(29%)	30(16%)
20%(140)	128(67%)	121(64%)	82(43%)	41(22%)
25%(175)	161(85%)	153(81%)	105(55%)	58(31%)
30%(209)	190(100%)	178(94%)	105(55%)	62(33%)

Table 2 is similar to table 1. From table 2, we can see that for the Annealing data set, GrC-based method performs the best among the four outlier detection methods. In fact, the performances of GrC-based and distance-based methods are very close, and they perform markedly better than the other two methods — FindCBLOF and KNN-based methods.

Furthermore, for the Annealing data set, the false alarm rates of GrC-based, distance-based, FindCBLOF and KNN-based method are 6%, 12%, 45% and 68%, respectively.

4.4   Wisconsin Breast Cancer Data

The Wisconsin breast cancer data set is found in the UCI machine learning repository [22]. The data set contains 699 instances with 9 continuous attributes. Here we follow the experimental technique of Harkins et al. by removing some of the *malignant* instances to form a very unbalanced distribution [17, 20-21]. The resultant data set had 39 (8%) *malignant* instances and 444 (92%) *benign* instances. Moreover, the 9 continuous attributes in the data set are transformed into categorical attributes, respectively [1] [17].

The experimental results are summarized in table 3.

Table 3. Experimental Results in Wisconsin Breast Cancer Data Set

Top Ratio (Number of Objects)	Number of Rare Class Included (Coverage)				
	GrC	DIS	FindCBLOF	RNN	KNN
1%(4)	4(10%)	4(10%)	4(10%)	3(8%)	4(10%)
2%(8)	7(18%)	5(13%)	7(18%)	6(15%)	7(18%)
4%(16)	14(36%)	11(28%)	14(36%)	11(28%)	13(33%)
6%(24)	21(54%)	18(46%)	21(54%)	18(46%)	20(51%)
8%(32)	28(72%)	24(62%)	27(69%)	25(64%)	27(69%)
10%(40)	32(82%)	29(74%)	32(82%)	30(77%)	32(82%)
12%(48)	37(95%)	36(92%)	35(90%)	35(90%)	38(97%)
14%(56)	39(100%)	39(100%)	38(97%)	36(92%)	39(100%)
16%(64)	39(100%)	39(100%)	39(100%)	36(92%)	39(100%)
18%(72)	39(100%)	39(100%)	39(100%)	38(97%)	39(100%)
20%(80)	39(100%)	39(100%)	39(100%)	38(97%)	39(100%)
28%(112)	39(100%)	39(100%)	39(100%)	39(100%)	39(100%)

From table 3, we can see that for the breast cancer data set, GrC-based method performs the best among the five outlier detection methods, except in the case when *Top Ratio* is 12%. In fact, the performances of GrC-based, FindCBLOF and KNN-based methods are very close, and they perform markedly better than the other two methods — RNN-based and distance-based methods.

---

[1]   The resultant data set is public available at:
  http://research.cmis.csiro.au/rohanb/outliers/breast-cancer/

Furthermore, for the Wisconsin breast cancer data set, the false alarm rates of GrC-based, distance-based, FindCBLOF, RNN-based and KNN-based method are 18%, 26%, 21%, 23% and 18%, respectively.

5   Conclusion

Finding outliers is an important task for many data mining applications. In this paper, we present a new method for outlier definition and outlier detection, which exploits the granular computing model using information tables proposed by Yao [1]. The main idea is that an object has more likelihood of being an outlier if the granules containing it have a high degree of outlierness. Experimental results on real data sets demonstrate the effectiveness of our method for outlier detection. In the next work, we may consider to further reduce the time complexity of our algorithm for finding GrC-based outliers.

Acknowledgements. This work is supported by the Natural Science Foundation (Grant Nos. 60475019 and 60775036), and the Specialized Research Fund for the Doctoral Program of Higher Education of China (Grant No. 20060247039)

References

1. Yao, Y.Y., Zhong, N.: Granular computing using information tables. In: Lin, T.Y., Yao, Y.Y., Zadeh, L.A. (eds.) Data Mining, Rough Sets and Granular Computing, pp. 102–124. Physica-Verlag (2002)
2. Zadeh, L.A.: Fuzzy sets and information granularity. In: Gupta, N., Ragade, R., Yager, R. (eds.) Advances in Fuzzy Set Theory and Applications, pp. 3–18. North-Holland, Amsterdam (1979)
3. Zadeh, L.A.: Some reflections on soft computing, granular computing and their roles in the conception, design and utilization of information/intelligent systems. Soft Computing 2(1), 23–25 (1998)
4. Skowron, A., Stepaniuk, J.: Towards discovery of information granules. In: Żytkow, J.M., Rauch, J. (eds.) PKDD 1999. LNCS (LNAI), vol. 1704, pp. 542–547. Springer, Heidelberg (1999)
5. Skowron, A., Stepaniuk, J.: Information Granules in Distributed Environment. In: Zhong, N., Skowron, A., Ohsuga, S. (eds.) RSFDGrC 1999. LNCS (LNAI), vol. 1711, pp. 357–366. Springer, Heidelberg (1999)
6. Miao, D.Q., Wang, G.Y., Liu, Q., et al.: Granular Computing Past, Present and Future Prospect. Science Press, Beijing (2007) (in Chinese)
7. Duan, Q.G., Miao, D.Q., Zhang, H.Y., Zheng, J.: Personalized Web Retrieval based on Rough-Fuzzy Method. Journal of Computational Information Systems 3(3), 1067–1074 (2007)
8. Duan, Q.G., Miao, D.Q., Wang, R.Z., Chen, M.: An Approach to Web Page Classification based on Granules. In: Proc. of 2007 IEEE/WIC/ACM Int. Conf. on Web Intelligence (WI 2007), Silicon Valley, USA, vol. 2-5, pp. 279–282 (2007)
9. Miao, D.Q., Chen, M., Wei, Z.H., Duan, Q.G.: A Reasonable Rough Approximation of Clustering Web Users. In: Zhong, N., Liu, J., Yao, Y., Wu, J., Lu, S., Li, K. (eds.) Web Intelligence Meets Brain Informatics. LNCS (LNAI), vol. 4845, pp. 428–442. Springer, Heidelberg (2007)

10. Yao, Y.Y.: A partition model of granular computing. LNCS Transactions on Rough Sets, vol. 1, pp. 232–253 (2004)
11. Knorr, E., Ng, R.: Algorithms for Mining Distance-based Outliers in Large Datasets. In: Proc. of the 24th VLDB Conf., New York, pp. 392–403 (1998)
12. Hawkins, D.: Identifications of Outliers. Chapman and Hall, London (1980)
13. Breunig, M.M., Kriegel, H.-P., Ng, R.T., Sander, J.: LOF: identifying density-based local outliers. In: Proc. of the 2000 ACM SIGMOD Int. Conf. on Management of Data, Dallas, pp. 93–104 (2000)
14. Jiang, F., Sui, Y.F., Cao, C.G.: Outlier Detection Using Rough Set Theory. In: Ślęzak, D., Yao, J., Peters, J.F., Ziarko, W., Hu, X. (eds.) RSFDGrC 2005. LNCS (LNAI), vol. 3642, pp. 79–87. Springer, Heidelberg (2005)
15. Johnson, S.C.: Hierarchical Clustering Schemes. Psychometrika 2, 241–254 (1967)
16. Nguyen, S.H., Nguyen, H.S.: Some efficient algorithms for rough set methods. In: Proc. of the 6th Int. Conf. on Information Processing and Management of Uncertainty (IPMU 1996), Granada, Spain, pp. 1451–1456 (1996)
17. He, Z.Y., Deng, S.C., Xu, X.F.: An Optimization Model for Outlier Detection in Categorical Data. In: Int. Conf. on Intelligent Computing (ICIC(1) 2005), Hefei, China, pp. 400–409 (2005)
18. He, Z.Y., Deng, S.C., Xu, X.F.: Discovering Cluster Based Local Outliers. Pattern Recognition Letters 24(9-10), 1651–1660 (2003)
19. Ramaswamy, S., Rastogi, R., Shim, K.: Efficient algorithms for mining outliers from large datasets. In: Proc. of the 2000 ACM SIGMOD Int. Conf. on Management of Data, Dallas, pp. 427–438 (2000)
20. Harkins, S., He, H.X., Willams, G.J., Baxter, R.A.: Outlier detection using replicator neural networks. In: Proc. of the 4th Int. Conf. on Data Warehousing and Knowledge Discovery, France, pp. 170–180 (2002)
21. Willams, G.J., Baxter, R.A., He, H.X., Harkins, S., Gu, L.F.: A Comparative Study of RNN for Outlier Detection in Data Mining. In: Proc. of the 2002 IEEE Int. Conf. on Data Mining (ICDM 2002), Japan, pp. 709–712 (2002)
22. Bay, S.D.: The UCI KDD repository (1999), http://kdd.ics.uci.edu
23. Aggarwal, C.C., Yu, P.S.: Outlier detection for high dimensional data. In: Proc. of ACM SIGMOD Int. Conf. on Managment of Data, California, pp. 37–46 (2001)
24. Liu, B., Hsu, W., Ma, Y.: Integrating Classification and Association Rule Mining. In: Proc.of the 4th Int. Conf. on Knowledge Discovery and Data Mining (KDD 1998), New York, pp. 80–86 (1998)

Implementing a Rule Generation Method Based on Secondary Differences of Two Criteria

Hidenao Abe and Shusaku Tsumoto

Shimane University
89-1 Enya-cho Izumo Shimane, 6938501, Japan
abe@med.shimane-u.ac.jp, tsumoto@computer.org

Abstract. In order to obtain valuable knowledge from stored data on database systems, rule mining is considered as one of the usable mining method. However, almost current rule mining algorithms only use primary difference of a criterion to select attribute-value pairs to obtain a rule set to a given dataset. In this paper, we implemented a rule generation method based on secondary differences of two criteria. Then, we performed a case study using UCI common datasets. With regarding to the result, we compared the accuracies of rule sets learned by our algorithm with that of three representative algorithms.

1 Introduction

In recent years, enormous amounts of data have been stored on information systems in natural science, social science, and business domains. People have been able to obtain valuable knowledge due to the development of information technology. Beside, data mining has been well known for utilizing data stored on database systems. In particular, if-then rules, which are produced by rule mining algorithms, are considered as one of the highly usable and readable outputs of data mining.

Considering tradeoff of two criteria when selecting an attribute-value pair for a closure of rules, primary difference is so naïve to obtain an adequate volume of rules. Since such rule mining method searches attribute-value space[1] exhaustibly [1], their outputs become enormous number of rules.

Considering above mentioned issue, Tsumoto [2] proposed a search strategy to obtain rules, which treat the tradeoff of accuracy and coverage using secondary differences. Therefore, we implemented the idea as a rule mining method.

In this paper, we describe the difference between our proposed method and other representative rule mining method in Section 2. Then, the detail of our method is described in Section 3. In Section 4, we show a result of case study using an implementation of our method. Finally, we conclude this paper in Section 5.

2 Related Work

There are many conventional studies about rule learning algorithms, which are most popular learning algorithms in the machine learning field.

---

[1] Maximum number of rules of dataset having n attribute-value pairs is 2^{n+1}.

C.-C. Chan et al. (Eds.): RSCTC 2008, LNAI 5306, pp. 293–298, 2008.

As rule mining algorithms, there are the following major approaches: separate-and-conquer [3], methods based on divide-and-conquer, reinforcement learning.

Their studies of separate-and-conquer algorithms, which are also called covering algorithms, include many famous algorithms such as AQ family of algorithms [4] and Version Space (VS) [5]. C4.5Rule [6] is based on the decision tree learned with information gain ratio called C4.5, which is classified as the divide-and-conquer approach.

Although separate-and-conquer approach has been developed for decades, many new algorithms are developed introducing ideas from the other viewpoints such as APRIORI-C [7]. These algorithms share the following top-level loop: an algorithm searches for a rule that explains a part of its training instances, separates these examples, and recursively conquers the remaining examples by learning more rules until no examples remain.

Focusing on the search strategy of rule learning algorithms, they use one simple criterion, such as precision as shown in VS and old AQ family of algorithms. Besides, to treat multiple criteria, other groups of algorithms use combined criterion such as strength of each rule and information gain as shown in Classifier Systems [8], ITRule [9], C4.5 Rule, and PART [10]. There is no algorithm handling two different criteria, because it is a hard work to treat the tradeoff between generality and specificity when an algorithm obtains each rule. APRIORI-C (or so-called predictive Apriori) can use two criteria to search rules from possible rule space. However, they do not treat the tradeoff, but searching the space exhaustively.

3 A Rule Mining Method Using Secondary Differences

Tsumoto proposed a rule generation algorithm using secondary differences of two different criteria, α and κ, to generate rules holding both of high accuracy and high coverage[2]. The search space of the algorithm is shown in Figure 1(a) as the gray colored region. Figure 1(b) shows the search space of exhaustive search. To similar, Figure 1(c) shows the search space of the algorithms, which don't consider the tradeoff.

Figure 2 shows the search strategies of our proposed algorithm. In this figure, $\Delta\alpha(i,i+1)$ and $\Delta\kappa(i,i+1)$ are the primary differences of α and κ. Also, $\Delta^2\alpha(i,i+1,i+2)$ and $\Delta^2\kappa(i,i+1,1+2)$ are the secondary differences of α and κ. For each R, these differences are calculated the following equations to dataset D, where i means the length of the consequents of R.

$$\Delta\alpha(i,i+1) = \alpha_{R(i+1)}(D) - \alpha_{R(i)}(D) \tag{1}$$

$$\Delta\kappa(i,i+1) = \kappa_{R(i+1)}(D) - \kappa_{R(i)}(D) \tag{2}$$

$$\Delta^2\alpha(i,i+1,i+2) = \Delta\alpha(i+1,i+2) - \Delta\alpha(i,i+1) \tag{3}$$

$$\Delta^2\kappa(i,i+1,i+2) = \Delta\kappa(i+1,i+2) - \Delta\kappa(i,i+1) \tag{4}$$

---

[2] Accuracy is also called precision or confidence. Coverage is also called recall.

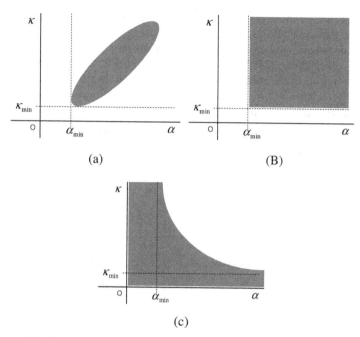

Fig. 1. Search spaces (gray colored) of the three search strategy

Strategy-I: Selecting rules with $\Delta\alpha > 0 \cap \Delta\kappa = \min(\Delta\kappa)$
Strategy-II: Selecting rules with $\Delta^2\alpha = \min(\Delta^2\alpha) \cap \Delta^2\kappa > 0$

Fig. 2. Search strategies of our rule learning algorithm using secondary differences

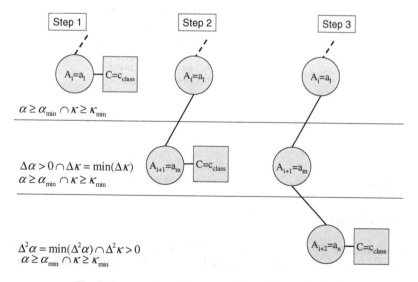

Fig. 3. An overview of the steps of the rule generation

The overview of the algorithm is shown in Figure 3. Using given two criteria and their lower threshold values, α and κ, by a user, the algorithm firstly obtains the rules with one clause in their consequent. Then, another clause is added to these rules. The rules, which don't satisfy the strategy-I in Figure 2, are pruned. The remaining rules are stored in to the rule set, and go to next step. In the next step, the rules are added another clause again. Then, the rules, which don't satisfy the strategy-II, are pruned. Storing rules, which satisfy α_{min} and κ_{min}, on each step, the algorithm iterates these steps for each attribute $i(i = 1,2,..., A-2)$ [3] and class value.

Figure 4 shows a pseudo code of the algorithm.

```
Input: Dataset,Attributes=A_{n-1}+Class, Alpha_{min}, Kappa_{min}
Output: Ruleset

Begin:
    for(class=0; class<ClassNum; class++){
        for(i=0; i<n-1; i++){
            inclementAntecedent(A_i, Ruleset, C_{class});
            calculateAlpha(Dataset, Ruleset);
            calculateKappa(Dataset, Ruleset);
            selectRules(Ruleset, =>Alphamin);
            selectRules(Ruleset, =>Kappamin);
            for(j=i+1; j<n-1; j++){
                inclementAntecedent(A_j, Ruleset, C_{class});
                calculateAlpha(Dataset, Ruleset);
                calculateKappa(Dataset, Ruleset);
                selectDelta(Ruleset, Alpha,>0);
                selectDelta(Ruleset, Kappa, min);
                selectRules(Ruleset, =>Alphamin);
                selectRules(Ruleset, =>Kappamin);
                for(k=i+2; k<n-1; k++){
                    inclementAntecedent(A_k, Ruleset, C_{class});
                    calculateAlpha(Dataset, Ruleset);
                    calculateKappa(Dataset, Ruleset);
                    selectDelta2(Ruleset, Alpha, min);
                    selectDelta2(Ruleset, Kappa, >0);

                    selectRules(Ruleset, =>Alphamin);
                    selectRules(Ruleset, =>Kappamin);

                }
            }
        }
    }
End;
```

Fig. 4. Pseudo code of the rule learning algorithm using secondary differences

4   Experiment

In this section, we describe about a case study of an implementation of the algorithm explained in Section 3. We implemented the algorithm in Java, combining a rule

---

[3] A: Number of Attributes of a dataset D.

evaluation index calculation module called COIN [11]. Using the implementation, we generated rule sets to the five datasets from UCI Machine Learning Repository [12]. The numerical attributes in these datasets, we discretized each attributes into ten bins with equalized width. For example, the number of possible rules of iris, which has four numerical attributes, is $2^{(44+1)}$. Then, the accuracies of the rule sets are compared with that of OneR [13], PART, and unpruned J4.8, which are implemented in Weka [14].

In this experiment, we specified precision and recall to search for rule sets. Precision shows the correct rate of the prediction of each rule as shown in Equation 5. In similar, recall shows the rate of correctly predicted instances in the dataset D for each class, as shown in Equation 6.

$$Precision_R = P(D|R) \qquad (5)$$

$$Recall_R = P(R|D) \qquad (6)$$

Our rule learning method also needs lower thresholds, α_{min} and κ_{min}. We set up these lower thresholds as $Precision_{min} = 0.5$ and $Recall_{min} = 0.3$ in this experiment.

Table 1 shows the averaged accuracies of each algorithm to the eight dataset. These accuracies are obtained with 100 times repeated 10-fold cross validation. The accuracy of the rule sets outperform their of OneR on several datasets.

Table 1. The average accuracies (%) of the four rule learning algorithms and standard deviations (SDs) of the accuracies

Dataset	Proposed Algorithm		OneR		PART		J4.8 (unpruned)	
	Acc.	SD	Acc.	SD	Acc.	SD	Acc.	SD
iris	68.5	17.0	96.0	4.7	94.8	5.5	96.0	4.7
balance-scale	75.1	3.6	57.7	3.5	76.3	4.7	64.5	4.6
glass	38.3	12.9	50.8	9.5	55.7	9.5	57.6	8.8
breast-cancer	73.8	7.7	67.1	5.9	69.7	7.1	73.9	5.6
diabetes	65.1	4.6	74.4	4.3	73.3	4.8	74.0	4.0

The disadvantage of the accuracies is caused by the two major reasons: lower threshold values and the confliction avoid strategy. The given lower threshold values were not optimized to obtain accurate rule sets. In addition, we avoid conflictions of rules using "better precision first", when predicting the class for each test instance. The strategy should be selected an adequate one to predict test instances more correctly.

5   Conclusion and Future Work

In this paper, we described a rule mining algorithm using secondary difference of two criteria. The result of the case study in Section 4 shows that our proposed algorithm can obtain the rule sets with lower correct rates, comparing with the four representative rule learning algorithms. However, our algorithm can obtain a kind of valid rule sets from different point of view, considering that our method searches for different space of the given datasets.

In the future, we will evaluate usefulness of the proposed method with actual medical data, comparing with the other metrics. Then, we will also obtain rule sets with pairs of objective rule evaluation indices, which have different functional behaviors [15].

References

1. Furnkranz, J., Flach, P.A.: ROC 'n' rule learning: towards a better understanding of covering algorithms. Machine Learning 58(1), 39–77 (2005)
2. Tsumoto, S.: Accuracy and coverage in rough set rule induction. In: 11th International Conference on Information Processing and Management of Uncertainty in Knowledge-Based Systems (2006)
3. Furnkranz, J.: Separate-and-Conquer Rule Learning. Artificial Intelligence Review 13(1), 3–54 (1999)
4. Michalski, R.S.: On the QuasiMinimal Solution of the Covering Problem. In: Proceedings of the 5th International Symposium on Information Processing (FCIP 1969) (Switching Circuits), vol. A3, pp. 125–128 (1969)
5. Mitchell, T.M.: Generalization as Search. Artificial Intelligence 18(2), 203–226 (1982)
6. Quinlan, J.R.: Programs for Machine Learning. Morgan Kaufmann, San Francisco (1992)
7. Jovanoski, V., Lavrac, N.: Classification Rule Learning with APRIORI-C. In: Brazdil, P.B., Jorge, A.M. (eds.) EPIA 2001. LNCS (LNAI), vol. 2258, pp. 44–51. Springer, Heidelberg (2001)
8. Booker, L.B., Holland, J.H., Goldberg, D.E.: Classifier Systems and Genetic Algorithms. Artificial Intelligence 40, 235–282 (1989)
9. Goodman, R.M., Smyth, P.: The induction of probabilistic rule sets—the Itrule algorithm. In: Proceedings of the sixth international workshop on Machine Learning, pp. 129–132 (1989)
10. Frank, E., Witten, I.H.: Generating accurate rule sets without global optimization. In: The Fifteenth International Conference on Machine Learning, pp. 144–151 (1998)
11. COIN Project, http://coin.sourceforge.jp/
12. Asuncion, A., Newman, D.J.: UCI Machine Learning Repository. University of California, School of Information and Computer Science, Irvine (2007), http://www.ics.uci.edu/~mlearn/MLRepository.html
13. Holte, R.C.: Very simple classification rules perform well on most commonly used datasets. Machine Learning 11, 63–91 (1993)
14. Witten, I.H., Frank, E.: Data Mining: Practical Machine Learning Tools and Techniques with Java Implementations. Morgan Kaufmann, San Francisco (2000)

Lower and Upper Approximations of Rules in Non-deterministic Information Systems

Hiroshi Sakai[1], Ryuji Ishibashi[1], and Michinori Nakata[2]

[1] Mathematical Sciences Section, Department of Basic Sciences
Faculty of Engineering, Kyushu Institute of Technology
Tobata, Kitakyushu 804, Japan
sakai@mns.kyutech.ac.jp
[2] Faculty of Management and Information Science
Josai International University
Gumyo, Togane, Chiba 283, Japan
nakatam@ieee.org

Abstract. A rule in a *Deterministic Information System* (*DIS*) is often defined by an implication τ such that both $support(\tau) \geq \alpha$ and $accuracy(\tau) \geq \beta$ hold for the threshold values α and β. In a *Non-deterministic Information System* (*NIS*), there are *derived DISs* due to the information incompleteness. The definition of a rule in a *DIS* is extended to the *lower* and *upper approximations* of a rule in a *NIS*. This definition explicitly handles non-deterministic information and incomplete information. To implement the utility programs for two approximations, *Apriori* algorithm is extended. Even though the number of derived *DISs* increases in exponential order, this extended algorithm does not depend upon the number of derived *DISs*. A prototype system is implemented, and this system is applied to some data sets.

Keywords: Rough sets, Non-deterministic information, Incomplete information, Rule generation, Lower and upper approximations, Apriori algorithm.

1 Introduction

We follow rule generation in *DISs* [12,13,17], and we describe rule generation in *NISs*. *NISs* were proposed by Pawlak [12], Orłowska [10,11] and Lipski [7,8] to handle information incompleteness in *DISs*, like null values, unknown values, missing values. Since the emergence of incomplete information research, *NISs* have been playing an important role.

The following shows some important research on rule generation from incomplete information. In [7,8], Lipski showed a question-answering system besides an axiomatization of logic. Orłowska established rough set analysis for non-deterministic information [10,11], and Grzymala-Busse developed a system named *LERS*, which depends upon *LEM*1 and *LEM*2 algorithms [3,4]. Stefanowski and Tsoukias also defined non symmetric similarity relations and valued tolerance relations for analyzing incomplete information [18]. Kryszkiewicz

C.-C. Chan et al. (Eds.): RSCTC 2008, LNAI 5306, pp. 299–309, 2008.

proposed a framework of rules in incomplete information systems [5,6]. According to authors' knowledge, these are the most important research on incomplete information. We have also focused on the semantic aspect for incomplete information [9], and proposed *Rough Non-deterministic Information Analysis* (*RNIA*) [14].

In this paper, we continue the framework of rule generation in $NISs$ [14,15,16], and propose the lower and upper approximations of rules in $NISs$. Here, we briefly survey the basic definitions in $RNIA$. A *Deterministic Information System* (*DIS*) is a quadruplet $(OB, AT, \{VAL_A | A \in AT\}, f)$. Let us consider two sets $CON \subseteq AT$ which we call *condition attributes* and $DEC \subseteq AT$ which we call *decision attributes*. An object $x \in OB$ is *consistent* (with any distinct object $y \in OB$), if $f(x, A) = f(y, A)$ for every $A \in CON$ implies $f(x, A) = f(y, A)$ for every $A \in DEC$.

A *Non-deterministic Information System* (*NIS*) is also a quadruplet $(OB, AT, \{VAL_A | A \in AT\}, g)$, where $g : OB \times AT \rightarrow P(\cup_{A \in AT} VAL_A)$ (a power set of $\cup_{A \in AT} VAL_A$). Every set $g(x, A)$ is interpreted as that there is an actual value in this set but this value is not known. For a $NIS = (OB, AT, \{VAL_A | A \in AT\}, g)$ and a set $ATR \subseteq AT$, we name a $DIS = (OB, ATR, \{VAL_A | A \in ATR\}, h)$ satisfying $h(x, A) \in g(x, A)$ a *derived DIS* (*for ATR*) from a NIS.

For a set $ATR = \{A_1, \cdots, A_n\} \subseteq AT$ and every $x \in OB$, let $PT(x, ATR)$ denote the Cartesian product $g(x, A_1) \times \cdots \times g(x, A_n)$. We name every element a *possible tuple* (*for ATR*) of x. For $\zeta = (\zeta_1, \cdots, \zeta_n) \in PT(x, ATR)$, let $[ATR, \zeta]$ denote a formula $\bigwedge_{1 \le i \le n} [A_i, \zeta_i]$. Let $PI(x, CON, DEC)$ ($x \in OB$) denote a set $\{[CON, \zeta] \Rightarrow [DEC, \eta] | \zeta \in PT(x, CON), \eta \in PT(x, DEC)\}$. We name an element of $PI(x, CON, DEC)$ a *possible implication* (*from CON to DEC*) of x. If $PI(x, CON, DEC)$ is a singleton set $\{\tau\}$, we say τ (from x) is *definite*. Otherwise, we say τ (from x) is *indefinite*.

2 An Example and Definitions

Let us consider Table 1. This is an exemplary NIS. In Table 1, every attribute value is not a value but a set. We usually interpret an indefinite attribute value as that "the real attribute value is in this set, but the real value is uncertain".

Table 1. A Table of NIS

OB	Temperature	Headache	Nausea	Flu
1	$\{high\}$	$\{yes, no\}$	$\{no\}$	$\{yes\}$
2	$\{high, very\_high\}$	$\{yes\}$	$\{yes\}$	$\{yes\}$
3	$\{normal, high, very\_high\}$	$\{no\}$	$\{no\}$	$\{no\}$
4	$\{high\}$	$\{yes\}$	$\{yes, no\}$	$\{yes\}$
5	$\{high\}$	$\{yes, no\}$	$\{yes\}$	$\{no\}$
6	$\{normal\}$	$\{yes\}$	$\{yes, no\}$	$\{no\}$
7	$\{normal\}$	$\{no\}$	$\{yes\}$	$\{no\}$
8	$\{normal, high, very\_high\}$	$\{yes\}$	$\{yes, no\}$	$\{yes\}$

In our previous research, we named a possible case in a NIS a *derived DIS* from a NIS, and extended several rough sets based concepts according to derived $DISs$. In Table 1, there are 576 ($2^6 \times 3^2$) derived $DISs$. For an implication τ, τ^x denotes an implication from object $x \in OB$. For $\tau : [Nausea, no] \Rightarrow [Flu, yes]$, τ^1 appears in 576 derived $DISs$, and τ^4 appears in 288 derived $DISs$.

Rules in a *Deterministic Information System* (DIS) are often defined by a set (for given threshold values α and β) as follows:

$$Rule(\alpha, \beta) = \{\tau | \tau \text{ is an implication}, support(\tau) \geq \alpha \text{ and } accuracy(\tau) \geq \beta\}.$$

We extend this definition to the lower and upper approximations of rules in $NISs$ as follows:

(1) $DD(\tau^x) = \{$derived $DIS | \tau^x$ appears in $DIS\}$.
(2) *Lower approximation of a set of rules (from x) in a NIS*:
$Rule(x, \alpha, \beta, LA) = \{\tau^x | support(\tau^x) \geq \alpha \text{ and } accuracy(\tau^x) \geq \beta$ hold
  for each derived DIS in $DD(\tau^x)\}$.
(3) *Lower approximation of a set of rules in a NIS*:
$Rule(\alpha, \beta, LA) = \cup_{x \in OB} Rule(x, \alpha, \beta, LA)$.
(4) *Upper approximation of a set of rules (from x) in a NIS*:
$Rule(x, \alpha, \beta, UA) = \{\tau^x | support(\tau^x) \geq \alpha \text{ and } accuracy(\tau^x) \geq \beta$ hold
  for some derived $DISs$ in $DD(\tau^x)\}$.
(5) *Upper approximation of a set of rules in a NIS*:
$Rule(\alpha, \beta, UA) = \cup_{x \in OB} Rule(x, \alpha, \beta, UA)$.

In order to divide definite and indefinite implications in $Rule(\alpha, \beta, LA)$, we may employ the following:

(6) *Lower approximation of a set of definite rules in a NIS*:
$Rule(\alpha, \beta, LA, def) = \{\tau \in Rule(\alpha, \beta, LA) | \tau \text{ is definite}\}$,
(7) *Lower approximation of a set of indefinite rules in a NIS*:
$Rule(\alpha, \beta, LA, indef) = \{\tau \in Rule(\alpha, \beta, LA) | \tau \text{ is indefinite}\}$,

Intuitively, every $\tau \in Rule(\alpha, \beta, LA, def)$ is a possible implication, which is not influenced by the information incompleteness at all. Similarly, every $\tau \in Rule(\alpha, \beta, UA)$ is a possible implication, which satisfies the conditions in some derived $DISs$. These two approximations depend upon the number of derived $DISs$. It increases in exponential order. Therefore, a method depending upon the number of derived $DISs$ will not be applicable to large data sets.

3   Background of This Work

Now, we describe the background of this work. We have already found some algorithms which do not depend upon the number of derived $DISs$ [15,16]. The most important definition is the following:

For a descriptor $[A_i, \zeta_i]$ and a conjunction $\wedge_i[A_i, \zeta_i]$,
$Descinf([A_i, \zeta_i])=\{x \in OB| \ g(x, A_i)=\{\zeta_i\}\}$.
$Descinf(\wedge_i[A_i, \zeta_i])=\cap_i Descinf([A_i, \zeta_i])$.
$Descsup([A_i, \zeta_i])=\{x \in OB| \ \zeta_i \in g(x, A_i)\}$.
$Descsup(\wedge_i[A_i, \zeta_i])=\cap_i Descsup([A_i, \zeta_i])$.

$Descinf$ and $Descsup$ are the minimum and the maximum sets for an equivalence class, respectively. In Table 1, the following holds:

$Descinf([Temperature, high])=\{1, 4, 5\}$,
$Descinf([Headach, yes] \wedge [Nausea, yes])=\{2, 4, 6, 8\} \cap \{2, 5, 7\}=\{2\}$,
$Descsup([Temperature, high])=\{1, 2, 3, 4, 5, 8\}$,
$Descsup([Headach, yes] \wedge [Nausea, yes])=\{1, 2, 4, 5, 6, 8\} \cap \{2, 4, 5, 6, 7, 8\}$.

Result 1. [15,16] For each τ^x there is an algorithm, which does not depend upon the size of $DD(\tau^x)$, to calculate the following:

$minsupp(\tau^x)=Min_{\psi \in DD(\tau^x)}\{support(\tau^x) \ in \ \psi\}$,
$maxsupp(\tau^x)=Max_{\psi \in DD(\tau^x)}\{support(\tau^x) \ in \ \psi\}$,
$minacc(\tau^x)=Min_{\psi \in DD(\tau^x)}\{accuracy(\tau^x) \ in \ \psi\}$,
$maxacc(\tau^x)=Max_{\psi \in DD(\tau^x)}\{accuracy(\tau^x) \ in \ \psi\}$.

For example, If $\tau^x : [CON, \zeta] \Rightarrow [DEC, \eta]$ is definite,

$minsupp(\tau^x)=|Descinf([CON, \zeta]) \cap Descinf([DEC, \eta])|//|OB|$,
$minacc(\tau^x)=\frac{|Descinf([CON,\zeta]) \cap Descinf([DEC,\eta])|}{|Descinf([CON,\zeta])|+|OUTACC|}$.

$OUTACC=[Descsup([CON, \zeta]) - Descinf([CON, \zeta])] - Descinf([DEC, \eta])$.
(A sketch of this proof) The proof of $minsupp(\tau^x)$ is trivial, so we show an overview of $minacc(\tau^x)$. The details are in [15,16]. Let $NUME$ be the amount of τ in a DIS, and $DENO$ be the amount of condition part in τ. Then, $accuracy(\tau^x)$ is a ratio $NUME/DENO$. Let us consider an object y $(y \neq x)$ satisfying $\tau^y \in PI(y, CON, DEC)$. If τ^y is definite, τ^y occurs in every derived $DISs$. Therefore, this object y belongs to both $NUME$ and $DENO$. On the other hand, if τ^y is indefinite, this object y influences the value of $accuracy(\tau^x)$. For an indefinite τ^y, there are three cases in the following:

(CASE 1) $[CON, \zeta] \Rightarrow [DEC, \eta'] \in PI(y, CON, DEC)$ $(\eta' \neq \eta)$,
(CASE 2) $[CON, \zeta'] \Rightarrow [DEC, \eta] \in PI(y, CON, DEC)$ $(\zeta' \neq \zeta)$,
(CASE 3) $[CON, \zeta'] \Rightarrow [DEC, \eta'] \in PI(y, CON, DEC)$ $(\eta' \neq \eta, \zeta' \neq \zeta)$.

Furthermore, $NUME/DENO \leq (NUME + K)/(DENO + K)$ $(K > 0)$ holds, so the occurrence of τ^y causes to increase $accuracy(\tau^x)$. Therefore, for every object y we do not select a possible implication τ^y. Instead of τ^y, we first select a possible implication in (CASE 1). If (CASE 1) does not hold, we select a possible implication in (CASE 2). Every object y in (CASE 1) belongs to just $DENO$, and a set $OUTACC$ defines such object y. Every object y in (CASE 2) does not belong to $NUME$ nor $DENO$. In this way, we obtain the formula of $minacc(\tau^x)$. This selection of cases specifies the attribute values in a NIS, therefore some derived $DISs$ are also specified as a side effect. Because there is no indefinite τ^y in every specified DIS, the numerator of $support(\tau^x)$ becomes $|Descinf([CON, \zeta]) \cap Descinf([DEC, \eta])|$. In every specified DIS, $support(\tau^x)$ becomes minimum. Similarly, it is possible to derive $maxsupp(\tau^x)$ and $maxacc(\tau^x)$, and we obtained the next result.

Result 2. [16] For each τ^x there is a derived DIS_{worst}, where both $support(\tau^x)$ and $accuracy(\tau^x)$ are minimum. There is also a derived DIS_{best}, where both $support(\tau^x)$ and $accuracy(\tau^x)$ are maximum.

We call DIS_{worst} a *derived DIS* with the *worst condition* for τ^x. We also call DIS_{best} a *derived DIS* with the *best condition* for τ^x. In Table 2, two derived $DISs$ with the worst condition for $\tau^1 : [Temperature, high] => [Flu, yes]$ are shown. These two $DISs$ are obtained as follows: Since $[Temperature, \zeta] \Rightarrow$

Table 2. For an implication $\tau^1 : [Temperature, high] => [Flu, yes]$ in Table 1, there exist two derived $DISs$ with the worst condition. The $*$ symbol shows the difference of two $DISs$. Here, $minsupp(\tau^1)=0.25$ and $minacc(\tau^1)=0.5$ hold.

OB	Temperature	Flu
1	high	yes
2	very_high	yes
3	high	no
4	high	yes
5	high	no
6	normal	no
7	normal	no
8	normal*	yes

OB	Temperature	Flu
1	high	yes
2	very_high	yes
3	high	no
4	high	yes
5	high	no
6	normal	no
7	normal	no
8	very_high*	yes

Table 3. For an implication $\tau^1 : [Temperature, high] => [Flu, yes]$ in Table 1, there exist two derived $DISs$ with the best condition. The $*$ symbol shows the difference of two $DISs$. Here, $maxsupp(\tau^1)=0.5$ and $maxacc(\tau^1)=0.8$ hold.

OB	Temperature	Flu
1	high	yes
2	high	yes
3	normal*	no
4	high	yes
5	high	no
6	normal	no
7	normal	no
8	high	yes

OB	Temperature	Flu
1	high	yes
2	high	yes
3	very_high*	no
4	high	yes
5	high	no
6	normal	no
7	normal	no
8	high	yes

$[Flu, \eta]$ is definite for objects 1,4,5,6 and 7, the attribute values of these objects are fixed. Objects 2,3 and 8 influence the value of $accuracy(\tau^1)$. In order to reduce it, we select a possible implication $[Temperature, high] \Rightarrow [Flu, no]$ in (CASE 1), and the attribute values of object 3 are fixed to $high$ and no. For objects 2 and 8, we select a possible implication $[Temperature, \zeta] \Rightarrow [Flu, yes]$ ($\zeta \neq high$) in (CASE 2). In this way, two derived $DISs$ with the *worst condition* for τ^1 are obtained. Generally, a DIS_{worst} may not be unique and a DIS_{best} may not be unique, either.

Proposition 1. The following holds.

(1) $\tau \in Rule(\alpha, \beta, LA)$ holds, if and only if there is an implication τ^x such that $support(\tau^x) \geq \alpha$ and $accuracy(\tau^x) \geq \beta$ hold in a DIS_{worst} for τ^x.
(2) $\tau \in Rule(\alpha, \beta, UA)$ holds, if and only if there is an implication τ^x such that $support(\tau^x) \geq \alpha$ and $accuracy(\tau^x) \geq \beta$ hold in a DIS_{best} for τ^x.

4   Extended Apriori Algorithms and a Real Execution

We follow *Apriori* algorithm in transaction data [1,2], and extend it to algorithms in $NISs$. *Apriori* algorithm employs a large item set, which corresponds to an equivalence class for a descriptor. On the other hand, Algorithm 1 employs two classes, i.e., $Descinf$ and $Descsup$. Due to Proposition 1 and the manipulation of $Descinf$ and $Descsup$, Algorithm 1 can pick up derived $DISs$ with the worst and the best conditions for τ^x and it calculates four values, i.e., $minsupp(\tau^x)$, $minacc(\tau^x)$, $maxsupp(\tau^x)$ and $maxacc(\tau^x)$. In this way, Algorithm 1 handles non-deterministic information as well as deterministic information. In Algorithm 1, it takes twice steps of *Apriori* algorithm for manipulating $Descinf$ and $Descsup$. Since the rest is the same, the complexity of Algorithm 1 is almost the same as *Apriori* algorithm.

The following shows a real execution about Table 1. Every program is implemented in C on a Windows PC with Pentium 4 (3.2GHz). We first apply Microsoft Excel to make the following data set *flu.csv* in Table 1. For handling indefinite attribute values, we employ a list notation, for example, $[high, very\_high]$.

```
high,"[yes,no]",no,yes                /* table data */
"[high,very_high]",yes,yes,yes
"[normal,high,very_high]",no,no,no
high,yes,"[yes,no]",yes
high,"[yes,no]",yes,no
normal,yes,"[yes,no]",no
normal,no,yes,no
"[normal,high,very_high]",yes,"[yes,no]",yes
```

Algorithm 1. Extended Apriori Algorithm for Lower Approximation: $Rule(\alpha, \beta, LA)$

Input : A NIS, a decision attribute DEC, threshold value α and β.
Output: Every rule defined by $Rule(\alpha, \beta, LA)$.
for (*every* $A \in AT$) **do**
| Generate $Descinf([A, \zeta])$ and $Descsup([A, \zeta])$;
end
For the condition $minsupp(\tau)=|SET|/|OB| \geq \alpha$, obtain the number NUM of
     elements in SET;
Generate a set $CANDIDATE(1)$, which consists of descriptors $[A, \zeta_A]$
     satisfying either ($CASE$ A) or ($CASE$ B) in the following;
     ($CASE$ A) $|Descinf([A, \zeta_A])| \geq NUM$,
     ($CASE$ B) $|Descinf([A, \zeta_A])| = (NUM - 1)$ and
            $(Descsup([A, \zeta_A]) - Descinf([A, \zeta_A])) \neq \{\}$.
Generate a set $CANDIDATE(2)$ according to the following procedures;
     (Proc 2-1) For every $[A, \zeta_A]$ and $[DEC, \zeta_{DEC}]$ ($A \neq DEC$) in
           $CANDIDATE(1)$, generate a new descriptor $[\{A, DEC\}, (\zeta_A, \zeta_{DEC})]$;
     (Proc 2-2) Examine condition ($CASE$ A) and ($CASE$ B) for each
           $[\{A, DEC\}, (\zeta_A, \zeta_{DEC})]$ and each object x;
           If either ($CASE$ A) or ($CASE$ B) holds and $minacc(\tau^x) \geq \beta$
             display $\tau^x : [A, \zeta_A] \Rightarrow [DEC, \zeta_{DEC}]$ as a rule;
           If either ($CASE$ A) or ($CASE$ B) holds and $minacc(\tau^x) < \beta$,
             add this descriptor to $CANDIDATE(2)$;
Assign 2 to n;
while $CANDIDATE(n) \neq \{\}$ **do**
| Generate $CANDIDATE(n + 1)$ according to the following procedures;
| (Proc 3-1) For $DESC_1$ and $DESC_2$ ($[DEC, \zeta_{DEC}] \in DESC_1 \cap DESC_2$)
| in $CANDIDATE(n)$, generate a new descriptor by using a
| conjunction of $DESC_1 \wedge DESC_2$;
| (Proc 3-2) Examine the same procedure as (Proc 2-2).
| Assign $n + 1$ to n;
end

Algorithm 2. Extended Apriori Algorithm for Upper Approximation: $Rule(\alpha, \beta, UA)$

Input : A NIS, a decision attribute DEC, threshold value α and β.
Output: Every rule defined by $Rule(\alpha, \beta, UA)$.
Algorithm 2 is proposed as Algorithm 1 with the following two revisions :
   1. ($CASE$ A) and ($CASE$ B) in Algorithm 1 are replaced with ($CASE$ C).
      ($CASE$ C) $|Descsup([A, \zeta_A])| \geq NUM$.
   2. $minacc(\tau^x)$ in Algorithm 1 is replaced with $maxacc(\tau^x)$.

In order to reduce the manipulation of string data, we translate this data to numerical data by using *trans.exe*. Then, we also make an attribute definition file in the following. In rule generation, we adjust values in this file. Finally, we execute *nis_apriori.exe* command.

```
8                                 /* the number of objects */
4                                 /* the number of attributes */
Temperature,Headache,Nausea,Flu   /* names of attributes */
3,1,2,3                           /* 3 candidates, condition attributes */
1,4                               /* 1 decision attribute, 4=Flu */
0.3                               /* threshold of support value */
0.8                               /* threshold of accuracy value */
```

```
===========================================
Lower Approximation Strategy
===========================================
CAN(1)=[Temperature,high],[Temperature,normal],[Headache,no],
    [Headache,yes],[Nausea,no],[Nausea,yes],[Flu,no],[Flu,yes](8)
CAN(2)=[Temperature,normal][Flu,no](<DEF>0.667,<INDEF>0.750),
    [Headache,no][Flu,no](<DEF>0.667,<INDEF>0.750),
    [Nausea,yes][Flu,no](<DEF>0.400,<INDEF>0.500),
    [Temperature,high][Flu,yes](<DEF>0.500,<INDEF>0.600),
    [Headache,yes][Flu,yes](<DEF>0.600,<INDEF>0.667)(5)
========== OBTAINED RULE ==========
EXEC_TIME=0.000(sec)
```

In the above execution, the constraint is $minsupp(\tau)=|SET|/|OB| \geq 0.3$. Thus, $|SET| \geq 3$ must hold. Therefore, we need to handle a descriptor $[A, \zeta_A]$ satisfying either (CASE A) or (CASE B) in the following:

(CASE A) $|Descinf([A, \zeta_A])| \geq 3$,
(CASE B) $|Descinf([A, \zeta_A])|=2$ and $(Descsup([A, \zeta_A]) - Descinf([A, \zeta_A])) \neq \{\}$.

A definite descriptor can be obtained in (CASE A), and an indefinite descriptor can be obtained in (CASE B). Like this, $CAN(1)$ is generated. Then, for descriptors $[A, \zeta_A], [Flu, \eta_{Flu}] \in CAN(1)$, if $minsupp([A, \zeta_A] \wedge [Flu, \eta_{Flu}]) \geq 3$, $minacc([A, \zeta_A] \wedge [Flu, \eta_{Flu}])$ is calculated according to Result 1. If this value is more than 0.8, the conjunction is a rule. Otherwise, we add this conjunction to $CAN(2)$. Algorithm 1 continues this process until $CAN(n)=\{\}$. The following is a case of the upper approximation.

```
===========================================
Upper Approximation Strategy
===========================================
CAN(1)=[Temperature,high],[Temperature,normal],[Headache,no],
    [Headache,yes],[Nausea,no],[Nausea,yes],[Flu,no],[Flu,yes](8)
CAN(2)=[Temperature,normal][Flu,no](<DEF>1.000,<INDEF>1.000),
    [Headache,no][Flu,no](<DEF>1.000,<INDEF>1.000),
    [Nausea,yes][Flu,no](<DEF>0.750,<INDEF>0.750),
    [Temperature,high][Flu,yes](<DEF>0.800,<INDEF>0.800),
    [Headache,yes][Flu,yes](<DEF>0.800,<INDEF>0.800)(5)
========== OBTAINED RULE ==========
```

```
[Temperature,normal]=>[Flu,no]
    maxsupp<DEF>=0.375,maxsupp<INDEF>=0.375,
    maxacc<DEF>=1.000,maxacc<INDEF>=1.000
    (<DEF>from 6,7) (<INDEF>from 3)
        :       :       :
[Headache,yes]=>[Flu,yes]
    maxsupp<DEF>=0.500,maxsupp<INDEF>=0.500,
    maxacc<DEF>=0.800,maxacc<INDEF>=0.800
    (<DEF>from 2,4,8) (<INDEF>from 1)
EXEC_TIME=0.000(sec)
```

Now, we briefly show the application to Mammographic data in UCI Machine Learning Repository [19]. This original data consists of 961 objects and 6 attributes, i.e., *BI-RADS assessment, Age, Shape, Margin, Density, Severity*. We obtained 150 objects from the top of the data. In this data, ? symbol is marked for every 76 missing value. We replaced these ? symbols with a list of possible values. There are $4^{55} \times 5^{21}$ derived $DISs$ for these 150 objects. Probably, it seems hard to enumerate all derived $DISs$ sequentially. For this data, it took 0.000(sec) for generating $Rule(0.2, 0.5, LA)$ and $Rule(0.2, 0.5, UA)$ in the following:

```
Rule(0.2,0.5,LA,def)=Rule(0.2,0.5,LA)
= {[SHAPE,1]=>[SEVERITY,0], [SHAPE,2]=>[SEVERITY,0],
    [MARGIN,1]=>[SEVERITY,0], [SHAPE,4]=>[SEVERITY,1]},
Rule(0.2,0.5,UA)-Rule(0.2,0.5,LA)
= {[DENSITY,3]=>[SEVERITY,0], [DENSITY,3]=>[SEVERITY,1]}.
```

If we employ a lower approximation strategy to this data, we may miss two implications on $DENSITY$ in $Rule(0.2, 0.5, UA) - Rule(0.2, 0.5, LA)$. By chance, these two implications are inconsistent, but this may occur in an upper approximation strategy. Because nearly 33 percent of the 150 objects are ? => *Any* in Table 4, and this percent is too large. Since $NUME/DENO \leq (NUME + K)/(DENO + K)$ $(K > 0)$ holds, τ_3 in Table 4 is identified with τ_1 in a DIS_{best} for τ_1. Thus, $maxacc(\tau_1) = (33+37)/(78+37) = 0.608$. As for τ_2, τ_4 is identified with τ_2, therefore $maxacc(\tau_2) = (45+10)/(78+10) = 0.625$. This example shows a characteristic aspect of an upper approximation strategy.

Table 4. Amount of each implication in 150 objects

Implication	Amount	Implication	Amount
$\tau_1 : [DENSITY, 3] => [SEVERITY, 0]$	33	$\tau_3 :? => [SEVERITY, 0]$	37
$\tau_2 : [DENSITY, 3] => [SEVERITY, 1]$	45	$\tau_4 :? => [SEVERITY, 1]$	10
$[DENSITY, 3] => Any$	78	$? => Any$	47

5   Concluding Remarks

We defined lower and upper approximations of rules in $NISs$. We employed $Descinf$, $Descsup$ and the concept of large item set in $Apriori$ algorithm, and proposed two extended $Apriori$ algorithms in $NISs$. The complexity of these extended algorithms is almost the same as $Apriori$ algorithm. Due to these utility programs, we can explicitly handle not only deterministic information but also non-deterministic information.

Acknowledgment. The authors would be grateful to anonymous referees for their useful comments. This work is partly supported by the Grant-in-Aid for Scientific Research (C) (No.18500214), Japan Society for the Promotion of Science.

References

1. Agrawal, R., Srikant, R.: Fast Algorithms for Mining Association Rules. In: Proceedings of the 20th Very Large Data Base, pp. 487–499 (1994)
2. Agrawal, R., Mannila, H., Srikant, R., Toivonen, H., Verkamo, A.: Fast Discovery of Association Rules. In: Advances in Knowledge Discovery and Data Mining, pp. 307–328. AAAI/MIT Press (1996)
3. Grzymala-Busse, J., Werbrouck, P.: On the Best Search Method in the LEM1 and LEM2 Algorithms. In: Incomplete Information: Rough Set Analysis, pp. 75–91. Physica-Verlag (1998)
4. Grzymala-Busse, J.: Data with Missing Attribute Values: Generalization of Indiscernibility Relation and Rule Induction. Transactions on Rough Sets 1, 78–95 (2004)
5. Kryszkiewicz, M.: Rules in Incomplete Information Systems. Information Sciences 113, 271–292 (1999)
6. Kryszkiewicz, M., Rybinski, H.: Computation of Reducts of Composed Information Systems. Fundamenta Informaticae 27, 183–195 (1996)
7. Lipski, W.: On Semantic Issues Connected with Incomplete Information Data Base. ACM Trans. DBS 4, 269–296 (1979)
8. Lipski, W.: On Databases with Incomplete Information. Journal of the ACM 28, 41–70 (1981)
9. Nakata, M., Sakai, H.: Lower and Upper Approximations in Data Tables Containing Possibilistic Information. Transactions on Rough Sets 7, 170–189 (2007)
10. Orłowska, E.: What You Always Wanted to Know about Rough Sets. In: Incomplete Information: Rough Set Analysis, vol. 13, pp. 1–20. Physica-Verlag (1998)
11. Orłowska, E., Pawlak, Z.: Representation of Nondeterministic Information. Theoretical Computer Science 29, 27–39 (1984)
12. Pawlak, Z.: Rough Sets. Kluwer Academic Publisher, Dordrecht (1991)
13. Pawlak, Z.: Some Issues on Rough Sets. Transactions on Rough Sets 1, 1–58 (2004)
14. Sakai, H., Okuma, A.: Basic Algorithms and Tools for Rough Non-deterministic Information Analysis. Transactions on Rough Sets 1, 209–231 (2004)
15. Sakai, H., Nakata, M.: On Possible Rules and Apriori Algorithm in Non-deterministic Information Systems. In: Greco, S., Hata, Y., Hirano, S., Inuiguchi, M., Miyamoto, S., Nguyen, H.S., Słowiński, R. (eds.) RSCTC 2006. LNCS (LNAI), vol. 4259, pp. 264–273. Springer, Heidelberg (2006)

16. Sakai, H., Ishibashi, R., Koba, K., Nakata, M.: On Possible Rules and Apriori Algorithm in Non-deterministic Information Systems 2. In: An, A., Stefanowski, J., Ramanna, S., Butz, C.J., Pedrycz, W., Wang, G. (eds.) RSFDGrC 2007. LNCS (LNAI), vol. 4482, pp. 280–288. Springer, Heidelberg (2007)
17. Skowron, A., Rauszer, C.: The Discernibility Matrices and Functions in Information Systems. In: Intelligent Decision Support - Handbook of Advances and Applications of the Rough Set Theory, pp. 331–362. Kluwer Academic Publishers, Dordrecht (1992)
18. Stefanowski, J., Tsoukias, A.: On the Extension of Rough Sets under Incomplete Information. In: Zhong, N., Skowron, A., Ohsuga, S. (eds.) RSFDGrC 1999. LNCS (LNAI), vol. 1711, pp. 73–82. Springer, Heidelberg (1999)
19. UCI Machine Learning Repository, http://mlearn.ics.uci.edu/MLRepository.html

A New Approach to Fuzzy-Rough Nearest Neighbour Classification

Richard Jensen[1] and Chris Cornelis[2]

[1] Dept. of Comp. Sci., Aberystwyth University, Ceredigion, SY23 3DB, Wales, UK
rkj@aber.ac.uk
[2] Dept. of Appl. Math. and Comp. Sci., Ghent University, Gent, Belgium
Chris.Cornelis@UGent.be

Abstract. In this paper, we present a new fuzzy-rough nearest neighbour (FRNN) classification algorithm, as an alternative to Sarkar's fuzzy-rough ownership function (FRNN-O) approach. By contrast to the latter, our method uses the nearest neighbours to construct lower and upper approximations of decision classes, and classifies test instances based on their membership to these approximations. In the experimental analysis, we evaluate our approach with both classical fuzzy-rough approximations (based on an implicator and a t-norm), as well as with the recently introduced vaguely quantified rough sets. Preliminary results are very good, and in general FRNN outperforms both FRNN-O, as well as the traditional fuzzy nearest neighbour (FNN) algorithm.

1 Introduction

The K-nearest neighbour (KNN) algorithm [6] is a well-known classification technique that assigns a test object to the decision class most common among its K nearest neighbours, i.e., the K training objects that are closest to the test object. An extension of the KNN algorithm to fuzzy set theory (FNN) was introduced in [8]. It allows partial membership of an object to different classes, and also takes into account the relative importance (closeness) of each neighbour w.r.t. the test instance. However, as Sarkar correctly argued in [11], the FNN algorithm has problems dealing adequately with insufficient knowledge. In particular, when every training pattern is far removed from the test object, and hence there are no suitable neighbours, the algorithm is still forced to make clear-cut predictions. This is because the predicted membership degrees to the various decision classes always need to sum up to 1.

To address this problem, Sarkar [11] introduced a so-called fuzzy-rough ownership function that, when plugged into the conventional FNN algorithm, produces class confidence values that do not necessarily sum up to 1. However, this method (called FRNN-O throughout this paper) does not refer to the main ingredients of rough set theory, i.e., lower and upper approximation. In this paper, therefore, we present an alternative approach, which uses a test object's nearest neighbours to construct the lower and upper approximation of each decision class, and then computes the membership of the test object to these approximations.

C.-C. Chan et al. (Eds.): RSCTC 2008, LNAI 5306, pp. 310–319, 2008.

The method is very flexible, as there are many options to define the fuzzy-rough approximations, including the traditional implicator/t-norm based model [10], as well as the vaguely quantified rough set (VQRS) model [3], which is more robust in the presence of noisy data.

This paper is structured as follows. Section 2 provides necessary details for fuzzy rough set theory, while Section 3 is concerned with the existing fuzzy (-rough) NN approaches. Section 4 outlines our algorithm, while comparative experimentation on a series of crisp classification problems is provided in Section 5. The paper is concluded in section 6.

2 Hybridization of Rough Sets and Fuzzy Sets

Rough set theory (RST) [9] provides a tool by which knowledge may be extracted from a domain in a concise way; it is able to retain the information content whilst reducing the amount of knowledge involved. Central to RST is the concept of indiscernibility. Let $(\mathbb{U}, \mathbb{A})$ be an information system[1], where $\mathbb{U}$ is a non-empty set of finite objects (the universe of discourse) and $\mathbb{A}$ is a non-empty finite set of attributes such that $a : \mathbb{U} \to V_a$ for every $a \in \mathbb{A}$. V_a is the set of values that attribute a may take. With any $B \subseteq \mathbb{A}$ there is an associated equivalence relation R_B:

$$R_B = \{(x, y) \in \mathbb{U}^2 | \forall a \in B, \ a(x) = a(y)\} \tag{1}$$

If $(x, y) \in R_B$, then x and y are indiscernible by attributes from B. The equivalence classes of the B-indiscernibility relation are denoted $[x]_B$. Let $A \subseteq \mathbb{U}$. A can be approximated using the information contained within B by constructing the B-*lower* and B-*upper* approximations of A:

$$R_B{\downarrow}A = \{x \in \mathbb{U} \mid [x]_B \subseteq A\} \tag{2}$$
$$R_B{\uparrow}A = \{x \in \mathbb{U} \mid [x]_B \cap A \neq \emptyset\} \tag{3}$$

The tuple $\langle R_B{\downarrow}A, R_B{\uparrow}A \rangle$ is called a rough set.

The process described above can only operate effectively with datasets containing discrete values. As most datasets contain real-valued attributes, it is necessary to perform a discretization step beforehand. A more intuitive and flexible approach, however, is to model the approximate equality between objects with continuous attribute values by means of a fuzzy relation R in $\mathbb{U}$, i.e., a $\mathbb{U} \to [0, 1]$ mapping that assigns to each couple of objects their degree of similarity. In general, it is assumed that R is at least a fuzzy tolerance relation, that is, $R(x, x) = 1$ and $R(x, y) = R(y, x)$ for x and y in $\mathbb{U}$. Given y in $\mathbb{U}$, its foreset Ry is defined by $Ry(x) = R(x, y)$ for every x in $\mathbb{U}$.

Given a fuzzy tolerance relation R and a fuzzy set A in $\mathbb{U}$, the lower and upper approximation of A by R can be constructed in several ways. A general definition [4,10] is the following:

---

[1] In the classification problems considered further on in this paper, $\mathbb{A} = \mathbb{C} \cup \{d\}$, where $\mathbb{C}$ represents the set of conditional attributes, and d is the decision or class attribute.

$$(R{\downarrow}A)(x) = \inf_{y \in \mathbb{U}} I(R(x,y), A(y)) \tag{4}$$

$$(R{\uparrow}A)(x) = \sup_{y \in \mathbb{U}} T(R(x,y), A(y)) \tag{5}$$

Here, I is an implicator[2] and T a t-norm[3]. When A is a crisp (classical) set and R is an equivalence relation in $\mathbb{U}$, the traditional lower and upper approximation are recovered.

Just like their crisp counterparts, formulas (4) and (5) (henceforth called the FRS approximations) are quite sensitive to noisy values. That is, a change in a single object can result in drastic changes to the approximations (due to the use of sup and inf, which generalize the existential and universal quantifier, respectively). In the context of classification tasks, this behaviour may affect accuracy adversely. Therefore, in [3], the concept of vaguely quantified rough sets (VQRS) was introduced. It uses the linguistic quantifiers "most" and "some", as opposed to the traditionally used crisp quantifiers "all" and "at least one", to decide to what extent an object belongs to the lower and upper approximation. Given a couple (Q_u, Q_l) of fuzzy quantifiers[4] that model "most" and "some", the lower and upper approximation of A by R are defined by

$$(R{\downarrow}A)(y) = Q_u \left(\frac{|Ry \cap A|}{|Ry|} \right) = Q_u \left(\frac{\sum_{x \in X} \min(R(x,y), A(x))}{\sum_{x \in X} R(x,y)} \right) \tag{6}$$

$$(R{\uparrow}A)(y) = Q_l \left(\frac{|Ry \cap A|}{|Ry|} \right) = Q_l \left(\frac{\sum_{x \in X} \min(R(x,y), A(x))}{\sum_{x \in X} R(x,y)} \right) \tag{7}$$

where the fuzzy set intersection is defined by the min t-norm and the fuzzy set cardinality by the sigma-count operation. As an important difference to (4) and (5), the VQRS approximations do not extend the classical rough set approximations, in a sense that when A and R are crisp, $R{\downarrow}A$ and $R{\uparrow}A$ may still be fuzzy.

3   Fuzzy Nearest Neighbour Classification

The fuzzy K-nearest neighbour (FNN) algorithm [8] was introduced to classify test objects based on their similarity to a given number K of neighbours (among the training objects), and these neighbours' membership degrees to (crisp or

---

[2] An implicator I is a $[0,1]^2 \to [0,1]$ mapping that is decreasing in its first and increasing in its second argument, satisfying $I(0,0) = I(0,1) = I(1,1) = 1$ and $I(1,0) = 0$.

[3] A t-norm T is an increasing, commutative, associative $[0,1]^2 \to [0,1]$ mapping satisfying $T(x,1) = x$ for x in $[0,1]$.

[4] By a fuzzy quantifier, we mean an increasing $[0,1] \to [0,1]$ mapping such that $Q(0) = 0$ and $Q(1) = 1$.

fuzzy) class labels. For the purposes of FNN, the extent $C(y)$ to which an un-classified object y belongs to a class C is computed as:

$$C(y) = \sum_{x \in N} R(x, y)C(x) \tag{8}$$

where N is the set of object y's K nearest neighbours, and $R(x, y)$ is the $[0,1]$-valued similarity of x and y. In the traditional approach, this is defined in the following way:

$$R(x, y) = \frac{||y - x||^{-2/(m-1)}}{\sum_{j \in N} ||y - j||^{-2/(m-1)}} \tag{9}$$

where $|| \cdot ||$ denotes the Euclidean norm, and m is a parameter that controls the overall weighting of the similarity. Assuming crisp classes, Figure 1 shows an application of the FNN algorithm that classifies a test object y to the class with the highest resulting membership. The complexity of this algorithm for the classification of one test pattern is $O(|\mathbb{U}| + K \cdot |C|)$.

FNN($\mathbb{U},\mathcal{C},y,K$).
$\mathbb{U}$, the training data; $\mathcal{C}$, the set of decision classes;
y, the object to be classified; K, the number of nearest neighbours.

(1) $N \leftarrow$ getNearestNeighbours(y,K);
(2) $\forall C \in \mathcal{C}$
(3) $C(y) = \sum_{x \in N} R(x, y)C(x)$
(4)   **output** $\arg\max_{C \in \mathcal{C}} (C(y))$

Fig. 1. The fuzzy KNN algorithm

Initial attempts to combine the FNN algorithm with concepts from fuzzy rough set theory were presented in [11,12]. In these papers, a fuzzy-rough own-ership function is constructed that attempts to handle both "fuzzy uncertainty" (caused by overlapping classes) and "rough uncertainty" (caused by insufficient knowledge, i.e., attributes, about the objects). The fuzzy-rough ownership func-tion τ_C of class C was defined as, for an object y,

$$\tau_C(y) = \frac{\sum_{x \in \mathbb{U}} R(x, y)C(x)}{|\mathbb{U}|} \tag{10}$$

In this, the fuzzy relation R is determined by:

$$R(x, y) = \exp\left(-\sum_{a \in C} \kappa_a(a(y) - a(x))^{2/(m-1)}\right) \tag{11}$$

where m controls the weighting of the similarity (as in FNN) and κ_a is a para-meter that decides the bandwidth of the membership, defined as

$$\kappa_a = \frac{|\mathbb{U}|}{2 \sum_{x \in \mathbb{U}} ||a(y) - a(x)||^{2/(m-1)}} \qquad (12)$$

$\tau_C(y)$ is interpreted as the confidence with which y can be classified to class C. The corresponding crisp classification algorithm, called FRNN-O in this paper, can be seen in Figure 2. Initially, the parameter κ_a is calculated for each attribute and all memberships of decision classes for test object y are set to 0. Next, the weighted distance of y from all objects in the universe is computed and used to update the class memberships of y via equation (10). Finally, when all training objects have been considered, the algorithm outputs the class with highest membership. The algorithm's complexity is $O(|\mathbb{C}||\mathbb{U}| + |\mathbb{U}| \cdot (|\mathbb{C}| + |\mathcal{C}|))$.

By contrast to the FNN algorithm, the fuzzy-rough ownership function considers all training objects rather than a limited set of neighbours, and hence no decision is required as to the number of neighbours to consider. The reasoning behind this is that very distant training objects will not influence the outcome (as opposed to the case of FNN). For comparison purposes, the K-nearest neighbours version of this algorithm is obtained by replacing line (3) with $N \leftarrow$ getNearestNeighbours(y, K).

FRNN-O$(\mathbb{U}, \mathbb{C}, \mathcal{C}, y)$.
$\mathbb{U}$, the training data; $\mathbb{C}$, the set of conditional features;
$\mathcal{C}$, the set of decision classes; y, the object to be classified.

(1) $\forall a \in \mathbb{C}$
(2) $\kappa_a = |\mathbb{U}|/2 \sum_{x \in \mathbb{U}} ||a(y) - a(x)||^{2/(m-1)}$
(3) $N \leftarrow |\mathbb{U}|$
(4) $\forall C \in \mathcal{C}, \tau_C(y) = 0$
(5) $\forall x \in N$
(6) $d = \sum_{a \in \mathbb{C}} \kappa_a (a(y) - a(x))^2$
(7) $\forall C \in \mathcal{C}$
(8) $\tau_C(y) += \frac{C(x) \cdot \exp(-d^{1/(m-1)})}{|N|}$
(9)   **output** $\arg\max_{C \in \mathcal{C}} \tau_C(y)$

Fig. 2. The fuzzy-rough ownership nearest neighbour algorithm

It should be noted that the algorithm does not use fuzzy lower or upper approximations to determine class membership. A very preliminary attempt to do so was described in [1]. However, the authors did not state how to use the upper and lower approximations to derive classifications.

4   Fuzzy-Rough Nearest Neighbour (FRNN) Algorithm

Figure 3 outlines our proposed algorithm, combining fuzzy-rough approximations with the ideas of the classical FNN approach. In what follows, FRNN-FRS and FRNN-VQRS denote instances of the algorithm where traditional, and VQRS, approximations are used, respectively. The rationale behind the algorithm is that the lower and the upper approximation of a decision class, calculated by means of the nearest neighbours of a test object y, provide good clues to predict the membership of the test object to that class.

In particular, if $(R{\downarrow}C)(y)$ is high, it reflects that all (most) of y's neighbours belong to C, while a high value of $(R{\uparrow}C)(y)$ means that at least one (some) neighbour(s) belong(s) to that class, depending on whether the FRS or VQRS approximations are used. A classification will always be determined for y due to the initialisation of $\mu_1(y)$ and $\mu_2(y)$ to zero in line (2). To perform crisp classification, the algorithm outputs the decision class with the resulting best fuzzy lower and upper approximation memberships, seen in line (4) of the algorithm. This is only one way of utilising the information in the fuzzy lower and upper approximations to determine class membership, other ways are possible (such as combining them into a single measure) but are not investigated in this paper. The complexity of the algorithm is $O(|\mathcal{C}| \cdot (2|\mathbb{U}|))$.

FRNN($\mathbb{U},\mathcal{C},y$).
$\mathbb{U}$, the training data; $\mathcal{C}$, the set of decision classes;
y, the object to be classified.

(1)	$N \leftarrow$ getNearestNeighbors(y,K)
(2)	$\mu_1(y) \leftarrow 0,\ \mu_2(y) \leftarrow 0,\ Class \leftarrow \emptyset$
(3)	$\forall C \in \mathcal{C}$
(4)	$\quad$ **if** $((R{\downarrow}C)(y) \geq \mu_1(y)$ && $(R{\uparrow}C)(y) \geq \mu_2(y))$
(5)	$\quad\quad Class \leftarrow C$
(6)	$\quad\quad \mu_1(y) \leftarrow (R{\downarrow}C)(y),\ \mu_2(y) \leftarrow (R{\uparrow}C)(y)$
(7)	**output** $Class$

Fig. 3. The fuzzy-rough nearest neighbour algorithm

When using FRNN-FRS, the use of K is not required in principle: as $R(x, y)$ gets smaller, x tends to have only have a minor influence on $(R{\downarrow}C)(y)$ and $(R{\uparrow}C)(y)$. For FRNN-VQRS, this may generally not be true, because $R(x, y)$ appears in the numerator as well as the denominator of (6) and (7).

Furthermore, the algorithm is dependent on the choice of the fuzzy tolerance relation R A general way of constructing R is as follows: given the set of conditional attributes $\mathbb{C}$, R is defined by

$$R(x,y) = \min_{a \in \mathbb{C}} R_a(x,y) \tag{13}$$

in which $R_a(x,y)$ is the degree to which objects x and y are similar for attribute a. Possible options include

$$R_a^1(x,y) = \exp\left(-\frac{(a(x) - a(y))^2}{2\sigma_a{}^2}\right) \tag{14}$$

$$R_a^2(x,y) = 1 - \frac{|a(x) - a(y)|}{|a_{\max} - a_{\min}|} \tag{15}$$

where $\sigma_a{}^2$ is the variance of attribute a, and $a_{\max}$ and $a_{\min}$ are the maximal and minimal occurring value of that attribute.

5   Experimentation

This section presents the initial experimental evaluation of the classification methods FNN, FRNN-O, FRNN-FRS and FRNN-VQRS for the task of pattern classification, over nine benchmark datasets from [2] and [11]. The details of the datasets used can be found in table 1. All of them have a crisp decision attribute.

Table 1. Dataset details

Dataset	Objects	Attributes
Cleveland	297	14
Glass	214	10
Heart	270	14
Ionosphere	230	35
Letter-dgoq	3114	17
Olitos	120	26
Water 2	390	39
Water 3	390	39
Wine	178	14

5.1   Experimental Setup

K is initialized as $|\mathbb{U}|$, the number of objects in the training dataset and then decremented by $1/30^{th}$ of $|\mathbb{U}|$ each time, resulting in 30 experiments for each dataset. For each choice of parameter K, $2\times$ 10-fold cross-validation is performed. For FNN and FRNN-O, m is set to 2. For the new approaches, the fuzzy relation given in equation (15) was chosen. In the FRNN-FRS approach, we used the min t-norm and the Kleene-Dienes implicator I defined by $I(x, y) = \max(1-x, y)$. The FRNN-VQRS approach was implemented using $Q_l = Q_{(0.1,0.6)}$ and $Q_u = Q_{(0.2,1.0)}$, according to the general formula

$$Q_{(\alpha,\beta)}(x) = \begin{cases} 0, & x \leq \alpha \\ \frac{2(x-\alpha)^2}{(\beta-\alpha)^2}, & \alpha \leq x \leq \frac{\alpha+\beta}{2} \\ 1 - \frac{2(x-\beta)^2}{(\beta-\alpha)^2}, & \frac{\alpha+\beta}{2} \leq x \leq \beta \\ 1, & \beta \leq x \end{cases}$$

5.2   Comparative Investigation

The results of the experiments are shown in Figure 4. Several interesting observations can be made from them. First, for all but one dataset (`letter-dgoq`, which was used in [11]), either FRNN-FRS or FRNN-VQRS yields the best results. Overall, FRNN-FRS produces the most consistent results. This is particularly

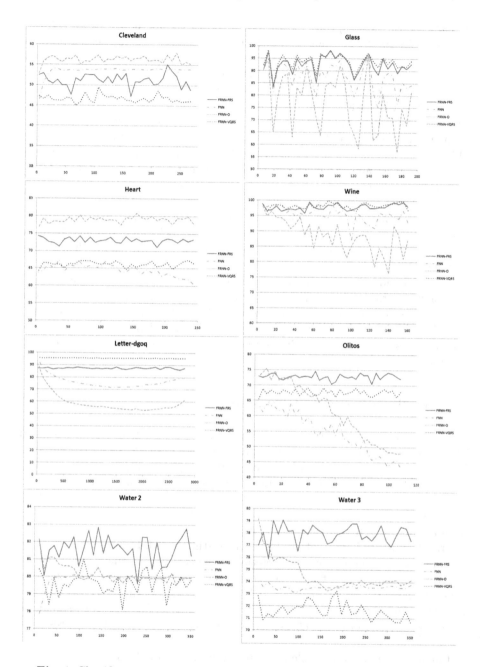

Fig. 4. Classification accuracy for the four methods and different values of K

remarkable considering the inherent simplicity of the method. FRNN-VQRS is best for `cleveland` and `heart`, which might be attributed to the comparative presence of noise in those datasets, but it performs rather disappointing for a number of other datasets (`glass`, `letter-dgoq`, `wine`, `olitos`).

It is also interesting to consider the influence of the number of nearest neighbours. Both FRNN-FRS and FRNN-O remain relatively unaffected by changes in K. This could be explained in that, for FRNN-FRS, an infimum and supremum are used which can be thought of as a worst case and best case respectively. When more neighbours are considered, $R(x, y)$ values decrease as these neighbours are less similar, hence $I(R(x, y), C(x))$ increases, and $T(R(x, y), C(x))$ decreases. In other words, the more distant a neighbour is, the more unlikely it is to change the infimum and supremum value. For FRNN-O, again $R(x, y)$ decreases when more neighbours are added, and hence the value $R(x, y)C(x)$ that is added to the numerator is also small. Since each neighbour has the same weight in the denominator, the ratios stay approximately the same when adding new neighbours.

For FNN and FRNN-VQRS, increasing K can have a significant effect on classification accuracy. This is most clearly observed in the results for the `olitos` data, where there is a clear downward trend. For FRNN-VQRS, the ratio $|Ry \cap C|/|Ry|$ has to be calculated. Each neighbour has a different weight in the denominator, so the ratios can fluctuate considerably even when adding distant neighbours.

6   Conclusion and Future Work

This paper has presented two new techniques for fuzzy-rough classification based on the use of lower and upper approximations w.r.t. fuzzy tolerance relations. The difference between them is in the definition of the approximations: while FRNN-FRS uses "traditional" operations based on a t-norm and an implicator, FRNN-VQRS uses a fuzzy quantifier-based approach. The results show that these methods are effective, and that they are competitive with existing methods such as the fuzzy K-nearest neighbour and the fuzzy-rough ownership function approach. Further investigation, however, is still needed, to adequately explain the impact of the choice of fuzzy relations, connectives and quantifiers.

Also, the impact of a feature selection preprocessing step upon classification accuracy needs to be investigated. It is expected that feature selectors that incorporate fuzzy relations expressing closeness of objects (see e.g. [5,7]) should be able to further improve the effectiveness of the classification methods presented here.

Finally, an important challenge is to adapt the algorithms so that they can deal with continuous decision attributes. In this case, we need to predict the membership of a test object to different, possibly overlapping classes. Such a prediction can be based on the test object's membership degrees to the lower and/or upper approximation (e.g., on the average of these two values).

Acknowledgment

Chris Cornelis would like to thank the Research Foundation—Flanders for funding his research.

References

1. Bian, H., Mazlack, L.: Fuzzy-Rough Nearest-Neighbor Classification Approach. In: Proceeding of the 22nd International Conference of the North American Fuzzy Information Processing Society (NAFIPS), pp. 500–505 (2003)
2. Blake, C.L., Merz, C.J.: UCI Repository of machine learning databases. Irvine, University of California (1998), http://archive.ics.uci.edu/ml/
3. Cornelis, C., De Cock, M., Radzikowska, A.M.: Vaguely Quantified Rough Sets. In: An, A., Stefanowski, J., Ramanna, S., Butz, C.J., Pedrycz, W., Wang, G. (eds.) RSFDGrC 2007. LNCS (LNAI), vol. 4482, pp. 87–94. Springer, Heidelberg (2007)
4. Cornelis, C., De Cock, M., Radzikowska, A.M.: Fuzzy Rough Sets: from Theory into Practice. In: Pedrycz, W., Skowron, A., Kreinovich, V. (eds.) Handbook of Granular Computing. Wiley, Chichester (2008)
5. Cornelis, C., Hurtado Martín, G., Jensen, R., Slezak, D.: Feature Selection with Fuzzy Decision Reducts. In: Proceedings of 3rd International Conference on Rough Sets and Knowledge Technology (RSKT2008) (2008)
6. Duda, R., Hart, P.: Pattern Classification and Scene Analysis. Wiley, New York (1973)
7. Jensen, R., Shen, Q.: Fuzzy-Rough Sets Assisted Attribute Selection. IEEE Transactions on Fuzzy Systems 15(1), 73–89 (2007)
8. Keller, J.M., Gray, M.R., Givens, J.A.: A fuzzy K-nearest neighbor algorithm. IEEE Trans. Systems Man Cybernet. 15(4), 580–585 (1985)
9. Pawlak, Z.: Rough Sets: Theoretical Aspects of Reasoning About Data. Kluwer Academic Publishing, Dordrecht (1991)
10. Radzikowska, A.M., Kerre, E.E.: A comparative study of fuzzy rough sets. Fuzzy Sets and Systems 126(2), 137–155 (2002)
11. Sarkar, M.: Fuzzy-Rough nearest neighbors algorithm. Fuzzy Sets and Systems 158, 2123–2152 (2007)
12. Wang, X., Yang, J., Teng, X., Peng, N.: Fuzzy-Rough Set Based Nearest Neighbor Clustering Classification Algorithm. In: Wang, L., Jin, Y. (eds.) FSKD 2005. LNCS (LNAI), vol. 3613, pp. 370–373. Springer, Heidelberg (2005)

Towards Approximation of Risk

Marcin Szczuka

Institute of Mathematics, Warsaw University
Banacha 2, 02-097 Warsaw, Poland
szczuka@mimuw.edu.pl

Abstract. We discuss the notion of risk in generally understood classification support systems. We propose a method for approximating the loss function and introduce a technique for assessing the empirical risk from experimental data. We discuss the general methodology and possible directions of development in the area of constructing compound classification schemes.

Keywords: risk assessment, loss function, granularity, approximation, neighbourhood, empirical risk.

1   Introduction

While constructing a decision support (classification) system for research purposes we usually rely on commonly used, convenient quality measures, such as success ratio (accuracy) on test set, coverage (support) and versatility of the classifier. While sufficient for the purposes of analysing classification methods in terms of their technical abilities, these measures sometimes fail to fit into a bigger picture.

In practical decision support applications the classifier is usually just a sprocket in a larger machine. The decision whether to construct and then use such system is taken by the user on the basis of his confidence in relative "safety" of his computer-supported decision. This confidence is closely related to the users' assessment of the *risk* involved in making the decision.

The overall topics of risk assessment, risk management and decision making in presence of risk constitute a separate field of science. The ubiquity of decision-making processes that involve risk is making risk assessment a crucial element in areas such as economy, investment, medicine, engineering and many others. Numerous approaches have been developed so far, and vast literature dedicated to these issues exist (see [1,2,3]). The topic or risk assessment and management is a topic of research in many fields of science, ranging from philosophy to seismology. In this article we restrict ourselves to a much narrower topic of calculating (assessing) the risk associated with the use of classifier in a decision-making process.

We focus on one commonly used method for calculating a risk of (using) a classifier, which is known from the basics of statistical learning theory [4]. In this approach the risk is measured as a *summarised chance* for creating a loss due to classifier's error. More formally, the risk is equal to the total *loss* (integral) over the probabilistic distribution of data. Loss is expressed in terms of a specialised function which compares the answer of classifier with the desired one and returns the numerical value corresponding

C.-C. Chan et al. (Eds.): RSCTC 2008, LNAI 5306, pp. 320–328, 2008.

to the amount of "damage" resulting from misclassification. Such a measure of risk is to a large extent intuitive in many situations. It is by no means the only scheme used by humans to judge the risk, but a popular one. It is quite common to make assessment of the involved risk by hypothesising the situations in which the gain/loss can be generated in our system, and then weighting them by the likelihood of their occurrence.

We investigate the possibilities for approximating the risk in a situation when the standard numerical, statistical learning methods cannot be applied to full extent. The real-life data is not always (verifiably) representative, large enough or sufficiently compliant with assumptions of underlying analytical model. Also, the information we posses about the amount of loss and its probabilistic distribution may be expressed in granular rather than crisp, numerical way. Nevertheless, we would like to be able to provide approximate assessment of risk associated with a classification method. For this purpose we put forward some ideas regarding the approximate construction of two crucial components in measuring risk i.e., the loss function and the summarisation method needed to estimate overall risk from the empirical, sample-dependant one.

This article is intended to pose some questions and provide suggestions in which direction we may search for answers, rather than deliver ready to use technical solutions. The paper starts with more formal introduction of risk functional, as known from statistical learning theory. Then, we discuss the possible sources of problems with such definition and suggest some directions, in particular an outline for a loss function approximation method. We also extend the discussion to the issue of finding the proper summarisation procedure for measuring the value of empirical risk functional. We conclude by pointing out several possible directions for further investigation.

2 Risk in Statistical Learning Theory

In the classical statistical learning approach, represented by seminal works of Vapnik [4,5], the risk associated with a classification method (classifier) α is defined as a functional (integral) of the *loss function* L_α calculated over an entire space with respect to probability distribution.

To put it more formally, let X^∞ be the complete (hypothetical) universe of objects from which we are drawing our finite sample $X \subset X^\infty$. In the analytical model of risk we are assuming that a probability distribution P is defined for entire σ-field of measurable subsets of X^∞.

Definition 1. *The risk value for a classifier α is defined as:*

$$R(\alpha) = \int_{X^\infty} L_\alpha dP$$

where $L_\alpha = L(x, f_\alpha(x))$ is the real-valued loss function defined for every point $x \in X^\infty$ where the classifier α returns the value $f_\alpha(x)$.

The classical definition of risk, as presented above, is heavily dependant on assumptions regarding the underlying analytical model of the space of discourse. While over the years several methods have been developed within the area of statistical learning in pursuit of practical means for calculating risk, there are still some important shortcomings in this approach. Some of them are:

1. Sensitivity to scarceness of the data sample. In real life experiments we may be very far from complete knowledge of our data universe. The sample we are given may be tiny in comparison with the range of possible outcomes.
2. Incomplete definition of loss function. We expect that $L(y, f_\alpha(x))$ is integrable wherever f_α takes value. Unfortunately, in practice all we are given is the set of points from the graph of L_α. From these few points we have to extend (approximate) the function L_α.
3. Incomplete knowledge of the distribution, which is closely related to the point 1 above. Even with large data sample X we may not be certain about its representativeness.

There are also several advantages of the classical risk functional definition. Thanks to solid mathematical grounding it is possible to provide answers with provable quality. As long as we can assure sufficient compliance to assumptions of the underlying statistical methodology the task of estimating the risk is equivalent to solving a numerical optimisation problem. For a given classifier α we search for the solution to:

$$\lim_{l \longrightarrow \infty} \Pr\left\{z \in (X^\infty)^l : |R(\alpha) - R_{emp}(\alpha)| > \varepsilon\right\} = 0$$

where z is a data sample of size l, probability $\Pr$ is calculated according to distribution P (see Def. 1), $\varepsilon \geq 0$, and $R_{emp}(\alpha)$ is the *empirical risk* measured for the classifier α on (labelled) sample z. The empirical risk is usually measured as an average over values of loss function. For a labelled sample z of length l

$$R_{emp}(\alpha) = \frac{\sum_{i=1}^{l} L(x, f_\alpha(x))}{l}.$$

It is visible, that the ability to calculate value of loss L_α, i.e., to compare the answer of classifier with the desired one is the key element in empirical risk assessment.

3   Approximation of Loss Function and Its Integral

The formal postulates regarding the loss function may be hard to meet, or even verify in practical situations. Nevertheless, we would like to be able to asses the loss. In this section we suggest a method for approximating the loss function from the available, finite sample. In the process we will consider the influence of granularity on our ability to make valid approximations of loss function.

First, we will attempt to deal with the situation when the value of loss function L_α for a classifier α is given as a set of positive real values defined for data points from a finite sample z. Let $z \in (X^\infty)^l$ be a sample consisting of l data points, by $\mathbb{R}_+$ we denote the set of non-negative reals (including 0). A function $\hat{L}_\alpha : z \mapsto \mathbb{R}_+$ is called a sample of loss function $L_\alpha : X^\infty \mapsto \mathbb{R}_+$ if L_α is an extension of $\hat{L}_\alpha$. For any $Z \subseteq X^\infty \times \mathbb{R}_+$ we introduce two projection sets as follows:

$$\pi_1(Z) = \{x \in X^\infty : \exists y \in \mathbb{R}_+ \ (x, y) \in Z\},$$

$$\pi_2(Z) = \{y \in \mathbb{R}_+ : \exists x \in X^\infty \ (x, y) \in Z\}.$$

We assume that we are also given a family $\mathcal{C}$ of neighbourhoods, i.e, non-empty, measurable subsets of $X^{\infty} \times \mathbb{R}_{+}$. These neighbourhoods shall be defined for a particular application.

Under the assumptions presented above the lower approximation of $\hat{L}_{\alpha}$ relative to $\mathcal{C}$ is defined by

$$\underline{\mathcal{C}}\hat{L}_{\alpha} = \bigcup \{c \in \mathcal{C} : \hat{L}_{\alpha}(\pi_1(c) \cap z) \subseteq \pi_2(c)\}. \tag{1}$$

Note, that the definition of lower approximation given by (1) is different from the traditional one, known from rough set theory [6,7]

One can define the upper approximation of f relative to $\mathcal{C}$ by

$$\overline{\mathcal{C}}\hat{L}_{\alpha} = \bigcup \{c \in \mathcal{C} : \hat{L}_{\alpha}(\pi_1(c) \cap z) \cap \pi_2(c) \neq \emptyset\}. \tag{2}$$

An illustration of the upper and lower approximations of a function given by a finite sample if provided in Fig.1.

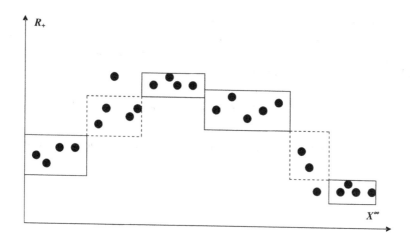

Fig. 1. Loss function approximation (neighbourhoods marked by solid lines belong to the lower approximation and with dashed lines - to the upper approximation)

Example 1. We present an illustrative example of a function $\hat{L}_{\alpha} : z \mapsto \mathbb{R}_{+}$ approximation in a simple situation where $z = \{1, 2, 4, 5, 7, 8\}$ is a sequence of $l = 6$ real numbers. Let $\hat{L}_{\alpha}(1) = 3$, $\hat{L}_{\alpha}(2) = 2$, $\hat{L}_{\alpha}(4) = 2$, $\hat{L}_{\alpha}(5) = 5$, $\hat{L}_{\alpha}(7) = 5$, $\hat{L}_{\alpha}(8) = 2$.

We consider a neighbourhood consisting of three indiscernibility classes $C_1 = [0, 3] \times [1.5, 4]$, $C_2 = [3, 6] \times [1.7, 4.5]$ and $C_3 = [6, 9] \times [3, 4]$.

We compute projections of indiscernibility classes: $\pi_1(C_1) = [0, 3]$, $\pi_2(C_1) = [1.5, 4]$, $\pi_1(C_2) = [3, 6]$, $\pi_2(C_2) = [1.7, 4.5]$, $\pi_1(C_3) = [6, 9]$ and $\pi_2(C_3) = [3, 4]$.

Hence, we obtain $\hat{L}_{\alpha}(\pi_1(C_1) \cap X) = \hat{L}_{\alpha}(\{1, 2\}) = \{2, 3\} \subseteq \pi_2(C_1)$, $\hat{L}_{\alpha}(\pi_1(C_2) \cap X) = \hat{L}_{\alpha}(\{4, 5\}) = \{2, 5\} \nsubseteq \pi_2(C_2)$ but $\hat{L}_{\alpha}(\pi_1(C_2) \cap X) \cap \pi_2(C_2) = \{2, 5\} \cap [1.7, 4.5] \neq \emptyset$, $\hat{L}_{\alpha}(\pi_1(C_3) \cap X) = \emptyset$.

As a result, we get the lower approximation $\underline{\mathcal{C}}\hat{L}_\alpha = C_1$ and the upper approximation $\overline{\mathcal{C}}\hat{L}_\alpha = C_1 \cup C_2$.

For the moment we have defined the approximation of loss function as a pair of sets created from the elements of neighbourhood family $\mathcal{C}$. From this approximation we would like to obtain an estimation of risk. For that purpose we need to define summarisation (integration) method that will return a value analogous to the one in Def. 1. We will define an integration functional that is based on the idea of probabilistic version of Lebesgue-Stieltjes integral [4,?].

In order to define our integral we need to make some additional assumptions. For the universe X^∞ we assume that μ is a measure on a Borel σ-field of subsets of X^∞ and that $\mu(X^\infty) < \infty$. By μ_0 we denote a measure on a σ-field of subsets of $\mathbb{R}_+$. We will also assume that $\mathcal{C}$ is a family of non-empty subsets of $X^\infty \times \mathbb{R}_+$ that are measurable relative to the product measure $\bar{\mu} = \mu \times \mu_0$. Finally, we assume that the value of loss function is bounded by some positive real B. Please, note that none of the above assumptions is unrealistic, and that in practical applications we are dealing with finite universes.

For the upper bound B we split the range $[0, B] \subset \mathbb{R}_+$ into $m > 0$ intervals of equal length $I_1, \ldots, I_m$, where $I_i = [\frac{(i-1)B}{m}, \frac{iB}{m}]$. This is a simplification of the most general definition, where the intervals do not have to be equal. For every interval I_i we consider the sub-family $\mathcal{C}_i \subset \mathcal{C}$ of neighbourhoods such that

$$\mathcal{C}_i = \left\{ c \in \mathcal{C} : \forall x \in (z \cap \pi_1(c)) \quad \hat{L}_\alpha(x) > \frac{(i-1)B}{m} \right\}.$$

With the previous notation the estimate for empirical risk functional is given by:

$$R_{emp}(\alpha) = \sum_{i=1}^{m} \frac{B}{m} \mu \left(\bigcup_{c \in \mathcal{C}_i} \pi_1(c) \right) \tag{3}$$

subject to limitation with $m \to \infty$. The parameter m in practical situation does not have to go to infinity. It is sufficient to find m such that for every pair of points $x_1 \neq x_2$ taken from sample z if $\hat{L}_\alpha(x_1) < \hat{L}_\alpha(x_2)$ then for some integer $i \leq m$ we have $\hat{L}_\alpha(x_1) < \frac{iB}{m} < \hat{L}_\alpha(x_2)$.

The notions of function approximations and risk functional that we have introduced are heavily dependant on the data sample z and decomposition of our domain into family of neighbourhoods $\mathcal{C}$. It is not yet visible, how the ideas we present may help in construction of better decision support (classification) systems. In the following section we discuss these matters in some detail.

4   Classifiers, Neighbourhoods and Granulation

Insofar we have introduced the approximation of loss and the measure of risk. To show the potential use of these entities, we intend to investigate the process of creation and evaluation (scoring) of classifier-driven decision support system as a whole.

The crucial component in all our definitions is the family of non-empty sets (neighbourhoods) $\mathcal{C}$. We have to know this family before we can approximate loss or estimate empirical risk. In practical situations the family of neighbourhoods have to be constructed in close correlation with classifier construction. It is quite common, especially for rough set approaches, to define these sets constructively by semantics of some formulas. An example of such formula could be the conditional part of decision rule or a template (in the sense of [9,10]). Usually the construction of proper neighbourhoods is a complicated search and optimisation task. The notions of approximation and empirical risk that we have introduced may be used to express requirements for this search/optimisation. For the purpose of making valid, low-risk decision by means of classifier α we would expect the family $\mathcal{C}$ to possess the following qualities:

1. *Precision.* In order to have really meaningful assessment of risk as well as good idea about the loss function we would like the elements of neighbourhood family to to be relative large in terms of universe X^{∞}, but at the same time having possibly low variation. These requirements translate to expectation that for the whole family $\mathcal{C}$ we want to minimise the *boundary region* in loss approximation, i.e., achieve possibly the lowest value of $\bar{\mu}(\overline{\mathcal{C}}\hat{L}_{\alpha} \setminus \underline{\mathcal{C}}\hat{L}_{\alpha})$. The minimisation of boundary region shall be constrained by requirements regarding the "shape" of elements of $\mathcal{C}$. We should try to find such a family $\mathcal{C}$ that for $c \in \mathcal{C}$ the value of $\mu(\pi_1(c))$ is relatively large while the value of $\mu_0(\pi_2(c))$ is relatively small. The neigbourhoods that fulfill these requirement correspond to ability of characterising large portions of the domain X^{∞} as being relatively uniform in terms of the value of loss function. The low variation of loss function on any given neighbourhood can be understood as equivalent to the requirement that this neighbourhood is well contained in a granule defined by application of the classifier α.

2. *Relevance.* This requirement is closely connected withe previous one (precision). While attempting to precisely dissect the domain into neighbourhoods we have to keep under control the relative quality (relevance) of neighbourhoods with respect to the data sample z. We are only interested in the neighbourhoods that contain sufficient number of elements of z. The actual threshold for this number is obviously dependant on the particular application and data set we are dealing with. It should be noted that without such threshold we are likely to produce neighbourhoods that are irrelevant to the data sample, hence potentially harmful for classifier learning process and risk assessment. The threshold for relevance of a neighbourhood is a subject to optimisation as well. If the threshold is too high, the resulting granularity of domain is to coarse and we are unable to make precise classification. On the other hand, if the threshold is too low and the resulting neighbourhoods contain too few elements from data sample, we may face the effect that is an equivalent of *overfitting* in classical classification systems.

3. *Coverage and adaptability.* One of the motivations that steer the process of creating the family of neighbourhoods and the classifier is the expectation regarding its ability to generalise and adapt the solution established on the basis of finite sample to a possibly large portion of the data domain. In terms of neighbourhoods it can be expressed as the requirement for minimisation of $\mu(X^{\infty} \setminus \bigcup_{c \in \mathcal{C}} \pi_1(c))$. This

minimisation has to be constrained by requirements expressed in points 1 and 2 above (precision and relevance). It shall also, to largest possible extent, provide for adaptability of our solution to classification problem. The adaptability requirement is understood as expectation that in the presence of newly collected data point the quality of loss approximation will not decrease dramatically as well as quality of our empirical risk assessment. In other words, the domain shall be covered by neighbourhoods in such a way that they are complaint with the expected behaviour of classifier on yet unseen examples. In terms of granulation of the domain we expect that the creation of neighbourhoods and learning of classifiers will be performed in some kind of feedback loop, that will make it possible to achieve possibly the highest compliance between the two. This requirement is formulated quite vaguely and obviously hard to turn into numerical optimisation criterion. It is, nevertheless, a crucial one, as we expect the system we are creating to be able to cater limited extensions of data sample with only small adjustments, definitely without the need for fundamental reconstruction of the entire system (classifier, neighbourhoods and loss function).

As discussed in points 1–3 above, the task of finding a family of neighbourhoods can be viewed as a multi-dimensional optimisation on meta-level. It is in par with the kind of procedure that has to be employed in construction of systems based on the *granular computing* paradigm [11,10].

The idea of granularity, impreciseness and limited expressiveness may surface in other places within the process of constructing the classifier, approximating the loss and assessing empirical risk. So far we have followed the assumption made at the beginning of Section 2, that the values of loss function are given as non-negative real numbers. In real application we may face the situation when the value of loss is given to us in less precise form. One such example is the loss function expressed in relative, qualitative terms. If the value of loss is given to us by the human expert, he/she may be unable to present us with precise, numerical values due to, e.g., imprecise or incompletely define nature of problem in discourse. We may then be confronted with situation when the loss is expressed in qualitative terms such as "big","negligible", "prohibitive","acceptable". Moreover, the value of loss may be expressed in relative terms, by reference to other, equally imprecise notions such as "greater than previous case" or "on the border between large and prohibitive". Such imprecise description of the loss function may in turn force us to introduce another training loop into our system, one that will learn how to convert the imprecise notions we have into concrete, numerical values of loss function. As a result, we may expect the final classification system to be compound, multi-stage and possibly hierarchical one.

Yet another axis for possible discussion and extensions of ideas presented in previous sections is associated with the choice of methodology for summation (integration). In order to introduce definition of risk measure (3) we have assumed that the the universe and corresponding sets are measurable in conventional sense. In other words, for finite samples we are dealing with, we have assumed that we ma rely on additivity of measure, and that the summation (integral) in our universe is a proper linear functional. That assumption may be challenged, as currently many authors (see [12]) bring examples that in several application domains we cannot easily make assumptions about linearity and sub-additivity of the mathematical structure of the universe. This is in particular the case of risk assessment. In complicated systems involving risk it may be necessary to perform a complex analysis that will lead to discovery of the actual properties of the domain. It is likely that such investigation will entail utilisation of expert's knowledge and domain theory/knowledge in addition to mathematical, analytic tools. The methodologies for proper utilisation of domain knowledge and experiences of human experts pose challenge that may be far greater and more general than the task of empirical risk assessment in classification systems we are dealing with in this paper.

5 Summary and Conclusion

In this paper we have discussed the issues that accompany the assessment of risk in classification systems on the basis of the finite set of examples. We have pointed out some sources of possible problems and outlined some directions, in which we may search for solutions that match our expectations sufficiently well.

In conclusion, we would like to go back to the more general issue of weighting the risk involved in computer-supported decision making. As we have mentioned in the introduction to this paper, in the real-life situations the human user may display various patterns in his/her risk assessment and aversion. In particular, even with a well established mathematical model that measures the risk in a given situation, we are frequently forced to change it as new information arrives. This is a natural phenomenon that we have to take into account in design of our solutions from the very beginning. In human terms, we can talk of evolution of the concept of risk in a given system as time passes and new information arrives. Humans may wish to change the way of perceiving the risk if they are able to use more information, form new experiences and make more informed judgement. In terms of computer-aided decision making the inflow of new information contributes to changes in parameters of the model. The challenge is to devise such a model that is flexible and far-fetching enough to be able to adjust for even significant changes resulting from changes generated by inflow of new information. It would be unrealistic to expect that it would be possible to devise and explicitly formulate a model, that would possess the extensibility as well as adaptability, and at the same time applicable in many different situations. It is much more likely that in practical situation we may need to learn (or estimate) not only the parameters, but the general laws governing its dynamics, at the same time attempting preserve its flexibility and ability to adapt for new cases.

Acknowledgements. The author wishes to thank Professor Andrzej Skowron, who's suggestions and guidance inspired creation of this article.

This research was supported by the grant N N516 368334 from Ministry of Science and Higher Education of the Republic of Poland and by the Innovative Economy Operational Programme 2007-2013 (Priority Axis 1. Research and development of new technologies) managed by Ministry of Regional Development of the Republic of Poland.

References

1. Warwick, B. (ed.): The Handbook of Risk. John Wiley & Sons, New York (2003)
2. Bostrom, A., French, S., Gottlieb, S. (eds.): Risk Assessment, Modeling and Decision Support. Risk, Governance and Society, vol. 14. Springer, Heidelberg (2008)
3. Cohen, M., Etner, J., Jeleva, M.: Dynamic decision making when risk perception depends on past experience. Theory and Decision 64, 173–192 (2008)
4. Vapnik, V.: Statisctical Learning Theory. John Wiley & Sons, New York (1998)
5. Vapnik, V.: Principles of risk minimization for learning theory. In: Proceedings of NIPS, pp. 831–838 (1991)
6. Pawlak, Z.: Rough Sets. Theoretical Aspects of Reasoning about Data. Kluwer Academic Publishers, Dordrecht (1991)
7. Pawlak, Z.: Rough sets, rough functions and rough calculus. In: Pal, S.K., Skowron, A. (eds.) Rough Fuzzy Hybridization: A New Trend in and Decision Making, pp. 99–109. Springer, Singapore (1999)
8. Halmos, P.: Measure Theory, Berlin. Springer, Berlin (1974)
9. Skowron, A., Synak, P.: Complex patterns. Fundamenta Informaticae 60, 351–366 (2004)
10. Jankowski, A., Peters, J.F., Skowron, A., Stepaniuk, J.: Optimization in discovery of compound granules. Fundamenta Informaticae 85 (in print, 2008)
11. Pedrycz, W., Skowron, A., Kreinovich, V. (eds.): Handbook of Granular Computing. John Wiley & Sons, New York (in print, 2007)
12. Kleinberg, J., Papadimitriou, C., Raghavan, P.: A microeconomic view of data mining. Data Mining and Knowledge Discovery 2, 311–324 (1998)

Business Aviation Decision-Making Using Rough Sets

Yu-Ping Ou Yang[1], How-Ming Shieh[1], Gwo-Hshiung Tzeng[2,3], Leon Yen[4],
and Chien-Chung Chan[5,6]

[1] Department of Business Administration, National Central University, Taiwan
ouyang.ping@msa.hinet.net
[2] Department of Business Administration, Kainan University, Taiwan
ghtzeng@mail.knu.edu.tw
[3] Institute of Management of Technology, National Chiao Tung University, Taiwan
ghtzeng@cc.nctu.edu.tw
[4] Department of International Business, Kainan University, Taiwan
leonyen@mail.knu.edu.tw
[5] Department of Computer Science, University of Akron, Akron, OH 44325, USA
chan@uakron.edu
[6] Department of Information Communications, Kainan University, Taiwan

Abstract. Business aviation has been popular in the United States, Europe, and South America, however, top economies in East Asia, including Japan, Korea, and Taiwan, have been more conservative and lag behind in the development of business aviation. In this paper, we hope to discover possible trends and needs of business aviation for supporting the government to make decision in anticipation of eventual deregulation in the near future. We adopt knowledge-discovery tools based on rough set to analyze the potential for business aviation through an empirical study. Although our empirical study uses data from Taiwan, we are optimistic that our proposed method can be similarly applied in other countries to help governments there make decisions about a deregulated market.

Keywords: Rough set theory, Information system, Decision rule, Business aviation, Data mining, Decision making.

1 Introduction

General aviation includes all aircrafts not flown by the airlines or the military, and business aviation is a segment of general aviation that consists of companies and individuals using aircrafts as tools in the conduct of their business. The largest market for business aviation is in North America, with Europe second in size. And this is not without reason. The U.S. government actively encourages and promotes the development of aviation. It invests a great deal of resources on the research of general aviation transportation and the establishment of the Advanced General Aviation Transport Experiments (AGATE) in 1994. Likewise, the British government offers special Customs, Immigration, Quarantine, Security (CIQS) to speed up immigration and customs procedures for business

C.-C. Chan et al. (Eds.): RSCTC 2008, LNAI 5306, pp. 329–338, 2008.

aviation passengers. It has also invested and constructed a Jet Center in 2002 that offers expedited passport control for business aviation passengers. But unlike most Western countries, business aviation has remained largely undeveloped in East Asian countries for three reasons: (1) lack of landing slots, parking spaces, and facilities at existing airports to support business aviation; (2) excessive regulations regarding the use of airports, pilot qualifications, and maintenance of aircrafts; (3) lack of understanding for the benefits and commercial value of business aviation. In Taiwan, since the completion of the second expressway and the new high-speed rail, it is expected that some airport capacity to become available for the development of business aviation. In addition, as more and more Taiwanese companies become multinational corporations, there should be substantial growth in demand for business aviation. The purpose of this study is to answer some pertinent questions in anticipation of deregulation in the business aviation industry in Taiwan. Specifically, we would like to address the third reason listed above, which is the lack of understanding for the economic benefits of business aviation, especially by the policymakers and the regulators. Is there a demand for business air travel in Taiwan? If so, which industries want to use business aviation? In addition, what are the most important criteria for business executives regarding business aviation? To find out the answer to these questions, we use the rough set approach to analyze the results from questionnaires given to Taiwanese companies on the subject of business aviation.

Rough Set Theory (RST) introduced by Pawlak in 1982 [7] is a knowledge-discovery tool that can be used to help induce logical patterns hidden in massive data. This knowledge can then be presented to the decision-maker as convenient decision rules. Its strength lies in its ability to deal with imperfect data and to classify, and its applications have grown in recent years. The rough set approach has already been applied with success in many applications, including data mining [6], business failure prediction [2,4], evaluation of bankruptcy risk [12], activity-based travel modeling [13], etc. However, RST has not been widely used in analyzing the airline industry, or for that matter, business aviation. Therefore, this study adopts RST to analyze above purposes, and the results demonstrate that the rough set approach is well suited for analyzing the market potential for business aviation and the needs of business aviation customers prior to the industry's deregulation. Likewise, the rough set approach can also be used to analyze the underlying market demand in any industry slated for deregulation.

The remainder of this paper is organized as follows: In Section 2, we present the basic concepts of RST. In Section 3, an empirical study is done using the rough set approach, and the results are presented and discussed. Section 4 concludes.

2   Basics of Rough Set Theory

The Rough Set Theory was first introduced by Pawlak in 1982 [7,8]. RST has been used by many researchers, and the theory has had a long list of achievements [10]. This section reviews the basic concepts of rough sets.

2.1 Information System and Approximations

RST is founded on the assumption that every object of the universe of discourse is associated with some information (data, knowledge). An *information system/table* can be represented as $S=(U, A, V, f)$, where U is the *universe* (a finite set of objects), A is a finite set of attributes (features, variables), $V = \underset{a \in A}{\cup} V_a$, where V_a is the set of values for each attribute a (called the domain of attribute a), and $f : U \times A \to V$ is an information function such that $f(x, a) \in V_a$, for all $x \in U$ and $a \in A$. Let $B \subseteq A$, and $x, y \in U$. We say x and y are indiscernible by the set of attributes B in S iff $f(x, b) = f(y, b)$ for every $b \in B$. Thus every $B \subseteq A$ generates a binary relation on U, called B *indiscernibility relation*, denoted by I_B.

In RST, the accuracy and the quality of approximations are very important in extracting decision rules. Let $B \subseteq A$ and $X \subseteq U$. The lower approximation of X in S by B, denoted by $\underline{B}X$, and the upper approximation of X in S by B, denoted by $\overline{B}X$ are defined as: $\underline{B}X = \{x \in U | I_B[x] \subseteq X\}$ and $\overline{B}X = \{x \in U | I_B[x] \cap X \neq \phi\}$ where the equivalence class of x in relation I_B is represented as $I_B[x]$. The boundary of X in S by B, is defined as $BN_B(X) = \overline{B}X - \underline{B}X$. An accuracy measure of the set X in S by $B \subseteq A$ is defined as:

$$\alpha_B(X) = card(\underline{B}X)/card(\overline{B}X),$$

where $card(\cdot)$ is the cardinality of a set. Let $F = \{X_1, X_2, \cdots, X_n\}$ be a classification of U, i.e., $X_i \cap X_j = \phi, \forall i, j \leq n, i \neq j$ and $\overset{n}{\underset{i=1}{\cup}} X_i = U$, X_i are called classes of F. The lower and upper approximations of F by $B \subseteq A$ are defined as: $\underline{B}F = \{\underline{B}X_1, \underline{B}X_2, \cdots, \underline{B}X_n\}$ and $\overline{B}F = \{\overline{B}X_1, \overline{B}X_2, \cdots, \overline{B}X_n\}$, respectively. The quality of approximation of classification F by the set B of attributes, or in short, quality of classification F is defined as: $\gamma_B(F) = \overset{n}{\underset{i=1}{\sum}} card(\underline{B}X_i)/card(U)$. It expresses the ratio of all B-correctly classified objects to all objects in the system.

2.2 Reductions and Core

An important issue in RST is about attribute reduction, which is performed in such a way that the reduced set of attributes $B, B \subseteq A$, provides the same quality of classification $\gamma_B(F)$ as the original set of attributes A. A minimal subset $C \subseteq B \subseteq A$ such that $\gamma_B(F) = \gamma_C(F)$ is called a F-*reduct* of B and is denoted by $RED_F(B)$. A *reduct* is a minimal subset of attributes that has the same classification ability as the whole set of attributes. Attributes that do not belong to a reduct are superfluous in terms of classification of elements of the universe. The *core* is the common part of all reducts. For example, $CORE_F(B)$ is called the F-core of B, if $CORE_F(B) = \cap RED_F(B)$.

2.3 Decision Rules

An information table $A = C \cup D$ can be seen as a *decision table*, where C and D are condition and decision attributes, respectively, and $C \cap D = \emptyset$. The set of decision attributes D induces an indiscernibility relation I_D that is independent of the conditional attributes C; objects with the same decision values are grouped together as decision elementary sets (decision classes). The reducts of the condition attribute set will preserve the relevant relationship between condition attributes and decision classes. And this relationship can be expressed by a decision rule.

A decision rule in S is expressed as $\Phi \rightarrow \Psi$, read *if Φ then Ψ* (a logical statement). Φ and Ψ are referred to as conditions and decisions of the rule, respectively. In data mining, we usually take into account relevant confirmation measures and apply them within RST to data analysis. They are presented as follows [8]. The *strength* of the decision rule $\Phi \rightarrow \Psi$ in S is expressed as: $\sigma_s(\Phi, \Psi) = supp_s(\Phi, \Psi)/card(U)$, where $supp_s(\Phi, \Psi) = card(\|\Phi \wedge \Psi\|_s)$ is called the *support* of the rule $\Phi \rightarrow \Psi$ in S and $card(U)$ is the cardinality of U. With every decision rule $\Phi \rightarrow \Psi$ we associate a *coverage factor/covering ratio (CR)* defined as: $cov_s(\Phi, \Psi) = supp_s(\Phi, \Psi)/card(\|\Psi\|_s)$.

CR is interpreted as the frequency of objects having the property Φ in the set of objects having the property Ψ. The strength of the decision rule can simply be expressed as the ratio - the number of facts that can be classified by the decision rule divided by the number of facts in the data table. Both CR and the *strength* of the decision rule are used to estimate the quality of the decision rules. They play an essential role for a decision-maker in considering which decision rule to use.

2.4   Flow Graphs

The study of flow graphs is not new [1, 5]. In this work, we use the approach introduced by Pawlak [8, 9]. The basic idea is that each branch of a flow graph is interpreted as a decision rule and the entire flow graph describes a decision algorithm. In flow graphs, the Bayesian factors, namely, support, strength, certainty, and coverage factors, associated with each decision rule as defined in previous Section 2.3 are formulated in terms of throughflows. More precisely, a flow graph is a directed acyclic finite graph $G = (V, E, w)$, where V is a set of nodes, $E \subseteq V \times V$, is a set of directed branches, and $w: E \rightarrow R^+$ is a flow function where R^+ is the set of non-negative real numbers. The *throughflow* of a branch (x, y) in E is denoted by $w(x, y)$. For each branch (x, y) in E, x is an input of y and y is an output of x. For x in V, let $I(x)$ denote the set of all inputs of x and $O(x)$ be the set of all outputs of x. The inputs and outputs of a graph G are defined by $I(G) = \{x$ in $V|I(x)$ is empty$\}$ and $O(G) = \{x$ in $V|O(x)$ is empty$\}$. For every node x in G, *inflow(x)* is the sum of throughflows from all its input nodes, and *outflow(x)* is the sum of throughflows from x to all its output nodes. The inflow and outflow of the whole flow graph can be defined similarly. It is assumed that for any node x in a flow graph, $inflow(x) = outflow(x) = throughflow(x)$. This is also true for the entire flow graph G.

Every branch (x, y) of a flow graph G is associated with the certainty *(cer)* and coverage *(cov)* factors defined as: $cer(x, y) = \sigma(x, y)/\sigma(x)$ and $cov(x, y) = \sigma(x, y)/\sigma(y)$, where $\sigma(x, y) = w(x, y)/w(G)$, $\sigma(x) = w(x)/w(G)$, and

$\sigma(y) = w(y)/w(G)$ are *normalized throughflows*, which are also called strength of a branch or a node, and we have $\sigma(x) \neq 0$, $\sigma(y) \neq 0$, and $0 \leq \sigma(x, y) \leq 1$. Properties of the coefficients and the certainty, coverage, and strength of paths and connections were defined and studied in Pawlak [8, 9].

3   An Empirical Case for Classifying the Business Aviation Prediction

This study adopts two stages for the rough set approach. Each stage focuses on the problem of classifying data sets into classes, and each stage follows the same analytical procedure: (1) Calculate the approximation; (2) Find the reductions and the core attributes; (3) Create the decision rules; (4) Arrange rules into a decision flow graph as the final decision algorithm induced from data. We use an example in business aviation to illustrate the strength of rough set approach. The results are used to identify and predict the willingness of companies to use business aviation; they can also be used to propose improved attributes levels through plans to satisfy customers' needs by the government agencies or airline companies. The background of the empirical study, the statement of problem, and the experimental setup are discussed below.

3.1   Background and the Statement of Problem

Taiwan's business aviation sector has not yet been deregulated. But with Taiwan's multinational businesses having production facilities strewn all over Asia, there is a clear demand for business aviation. Our research aims to find out the underlying demand for business aviation if the industry were to be deregulated in the future. We will use the rough set approach to analyze the demand for business aviation and make suggestions for government agencies to start making plans for a deregulated future today.

3.2   Experimental Setup

In order to investigate the possibility of utilizing business aviation in the future, we randomly select 200 companies from the top 500 corporations in Taiwan (according to the annual ranking by a notable business magazine in Taiwan) as our survey subjects and mail out a questionnaire to these selected companies. The content of the questionnaire, the criteria or attribute about business aviation that we are interested in, is based on a research report done by Taiwan's Civil Aeronautics Administration (Research Program of CAA in 2004). Of the 88 questionnaires returned, 76 of them are qualified replies. Among the 76 replies, 2 (2.6%) companies said that they would definitely use business aviation when business aviation becomes available, 40 (52.6%) companies said that they are considering the possibility of using business aviation, and 34 (44.7%) companies said they would not use business aviation. In order to operate multiple-choice question value and to improve the classification rate, the value class has been redefined by the Research Program of CAA in 2004.

Our analysis includes two stages based on the rough set approach. The first stage is to find out the market potential for business aviation. The companies' basic attributes and their present experiences of flying are stated as conditional attributes and the classification is defined as a decision attribute (D1). The decision classes are *Definitely yes, Considering the possibility,* and *No.* The second stage is to understand the factors elaborated by companies in considering the use of business aviation. These factors are the six condition attributes. They include: (B1) the purpose of utilizing business aviation; (B2) the estimated flying trips in a year; (B3) what special services are needed; (B4) which Taiwan airports are preferred; (B5) what is the maximum price the company is willing to pay for a business chartered plane; (B6) whether the business is willing to buy an aircraft or rent instead. Four factors that influence the companies in considering whether to use business aviation are: (D2-1) direct flight between Taiwan and China; (D2-2) whether the cost is too expensive or not; (D2-3) the quality of ground services and maintenance; (D2-4) the scheduled flights cannot meet your needs. These 4 factors serve as decision attributes. The details of our analysis of this survey data are as follows.

3.3   First Stage Results

In the first stage, we use question D1 to define a classification on our sample data, and we calculate the accuracy of approximations of each class. Here class 1 denotes companies who answered "Definitely yes" on question D1, class 2 denotes companies who answered "Considering the possibility", and class 3 denotes companies who answered "No". Both the accuracy and the quality of the classification are equal to 1, which means that the results of the classification are very satisfactory.

In the next step, we used the ROSE2 (Rough Set Data Explorer) tool [11] to find all potential reducts and the core attributes. Two reducts were found: Set1 = {A1, A2, A3, A4, A5} and Set2 = {A2, A3, A4, A5, A7}. Therefore, the core is the set of attributes {A2, A3, A4, A5}. These attributes are the most significant attributes for classification purposes. Dimitras et al. [4] proposed some criteria for selecting reducts: (1) the reduct should contain as small a number of attributes as possible, (2) the reduct should not miss the attributes judged by the decision maker as the most significant for evaluation of the objects. Following the criteria, the reduct Set1 was selected for further analysis. We used the BLEM2 tool [3] to generate decision rules from data set reduced to attributes in Set1. There are 51 rules generated by BLEM2 for classes 1, 2, and 3. There are 2 rules with total support of 2 for class 1 (D1 = "1"), which denotes the companies definitely will utilize business aviation, there are 26 rules with total support of 40 for class 2 (D1 = "2"), which denotes companies that are considering the possibility of utilizing it, and there are 23 rules with total support of 34 for class 3 (D1 = "3"), denoting companies have no willingness at all to use business aviation.

We have applied a minimum support threshold for further reduction of rules. For class 2 rules, with minimum support value ≥ 2, the number of rules is reduced from 26 to 7 as shown in Table 1. In addition, condition attribute A4 is no longer used among these rules as shown in the table that entries in A4 are all blanks.

Table 1. Stage 1 rules for class 2 with minimum support value ≥ 2

A1	A2	A3	A4	A5	D1	Support	Certainty	Strength	Coverage
4					2	5	1	0.0658	0.125
	5	3		2	2	3	1	0.0395	0.075
	2	4			2	2	1	0.0263	0.05
	4			1	2	3	1	0.0395	0.075
1	3	1			2	4	1	0.0526	0.1
1	5	1			2	2	1	0.0263	0.05
1	3	3			2	2	1	0.0263	0.05

Rules in Table 1 can be translated into one decision algorithm represented by the decision flow graph shown in Figure 1. For simplicity, certainty, strength, and coverage factors associated with each branch are omitted, only throughflows (supports) are shown in the figure. The total inflow of the graph is 21, which is the sum of the supports corresponding to rules appeared in Table 1.

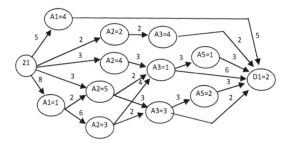

Fig. 1. Decision flow graph for rule-set shown in Table 1

3.4   Second Stage Results

In the second stage, we isolate the 40 companies who said they would consider using business aviation and analyze them separately. Four decision tables are generated, with each decision table made up of the same six conditional attributes (B1~B6) and one decision attribute D2-X (where X is assigned 1, 2, 3, 4, respectively). Similarly, the approximations of decision classes are calculated the same way as in the first stage, and the results are: the classification accuracy of D2-1 and D2-4 is 0.6667 and their classification quality is 0.8; both of the classification accuracy and quality of D2-2 are 1; the classification accuracy of

Table 2. The support value more than 2 from Stage 2 (D2-1=1)

B1	B2	B3	B4	B5	D2-1	support	certainty	strength	coverage
3					1	3	1	0.075	0.1364
2	1	H	A		1	2	1	0.05	0.0909
1	1	H		1	1	3	0.75	0.075	0.1364

Table 3. The support value more than 2 from Stage 2 (D2-2=1)

B2	B3	B4	D2-2	support	certainty	strength	coverage
	J		1	9	1	0.225	0.2432
	K		1	4	1	0.1	0.1081
	N		1	3	1	0.075	0.0811
N	H	A	1	4	1	0.1	0.1081
1	H		1	11	1	0.275	0.2973
2	H	B	1	2	1	0.05	0.0541

Table 4. The support value more than 2 from Stage 2 (D2-3=1)

B1	B2	B3	B4	B5	D2-3	support	certainty	strength	coverage
		N			1	3	1	0.075	0.1364
2	N				1	2	1	0.05	0.0909
2	1	H	A		1	2	1	0.05	0.0909
1	1	H		1	1	4	1	0.1	0.1818

Table 5. The support value more than 2 from Stage 2 (D2-4=1)

B1	B2	B3	B4	B5	D2-4	support	certainty	strength	coverage
	2	N			1	2	1	0.05	0.0952
2	N				1	2	1	0.05	0.0952
	2			N	1	2	1	0.05	0.0952
2	1	H	A		1	2	1	0.05	0.0952
1	1	H		1	1	3	0.75	0.075	0.1429

D2-3 is 0.8182 and its classification quality is 0.9. These results of the classification are acceptable, since all the accuracies for D2-1 through D2-4 are better than 0.6, and their qualities are close to 1. Next, we find the reductions and the core attributes. As a result, the reduct and the core for D2-2 obtained are { B2, B3, B4}; the reduct and the core for the others are {B1, B2, B3, B4, B5}. The next step of data analysis is to generate decision rules by using BLEM2 based on the selected reducts. The resulting rule sets with minimum support value $\geq$ 2 are shown in Tables 2 - 5.

4   Discussions

Our empirical study on the market potential for business aviation using the rough set approach has uncovered some important facts. In the first stage, by correlating their decision to use business aviation with their basic attributes, including type of industry and the regions they are doing business with, our results match up closely to the present economic realities in Taiwan: (1) the majority of businesses who replied to our survey said they would consider using business aviation if it became available; (2) the regions that these companies are doing business with are mainly China region, North America, and Northeast Asia; (3) the companies' type of industry is mainly traditional industries, manufacturing industries, and information technology industries. The RST analysis clearly shows that there is a market potential for business aviation in Taiwan. Therefore, if the Taiwan government wants to develop its business aviation industry, it should start thinking about how to satisfy companies' flying needs before business aviation is deregulated in the future. Our study has also addressed this issue through a second stage RST analysis, and we found five common patterns which could be of interest to government policymakers as they plan relative resources distribution, provide special services, select available airports, and set reasonable price, etc. Based on Tables 2 - 5 with higher strength, we find several proposals could be made:

1. Government agency should consider the regions that the companies are doing business with when they plan air traffic rights in the future.
2. Government agency should plan a way for these passengers to pass through the immigration & customs quickly. In addition, airports need to construct a business aviation center that can deal with related services such as transportation and hotel accommodation.
3. The most preferred airport for business aviation is Taipei Songshan Airport. Previously, Songshan Airport was an international airport. Nowadays, it is a domestic airport. Although the land this airport sits on isn't that big, it is big enough for smaller aircrafts. And it is situated right in the heart of Taipei, the capital of Taiwan, where most Taiwanese companies are headquartered. Also, Taipei's subway system is being extended to Songshan Airport. This makes Taipei Songshan Airport an ideal airport for business aviation.
4. The maximum price that businesses are willing to pay for business chartered flights should be limited to no more than double the price of first-class tickets in future planning.

The above proposals are made for the consideration of Taiwan government agencies. Moreover, government agencies should consider the resource distribution between existing scheduled commercial flights and business aviation, so that they can both thrive in the market. Currently, business aviation has not yet been deregulated in many countries. The rough set approach should be able to help these countries predict the underlying market demand and trends for business aviation and help their government agencies plan for a future where business aviation is allowed to thrive.

5   Conclusions

This paper uses the rough set approach to analyze the underlying market demand and needs for business aviation in Taiwan. Discovered patterns are interesting and corresponding well to current local economical reality. Therefore, we have shown that rough set approach is a promising method for discovering important facts hidden in data and for identifying minimal sets of relevant data (data reduction) for the business aviation. Our results provide several useful proposals for local government agencies. In addition, although our empirical study uses data from Taiwan, we are optimistic that the method can be similarly applied in other countries to help governments there make decisions about a deregulated market.

References

1. Berthold, M., Hand, D.J.: Intelligent Data Analysis – An Introduction. Springer, Heidelberg (1999)
2. Beynon, M.J., Peel, M.J.: Variable precision rough set theory and data discretisation: an application to corporate failure prediction. Omega, International Journal of Management Science 29(6), 561–576 (2001)
3. Chan, C.-C., Santhosh, S.: BLEM2: Learning Bayes' rules from examples using rough sets. In: Proc. NAFIPS 2003, 22nd Int. Conf. of the North American Fuzzy Information Processing Society, Chicago, Illinois, July 24 – 26, 2003, pp. 187–190 (2003)
4. Dimitras, A.I., Slowinski, R., Susmaga, R., Zopounidis, C.: Business failure prediction using rough sets. European Journal of Operational Research 114(2), 263–280 (1999)
5. Ford, L.R., Fulkerson, D.R.: Flows in Networks. Princeton University Press, Princeton (1962)
6. Hu, Y.C., Chen, R.S., Tzeng, G.H.: Finding fuzzy classification rules using data mining techniques. Pattern Recognition Letters 24(1-3), 509–519 (2003)
7. Pawlak, Z.: Rough sets. International Journal of Computer and Information Sciences 11(5), 341–356 (1982)
8. Pawlak, Z.: Rough sets, decision algorithms and Bayes' theorem. European Journal of Operational Research 136(1), 181–189 (2002)
9. Pawlak, Z.: Flow graphs and intelligent data analysis. Fundamenta Informaticae 64(1/4), 369–377 (2005)
10. Pawlak, Z., Skowron, A.: Rudiments of rough sets. Information Sciences 177(1), 3–27 (2007)
11. Predki, B., Slowinski, R., Stefanowski, J., Susmaga, R., Wilk, Sz.: ROSE - Software Implementation of the Rough Set Theory. In: Polkowski, L., Skowron, A. (eds.) RSCTC 1998. LNCS (LNAI), vol. 1424, pp. 605–608. Springer, Heidelberg (1998)
12. Slowinski, R., Zopounidis, C.: Application of the rough set approach to evaluation of bankruptcy risk. International Journal of Intelligent Systems in accounting, Finance and Management 4(1), 27–41 (1995)
13. Witlox, F., Tindemans, H.: The application of rough sets analysis in activity-based modeling, opportunities and constraints. Expert Systems with Application 27(2), 171–180 (2004)

Phase Transition in SONFIS and SORST

Hamed O. Ghaffari[1], Mostafa Sharifzadeh[1], and Witold Pedrycz[2]

[1] Department of Mining & Metallurgical Engineering, Amirkabir University of Technology,
Tehran, Iran
[2] Department of Electrical and Computer Engineering
University of Alberta
Alberta, Canada
h.o.ghaffari@gmail.com, Sharifzadeh@aut.ac.ir,
pedrycz@ece.ualberta.ca

Abstract. This study introduces new aspects of phase transition in two new hybrid intelligent systems called Self-Organizing Neuro-Fuzzy Inference System (SONFIS) and Self-Organizing Rough SeT (SORST). We show how our algorithms can be taken as a linkage of government-society interaction, where government catches various states of behaviors: "solid (absolute-oppressive) or flexible (democratic)". So, transition of such System, by changing of connectivity parameters (noise) and using a simple linear relation, from order to disorder states is inferred.

Keywords: phase transition, SONFIS, SORST, Nations-Government interactions.

1 Introduction

Social systems as a type of the Complex systems are often coincided with uncertainty and order-disorder transitions (or reverse) so that a small event can trigger an impressive transition. Apart of uncertainty, fluctuations forces due to competition of between constructive particles of the system drive the system towards order and disorder [1], [12]. For example a social system (ranging from stock markets to political systems, and traffic flow) whose individuals have an inclination to conform to each other may show a phase transition step [2]. In other view, in monitoring of most complex systems, there are some generic challenges for example sparse essence, conflicts in different levels of a system, inaccuracy and limitation of measurements are the real obstacles in analysis and predication of the possible emerged behaviors.

There are many methods to analyzing of systems include many particles that are acting on each other, for example statistical methods [3], [4], Vicsek model [5]. Other solution is finding out of main nominations of each distinct behavior which may has overlapping- in part-to others. This advance is to bate some mentioned difficulties that can be concluded in the *"information granules"* proposed by Zadeh [6]. In fact, more complex systems in their natural shapes can be described in the sense of networks, which are made of connections among the units. These units are several facets of information granules as well as clusters, groups, communities, modules [7].

Regarding mentioned aspects and the role of information flow in mass behavior, analysis and evaluation of social systems are accomplished typically in the context of

C.-C. Chan et al. (Eds.): RSCTC 2008, LNAI 5306, pp. 339–348, 2008.
© Springer-Verlag Berlin Heidelberg 2008

the specific system. For example consider a society involving political groups, parts, social networks and so forth. A stimulated wave – or information flows- in to this society, may amplify (or abridge) the range of distinguished nodes which are the nominations of a complex system. Evolving of social cores –as well as semi-self organizing system - is associated with the responses of a government or external forces. Definitely, when an oppressive government slightly worsens the living condition – with their oppressive instruments-this task may leads to increase of the opposition to the government. Depending on the information statics, the impact of the pervious states (background mind) of groups or government or even a small parameter, a dramatic transition can be occurred. So, the role of significant external trigger can't be denied. Phase transition modeling in specific social systems (such political revolution, fragility of mass behavior, social cooperation, traffic flow, and stock market crashes so forth) has been reviewed in [2].

Phase transitions in some intelligent systems especially in neural networks, stochastic self organizing maps and cellular neural network regarding statistical physics theorems have been applied [8-11]. In this study, we reproduce two hybrid intelligent systems called Self-Organizing Neuro-Fuzzy Inference System (SONFIS) and Self-Organizing Rough SeT (SORST), and investigate several levels of responses against the real information. The motivations of our methods are considering of several states of society, government, their effects on each other and the role of other parameters (such external forces) which are concluded behind transformations of information nodes. In [18] we showed how relatively such our simple methods can produce (mimic) government-nation interactions where the adaptive scaling scheme of SORST (SORST-AS) and SONFIS with the changing of the highlighted parameters have been utilized. In this paper the aim is to analysis and inferring new contexts upon the regarding of more rules and external forces in SONFIS and direct transition evaluation of SORST (with constant scaling). In fact, Mutual relations between algorithms layers identify order-disorder transferring of such systems. Developing of such intelligent hierarchical networks, investigations of their performances on the noisy information and exploration of possible relate between phase transition steps and flow of information in to such systems are new interesting fields, as well in various fields of science and economy.

2   The Proposed Procedure

In this section based upon self organizing feature map [13], adaptive neuro fuzzy inference system [14] and rough set theory (RST) [15], we reproduce: Self Organizing Neuro-Fuzzy Inference System (SONFIS) and Self Organizing Rough SeT (SORST) [9], [10]. In this study our aim is to investigate order-disorder transition in the mentioned systems. The mentioned algorithms use four basic axioms upon the balancing of the successive granules assumption:

- Step (1): dividing the monitored data into groups of training and testing data
- Step (2): first granulation (crisp) by SOM or other crisp granulation methods
  Step (2-1): selecting the level of granularity randomly or depending on the obtained error from the NFIS or RST (regular neuron growth)
  Step (2-2): construction of the granules (crisp).

- Step (3): second granulation (fuzzy or rough granules) by NFIS or RST
  Step (3-1): crisp granules as a new data.
  Step (3-2): selecting the level of granularity; (Error level, number of rules, strength threshold...)
  Step (3-3): checking the suitability. (Close-open iteration: referring to the real data and reinspect closed world)
  Step (3-4): construction of fuzzy/rough granules.
- Step (4): extraction of knowledge rules

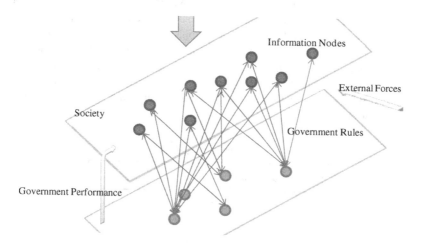

Fig. 1. A general schematic of Society-Government network

Mutual relations of the two main layers of the algorithms can be imagined as society – government interactions, where reactions of a dynamic community to an "absolute (solid) or flexible" government (regulator) are controlled by correlation (noise) factors of the two simplified systems. In absolute case, the second layer (government\ regulator) has limited rules with stable learning iteration for all of matters. In first layer, society selects the main structures of the stimulator where these clusters upon the reaction of government and the pervious structures will be adjusted. Flexible regulator (democratic government) has ability of adapting with the evolution of society. This situation can be covered by two discrete alternatives: evolution of constitutive rules (policies) over time passing or a general approximation of the dominant rules on the emerged attitudes. In latter case the legislators can considers being conflicts of the emerged states. Other mode can be imagined as poor-revealing structures (information nodes) of the society due to poor-learning or relatively high disturbances within inner layers of the community. In other view and using complex networks theory [19], we may make a hierarchical complex network in which information nodes in a society depend on their requirements (or in a compulsory way) are connected to the government's rule(s) (Fig.1). Considering of several aspects of intellectuality, uncertainty, evolution of nodes (society and/or government) and other possible fascinating results are some of the motivations of such structures.

With variations of correlation factors of the two sides (or more connected intelligent particles), one can identify point (or interval) of changes behavior of society or overall system and then controlling of society may be satisfied. Obviously, considering of simple and reasonable phase transition measure in the mentioned systems will be necessary so that regarding of entropy and order parameter are two distinguished criteria [1]. In this study we use crisp granules level to emulation of phase passing. So, we consider both "absolute and flexible" government while in latter case the approximated rules are contemplated and the presumed society will cope to the disturbances. As before mentioned, in this macroscopic sight of the interactions, the reaction of first layer to the current stimulator is upon the batching inserting of information, and so cannot take in to account of "historical memory decadence" of individuals of the community. In other sense all of information gets an equal effect on the society whereas in actual case *time streamlet bears other generations and new worlds, so new politics.*

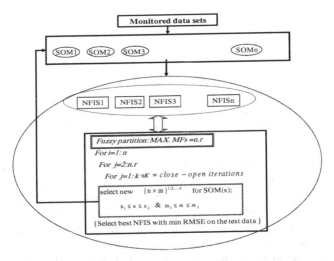

Fig. 2. Self Organizing Neuro-Fuzzy Inference System (SONFIS)

Balancing assumption is satisfied by the close-open iterations: this process is a guideline to balancing of crisp and sub fuzzy/rough granules by some random/regular selection of initial granules or other optimal structures and increment of supporting rules (fuzzy partitions or increasing of lower /upper approximations), gradually. The overall schematic of Self Organizing Neuro-Fuzzy Inference System -Random: SONFIS-R has been shown in Fig.2. In first granulation step, we use a linear relation is given by:

$$N_{t+1} = \alpha N_t + \Delta_t ; \Delta_t = \beta E_t + \gamma, \tag{1}$$

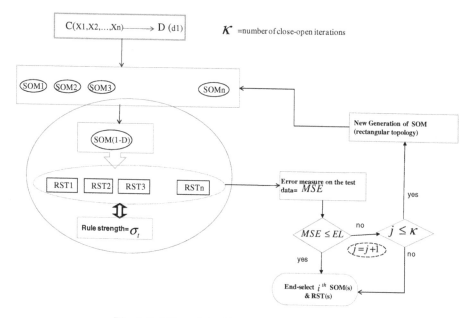

Fig. 3. Self Organizing Rough Set Theory (SORST)

where $N_t = n_1 \times n_2; |n_1 - n_2| = Min.$ is the number of neurons in SOM or Neuron Growth (NG); E_t is the obtained error (measured error) from second granulation on the test data and coefficients must be determined, depend on the used data set. Obviously, one can employ like manipulation in the rule (second granulation) generation part, i.e., number of rules (as a pliable regulator). Determination of granulation level is controlled with three main parameters: range of neuron growth, number of rules and/or error level. The main benefit of this algorithm is to looking for best structure and rules for two known intelligent system, while in independent situations each of them has some appropriate problems.

In second algorithm RST instead of NFIS has been proposed (Fig. 3). Because of the generated rules by a rough set are coarse and therefore need to be fine-tuned, here, we have used the preprocessing step on data set to crisp granulation by SOM (close world assumption).

3 Phase Transition on the "Lugeon Data Set"

In this part of paper, we ensue our algorithms on the "lugeon data set" [16-18]. So, a similar procedure has been accomplished in other data sets [20, 21]. To evaluate the interactions due to the lugeon values we follow two situations where phase transition measure is upon the crisp granules (here NG): 1) second layer gets a few limited rules by using NFIS; 2) second layer gets all of the extracted rules by RST and under an approximated progressing.

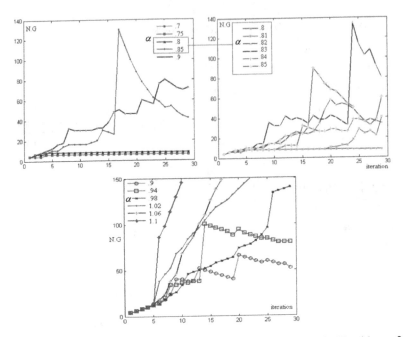

Fig. 4. Effect of Alpha variations in the neuron growth (N.G) of SONFIS-AR with $n.r$=2 over 30 iterations; β =.001, γ=.5 a) .7=< α <=.9; b). 8=< α <=.85 and c).9=<α <=1.1

Analysis of first situation is started off by setting number of close-open iteration and maximum number of rules equal to 30 and 3 in SONFIS respectively. The error measure criterion in SONFIS is Root Mean Square Error (RMSE), given as below:

$$RMSE = \sqrt{\frac{\sum\limits_{i=1}^{m} (t_i - t_i^*)^2}{m}} , \qquad (2)$$

where t_i is output of SONFIS and t_i^* is real answer; m is the number of test data (test objects). In the rest of paper, let m=93 and number of inserting data set =600. By employing of Eq.1 in SONFIS (β =.001 and γ=.5); the general patterns of NG and RMSE vs. time steps and variations of α can be observed (Fig.4). It must be noticed for two like process (i.e., α=.9), we have different situation of neuron growth. The main reason of such behavior is on the regulation of weight neurons in SOM thank to initial random selection and fall in to the "dead neurons state". However, this will be interesting if we see real case, as is appeared in a real society, in order to "in an identical cases (but in an unlike iteration) society may shows other behavior, not completely different from other mate". The neurons fluctuations with the time passing reveal more chaos while the phase transition step can be transpired in α=.8-.85(Fig.5).

Increasing of rules number (*n.r*) in second layer to 3, system doesn't disclose a significant change in phase transition step (starting of rush after α =.8) where the initial values of NFIS and an upper limit for N.G, i.e., 150 have been employed (Fig.5). It must be noticed that the variation of RMSE (or deduction of the government from the society) is not coincided with the N.G whereas with fixing of β and changing α, the assumed government considers a constant impact for repression of a dynamic society [18]. In the former and latter options, the phase transition has been occurred gradationally likewise one can consider three discrete steps to these conversions: society with "silent dead (laminar)", in transition and in triggering of revolutionary community.

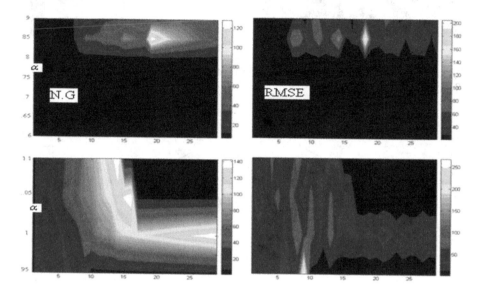

Fig. 5. Effect of Alpha variations in neuron growth –Beta=.001(N.G)-left- &RMSE-right- of SONFIS with *n.r*=3 over 30 iterations; a) .6=< Alpha<=.9; b) .95=< Alpha <=1.1

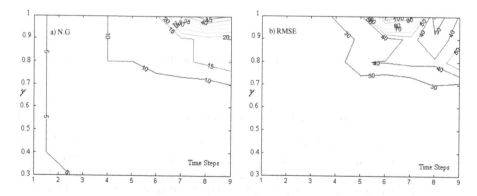

Fig. 6. Diagram contour of γ variations on the N.G and RMSE ($\alpha = .8, \beta = 10^{-3}$) over 9 time steps

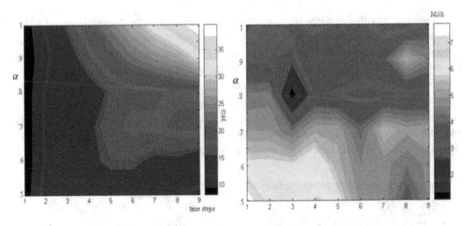

Fig.7. Color code of alpha variations in SORST on the N.G and MSE over 9 time steps- β =1.01

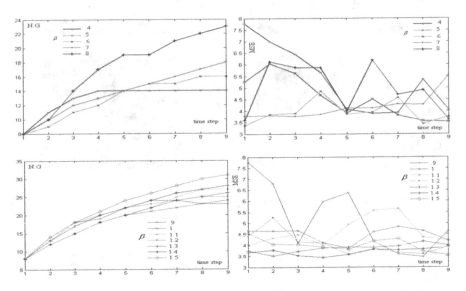

Fig. 8. Effect of β variations in neuron growth (N.G) &MSE of SORST over 9 iterations-; α =.8

In other process and to investigate the position of γ as a small parameter (we call external forces), we considered SONFIS with α = .8, β = 10^{-3} over 9 time steps (Fig.6). As if system shows a transition step after .9, one can recognize the disordered zones.

In second situation instead of NFIS, we employ RST upon this assumption that the government based on history, experience and other like fashions in the world, can has ability to elicitation of relatively approximated rules of the observed and distinguished behaviors (by transferring of attributes to three scaled classes using 1-D SOM, as well as low, middle and high). The applied error measure for measure of performance of SORST is given by:

$$MSE = \frac{\sum\limits_{i=1}^{m} (d_i{}^{real} - d_i{}^{classified})^2}{m}, \tag{3}$$

In deducing of decision for approximated rules (not unique decision part), we select highest value (largest ambiguities) for such decisions. By repeating of steps as well pervious situation, we obtain other behavior of SORST where we employ $\gamma = 1$, in Eq.1 (Fig.7&Fig.8). Fig.7 shows how with keeping of β as a fixed value, N.G after $\alpha = .8$ gets in to the steeper rate while it has endured relatively high transition time. In fact government without changing of his affecting power (β) preserve, in a long time society between "tranquil and rushing". Other guessed difference with first option, is on the fast change of the society over passing time and for high α values as though MSE exhibits a low range of variations.

4 Conclusion

In this study we proposed two new algorithms in which SOM, NFIS and RST make SONFIS and SORST. Mutual relations between algorithms layers identify order-disorder transferring of such systems. So, we found our proposed methods have good ability in mimicking of government-nation interactions while government and society can take different states of responses. Developing of such intelligent hierarchical networks, investigations of their performances on the noisy information and exploration of possible relate between phase transition steps and flow of information are new interesting fields, as well in various fields of science and economy.

References

1. Wicks, R.T., Chapman, S.C., Dendy, R.O.: Mutual Information as a Tool for Identifying Phase Transition in Dynamical Complex Systems with Limited Data. Phys. Rev. E. 75, 051125 (2007)
2. Levy, M.: Social Phase Transitions. Journal of Economic Behavior& Organization 57, 71–87 (2005)
3. Convey, P.V.: The Second Law of Thermodynamics: Entropy, Irreversibility and Dynamics. Nature 333, 409 (1988)
4. Hakan, H.: Synergetics: An Introduction. Springer, New York (1983)
5. Vicsek, T., Czirok, A., Ben-Jacob, E., Cohen, I., Shochet, O.: Novel Type of Phase Transition in a System of Self-Driven Particles. Phys. Rev. Lett. 75, 1226 (1995)
6. Zadeh, L.A.: Toward a Theory of Fuzzy Information Granulation and Its Centrality in Human Reasoning and Fuzzy Logic. Fuzzy sets and systems 90, 111–127 (1997)
7. Palla, G., Derenyi, I., Farkas, I., Vicsek, T.: Uncovering the Overlapping Community Structure of Complex Networks in Nature and Society. Nature 435, 814–818 (2005)
8. Puljic, M., Kozma, R.: Phase Transition in a Probabilistic Cellular Neural Network Modeling Having Local and Remote Connections. In: International Joint Conference on Neural Networks, Portland, OR, pp. 831–835 (2003)

9. Biehl, M., Schlosser, E., Ahr, M.: Phase Transitions in Soft Committee Machines (1998), http://www.arxiv.org/cond-mat/9805182v2

10. Kinzel, W.: Phase Transitions of Neural Networks (1997), http://www.arxiv.org/cond-mat/9704098v1

11. Huepe, C., Aldana-Gonzalez, M.: Dynamical Phase Transition in a Neural Network Model with Noise: An Exact Solution (2002), http://www.arxiv.org/cond-mat/0202411v1

12. Vicsek, T.: The Bigger Picture. Nature 418, 131 (2002)

13. Kohonen, T.: Self-Organization and Associate Memory, 2nd edn. Springer, Berlin (1983)

14. Roges Jang, J.-S., Sun, C.T., Mizutani, E.: Neuro-Fuzzy and Soft Computing, Newjersy. Prentice Hall, Englewood Cliffs (1997)

15. Pawlak, Z.: Rough Sets: Theoretical Aspects Reasoning about Data. Kluwer academic, Boston (1991)

16. Ghaffari, O.H., Sharifzadeh, M., Shahrair, K.: Rock Engineering Modeling based on Soft Granulation Theory. In: 42nd U.S. Rock Mechanics Symposium & 2nd U.S.-Canada Rock Mechanics Symposium, San Francisco, CD-ROM (2008)

17. Ghaffari, O.H., Shahriar, K., Pedrycz, W.: Graphical Estimation of Permeability Using RST&NFIS. In: NAFIPS 2008, New York (2008)

18. Ghaffari, O.H., Pedrycz, W., Sharifzadeh, M.: Order - Disorder Transition in Hybrid Intelligent Systems: a Hatch to the Interactions of Nations –Governments. In: The 2008 IEEE International Conference on Granular Computing (GrC 2008), China (August 2008)

19. Albert, R., Barabasi, A.-L.: Statistical Mechanics of Complex Networks. Review of Modern Physics 74, 47–97 (2002)

20. Ghaffari, O.H.: Applications of Intelligent Systems in Analysis of Hydro- Mechanical Coupling Behavior of a Single Rock Joint under Normal and Shear Load. M.Sc Thesis, Amirkabir University of Technology (2008)

21. Ghaffari, O.H., Ejtemaei, M., Iranajad, M.: Fusion of Intelligent Systems based on Information Granulation Approach to Analysis of Hydro Cyclone Performance. In: 8thWorld Congress on Computational Mechanics (WCCM8)-5th European Congress on Computational Methods in Applied Sciences and Engineering (ECCOMAS 2008), Venice, Italy, June 30 – July 5 (2008)

A New Rough Sets Decision Method Based on PCA and Ordinal Regression

Dun Liu[1], Tianrui Li[2], and Pei Hu[1]

[1] School of Economics and Management, Southwest Jiaotong University
Chengdu 610031, P.R. China
newton83@163.com, huhupei@126.com
[2] School of Information Science and Technology, Southwest Jiaotong University
Chengdu 610031, P.R. China
trli@swjtu.edu.cn

Abstract. The classical multivariate statistical method can only discuss the effectiveness of result, but can't explain the cause and intrinsic mechanism when dealing with classification problems. In this paper, a new rough sets decision method based on the Principal Component Analysis (PCA) and the ordinal regression is proposed which may help to explain the cause and the intrinsic mechanism of classification problems. An empirical study is employed to validate the reasonability and effectiveness of the proposed method.

Keywords: Principal component analysis, rough sets theory, ordinal regression, classification.

1 Introduction

In the process of decision-makings, people may meet many classification problems, *e.g.*, the financial risk early warning of listed companies, the credit evaluation and performance evaluation. The traditional statistic methodologies such as the logistic regression, discriminant analysis and cluster analysis can be used for classification problems. For the two former methods, they need the prior information to construct the discriminant function and the final discriminant rule relies on the probability distributing or the measurement. However, the prior information sometimes may not be available in our real applications. In addition, the number of clusters has to be given at first when using the k-mean algorithm in the cluster analysis. Moreover, the only standard, the precision of classification, is employed to judge the classification quality of the above methods while the reasons for generating all labels are not considered, which may not help us explore the latent rules in complex systems.

The Rough Sets Theory (RST), proposed by Pawlak in 1982, has been extensively studied in recent years. It is a mathematical tool to deal with vagueness and uncertainty and has been applied successfully in data mining [1,2,8,9,10,11,12]. For example, RST has been used to mining classification rules in databases, which helps people to understand the inner mechanism among objects. In addition, when

C.-C. Chan et al. (Eds.): RSCTC 2008, LNAI 5306, pp. 349–358, 2008.

we deal with practical problems, *e.g.*, in the case that the prior information does not exist, we only obtain the information table and it is difficult to acquire the values of the decision attribute of the decision table which is vital to our decision-makings.

As a classical method in statistics, the Principal Component Analysis (PCA) is usually used to reduce the dimensionality of the orthogonal space. Swiniarski firstly used RST and PCA in data model building and classification as well as feature selection and recognition [3,4]. Swiniarski and Skowron presented a description of the algorithm for feature selection and reduction based on the rough sets method proposed jointly with PCA, and then a description of hybrid methods of face recognition which are based on independent component analysis, PCA and RST. The feature extraction and pattern forming from face images have been provided using Independent Component Analysis and PCA. The feature selection/reduction has been realized using the rough sets technique. The face recognition system was designed as rough-sets rule based classifier [5,6]. Zeng, et al advocated a knowledge acquisition approach based on RST and PCA (KA-RSPCA) to acquire rules with stronger generalization capabilities. KA-RSPCA used a collective correlation coefficient as heuristic knowledge to assist attribute reduction and attribute value reduction. The coefficient was a PCA-based quantitative index to measure every condition attribute's contributions to the state space constructed by the whole of the condition attributes in a decision table [7]. However, in the domain of management, PCA can be used to construct the principal component function and obtain the estimation and order of every object, which may form a natural values of the decision attribute. In this paper, PCA is used to generate the values of the decision attribute and RST is used to acquire the decision rules.

The remaining of the paper is organized as follows. Some basic concepts are reviewed in Section 2. A new rough sets decision method based on PCA and ordinal regression is proposed to deal with classification problems in Section 3. A case study is given to validate the proposed model in Section 4. Section 5 concludes the research work of this paper.

2   Preliminaries

The basic concepts, correlative notations of rough sets and PCA are briefly reviewed [1,9,10].

Definition 1. *RST is based on an information system $S = (U, A, V, f)$, where U is a non-empty finite set of objects, $A = C \cup D$ is a non-empty finite set of attributes, C denotes the set of condition attributes and D denotes the set of decision attributes, $C \cap D = \emptyset$. $V = \bigcup_{a \in A} V_a$ and V_a is a domain of the attribute a, and $f : U \times A \to V$ is an information function such that $f(x, a) \in V_a$ for every $x \in U$, $a \in A$. When $D = \emptyset$, the system is an information table.*

For PCA, U stands for the object sets, A stands for the factor sets. PCA uses the linear combination of target sets A to construct the comprehensive factors

$F = \{F_1, F_2, \cdots, F_p\}$. If the accumulative contribute rate of the anterior p principal components exceeds 80%, then we think that $F_1, F_2, \cdots, F_p$ contain nearly all of the original information of the system. Suppose that the information system S is constituted by the m objects $u_1, u_2, \cdots, u_m$ and n factors $a_1, a_2, \cdots, a_n$, which form the data matrix as follows:

$$U = (u_1, u_2, \cdots, u_m) = \begin{pmatrix} v_{11} & v_{12} & \cdots & v_{1n} \\ v_{21} & v_{22} & \cdots & v_{2n} \\ \vdots & \vdots & \vdots & \vdots \\ v_{m1} & v_{m2} & \cdots & v_{mn} \end{pmatrix}$$

where, $u_i \in U$, $v_{ij} \in V$. Then,

$$\begin{cases} F_1 = w_{11}a_1 + w_{21}a_2 + \cdots + w_{n1}a_n \\ F_2 = w_{12}a_1 + w_{22}a_2 + \cdots + w_{n2}a_n \\ \qquad \cdots \\ F_p = w_{1p}a_1 + w_{2p}a_2 + \cdots + w_{np}a_n \end{cases}$$

where, $p < n$, $w_{1i}^2 + w_{2i}^2 + \cdots + w_{ni}^2 = 1$. If we denote $Var(F) = Var(w'X) = w'\Sigma w$, then $Var(F_1) = max\,Var(F_i)$ and $Var(F_1) \geq Var(F_2) \geq \cdots \geq Var(F_p)$, $i = 1, 2, \cdots, p$. In addition, the characteristic roots of the characteristic equation $|\lambda\Sigma - I| = 0$ are denoted as $\lambda_1, \lambda_2, \cdots, \lambda_p$, where $\lambda_1 \geq \lambda_2 \geq \cdots \geq \lambda_p$.

Definition 2. *Suppose $S = (U, C \bigcup D, V, F)$ is a complete information system, we denote $U/C = \{X_1, X_2, \cdots, X_m\}$, $U/D = \{D_1, D_2, \cdots, D_n\}$. the support, accuracy and coverage of $X_i \rightarrow D_j$ are defined as: Support of $X_i \rightarrow D_j : Sup(D_j \rightarrow X_i) = |X_i \bigcap D_j|$; Accuracy of $X_i \rightarrow D_j : Acc(D_j \rightarrow X_i) = |X_i \bigcap D_j|/|X_i|$; Coverage of $X_i \rightarrow D_j : Cov(D_j \rightarrow X_i) = |X_i \bigcap D_j|/|D_j|$, where $X_i \in U/C, i = 1, 2, \cdots, m$ and $D_j \in U/D, j = 1, 2, \cdots, n$.*

3 A New Rough Sets Decision Method Based on PCA and Ordinal Regression

For the RST, it is mainly based on the equivalence class, in other words, it uses an equivalence relation to obtain a partition of the universe. For the PCA, it usually use the information contribution maximization principle to select the principal component in turn, then construct the comprehensive principal component function by the weighted method. Finally, we can get the comprehensive evaluation value for every object. It is easy to see that we can use the comprehensive evaluation value to generate the values of the decision attribute. Then it can also form a partition of the universe after its values are discretized. Therefore, both RST and PCA can give a partition of the universe, we can combine the two methodologies together. Above all, we can expand the information table into a decision table by using PCA and ordinal regression.

In the following, we construct the rough sets decision method based on PCA and ordinal regression. Following is the detailed process (See Fig. 1).

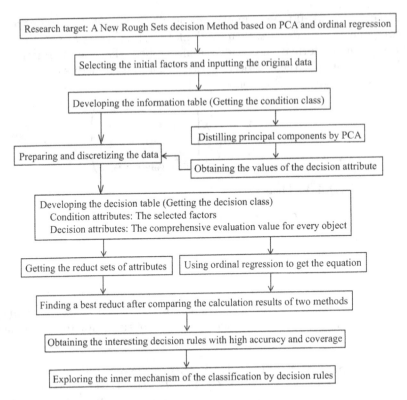

Fig. 1. The rough sets decision method based on PCA and ordinal regression

Step 1: Selecting the investigated objects, and using the objects and factors to construct the information table.

Step 2: Using PCA, obtaining the principal components from the information table and constructing the comprehensive principal component function.

Step 3: Calculating the comprehensive evaluation value for every object, and using it to generate the decision attribute and construct the decision table.

Step 4: Using RST and ordinal regression to obtain the interesting rules and available knowledge from the decision table and explore the inner reason of the classification.

4   Case Study

In this section, we consider the economic benefit conditions of 31 areas of P.R. China in 2000, and we select GDP (a_1), industrial added value (a_2), total capital contribution ratio (a_3), asset-liability ratio (a_4), current assets turnover times (a_5), profit rate of industrial costs and expenses (a_6), all laborproductivity (a_7)

Table 1. The economic benefit conditions of 31 areas in 2000

Area	a_1	a_2	a_3	a_4	a_5	a_6	a_7	a_8
1	2478.76	28.17	6.94	57.65	1.43	5.02	6.39	95.95
2	1639.36	24.18	8.50	60.83	1.79	6.48	5.24	98.39
3	5088.96	33.06	9.00	61.73	1.75	5.75	4.20	97.64
4	1643.81	35.23	5.52	63.92	0.98	1.94	2.34	96.88
5	1401.01	37.32	6.19	59.25	1.24	2.23	3.28	98.45
6	4669.06	28.10	6.91	60.46	1.46	4.27	4.05	97.79
6	1821.19	29.54	7.91	65.82	1.27	5.70	3.68	98.08
8	3253.00	49.29	21.60	58.07	1.60	31.39	6.22	98.12
9	4551.15	27.19	8.87	49.80	1.56	6.43	8.23	99.02
10	8582.73	24.92	8.92	61.80	1.91	3.87	5.03	97.15
11	6036.34	23.62	11.26	57.11	2.02	5.77	4.83	96.87
12	3038.24	30.54	7.09	63.19	1.43	2.40	3.12	98.48
13	3920.07	30.47	8.60	57.52	1.89	4.76	5.12	96.95
14	2003.07	28.93	6.20	68.30	1.27	1.43	2.48	97.27
15	8542.44	30.67	11.86	62.54	2.07	7.31	4.88	97.86
16	5137.66	31.94	8.28	66.43	1.52	4.47	3.23	98.00
17	4276.32	33.02	7.86	63.50	1.50	3.95	4.39	99.14
18	3691.88	32.44	8.86	67.60	1.45	2.38	3.17	98.61
19	9662.23	27.43	8.55	57.56	1.87	4.82	5.98	97.40
20	2050.14	32.28	7.82	68.39	1.49	3.86	3.55	97.85
21	518.48	31.18	5.54	66.48	1.16	2.73	5.27	94.41
22	1589.34	29.48	6.09	64.80	1.18	1.67	3.13	99.09
23	4010.25	31.89	6.85	64.45	1.18	3.60	3.18	98.21
24	993.53	34.35	7.17	69.29	0.91	2.31	3.18	96.61
25	1955.09	49.98	15.80	55.48	1.31	8.39	6.90	98.79
26	117.46	56.29	5.31	26.97	0.71	21.97	3.16	92.11
27	1660.92	34.71	7.6	68.17	1.12	6.06	3.29	96.70
28	983.36	29.11	5.10	65.05	1.08	1.34	2.68	95.15
29	263.59	33.32	4.77	72.90	0.63	0.41	4.12	97.22
30	265.57	30.81	5.50	60.01	1.20	1.87	3.29	96.26
31	1364.36	41.86	11.01	62.71	1.55	12.83	7.67	97.62

and product sales rate (a_8) as the evaluation factors (see Table 1). We hope to obtain the causes which affect the economic benefit in Table 1.

Firstly, we employ z-score transformation to standardize the continuous data in Table 1, then we calculate 3 characteristic roots: $\lambda_1 = 2.887$, $\lambda_2 = 2.636$, $\lambda_3 = 1.094$ (shown in Table 2), and the accumulative contribute rate of the 3 principal components exceeds to 82.7%. The component matrix is shown in Table 3.

Secondly, we use the component score (see Table 3) to divide the corresponding square root of principal components and obtain the coefficient of each principal component (for example, the coefficient of a_1 in Formula (1) can be calculated

Table 2. The accumulative contribute rate of 3 principal components

Component	Total	% of Variance	Cumulstive %
1	2.887	36.114	36.114
2	2.636	32.942	69.056
3	1.092	13.653	82.710

Table 3. The component matrix of the 3 principal components

	1	2	3		1	2	3
a_1	0.489	0.649	-0.382	a_5	0.627	0.667	-0.254
a_2	0.346	-0.827	0.288	a_6	0.745	-0.568	0.077
a_3	0.879	0.018	0.386	a_7	0.744	0.102	-0.037
a_4	-0.455	0.571	0.551	a_8	0.217	0.654	0.584

as: $0.489/\sqrt{2.887} = 0.2875$). Then, we construct the 3 principal components as follows:

$F_1 = 0.2875 a_1 + 0.2034 a_2 + 0.5176 a_3 - 0.2676 a_4 + 0.3690 a_5 + 0.4387 a_6 + 0.4376 a_7 + 0.1278 a_8;$

$F_2 = 0.3997 a_1 - 0.5092 a_2 + 0.0110 a_3 - 0.3517 a_4 + 0.4108 a_5 - 0.3496 a_6 + 0.0627 a_7 + 0.4026 a_8;$

$F_3 = -0.3653 a_1 + 0.2756 a_2 + 0.3689 a_3 + 0.5266 a_4 - 0.2424 a_5 + 0.0738 a_6 - 0.0351 a_7 + 0.5581 a_8.$

Thirdly, the comprehensive evaluation value for each object can be calculated by the following formula.

$$F = \sigma_1 F_1 + \sigma_2 F_2 + \sigma_3 F_3 \tag{1}$$

where, $\sigma_i = \lambda_i/(\lambda_1 + \lambda_2 + \lambda_3)$, $i = 1, 2, 3$. That is, $\sigma_1 = 2.887/(2.887 + 2.636 + 1.094) = 0.4363$, $\sigma_2 = 0.3984$, $\sigma_3 = 0.1653$.

Fourthly, we calculate the economic benefit comprehensive evaluation value F of 31 areas in P.R. China by Formula (1) (see Table 4).

Fifthly, we use the comprehensive evaluation value F to generate the values of the economic benefit evaluation and construct a decision table. In addition, the data in Table 1 and Table 4 are continuous. We discretize the original data in order to obtaining reducts by RST. The criterion of discretization by the expert evaluation method is shown in Table 5 and the final decision table is shown in Table 6.

Sixthly, we obtain the the reduct sets of Table 6 by RST. Here we directly use the software Rosetta produced by Warsaw University in Poland to get the 17 reduction sets: $\{a_3, a_7, a_8\}$, $\{a_1, a_4, a_7, a_8\}$, $\{a_1, a_2, a_3, a_7\}$, $\{a_1, a_2, a_3, a_8\}$, $\{a_1, a_4, a_6, a_8\}$, $\{a_1, a_4, a_6, a_7\}$, $\{a_1, a_2, a_4, a_8\}$, $\{a_2, a_3, a_6, a_8\}$, $\{a_2, a_4, a_5, a_8\}$, $\{a_2, a_3, a_4, a_8\}$, $\{a_2, a_4, a_6, a_8\}$, $\{a_3, a_4, a_6, a_8\}$, $\{a_4, a_6, a_7, a_8\}$, $\{a_4, a_5, a_6, a_8\}$, $\{a_1, a_2, a_4, a_5, a_7\}$, $\{a_1, a_3, a_5, a_6, a_7\}$, $\{a_2, a_3, a_5, a_6, a_7\}$.

Table 4. The comprehensive evaluation value in 31 areas

Area	F	Area	F	Area	F	Area	F
1	-0.2731	9	0.909	17	0.5297	25	0.945
2	0.5538	10	0.8335	18	0.3433	26	-2.9049
3	0.4715	11	0.783	19	0.9335	27	-0.5105
4	-0.9815	12	0.0232	20	0.0698	28	-1.2949
5	-0.3986	13	0.3297	21	-1.0633	29	-0.9145
6	0.0763	14	-0.4843	22	-0.1602	30	-1.0071
7	-0.0315	15	1.3466	23	-0.1227	31	0.7023
8	1.7663	16	0.3087	24	-0.7781		

Table 5. The criterions of data discretization

	a_1	a_2	a_3	a_4	a_5	a_6	a_7	a_8	F
A	≥ 5000	≥ 40	≥ 10	≤ 55	≥ 2	≥ 8	≥ 6	≥ 99	≥ 0.6
B	3500 to 5000	35 to 40	8 to 10	55 to 60	1.5 to 2	6 to 8	5 to 6	98 to 99	0.1 to 0.6
C	2000 to 3500	30 to 35	6 to 8	60 to 65	1 to 1.5	4 to 6	4 to 5	97 to 98	-0.4 to 0.1
D	500 to 2000	25 to 30	≤ 6	≥ 65	≤ 1	2 to 4	3 to 4	96 to 97	-0.9 to -0.4
E	≤ 500	≤ 25				≤ 2	≤ 3	≤ 6	≤ -0.9

After that, we will validate which reduct is more important from 17 reduction sets in our case. By considering the decision attribute value F in Table 6 is induced by the comprehensive evaluation value of PCA, we can choose it as the dependent variable, and we also can choose $a_1, \cdots, a_8$ as independent variables. Due to F value in Table 6 is the string variable, we transform them into the numeric variable (A → 1, B → 2, and so on). In addition, we can obviously obtain that F is also a ordinal variable by the former analysis (which means the value A is better than B, $\cdots$, D is better than E). So we will use the ordinal regression (F is treated as a dummy variable) to find the relations between the dependent variable and independent variables, and the result is shown in Table 7.

From Table 7, we get the 2 variables a_7 and a_8 with the Wald value are more than 2 and p value are less than 0.1 simultaneously, which indicate the coefficient of the 2 variables are statistically significant when the confidence level is 90%. So, the 2 variables will enter into the final regression equation.

After getting the equation, we discover that the set $\{a_3, a_7, a_8\}$ will be the best reduct in our information system by comparing the computing results of RST and ordinal regression. So, we can choose the 3 attributes to generate decision rules. In addition, the rules acquired by the system should have high accuracy and coverage (the accuracy is equal to 1 and the coverage is no less than 0.15). Table 8 shows the 10 main and valuable rules.

According to Table 8, we find that the total capital contribution ratio (a_3), all laborproductivity (a_7) and the product sales rate (a_8) are the most important attributes, and $\{a_3, a_7, a_8\}$ is the shortest of all the 17 reduct sets. However, the all laborproductivity stands for the labor market, the total capital contribution

Table 6. Final decision table of the economic benefit in 31 regions

Area	a_1	a_2	a_3	a_4	a_5	a_6	a_7	a_8	F	Area	a_1	a_2	a_3	a_4	a_5	a_6	a_7	a_8	F
1	C	D	C	B	C	C	A	E	C	17	B	C	C	C	C	D	C	A	B
2	E	E	B	C	B	B	B	B	B	18	B	C	B	D	C	D	D	B	B
3	A	C	B	C	B	C	C	D	B	19	A	D	B	B	B	C	B	C	A
4	D	B	D	C	D	E	E	D	E	20	C	C	C	D	C	D	D	C	C
5	D	B	C	B	C	D	D	B	C	21	D	C	D	D	C	D	B	E	E
6	B	D	C	C	C	C	C	C	C	22	D	D	C	C	C	E	D	A	C
7	D	D	C	D	C	C	D	B	C	23	B	C	C	C	C	D	D	B	C
8	C	A	A	B	B	A	A	B	A	24	D	C	C	D	D	D	D	D	D
9	B	D	B	A	B	B	A	A	A	25	D	A	A	B	C	A	A	B	A
10	A	E	B	C	B	D	B	C	A	26	E	A	D	A	D	A	D	E	E
11	A	E	A	B	A	C	C	D	A	27	D	C	C	D	C	B	D	D	D
12	C	C	C	C	C	D	D	B	C	28	D	D	D	D	C	E	E	E	E
13	B	C	B	B	B	C	B	D	B	29	E	C	D	D	D	E	C	C	E
14	C	D	C	D	C	E	E	C	D	30	E	C	D	C	C	E	D	D	E
15	A	C	A	C	A	B	C	C	A	31	D	A	A	C	B	A	A	C	A
16	A	C	B	D	B	C	D	C	B										

Table 7. The ordinal regression result

		Estimate	Std.Error	Wald	df	Sig.
Threshold	[F=1]	-8.886	3.469	6.561	1	.010
	[F=2]	-1.098	2.282	.232	1	.630
	[F=3]	7.747	2.786	7.736	1	.005
	[F=4]	12.776	4.059	9.908	1	.002
Location	a_1	-2.830	1.895	2.231	1	.135
	a_2	-.921	2.639	.122	1	.727
	a_3	-9.411	5.904	2.541	1	.111
	a_4	.118	2.888	.002	1	.967
	a_5	-1.968	3.208	.376	1	.540
	a_6	3.435	5.079	.458	1	.499
	a_7	-3.423	2.067	2.743	1	.098
	a_8	-4.044	1.853	4.764	1	.029

ratio can be considered as money market, and the product sales rate is an important factor in the product market. The rules generated by Table 8 shows the the economic benefit conditions is decided by the labor market, money market and product market, which is consistent with the Keynesian model in macro-economy theory.

In addition, the 10 rules obtained from the system are not only intuitive and clear and conforms with the reality and the people's experience, but also show the internal essence of the information table. For example, the attribute value performs A in a_2, a_4 and a_6 in the area of 26, but the value performs D, E, E in a_3, a_7 and a_8 respectively, and we can use the 10th rule in Table 8 to judge

Table 8. The 10 main rules

No.	The detailed rules	Support	Accuracy	Coverage
1	$(a_3, A) \wedge (a_7, A) \wedge (a_8, B) \to (F, A)$	2	1	0.25
2	$(a_3, C) \wedge (a_7, D) \wedge (a_8, B) \to (F, C)$	4	1	0.5
3	$(a_3, B) \wedge (a_7, B) \wedge (a_8, C) \to (F, A)$	2	1	0.25
4	$(a_3, C) \wedge (a_7, D) \wedge (a_8, D) \to (F, D)$	2	1	0.667
5	$(a_3, C) \wedge (a_7, E) \wedge (a_8, C) \to (F, D)$	1	1	0.333
6	$(a_3, B) \wedge (a_7, B) \wedge (a_8, D) \to (F, B)$	1	1	0.167
7	$(a_3, B) \wedge (a_7, D) \wedge (a_8, C) \to (F, B)$	1	1	0.167
8	$(a_3, D) \wedge (a_7, D) \wedge (a_8, D) \to (F, E)$	1	1	0.167
9	$(a_3, D) \wedge (a_7, D) \wedge (a_8, E) \to (F, E)$	1	1	0.167
10	$(a_3, D) \wedge (a_7, E) \wedge (a_8, E) \to (F, E)$	1	1	0.167

the comprehensive evaluation value of the 26th area is E. So, the example shows that a_3, a_7 and a_8 play a more important role than other condition attributes, although there are no core attributes in our information table. Moreover, the process of case study shows that the proposed model not only can be used for classification tasks, but also can explain the inner mechanism of the classification and explore the interesting rules in complex systems, which is helpful for the knowledge discover tasks in information system.

5   Conclusions

In this paper, PCA and ordinal regression was induced into RST to obtain the rules of information systems firstly. Then, a new rough sets decision method was proposed and a case study was given to validate the rationality and validity of the proposed method. Our future research work is focused on the incompletely information system and extend the basic relation to a dynamic environment. It also seems worthwhile to explore if the proposed approach can be extended to other generalized rough set models like fuzzy rough set theory.

Acknowledgements

This work is partially supported by the Research Fund for the Doctoral Program of Higher Education (No. 20060613019) and the Basic Science Foundation of Southwest Jiaotong University (No. 2007B13).

References

1. Pawlak, Z.: Rough sets. International Journal of Computer and Information Science 11, 341–356 (1982)
2. Pawlak, Z.: Rough set theory and its application to data analysis. Cybernetics and Systems 29, 661–688 (1998)

3. Swiniarski, R.: Rough sets and principal component analysis and their applications in data model building and classification. In: Pal, S.K. (ed.) Rough Fuzzy Hybridization: New Trend in Decision Making, pp. 275–300. Springer, Singapore (1999)
4. Swiniarski, R.: Rough sets methods in feature reduction and selection. International Journal of Applied Mathematical and Computer Science 11, 565–582 (2001)
5. Swiniarski, R., Skowron, A.: Rough set methods in feature selection and recognition. Pattern Recognition Letters 24, 833–849 (2003)
6. Swiniarski, R., Skowron, A.: Independent component analysis, principal component analysis and rough sets in face recognition. Transactions on Rough Sets I, 392–404 (2004)
7. Zeng, A., Pan, D., Zheng, Q.L.: Knowledge acquisition based on rough set theory and principal analysis. IEEE Intelligent Systems, 78–85 (2006)
8. Zhang, W.X., Wu, W.Z.: Rough sets theory and methodology. The Press of Science, Beijing (2001) (in Chinese)
9. Wang, G.Y.: Rough sets theory and knowledge discovery. Xi'an Jiaotong University Press, Xi'an (2001) (in Chinese)
10. Liu, D., Hu, P., Jiang, C.Z.: The methodology of the variable precision rough set increment study based on completely information system. In: Wang, G.Y., Li, T., Grzymała-Busse, J.W., Miao, D., Skowron, A., Yao, Y. (eds.) RSKT 2008. LNCS (LNAI), vol. 5009, pp. 276–283. Springer, Heidelberg (2008)
11. Grzymala-Busse, J.W.: A comparison of three strategies to rule induction from data with numerical attributes. Electronic Notes in Theoretical Computer Science 82, 132–140 (2003)
12. Chan, C.C.: A rough set approach to attribute generalization in data mining. Information Sciences 107, 177–194 (1998)

Rough Set Flow Graphs and $Max - *$ Fuzzy Relation Equations in State Prediction Problems

Zofia Matusiewicz[1] and Krzysztof Pancerz[1,2]

[1] University of Information Technology and Management
Sucharskiego Str. 2, 35-225 Rzeszów, Poland
{zmatusiewicz,kpancerz}@wsiz.rzeszow.pl
[2] College of Management and Public Administration
Akademicka Str. 4, 22-400 Zamość, Poland

Abstract. The paper makes the first attempt to combine two methodologies concerning uncertainty and fuzzy reasoning, namely rough set flow graphs and fuzzy relation equations. Rough set flow graphs proposed by Z. Pawlak are a useful tool for the knowledge representation. In this paper, we use them to represent the knowledge of transitions between states included in multistage dynamic information systems. The knowledge represented by flow graphs is a basis for determining possibilities of appearances of states in the future using the $max - *$ fuzzy composition. In the approach proposed in the paper, we take advantage of some properties of the $max - *$ fuzzy relation equations.

Keywords: rough sets, flow graphs, fuzzy relations, prediction.

1 Introduction

One of the important aspects of data mining is the analysis of data changing in time (temporal data). Different methodologies of soft computing are used for prediction with temporal data, e.g., fuzzy reasoning, neural networks, rough sets. In this paper, we use rough set flow graphs proposed by Z. Pawlak [10] as a tool for representing the knowledge of transitions between states. We assume that the knowledge is included in the multistage dynamic information systems introduced in [8]. Multistage dynamic information systems are generalization of dynamic information systems proposed by Z. Suraj in [12]. Having the knowledge represented in rough set flow graphs, we use the $max - *$ fuzzy relation equations to predict possibilities of the appearance of states in the future. In our approach, we treat the flow graph as a "black box". The knowledge included in the flow graph is described by means of a matrix with fuzzy elements (numbers), called a certainty matrix. We are interested only in input and output of the flow graph. This leads to the $max - *$ composition. Input is determined as a possibility distribution of current states of a given system. As a result of the $max - *$ composition we obtain a possibility distribution of states of the system in the future, i.e., after several steps (time units). This information can be useful in the state prediction problems. The usage of fuzzy relation equations

C.-C. Chan et al. (Eds.): RSCTC 2008, LNAI 5306, pp. 359–368, 2008.
© Springer-Verlag Berlin Heidelberg 2008

is very interesting because they have the solid theoretical basis. Moreover, a lot of properties of fuzzy relation equations have been proved. We recall and take advantage of some of them.

The rest of the paper is organized as follows. In Section 2, a brief review of the basic concepts of information systems and rough set flow graphs is presented. Section 3 presents multistage dynamic information systems describing transitions between states of considered systems. Section 4 includes essential information on $max - *$ fuzzy relation equations and their properties. Section 5 shows how to apply rough set flow graphs and fuzzy relation equations in state prediction problems. Finally, Section 6 consists of some conclusions.

2   Preliminaries

In this section, we recall the basic concepts concerning information systems and rough set flow graphs which are crucial to understand the approach proposed in the paper.

Information Systems. An *information system* is a pair $S = (U, A)$, where U is a set of *objects*, A is a set of *attributes*, i.e., $a : U \rightarrow V_a$ for $a \in A$, where V_a is called a value set of a. A *decision system* is a pair $DS = (U, A)$, where $A = C \cup D$, $C \cap D = \emptyset$, and C is a set of *condition attributes*, D is a set of *decision attributes*. Any information (decision) system can be represented as a data table, whose columns are labeled with attributes, rows are labeled with objects, and entries of the table are attribute values.

Rough Set Flow Graphs. Rough set flow graphs have been defined by Z. Pawlak [10] as a tool for reasoning from data. A flow graph is a directed, acyclic, finite graph $G = (N, B, \sigma)$, where N is a set of nodes, $B \subseteq N \times N$ is a set of directed branches and $\sigma : B \rightarrow [0, 1]$ is a flow function.

An input of a node $x \in N$ is the set $I(x) = \{y \in N : (y, x) \in B\}$, whereas an output of a node $x \in N$ is the set $O(x) = \{y \in N : (x, y) \in B\}$. $\sigma(x, y)$ is called a strength of a branch $(x, y) \in B$ and it is also denoted as $str(x, y)$. We define the input and the output of the graph G as $I(G) = \{x \in N : I(x) = \emptyset\}$ and $O(G) = \{x \in N : O(x) = \emptyset\}$, respectively. The input and the output of G consist of external nodes of G. The remaining nodes of G are its internal nodes.

With each node $x \in N$ we associate its inflow $\delta_+(x)$ and outflow $\delta_-(x)$ defined by:

$$- \ \delta_+(x) = \sum_{y \in I(x)} \sigma(y, x),$$
$$- \ \delta_-(x) = \sum_{y \in O(x)} \sigma(x, y).$$

For each node $x \in N$, its throughflow $\delta(x)$ is defined as follows:

$$\delta(x) = \begin{cases} \delta_-(x) & \text{if } x \in I(G) \\ \delta_+(x) & \text{if } x \in O(G) \\ \delta_-(x) = \delta_+(x) & \text{otherwise.} \end{cases} \qquad (1)$$

With each branch $(x, y) \in B$ we also associate:

- certainty: $cer(x, y) = \frac{\sigma(x,y)}{\delta(x)}$,
- coverage: $cov(x, y) = \frac{\sigma(x,y)}{\delta(y)}$,

where $\delta(x) \neq 0$ and $\delta(y) \neq 0$.

A directed path $[x \dots y]$ between nodes x and y in G, where $x \neq y$, is a sequence of nodes $x_1, x_2, \dots, x_n$ such that $x_1 = x$, $x_n = y$ and $(x_i, x_{i+1}) \in B$, where $1 \leq i \leq n - 1$. For each path $[x_1 \dots x_n]$, we define:

- certainty: $cer[x_1 \dots x_n] = \prod\limits_{i=1}^{n-1} cer(x_i, x_{i+1})$,
- coverage: $cov[x_1 \dots x_n] = \prod\limits_{i=1}^{n-1} cov(x_i, x_{i+1})$,
- strength: $str[x_1 \dots x_n] = \delta(x_1)cer[x_1 \dots x_n] = \delta(x_n)cov[x_1 \dots x_n]$.

A connection $\langle x, y \rangle$ from the node x to the node y is a set of all paths from x to y in G. For each connection $\langle x, y \rangle$, we define:

- certainty: $cer \langle x, y \rangle = \sum\limits_{[x \dots y] \in \langle x,y \rangle} cer[x \dots y]$,
- coverage: $cov \langle x, y \rangle = \sum\limits_{[x \dots y] \in \langle x,y \rangle} cov[x \dots y]$,
- strength: $str \langle x, y \rangle = \sum\limits_{[x \dots y] \in \langle x,y \rangle} str[x \dots y]$.

For a given rough set flow graph G, we can define the so-called certainty matrix $C(G)$ as follows.

Definition 1. *A certainty matrix $C(G) = [c_{ij}] \in [0, 1]^{m \times n}$ of the rough set flow graph G, where $m = card(O(G))$ and $n = card(I(G))$, is a matrix such that $c_{ij} = cer \langle x_j, y_i \rangle$, where $x_j \in I(G)$ and $y_i \in O(G)$.*

3   Multistage Dynamic Information Systems

A notion of dynamic information systems was introduced by Z. Suraj in [12] to represent dynamic behavior of systems (transitions between states). We can extend a notion of dynamic information systems to the so-called multistage dynamic information systems (in short, MDISs) (cf. [8]). If we are interested in sequences of changes of states, then we should represent such changes by means of polyadic relations over the sets of states. In this section, we give some crucial notions concerning multistage dynamic information systems.

Definition 2. *A multistage transition system is a pair $MTS = (U, T)$, where U is a nonempty set of states and $T \subseteq U^k$ is a multistage transition relation, where $k > 2$.*

A multistage transition relation $T \subseteq U^k$ is a polyadic relation defined over the set U.

Definition 3. *A multistage dynamic information system is a tuple* $MDIS = (U, A, T)$, *where* $S = (U, A)$ *is an information system called the underlying system of* $MDIS$ *and* $MTS = (U, T)$ *is a multistage transition system.*

Each element of a multistage transition relation T in a multistage dynamic information system $MDIS = (U, A, T)$ is a sequence of global states (from the set U) which can be referred to as an episode.

Definition 4. *Let* $MDIS = (U, A, T)$ *be a multistage dynamic information system, where* $T \subseteq U^k$. *Each element* $(u^1, u^2, \ldots, u^k) \in T$, *where* $u^1, u^2, \ldots, u^k \in U$, *is called an episode in* $MDIS$.

A dynamic information system can be presented by means of data tables representing information systems in the Pawlak's sense. In this case, each dynamic information system is depicted by means of two data tables. The first data table represents an underlying system that is an information system. The second one represents a decision system that is further referred to as a decision transition system. This table represents transitions determined by a transition relation. Analogously, we can use a suitable data table to represent a multistage transition system. Such a table will represent the so-called multistage decision transition system.

Definition 5. *Let* $MTS = (U, T)$ *be a multistage transition system. A multistage decision transition system is a pair* $MDTS = (U_T, A^1 \cup A^2 \cup \ldots \cup A^k)$, *where each* $t \in U_T$ *corresponds exactly to one element of the polyadic transition relation* T *whereas attributes from the set* A^i *determine global states of the i-th domain of* T, *where* $i = 1, 2, \ldots, k$.

Each object in a multistage decision transition system represents one episode in a given multistage dynamic information system.

By Inf_{A^i} we denote the set of all information vectors (appearing in the i-th stage in the multistage decision transition system $MDTS$) in the form:

$$(a_1^i(u_t), a_2^i(u_t), \ldots, a_m^i(u_t)), \tag{2}$$

where $a_1^i, a_2^i, \ldots, a_m^i \in A^i$, $i = 1, 2, \ldots, k$, and $u_t \in U_T$.

Let $MDTS = (U_T, A^1 \cup A^2 \cup \ldots \cup A^k)$ be a multistage decision transition system. By $|(v_1, v_2, \ldots, v_m)|^i_{MDTS}$ we denote the set of objects in U_T for which an information vector $(v_1, v_2, \ldots, v_m)$ appears in the i-th stage in $MDTS$, i.e.,

$$|(v_1, v_2, \ldots, v_m)|^i_{MDTS} = \{u_t \in U_T : a_j^i(u_t) = v_j, a_j^i \in A^i, j = 1, 2, \ldots, m\}.$$

Each information vector in the form of (2) can be interpreted as a global state observed in the time instant i. In that case, each episode in $MDIS$ can be described as a sequence of information vectors:

$$\left[(a_1^1(u_t), \ldots, a_m^1(u_t)), (a_1^2(u_t), \ldots, a_m^2(u_t)), \ldots, (a_1^k(u_t), \ldots, a_m^k(u_t)) \right], \tag{3}$$

where $a_1^i, \ldots, a_m^i \in A^i$, $i = 1, 2, \ldots, k$ and $u_t \in U_T$.

By $\langle (a_1^1(u_t), \ldots, a_m^1(u_t)), (a_1^k(u_t), \ldots, a_m^k(u_t)) \rangle$ we denote an episode starting from the global state described by the information vector $(a_1^1(u_t), \ldots, a_m^1(u_t))$ and ending in the global state described by the information vector $(a_1^k(u_t), \ldots, a_m^k(u_t))$.

4 $Max - *$ Fuzzy Relation Equations and Their Properties

In this section, we recall some properties of the $max - *$ composition with necessary assumptions on the operation $*$ important from our point of view. We assume the following notation: sup denotes supremum whereas inf denotes infimum.

By $max - *$ composition of a matrix $A = [a_{ij}] \in [0, 1]^{m \times n}$ and a vector $B = [b_i] \in [0, 1]^n$ we understand a vector $C = A \circ B = [c_i] \in [0, 1]^m$ such that:

$$c_i = \sup_{k=1,2,\ldots,n} (a_{ik} * b_k), \tag{4}$$

where $i = 1, 2, \ldots, m$.

Let $m, n \in \mathbb{N}$, $A = [a_{ij}] \in [0, 1]^{m \times n}$, and $B = [b_i] \in [0, 1]^m$. We are interested in solutions $X = [x_i] \in [0, 1]^n$ to the $max - *$ system $A \circ X = B$, i.e.,

$$\sup_{j=1,2,\ldots,n} (a_{ij} * x_j) = b_i, \tag{5}$$

where $i = 1, 2, \ldots, m$. The family of all such solutions will be denoted by $S(A, B)$.

Theorem 1. *If operation $*$ with neutral element $e = 1$ is infinitely sup-distributive, then $U = A \to B$, where*

$$u_j = \inf_{k=1,2,\ldots,m} (a_{kj} \xrightarrow{*} b_k)$$

for $j = 1, 2, \ldots, n$ is the greatest solution of inequality $A \circ X \leq B$, where $a \xrightarrow{*} b = max\{t \in [0, 1] : a * t \leq b\}$ is residuated implication induced by $*$.

Proof. The proof can be found in [11].

The product operation satisfies, among others, assumptions given in Theorem 1.

If an operation $*$ fulfills assumptions from Theorem 1, then for a matrix $A = [a_{ij}]$, the matrix $A^r = [a_{ij}^r]$, where

$$a_{ij}^r = \begin{cases} a_{ij} & \text{if } a_{ij} * u_j = b_i \\ 0 & \text{otherwise.} \end{cases} \tag{6}$$

is called a reduced matrix of A.

Analogously, we can reduce a matrix A for each fixed solution X less than or equal to U. Such a reduced matrix will be denoted by A_X^r.

Matrices $A, C \in [0,1]^{m \times n}$ are solution invariant about $B \in [0,1]^m$ if $S(A,B) = S(A,C) \neq \emptyset$. Then we write $A \sim_B C$. It is clear that relation $\sim_B$ is an equivalence in $[0,1]^{m \times n}$ and we can consider equivalence classes $[A]_{\sim_B}$.

One can prove that under the assumptions of Theorem 1 we have $A^r \in [A]_{\sim_B}$. Therefore, $S(A,B) = S(A^r, B)$. For the fixed solution X, if we have the reduced matix A_X^r, then we obtain that $A \circ X = A_X^r \circ X = B$.

5   State Prediction

To represent an information flow distribution in multistage decision transition systems we can use rough set flow graphs proposed by Z. Pawlak. Let $MDTS = (U_T, A^1 \cup A^2 \cup \ldots \cup A^k)$ be a multistage decision transition system. A rough set flow graph G corresponding to $MDTS$ consists of k layers. Nodes in the i-th layer of G represent information vectors (global states) determined by attributes from the set A^i, where $i = 1, 2, \ldots, k$. Since attribute sets A^1, A^2, ..., A^k are ordered in time, we can call the graph G a temporal rough set flow graph. It represents temporal flow distribution in the multistage decision transition system. In order to construct a rough set flow graph G corresponding to a multistage decision transition system $MDTS$ we may perform Algorithm 1. A graph constructed using this algorithm is supplemented with a certainty function which assigns a number called certainty from the interval $[0,1]$ to each branch in G. So, we have $G = (N, B, \sigma, cer)$, where N is a set of nodes, B is a set of directed branches, $\sigma : B \to [0,1]$ is a flow function, and $cer : B \to [0,1]$ is a certainty function. Certainty factors of branches will be used in our approach presented here.

If we have a possibility distribution (cf. [5]) of global states (information vectors) being the start points of episodes (global states determined by attributes from A^1) and the knowledge included in a multistage decision transition system and expressed by the temporal rough set flow graph, then we can determine a possibility distribution of global states (information vectors) being the end points of episodes (global states determined by attributes from A^k). Thus we can determine a possibility distribution of global states after $k-1$ transitions between global states (in $k-1$ time units).

Let $MDTS = (U_T, A^1 \cup A^2 \cup \ldots \cup A^k)$ be a multistage decision transition system and G be a rough set flow graph corresponding to $MDTS$. The possibility distribution of global states determined by attributes from A^1 defines the fuzzy set over the set of information vectors from Inf_{A^1}:

$$\frac{\mu_1}{(v_{1_1}^1, \ldots, v_{m_1}^1)} + \ldots + \frac{\mu_q}{(v_{1_q}^1, \ldots, v_{m_q}^1)},$$

where $(v_{1_1}^1, \ldots, v_{m_1}^1), \ldots, (v_{1_q}^1, \ldots, v_{m_q}^1) \in Inf_{A^1}$ and $\mu_1, \ldots, \mu_q \in [0,1]$. This possibility distribution can be presented as a column vector $[\mu_1, \ldots, \mu_q]^T$. Analogously the possibility distribution of global states determined by attributes from A^k defines the fuzzy set over the set of information vectors from Inf_{A^k}:

$$\frac{\mu_1}{(v_{1_1}^k, \ldots, v_{m_1}^k)} + \ldots + \frac{\mu_r}{(v_{1_r}^k, \ldots, v_{m_r}^k)},$$

Algorithm 1. Algorithm for creating a temporal rough set flow graph corresponding to a multistage decision transition system

---

Input : A multistage decision transition system
$$MDTS = (U_T, A^1 \cup A^2 \cup \ldots \cup A^k).$$
Output: A temporal rough set flow graph $G = (N, B, \sigma, cer)$ corresponding to $MDTS$.

$N \longleftarrow \emptyset;$
$B \longleftarrow \emptyset;$
for *each* $i = 1, 2, \ldots, k$ **do**
   Create an epmty set N_i of nodes;
   for *each information vector* $inf \in Inf_{A^i}$ **do**
     | Create a node n_{inf} representing inf and add it to N_i;
   end
   $N \longleftarrow N \cup N_i;$
end
for *each* $i = 1, 2, \ldots, k - 1$ **do**
   for *each node* $n_{inf_x} \in N_i$ **do**
     for *each node* $n_{inf_y} \in N_{i+1}$ **do**
       Create a branch $b = (n_{inf_x}, n_{inf_y});$
       $\sigma(b) \longleftarrow \dfrac{card(|inf_x|^i_{MDTS} \cap |inf_y|^{i+1}_{MDTS})}{card(U_T)};$
       $cer(b) \longleftarrow \dfrac{card(|inf_x|^i_{MDTS} \cap |inf_y|^{i+1}_{MDTS})}{card(|inf_x|^i_{MDTS})};$
       $B \longleftarrow B \cup \{b\};$
     end
   end
end

---

where $(v^k_{1_1}, \ldots, v^k_{m_1}), \ldots, (v^k_{1_r}, \ldots, v^k_{m_r}) \in Inf_{A^k}$ and $\mu_1, \ldots, \mu_r \in [0, 1]$. This possibility distribution can be presented as a column vector $[\mu_1, \ldots, \mu_r]^T$.

An element c_{ij} of the certainty matrix $C(G)$ determines certainty of appearing of the episode $\left\langle (v^1_{1_j}, \ldots, v^1_{m_j}), (v^k_{1_i}, \ldots, v^k_{m_i}) \right\rangle$, where $(v^1_{1_j}, \ldots, v^1_{m_j}) \in Inf_{A^1}$, $(v^k_{1_i}, \ldots, v^k_{m_i}) \in Inf_{A^k}$, $j = 1, 2, \ldots, card(Inf_{A^1})$, $i = 1, 2, \ldots, card(Inf_{A^k})$, and $m = card(A^1) = \ldots = card(A^k)$.

Suppose, we have determined a possibility distribution of global states (information vectors) being the start points of episodes given as a vector $\Pi(Inf_{A^1}) = [\pi^1_i]^n$, where $n = card(Inf_{A^1})$, and a certainty matrix $C(G)$ of the flow graph G. We can obtain a possibility distribution of global states (information vectors) being the end points of episodes and expressed as a vector $\Pi(Inf_{A^k}) = [\pi^k_i]^p$, where $p = card(Inf_{A^k})$, in the following way $\Pi(Inf_{A^k}) = C(G) \circ \Pi(Inf_{A^1})$, where $\circ$ denotes $max - product$ composition. Properties of $max - *$ fuzzy relation equations recalled in Section 4 enable us to obtain some useful information what will be shown in an example.

Example 1. Now, we give a simple example enabling readers to understand the approach proposed in this section. Let us take daily exchange rates between the Polish zloty and two currencies: the US dollar (marked with usd) and the euro (marked with *euro*). The meaning of values of attributes is the following: -1 denotes decreasing a given exchange rate in relation to the previous exchange rate, 0 denotes remaining a given exchange rate on the same level in relation to the previous exchange rate, 1 denotes increasing a given exchange rate in relation to the previous exchange rate. A multistage decision transition system $MDTS$ is shown in Table 1. $MDTS$ contains five episodes of length 3. Each global state belonging to these episodes is described by two attributes *usd* and *euro*.

Table 1. A multistage decision transition system $MDTS$

U_T	A^1		A^2		A^3	
	usd^1	$euro^1$	usd^2	$euro^2$	usd^3	$euro^3$
e_1	-1	1	0	0	0	0
e_2	-1	1	-1	0	0	0
e_3	0	0	-1	0	1	0
e_4	0	1	-1	0	1	1
e_5	-1	0	-1	0	1	0

A rough set flow graph G corresponding to the multistage decision transition system $MDTS$ and built using Algorithm 1 is shown in Figure 1. For legibility, branches are labeled only with certainties.

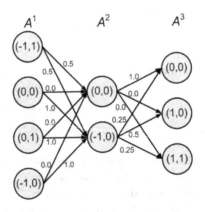

Fig. 1. A flow graph G corresponding to $MDTS$

The certainty matrix $C(G)$ of the rough set flow graph G corresponding to $MDTS$ has the following form:

$$C(G) = \begin{bmatrix} 0.625 & 0.25 & 0.25 & 0.25 \\ 0.25 & 0.5 & 0.5 & 0.5 \\ 0.125 & 0.25 & 0.25 & 0.25 \end{bmatrix}$$

Let us assume the following possibility distribution of current global states:

$$\frac{1.0}{(-1,1)} + \frac{0.5}{(0,0)} + \frac{0.2}{(0,1)} + \frac{0.8}{(-1,0)}.$$

Solving $max - product$ composition we obtain the following possibility distribution of global states after two transitions between states:

$$\frac{0.625}{(0,0)} + \frac{0.4}{(1,0)} + \frac{0.2}{(1,1)}.$$

The reduced matrix $C(G)^r_{\Pi(Inf_{A1})}$ of $C(G)$ has the form:

$$C(G)^r_{\Pi(Inf_{A1})} = \begin{bmatrix} 0.625 & 0 & 0 & 0 \\ 0 & 0 & 0 & 0.5 \\ 0 & 0 & 0 & 0.25 \end{bmatrix}$$

It means that:

- the appearance of the global state $(0,0)$ in two time units in the future is announced, first of all, by the global state $(-1,1)$ currently appearing,
- the appearance of the global state $(1,0)$ in two time units in the future is announced, first of all, by the global state $(-1,0)$ currently appearing,
- the appearance of the global state $(1,1)$ in two time units in the future is announced, first of all, by the global state $(-1,0)$ currently appearing.

The greatest solution has the following form:

$$U = \begin{bmatrix} 1 \\ 0.8 \\ 0.8 \\ 0.8 \end{bmatrix}$$

The greatest solution determines a maximal possibility distribution of current global states leading to a given possibility distribution of future global states. It means that having the following possibility distribution of current global states:

$$\frac{1.0}{(-1,1)} + \frac{0.8}{(0,0)} + \frac{0.8}{(0,1)} + \frac{0.8}{(-1,0)}$$

we also obtain the same possibility distribution of global states after two transitions between states, like previously.

6 Conclusions

The paper has shown how to combine rough set flow graphs and $max - *$ fuzzy relation equations in the state prediction problems. It is worth noting that the approach proposed in the paper provides only one of possible solutions. In the future works we will study deeper, incorporating fuzzy relation equations in reasoning by means of the knowledge represented in the rough set flow graphs. It seems to be very interesting because fuzzy relation equations have the solid theoretical basis and a lot of their important properties have been proved. It is also necessary to examine our approach on real-life data.

Acknowledgments

This paper has been partially supported by the grant from the University of Information Technology and Management in Rzeszów, Poland.

References

1. Butz, C.J., Yan, W., Yang, B.: The Computational Complexity of Inference Using Rough Set Flow Graphs. In: Ślęzak, D., Wang, G., Szczuka, M.S., Düntsch, I., Yao, Y. (eds.) RSFDGrC 2005. LNCS (LNAI), vol. 3641, pp. 335–344. Springer, Heidelberg (2005)
2. Drewniak, J.: Fuzzy Relation Equations and Inequalities. Fuzzy Sets and Systems 14(3), 237–247 (1984)
3. Drewniak, J.: Fuzzy Relation Calculus. University of Silesia, Katowice (1989)
4. Greco, S., Pawlak, Z., Słowiński, R.: Generalized Decision Algorithms, Rough Inference Rules, and Flow Graphs. In: Alpigini, J.J., et al. (eds.) RSCTC 2002. LNCS (LNAI), vol. 2475, pp. 93–104. Springer, Heidelberg (2002)
5. Lee, K.H.: First Course on Fuzzy Theory and Applications. Springer, Heidelberg (2005)
6. Matusiewicz, Z.: Fuzzy relation equations of type max-* for Lukasiewicz conjunction. Scientific Bulletin of Chelm, Section of Mathematics and Computers Science 1/07, 83–86 (2007)
7. Matusiewicz, Z.: Properties of fuzzy relation equations of type max-* for some classes of triangular norm conjunctions. In: Klement, E.P., et al. (eds.) Proc. of the FSTA 2008, Liptovsky Jan, Slovak Republic, p. 80 (2008)
8. Pancerz, K.: Extensions of Dynamic Information Systems in State Prediction Problems: the First Study. In: Magdalena, L., Ojeda-Aciego, M., Verdegay, J.L. (eds.) Proc. of the IPMU 2008, Malaga, Spain, pp. 101–108 (on CD, 2008)
9. Pawlak, Z.: Rough Sets - Theoretical Aspects of Reasoning About Data. Kluwer Academic Publishers, Dordrecht (1991)
10. Pawlak, Z.: Flow Graphs and Data Mining. In: Peters, J.F., Skowron, A. (eds.) Transactions on Rough Sets III. LNCS, vol. 3400, pp. 1–36. Springer, Heidelberg (2005)
11. Sanchez, E.: Resolution of composite fuzzy relation equations. Information Control 30, 38–48 (1976)
12. Suraj, Z.: The Synthesis Problem of Concurrent Systems Specified by Dynamic Information Systems. In: Polkowski, L., Skowron, A. (eds.) Rough Sets in Knowledge Discovery 2, pp. 418–448. Physica-Verlag, Berlin (1998)
13. Suraj, Z., Pancerz, K.: Flow Graphs as a Tool for Mining Prediction Rules of Changes of Components in Temporal Information Systems. In: Yao, J., et al. (eds.) RSKT 2007. LNCS (LNAI), vol. 4481, pp. 468–475. Springer, Heidelberg (2007)
14. Zhang, C., Lu, C., Li, D.: On perturbation properties of fuzzy relation equations. J. Fuzzy Math. 14(1), 53–63 (2006)

Precision of Rough Set Clustering

Pawan Lingras[1], Min Chen[1,2], and Duoqian Miao[2]

[1] Department of Mathematics & Computing Science, Saint Mary's University,
Halifax, Nova Scotia, B3H 3C3, Canada
[2] School of Electronics and Information Engineering, Tongji University, Shanghai,
201804, P.R. China
pawan.lingras@smu.ca

Abstract. Conventional clustering algorithms categorize an object into precisely one cluster. In many applications, the membership of some of the objects to a cluster can be ambiguous. Therefore, an ability to specify membership to multiple clusters can be useful in real world applications. Fuzzy clustering makes it possible to specify the degree to which a given object belongs to a cluster. In Rough set representations, an object may belong to more than one cluster, which is more flexible than the conventional crisp clusters and less verbose than the fuzzy clusters. The unsupervised nature of fuzzy and rough algorithms means that there is a choice about the level of precision depending on the choice of parameters. This paper describes how one can vary the precision of the rough set clustering and studies its effect on synthetic and real world data sets.

Keywords: Rough sets, K-means clustering algorithm, precision.

1   Introduction

In addition to clearly identifiable groups of objects, it is possible that a data set may consist of several objects that lie on the fringes. The conventional clustering techniques will mandate that such objects belong to precisely one cluster. Such a requirement is found to be too restrictive in many data mining applications. In practice, an object may display characteristics of different clusters. In such cases, an object should belong to more than one cluster, and as a result, cluster boundaries necessarily overlap. Fuzzy set representation of clusters, using algorithms such as fuzzy C-means, make it possible for an object to belong to multiple clusters with a degree of membership between 0 and 1 [11]. In some cases, the fuzzy degree of membership may be too descriptive for interpreting clustering results. Rough set based clustering provides a solution that is less restrictive than conventional clustering and less descriptive than fuzzy clustering.

Rough set theory has made substantial progress as a classification tool in data mining [1,14]. The basic concept of representing a set as lower and upper bounds can be used in a broader context such as clustering. Clustering in relation to rough set theory is attracting increasing interest among researchers [4,2,8,9,10,15,13]. Lingras [5] described how a rough set theoretic classification scheme can be represented using a rough set genome. In subsequent publications

C.-C. Chan et al. (Eds.): RSCTC 2008, LNAI 5306, pp. 369–378, 2008.

[6,7], modifications of K-means and *Kohonen Self-Organizing Maps* (SOMs) were proposed to create intervals of clusters based on rough set theory.

Clustering is an unsupervised learning process. That means there is no correct solution prescribed by an expert. For example, in a multidimensional space with a large number of objects, one cannot easily identify the number of clusters an algorithm should aim for. Researchers have proposed various cluster quality measures that make it possible to arrive at the appropriate number of clusters. The rough clustering has an additional issue that one needs to consider, namely, the precision of the clusters. Precision of the clusters refers to the number of objects that are precisely assigned to a cluster. An object in rough set clustering may be assigned to exactly one cluster or it may be assigned to multiple clusters. The objects that are assigned to multiple clusters are said to belong to the boundary region. Percentage of objects in boundary region is inversely proportional to the precision of rough clustering. This paper demonstrates how the size of boundary region can be varied with the help of *threshold* in rough set clustering. Experiments with a synthetic data set and a real world data set also suggest a procedure for choosing an appropriate precision.

2   Adaptation of Rough Set Theory for Clustering

Due to space limitations, some familiarity with rough set theory is assumed [14]. Rough sets were originally proposed using equivalence relations. However, it is possible to define a pair of upper and lower bounds $(\underline{A}(C), \overline{A}(C))$ or a rough set for every set $C \subseteq U$ as long as the properties specified by Pawlak [14] are satisfied. Yao *et al.* [16] described various generalizations of rough sets by relaxing the assumptions of an underlying equivalence relation. Such a trend towards generalization is also evident in rough mereology proposed by Polkowski and Skowron [12] and the use of information granules in a distributed environment by Skowron and Stepaniuk. The present study uses such a generalized view of rough sets. If one adopts a more restrictive view of rough set theory, the rough sets developed in this paper may have to be looked upon as interval sets.

Let us consider a hypothetical classification scheme

$$U/P = \{C_1, C_2, \dots, C_k\} \tag{1}$$

that partitions the set U based on an equivalence relation P. Let us assume due to insufficient knowledge that it is not possible to precisely describe the sets $C_i, 1 \leq i \leq k$, in the partition. Based on the available information, however, it is possible to define each set $C_i \in U/P$ using its lower $\underline{A}(C_i)$ and upper $\overline{A}(C_i)$ bounds. We will use m-dimensional vector representations, $\mathbf{u}, \mathbf{v}$ for objects and $\mathbf{c_i}$ for cluster C_i.

We are considering the upper and lower bounds of only a few subsets of U. Therefore, it is not possible to verify all the properties of the rough sets [14]. However, the family of upper and lower bounds of $\mathbf{c_i} \in \mathbf{U/P}$ are required to follow some of the basic rough set properties such as:

(P1) An object $\mathbf{x}$ can be part of at most one lower bound

(P2) $\mathbf{x} \in \underline{A}(\mathbf{c}_i) \Longrightarrow \mathbf{x} \in \overline{A}(\mathbf{c}_i)$

(P3) An object $\mathbf{x}$ is not part of any lower bound $\Longleftrightarrow$

$\mathbf{x}$ belongs to two or more upper bounds.

Property (P1) emphasizes the fact that a lower bound is included in a set. If two sets are mutually exclusive, their lower bounds should not overlap. Property (P2) confirms the fact that the lower bound is contained in the upper bound. Property (P3) is applicable to the objects in the boundary regions, which are defined as the differences between upper and lower bounds. The exact membership of objects in the boundary region is ambiguous. Therefore, property (P3) states that an object cannot belong to only a single boundary region. Their discussion can provide more insight into the essential properties for a rough set model. Note that (P1)-(P3) are not necessarily independent or complete. However, enumerating them will be helpful later in understanding the rough set adaptation of evolutionary, neural, and statistical clustering methods. In the context of decision-theoretic rough set model, Yao and Zhao [17] provide a more detailed discussion on the important properties of rough sets and positive, boundary, and negative regions.

3   Adaptation of K-Means to Rough Set Theory

Here, we refer readers to [3] for discussion on conventional K-means algorithm. Incorporating rough sets into K-means clustering requires the addition of the concept of lower and upper bounds. Calculation of the centroids of clusters from conventional K-Means needs to be modified to include the effects of these bounds. The modified centroid calculations for rough sets are then given by:

if $\quad \underline{A}(\mathbf{c}) \neq \emptyset$ and $\overline{A}(\mathbf{c}) - \underline{A}(\mathbf{c}) = \emptyset$

$$c_j = \frac{\sum_{\mathbf{x} \in \underline{A}(\mathbf{c})} x_j}{|\underline{A}(\mathbf{c})|}$$

else if $\underline{A}(\mathbf{c}) = \emptyset$ and $\overline{A}(\mathbf{c}) - \underline{A}(\mathbf{c}) \neq \emptyset$

$$c_j = \frac{\sum_{\mathbf{x} \in (\overline{A}(\mathbf{c}) - \underline{A}(\mathbf{c}))} x_j}{|\overline{A}(\mathbf{c}) - \underline{A}(\mathbf{c})|} \qquad (2)$$

else

$$c_j = w_{lower} \times \frac{\sum_{\mathbf{x} \in \underline{A}(\mathbf{c})} x_j}{|\underline{A}(\mathbf{c})|} + w_{upper} \times \frac{\sum_{\mathbf{x} \in (\overline{A}(\mathbf{c}) - \underline{A}(\mathbf{c}))} x_j}{|\overline{A}(\mathbf{c}) - \underline{A}(\mathbf{c})|},$$

where $1 \leq j \leq m$. Here, m is the dimensions of the vectors $\mathbf{c}$ and $\mathbf{x}$. The parameters w_{lower} and w_{upper} correspond to the relative importance of lower and upper bounds, and $w_{lower} + w_{upper} = 1$. If the upper bound of each cluster were equal to its lower bound, the clusters would be conventional clusters. Therefore, the boundary region $\overline{A}(\mathbf{c}) - \underline{A}(\mathbf{c})$ will be empty, and the second term in the equation will be ignored. Thus, Eq. (2) will reduce to conventional centroid calculations.

The next step in the modification of the K-means algorithms for rough sets is to design criteria to determine whether an object belongs to the upper or lower bound of a cluster given as follows. For each object vector $\mathbf{x}$, let $d(\mathbf{x}, \mathbf{c}_j)$ be the distance between itself and the centroid of cluster $\mathbf{c}_j$. Let $d(\mathbf{x}, \mathbf{c}_i) = \min_{1 \leq j \leq k} d(\mathbf{x}, \mathbf{c}_j)$. The ratio $d(\mathbf{x}, \mathbf{c}_i)/d(\mathbf{x}, \mathbf{c}_j)$, $1 \leq i, j \leq k$, are used to determine the membership of $\mathbf{x}$. Let $T = \{j : d(\mathbf{x}, \mathbf{c}_i)/d(\mathbf{x}, \mathbf{c}_j) \leq threshold$ and $i \neq j\}$.

1. If $T \neq \emptyset$, $\mathbf{x} \in \overline{A}(\mathbf{c}_i)$ and $\mathbf{x} \in \overline{A}(\mathbf{c}_j), \forall j \in T$. Furthermore, $\mathbf{x}$ is not part of any lower bound. The above criterion guarantees that property (P3) is satisfied.
2. Otherwise, if $T = \emptyset$, $\mathbf{x} \in \underline{A}(\mathbf{c}_i)$. In addition, by property (P2), $\mathbf{x} \in \overline{A}(\mathbf{c}_i)$.

It should be emphasized that the approximation space A is not defined based on any predefined relation on the set of objects. The upper and lower bounds are constructed based on the criteria described above.

4   Refinements of Rough Set Clustering

Rough clustering is gaining increasing attention from researchers. The rough K-means approach, in particular, has been a subject of further research. Peters [15] discussed various deficiencies of Lingras and West's original proposal [6]. The first set of independently suggested alternatives by Peters are similar to the Eq. (2). Peters also suggest the use of ratios of distances as opposed to differences between distances similar to those used in the rough set based Kohonen algorithm described in [7]. The use of ratios is a better solution than differences. The differences vary based on the values in input vectors. The ratios, on the other hand, are not susceptible to the input values. Peters [15] have proposed additional significant modifications to rough K-means that improve the algorithm in a number of aspects. The refined rough K-means algorithm simplifies the calculations of the centroid by ensuring that lower bound of every cluster has at least one object. It also improves the quality of clusters as clusters with empty lower bound have a limited basis for its existence. Peters tested the refined rough K-means for various datasets. The experiments were used to analyze the convergence, dependency on the initial cluster assignment, study of Davies-Boulden index, and to show that the boundary region can be interpreted as a security zone as opposed to the unambiguous assignments of objects to clusters in conventional clustering. Despite the refinements, Peters concluded that there are additional areas in which the rough K-means needs further improvement, namely in terms of selection of parameters.

By its very definition, unsupervised learning is an exercise with no known solution. Clustering is one of the primary examples of unsupervised clustering, which attempts to find groups of objects with similar characteristics. There are a number of unknowns involved in the process. The appropriate number of groups is not known apriori. Measures such as Davies-Boulden index have been used to identify the most appropriate number of clusters. As mentioned previously, even if there were clearly identifiable clusters of objects, it is quite often likely that some of the objects may be straying from these clusters. In that case, the next issue is how to decide what percentage of objects are straying from the neatly formed clusters. These stray objects will then be assigned to boundary regions of multiple clusters using the rough K-means algorithm. This paper experiments with the issue of determining the appropriate number of boundary region objects using two data sets. The first data set is a two dimensional set of objects artificially created with clearly identifiable clusters and stray objects. Since we can visualize the appropriate rough set clustering, we can test the behavior of the rough K-means algorithm for different values of *threshold*. The *threshold* parameter helps us control the size of the boundary region. We define the percentage of boundary region as a ratio of cardinality of the union of all the boundary regions divided by the total number of objects expressed as percentages, given by:

$$BoundarySize = \frac{\| \bigcup_{\mathbf{c} \in \mathbf{U}/\mathbf{P}} (\overline{A}(\mathbf{c}) - \underline{A}(\mathbf{c})) \|}{\|U\|} \times 100 \tag{3}$$

The following section studies the variation in *BoundarySize* along with qualitative analysis of changing memberships to suggest a procedure for identifying appropriate value of the *threshold* in the rough K-means algorithm.

5   Study Data and Experimental Analysis

We use two kinds of data, synthetic data and real data, to demonstrate how to choose an appropriate *threshold* for rough clustering.

5.1   Synthetic Data

The synthetic data set has been developed to study how the *BoundarySize* varies with *threshold* for rough clustering. In order to visualize the data set, we restrict it to two dimensions as can be seen in Fig. 1. There are a total of 65 objects. It is obvious that there are three distinct clusters, denoted by C_1, C_2 and C_3. However, five objects, identified as x_i $(1 \leq i \leq 5)$, do not belong to any particular cluster. We performed rough clustering on the synthetic data set for different values of *threshold*.

Fig. 2 shows how changing the value of *threshold* can affect the *BoundarySize* of rough clustering with $k = 3$ and $w_{lower} = 0.75$. In the inset figure, we can see a slow increase in the *BoundarySize* until the *threshold* reaches a value of 1.4, since the higher values lead to larger boundary regions. While the *threshold* values were changed from 1.4 to 2, the *BoundarySize* remained constant at 7.7%. However, the re-distribution of objects in the boundary region did occur.

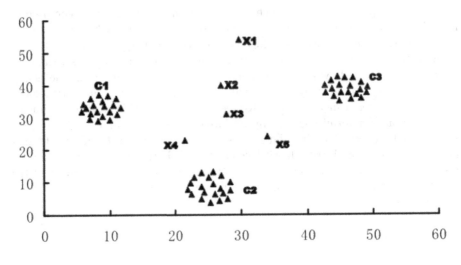

Fig. 1. Synthetic data

For example, x_5, which was in the boundary region of c_2 and c_3, was also added to the boundary region of c_1, when *threshold* changed the value from 1.6 into 1.7. It is obvious from Fig. 1 that x_5 should only belong to the boundary region of c_2 and c_3. That means increasing the value of *threshold* beyond a certain value can lead to unreasonable addition of some objects to boundary regions of some of the clusters. Moreover, one should not increase the boundary region too much as it will lead to fairly indecisive and uninformative rough clustering. Fig. 2 shows a sudden and sharp increase in the *BoundarySize* after *threshold* reaches a value of 2. The *BoundarySize* goes up to a value of more than 50% when *threshold* reaches the value of 2.5. Therefore, it is reasonable to consider *threshold* = 1.4 as an appropriate value in terms of the variance in *BoundarySize*. This value of *threshold* can be identified by the fact that further number of increases in *threshold* do not lead to net change in *BoundarySize*.

5.2   Real Data

This section reports experiments with a real world data set belonging to a small retail chain. The data consists of all the customer transactions in 2006. There were a total of 68716 transactions, one transaction per item purchased. 40260 of these transactions can be associated with 5878 identified customers. The objective of the experiment is to cluster the customers based on their spending habits. Each customer is represented by his monthly spending patterns. The monthly spending pattern gives a better understanding of a customer's spending habits than total spending. A customer who spends $100 regularly may be a little more loyal than one who spends $1000 during a single visit. The chronological ordering of spending does not help us understand the propensity of a customer to spend. For example, a person spending $100, $200, $300 in three months will look different from the one

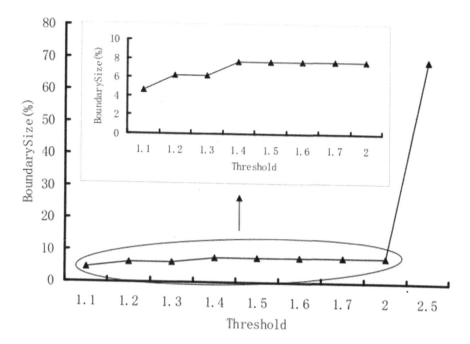

Fig. 2. Synthetic data: Change in *BoundarySize* with *threshold*

who spends $300, $100, $200 during the same three months. Therefore, we sort the spending values, which makes the two customers identical in terms of their revenue generation potential. Instead of using twelve monthly spending and visit values, which may be too detailed for the purpose of grouping, we will represent the patterns using the lowest, highest and average spending. However, in some cases, lowest and highest values can be outliers. Therefore, we use second highest, second lowest and median values as a representative of the pattern.

313 customers visited in only one month. These customers were termed as infrequent customers. It was decided that there was no further need for grouping these customers. After eliminating the 313 customers, the number of customers was 5565. After experimenting with different number of clusters we set $k = 5$. w_{lower} was set at 0.75.

Fig. 3 describes the *BoundarySize* changes with the *threshold*, which is similar to the one found for the synthetic data. The *BoundarySize* goes up a little slowly until the *threshold* reaches a value of 1.4, where there is a marked increase. This suggests that 1.4 may be an appropriate value for the *threshold*. We can also see a sudden and sharp jump at *threshold* = 2.5. This reinforces our earlier observation that high values of *threshold* may lead to inconclusive rough clustering. Fig. 4 presents the rough centroids as the representative patterns for each cluster. Cluster c_1 is the largest cluster consisting of moderate spenders who spend $0 to $52 in a month. The next cluster, c_2, is about the quarter the size of c_3 with spending ranging from $0 to $100. Third cluster ($c_3$) is even smaller

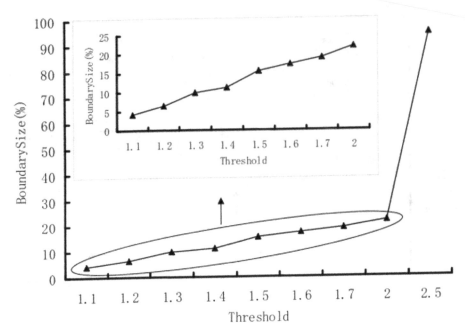

Fig. 3. Real data: Change in *BoundarySize* with *threshold*

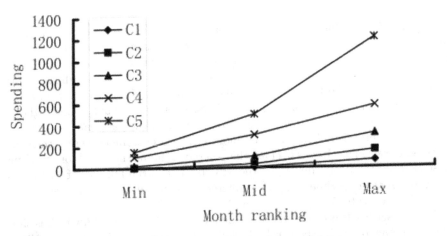

Fig. 4. Rough centroids for the retail data

with spending ranging from \$10 to \$250. Fourth cluster has approximately 70 to 100 customers who spend \$120 to \$500. The last cluster is the smallest with spending ranging from \$137 to \$1330. The overlap between different clusters for *threshold* = 1.4 and *threshold* = 2 are shown in Table 1. It can be seen in Table 1(a) that the intermediate clusters, i.e. c_2, c_3, and c_4 have overlaps with two clusters on either side. For example, c_2 overlaps with c_1 and c_3, while c_3

Table 1. The number of objects in the intersection of clusters

	C1	C2	C3	C4	C5
C1	–	403	0	0	0
C2	403	–	177	0	0
C3	0	177	–	41	0
C4	0	0	41	–	9
C5	0	0	0	9	–

(a)threshold=1.4

	C1	C2	C3	C4	C5
C1	–	809	81	11	8
C2	809	–	388	28	9
C3	81	388	–	163	18
C4	11	28	163	–	59
C5	8	9	18	59	–

(b)threshold=2.0

overlaps with c_2 and c_4, and c_4 overlaps with c_3 and c_5. Clusters c_1 and c_5 have overlap with only one cluster: c_1 with c_2 and c_5 with c_4. When the *threshold* is raised to 2.0, we can see from Table 1(b) that each cluster overlaps with other four clusters. That means many objects have now moved to boundary regions of all the clusters. This makes any conclusion about their membership impossible.

6 Conclusions

Rough set clustering makes it possible to assign stray objects - that may not belong to a precise cluster - to boundary regions of two or more clusters. This aspect of rough set clustering adds a degree of imprecision to the clustering scheme. The degree of imprecision is an additional unknown in the unsupervised learning based on rough set theory. The experiments with a synthetic data set and a real world data set show that it is important to choose a right balance between rough and precise cluster assignments. The paper describes a procedure that can be used to control the imprecision in rough set clustering for the rough K-means algorithm by varying the *threshold* parameter. The results presented here lay foundations for a more comprehensive study of the quality of rough set clustering, which will be presented in a subsequent publication.

Acknowledgement

The authors would like to thank China Scholarship Council and NSERC Canada for their financial support.

References

1. Banerjee, M., Mitra, S., Pal, S.K.: Rough fuzzy MLP: knowledge encoding and classification. IEEE Transactions on Neural Networks 9(6), 1203–1216 (1998)
2. Ho, T.B., Nguyen, N.B.: Nonhierarchical Document Clustering by a Tolerance Rough Set Model. International Journal of Intelligent Systems 17(2), 199–212 (2002)
3. Hartigan, J.A., Wong, M.A.: Algorithm AS136: A K-Means Clustering Algorithm. Applied Statistics 28, 100–108 (1979)

4. Hirano, S., Tsumoto, S.: Rough Clustering and Its Application to Medicine. Information Sciences 124, 125–137 (2000)
5. Lingras, P.: Unsupervised Rough Set Classification using GAs. Journal Of Intelligent Information Systems 16(3), 215–228 (2001)
6. Lingras, P., West, C.: Interval Set Clustering of Web Users with Rough K-means. Journal of Intelligent Information Systems 23(1), 5–16 (2004)
7. Lingras, P., Hogo, M., Snorek, M.: Interval Set Clustering of Web Users using Modified Kohonen Self-Organizing Maps based on the Properties of Rough Sets. Web Intelligence and Agent Systems: An International Journal 2(3), 217–230 (2004)
8. Mitra, S.: An evolutionary rough partitive clustering. Pattern Recognition Letters 25, 1439–1449 (2004)
9. Mitra, S., Bank, H., Pedrycz, W.: Rough-Fuzzy Collaborative Clustering. IEEE Trans. on Systems, Man and Cybernetics 36(4), 795–805 (2006)
10. Nguyen, H.S.: Rough Document Clustering and the Internet. Handbook on Granular Computing (2007)
11. Pedrycz, W., Waletzky, J.: Fuzzy Clustering with Partial Supervision. IEEE Trans. on Systems, Man and Cybernetics 27(5), 787–795 (1997)
12. Polkowski, L., Skowron, A.: Rough Mereology: A New Paradigm for Approximate Reasoning. International Journal of Approximate Reasoning 15(4), 333–365 (1996)
13. Peters, J.F., Skowron, A., Suraj, Z., Rzasa, W., Borkowski, M.: Clustering: A rough set approach to constructing information granules. In: Proceedings of 6th International Conference on Soft Computing and Distributed Processing, Rzeszow, Poland, June 24-25, 2002, pp. 57–61 (2002)
14. Pawlak, Z.: Rough Sets: Theoretical Aspects of Reasoning about Data. Kluwer Academic Publishers, Dordrecht (1992)
15. Peters, G.: Some Refinements of Rough k-Means. Pattern Recognition 39(8), 1481–1491 (2006)
16. Yao, Y.Y.: Constructive and algebraic methods of the theory of rough sets. Information Sciences 109, 21–47 (1998)
17. Yao, Y.Y., Zhao, Y.: Attribute reduction in decision-theoretic rough set models. Information Sciences (to appear, 2008)

A Dynamic Approach to Rough Clustering

Georg Peters[1] and Richard Weber[2]

[1] University of Applied Sciences - München
Department of Computer Science and Mathematics
80335 Munich, Germany
georg.peters@cs.hm.edu
[2] University of Chile
Department of Industrial Engineering
República 701, Santiago, Chile
rweber@dii.uchile.cl

Abstract. Many projects in data mining face, besides others, the following two challenges. On the one hand concepts to deal with uncertainty - like probability, fuzzy set or rough set theory - play a major role in the description of real life problems. On the other hand many real life situations are characterized by constant change - the structure of the data changes. For example, the characteristics of the customers of a retailer may change due to changing economical parameters (increasing oil prices etc.). Obviously the retailer has to adapt his customer classification regularly to the new situations to remain competitive. To deal with these changes dynamic data mining has become increasingly important in several practical applications. In our paper we utilize rough set theory to deal with uncertainty and suggest an engineering like approach to dynamic clustering that is based on rough k-means.

Keywords: Rough Clustering, Dynamic Clustering, Dynamic Data Mining.

1 Introduction

Rough clustering approaches have gained increasing attention since Lingras et al. introduced rough k-means [1,2]. Several extensions and modifications have been suggested in the meantime (e.g. in [3,4,5,6,7]). In applications it has been shown that rough clustering is a successful method for many real life problems (for example [2,4]).

Up to now, rough clustering algorithms only perform in stable environments, in environments where the structure of the data remains unchanged. However, Crespo and Weber [8] pointed out that many real life situations are characterized by changing data structures. They identified three strategies to deal with such situations: (1) Neglect the changes and do not change the classifier[1]. (2) Perform

---

[1] Note, we use the terms classify and classifier in the sense that new data are assigned to the clusters obtained by a cluster analysis. Both terms are also used in supervised learning and have a different meaning there.

C.-C. Chan et al. (Eds.): RSCTC 2008, LNAI 5306, pp. 379–388, 2008.
© Springer-Verlag Berlin Heidelberg 2008

the cluster algorithm again using all data and build a new classifier. (3) Dynamically adapt the initial classifier with respect to the changes in the data structure. Crespo and Weber [8] argued that the last approach is very adequate for dynamic data structures and therefore suggested a method to dynamically adapt the classifier derived by fuzzy c-means [9].

The objective of our paper is to develop an engineering like method[2] how to dynamically adapt the classifier of Lingras rough k-means.

The remaining paper is organized as follows. In the next Section we discuss related literature. In Section 3 we present our approach how to engineer the changes in rough clustering. The paper ends with a conclusion in Section 4.

2   Related Literature

In literature several approaches for dynamic data mining have been proposed. In our context the paper of Crespo and Weber [8] is of special interest. They suggested a dynamic approach to fuzzy c-means (FCM) and applied it to traffic data besides others.

Basically they proposed to check the results of the objects' classification for changes in the data structure after each cycle of new data. Depending on the changes they proposed a five step approach with three different actions (see Figure 1): (1) create new clusters, (2) move clusters or (3) eliminate clusters.

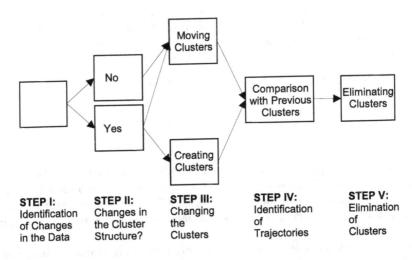

Fig. 1. Crespo's and Weber's Methodology for Dynamic Data Mining [8]

For a more detailed overview and a discussion on related literature on dynamic clustering the reader is referred to Weber [10].

---

[2] We define an "engineering like method" as an approach that operates in many but not necessarily in all real life situations properly.

Dynamic approaches can also be seen in the context of the determination of the initial parameters of partitive clustering algorithms. Several cluster validity indexes have been suggested (e.g. [11,12,13]) which can also be used to determine the optimal number of clusters. However, Mitra [5] pointed out the determination of optimal initial parameters remains one of the main challenges especially in rough clustering where the weights of the approximations and a threshold have to be determined besides the number of clusters.

A dynamic, engineering-like approach can be a promising method to support this process since the parameters are questioned and discussed after each cycle of new data again.

3   Engineering Dynamic Cluster Structures

3.1   Basic Notations

For our dynamic rough clustering approach we use notations as follows:

Notation	Meaning
i	Cycle i
K^i	Number of clusters in cycle i
$M^i = \overline{M^i}$	Total number of objects respectively total number of objects in upper approximations after cycle i
$\underline{M^i} = \sum_{k=1}^{K^i} \underline{M_k^i}$	Total number of objects in lower approximations after cycle i
$\overline{M_k^i}$, $\underline{M_k^i}$	Number of objects in the upper respectively lower approximation of cluster k after cycle i
N	Number of new objects in cycle i
$\overline{N_k^i}$, $\underline{N_k^i}$	Number of new objects in the upper respectively lower approximation of cluster k in cycle i
w_B, w_L	Weights of the boundary region respectively lower approximation (with $w_B + w_L = 1$)
ϵ	Threshold in rough k-means
F	Number of objects that are far away from existing cluster centers

Details on the underlying rough k-means can be found in e.g. [14,2].

3.2   Setup of the Dynamic Rough Clustering System

The setup of the dynamic rough clustering system is depicted in Figure 2. At the beginning of each cycle the rough k-means algorithm is performed. The obtained parameters are used to classify new objects.

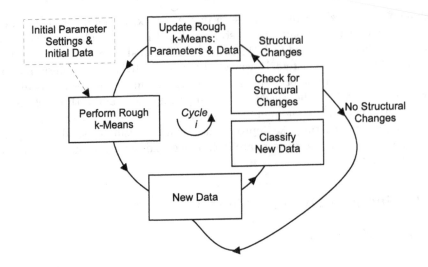

Fig. 2. Dynamic Rough Clustering Loop

Then we check for structural changes in the new data. If there are none the classifier remains unchanged. If there are structural changes the current parameters of the rough k-means are updated and, in the case of dying clusters, old objects belonging to those are eliminated. Finally the rough cluster algorithm is performed again. The method's details are described in the following Section.

3.3   Possible Changes in the Cluster Structure

In our engineering approach we consider three possible changes in the data structure which lead to the adaption of the current parameters of the rough k-means algorithm or an elimination of objects.

- **Dying Clusters.** A cluster is dying when it won't be refreshed sufficiently by new objects. As a consequence we proposed to eliminate the cluster.
- **Emerging Clusters.** If new objects do not sufficiently fit to the current cluster structure we propose to create a new cluster.
- **Changing Uncertainty.** When the fraction between the number of objects in the lower approximation and the number of objects in the boundary region changes significantly we consider this as a change in the uncertainty of the clusters. Here we propose to adapt the threshold parameter ϵ to the new situation.

In this phase of our research we do not consider a possible tuning of the rough k-means by changing the initial parameters w_B and w_L. Please note, that we also consider the feature space as stable in our analysis. So we exclude any changes in the dimensions of the feature space and the replacement of features.

3.4 Dying Clusters

Condition. We propose that a cluster should be refreshed - that is to get a reasonable number of new *sure* objects in every cycle - to stay alive. So we only consider new objects that are assigned to lower approximations.

We propose the following criterion. The cluster strength, in terms of its sure members, should not decline significantly in relation to the overall number of objects:

$$\frac{N_k^{i+1}/M^{i+1}}{N_k^i/M^i} = r_1 < \tau_1 \tag{1}$$

For $r_1 = 1$ the relative significance of cluster k remains unchanged from cycle i to $i+1$ while $r_1 > 1$ indicates an increasing importance of that cluster. If $r_1 < 1$ the importance of cluster k decreases.

The parameter τ_1 is a user defined threshold. In our context, the identification of dying clusters, τ_1 must be selected in a range of $0 < \tau_1 < 1$.

Action. If a cluster is to be eliminated the members of its lower approximation will be eliminated. For the objects in its upper approximation we have to distinguish two cases:

- An object still belongs to at least two other upper approximations. Since this complies with the properties of interval based rough set clustering no further action is required.
- An object belongs to only one upper approximation after the elimination of the cluster. Since this would violate a property of interval based rough set clustering we suggest to assign this object to the lower approximation of the corresponding cluster.

The obtained solution is already quite good but might not be optimal in the sense of rough clustering. Therefore, we suggest two alternative strategies here:

- The obtained solution is good enough for the needs of a data analyst (in our engineering like approach the "good criterion" is based on a subjective evaluation of an expert). Therefore, we do not need to perform the rough k-means again.
- After the elimination of the objects the rough k-means will be performed again with a cluster number of $K^{i+1} = K^i - 1$. The initial assignment of the objects to the clusters should equal the ones obtained after the elimination of the objects in iteration i.

3.5 Emerging Clusters

Condition. For fuzzy c-means Crespo and Weber [8] suggested to create a new cluster when the following two conditions hold:

- The membership values of *many new objects* are close to $1/c$ (with c the number of fuzzy clusters).
- These *many new objects* are far away from current cluster centers v_k.

In the case of rough k-means we can only apply the latter condition since new groups of objects can emerge in upper as well as in lower approximations. So we cannot transfer the membership criterion to rough approximations.

Therefore, along the lines with Crespo and Weber [8], we suggest the following indicator when to create a new cluster:

$$\hat{d}_{jh} > \tfrac{1}{2}min\{d(v_j, v_k)\} \forall h \in \{M^i + 1, ..., M^i + N^{i+1}\} \forall j \neq k \in \{1, ..., K^i\} \quad (2)$$

Whenever we discover *many* new objects that are far away from existing cluster centers we assume that they might form one new cluster[3] (see Figure 3). The term *many* should be defined use-dependent. We propose, for example, that the number F of this kind of new objects should at least exceed the average number of objects in the lower approximations of the existing clusters:

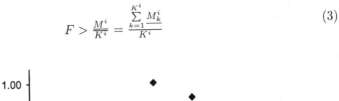

$$F > \frac{M^i}{K^i} = \frac{\sum_{k=1}^{K^i} M_k^i}{K^i} \quad (3)$$

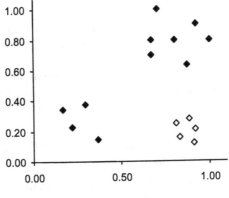

◆ Old Objects   ◇ New Objects

Fig. 3. Required Action: Creating a New Cluster

Action. In the given case we set the initial parameter $K^{i+1} = K^i + 1$ and leave the remaining initial parameters unchanged. Then we perform the rough k-means again.

3.6   Changing Uncertainty

There are several possibilities to define criteria that indicate a significant change in the data's structure. In our context we emphasize on the relation between the

---

[3] In general, one or more new clusters may have emerged. Preliminary, for simplicity, we restrict our analysis to one new cluster here.

number of objects in the lower approximations and the number of objects in the upper approximations of the data set respectively all objects.

In many real-life situations a fraction between sure objects and objects that "need a second look" is defined. Ideally this fraction can be derived intrinsically - the fraction is within the data. However, in many cases the fraction is defined extrinsically, for example by the famous 80/20 proportion: 80% of the cases go through immediately while 20% need to have a second look.

Applying this idea to rough clustering this means 80% (or any other user defined percentage) of the objects should belong to lower approximations since their assignment to a cluster is clear. However, the remaining 20% are members of boundary regions since their membership is unclear. Therefore, they need a second look.

As a criterion for the change in the clusters' uncertainty we suggest the following quotient:

$$\frac{\underline{\underline{M^i}}}{M^i} \tag{4}$$

that is the relation of the number of objects in all lower approximations to the number of all objects[4].

We distinguish two cases here - whether the uncertainty of the clusters increase or decrease:

I. Increase of Uncertainty: $\frac{\underline{\underline{M^i}}}{M^i} > \frac{\underline{\underline{M^{i+1}}}}{M^{i+1}}$,

II. Decrease of Uncertainty: $\frac{\underline{\underline{M^i}}}{M^i} < \frac{\underline{\underline{M^{i+1}}}}{M^{i+1}}$.

We discuss these two cases of uncertainty in the following paragraphs in more detail.

I. Increase of Uncertainty

Condition. As discussed above we define an increase of uncertainty as follows: $\frac{\underline{\underline{M^i}}}{M^i} > \frac{\underline{\underline{M^{i+1}}}}{M^{i+1}}$. The fraction between objects surely belonging to clusters and all objects decreases. So relatively more objects are in the boundary region. One or more clusters become blurred.

A possible reason is that new objects emerge around the core cluster as depicted in Figure 4. Therefore, they are possibly not assigned to the lower approximation of the cluster but belong to the boundary region[5].

Action. As already mentioned above we assume that our goal is to keep the quotient of the number of objects in the lower approximation and the total number of objects constant. So the threshold ϵ must be decreased.

One way to optimize the threshold ϵ according to our objective is to implement a genetic algorithm [15]. A simpler way is to apply an iterative method as defined as follows:

---

[4] Note, that this number equals the number of objects in all upper approximations.

[5] Please note, that the upper approximations are not necessarily spatial, which might be implied by Figure 4.

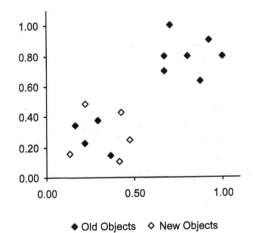

◆ Old Objects   ◇ New Objects

Fig. 4. Required Action: Adaption of the Parameters

- *Initial Settings:*
  Set $i = 0$, $\epsilon_i^{max} = \epsilon^{initial}$ and $\epsilon_i^{min} = 0$.

- *Iteration:*

  1. Set $i = i + 1$.
  2. Let $\epsilon_i = \epsilon_i^{min} + \frac{\epsilon_i^{max} - \epsilon_i^{min}}{2}$.
  3. Conduct rough k-means with the new initial parameter ϵ_i.
  4. IF $[\frac{\underline{M^i}}{M^i} \approx \frac{\underline{M^{i+1}}}{M^{i+1}}]$ THEN [Stop].
     ELSE IF $[\frac{\underline{M^i}}{M^i} > \frac{\underline{M^{i+1}}}{M^{i+1}}]$ THEN $[\epsilon_{i+1}^{max} = \epsilon_i$ and $\epsilon_{i+1}^{min} = \epsilon_i^{min}$
     and continue with Step 1].
     ELSE $[\epsilon_{i+1}^{min} = \epsilon_i$ and $\epsilon_{i+1}^{max} = \epsilon_i^{max}$
     and continue with Step 1].

When the algorithm has converged *sufficiently*, new data of the next cycle will be sent on the obtained rough classifier.

The *sufficiently* criterion is indicated by the $\approx$ symbol in the iteration above. The threshold - when both fractions are approximately equal - must be defined by the data analyst context dependently.

II. Decrease of Uncertainty

Condition. As discussed above we define a decrease of uncertainty as follows: $\frac{\underline{M^i}}{M^i} < \frac{\underline{M^{i+1}}}{M^{i+1}}$. The fraction between objects surely belonging to clusters and all objects increases. So relatively more objects are in lower approximations in comparison to the total number of objects.

Action. To increase the number of objects in the boundary region the threshold criterion has to be relaxed, i.e. the value ϵ must be increased. Our approach is similar to the approach we have presented already in the previous Section *I. Increase of Uncertainty*. The main difference is that the parameter ϵ will be increased instead of decreased. Therefore, we have to define an upper limit for ϵ. For example, we suggest to select ϵ_0^{max} ten-time higher than $\epsilon^{initial}$ which will be more than sufficient for most real life applications. Then the algorithms proceeds as follows:

– *Initial Settings:*
  Set $i = 0$, $\epsilon_i^{max} = 10 \cdot \epsilon^{initial}$ and $\epsilon_i^{min} = \epsilon^{initial}$.

– *Iteration:*

1. Set $i = i + 1$.
2. Let $\epsilon_i = \epsilon_i^{min} + \frac{\epsilon_i^{max} - \epsilon_i^{min}}{2}$.
3. Conduct rough k-means with the new initial parameter ϵ_i.
4. IF $[\frac{M^i}{\underline{M^i}} \approx \frac{M^{i+1}}{\underline{M^{i+1}}}]$ THEN [Stop].
   ELSE IF $[\frac{M^i}{\underline{M^i}} > \frac{M^{i+1}}{\underline{M^{i+1}}}]$ THEN $[\epsilon_{i+1}^{max} = \epsilon_i$ and $\epsilon_{i+1}^{min} = \epsilon_i^{min}$
   and continue with Step 1].
   ELSE $[\epsilon_{i+1}^{min} = \epsilon_i$ and $\epsilon_{i+1}^{max} = \epsilon_i^{max}$
   and continue with Step 1].

Again, as in Section *I. Increase of Uncertainty*, when the algorithm has converged sufficiently, new data of the next cycle will be sent on the obtained rough classifier.

4   Conclusion

In this paper we presented a novel dynamic approach for rough clustering. Our method should be applied in real life situation that are characterized by changing structures in the underlying data. It can also be utilized to determine the initial parameters of rough set clustering in an engineering like approach.

Presently, we are conducting experiments with synthetic as well as real data and developing methods to extend and further automate the dynamic rough cluster algorithm.

Acknowledgment

This paper has been funded by the Millenium Nucleus on Complex Engineering Systems (www.sistemasdeingenieria.cl).

References

1. Lingras, P., West, C.: Interval set clustering of web users with rough k-means. Technical Report 2002-002, Department of Mathematics and Computer Science, St. Mary's University, Halifax, Canada (2002)
2. Lingras, P., West, C.: Interval set clustering of web users with rough k-means. Journal of Intelligent Information Systems 23, 5–16 (2004)

3. Lingras, P., Hogo, M., Snorek, M.: Interval set clustering of web users using modified Kohonen self-organizing maps based on the properties of rough sets. Web Intelligence and Agent Systems 2(3), 217–225 (2004)
4. Lingras, P.: Applications of rough set based k-means, Kohonen SOM, GA clustering. In: Peters, J., Skowron, A., Marek, V., Orlowska, E., Slowinski, R., Ziarko, W. (eds.) Transactions on Rough Sets VII. LNCS, vol. 4400, pp. 120–139. Springer, Heidelberg (2007)
5. Mitra, S.: An evolutionary rough partitive clustering. Pattern Recognition Letters 25, 1439–1449 (2004)
6. Peters, G.: Some refinements of rough k-means. Pattern Recognition 39, 1481–1491 (2006)
7. Peters, G., Lampart, M., Weber, R.: Evolutionary rough k-medoids clustering. In: Peters, J.F., Skowron, A. (eds.) Transactions on Rough Sets VIII. LNCS, vol. 5084, pp. 289–306. Springer, Heidelberg (2008)
8. Crespo, F., Weber, R.: A methodology for dynamic data mining based on fuzzy clustering. Fuzzy Sets and Systems 150(2), 267–284 (2005)
9. Bezdek, J.: Pattern Recognition with Fuzzy Objective Algorithms. Plenum Press, New York (1981)
10. Weber, R.: Fuzzy clustering in dynamic data mining - techniques and applications. In: Valente de Oliveira, J., Pedrycz, W. (eds.) Advances in Fuzzy Clustering and Its Applications, pp. 315–332. John Wiley and Sons, Hoboken (2007)
11. Bezdek, J., Pal, N.: Some new indexes of cluster validity. IEEE Transactions on Systems, Man, and Cybernetics 28, 301–315 (1998)
12. Davies, D., Bouldin, D.: A cluster separation measure. IEEE Transactions on Pattern Analysis and Machine Intelligence 1, 224–227 (1979)
13. Windham, M.: Cluster validity for fuzzy clustering algorithms. Fuzzy Sets and Systems 5, 177–185 (1981)
14. Lingras, P., Yan, R., West, C.: Comparison of conventional and rough k-means clustering. In: Wang, G., Liu, Q., Yao, Y., Skowron, A. (eds.) RSFDGrC 2003. LNCS (LNAI), vol. 2639, pp. 130–137. Springer, Heidelberg (2003)
15. Goldberg, D.: Genetic Algorithms in Search, Optimization, and Machine Learning. Addison-Wesley Professional, Boston (1989)

Learning Patterns from Clusters Using Reduct

Alka Arora[1], Shuchita Upadhyaya[2], and Rajni Jain[3]

[1] Indian Agricultural Statistics Research Institute, Library Avenue, Pusa,
New Delhi-110012, India
alkak@iasri.res.in
[2] Department of Computer Science and Application, Kurukshetra University,
Kurukshetra, India
shuchita-bhasin@yahoo.com
[3] National Centre for Agricultural Economics and Policy Research Pusa Campus,
DPS Marg, New Delhi 110012, India
rajni@ncap.res.in

Abstract. Usual clustering algorithms just generate general description of the clusters like which entities are member of each cluster and lacks in generating cluster description in the form of pattern. Pattern is defined as a logical statement describing a cluster structure in terms of relevant attributes. In the proposed approach reduct from rough set theory is employed to generate pattern. Reduct is defined as the set of attributes which distinguishes the entities in a homogenous cluster, therefore these can be clear cut removed from the same. Remaining attributes are ranked for their contribution in the cluster. Cluster description is then formed by conjunction of most contributing attributes. Proposed approach is demonstrated using benchmarking mushroom dataset from UCI repository.

Keywords: Rough set theory, Reduct, Indiscernibility, Clustering, Cluster description, Mushroom, Pattern.

1 Introduction

Clustering partitions a given dataset into clusters such that entities in a cluster are more similar to each other than entities in different clusters [3]. Description of clusters helps in understanding these different clusters. Cluster description is able to approximately describe the cluster in the form that ' this cluster consists just of all the entities having the pattern P, where the pattern is formulated using the attribute and values of the given many valued context' [2]. Clustering algorithms in literature are divided into different categories [3,8]. Partitional clustering algorithms are commonly used clustering algorithms. K-Means and Expectation Maximization (EM) algorithms are the widely known partitional algorithms. These clustering algorithms just generate general description of the clusters like which entities are member of each cluster and lacks in generating pattern. This is because classical approach has no mechanism for selecting and evaluating the attributes in the process of generating clusters [6]. Therefore post processing of clusters is required to extract patterns from the same. The

C.-C. Chan et al. (Eds.): RSCTC 2008, LNAI 5306, pp. 389–398, 2008.

problem of finding pattern of a single cluster is relatively new and is beneficial in situation where interpretation of clusters is required in meaningful and user understandable format. From an intelligent data analysis perspective, pattern learning is as important in clustering as cluster finding.

In this paper, an attempt is being made to describe the clusters using Rough Set Theory (RST). Indiscernibility relation is core concept of RST. Indiscernibility relation partitions the set of entities into equivalence/ indiscernible classes, that's why it has a natural appeal to be applied in clustering as every indiscernible class can be considered as natural cluster. Moreover RST performs automatic concept approximation by producing minimal subset of attributes (Reduct) which distinguishes entities in the indiscernible class. Mostly reduct is computed relative to decision attribute in the dataset [4]. However, our approach of reduct computation is different. Clustering is done on unsupervised data where decision/class information is not present, hence reduct computation is purely on the basis of indiscernibility. Such reducts are referred as unsupervised reducts in this paper. We have computed unsupervised reduct for individual cluster as compared to reduct computation for dataset because our aim is to generate patterns of individual clusters.

Removal of reduct attributes from the cluster will lead to attributes which will have same attribute value pair for majority of its instances in the cluster. These remaining attributes play significant role in pattern formulation. These attributes are ranked on Precision Error for their significance in the cluster. Pattern formulation is then carried out with the conjunction of major contributing attributes. The efficacy of the proposed approach is demonstrated with the help of benchmarking mushroom dataset from the UCI repository [9]. Objective of applying the proposed approach on mushroom dataset is to study the relationship of attributes with edible and poisonous nature of mushrooms.

The paper is organized as follows. In section 2 the basic notions of rough set theory and cluster description is described. Section 3 presents the cluster description approaches including the proposed approach. In section 4, application of proposed approach is demonstrated on mushroom dataset followed by conclusions in Section 5.

2   Basic Notions

2.1   Rough Set Theory Concepts

In RST data is represented as an information system $X = (U, A)$ [5,11]. In this U is non-empty finite set of entities and A is a non-empty, finite set of attributes. With every attribute $a \in A$, we associate a set V_a such that $a : U \to V_a$. The set V_a is called the domain or value set of attribute a. Every entity x in the information system X is characterized by its information vector:

$$InfX(x) = \{(a, a(x)) : a \in A\} \tag{1}$$

Relationship between entities is described by their attribute values. Indiscernibility relation $IND(B)$, for any set $B \subseteq A$ is defined by:

$$xIND(B)y \Leftrightarrow \forall_a \in_B (a(x) = a(y)) \tag{2}$$

Two entities are considered to be indiscernible by the attributes in B, if and only if they have the same value for every attribute in B. Entities in the information system about which we have the same knowledge form an equivalence relation. Indiscernibility relation on B, $IND(B)$ is an equivalence relation that partitions U into set of indiscernible classes. Set of such partitions are denoted by $U/IND(B)$.

Reduct is the set of attributes that can differentiate all indiscernible classes. More formally reduct(R) is a set of attributes such that:

$$
\begin{aligned}
&R \subseteq A \\
&IND_R(U) = IND_A(U) \\
&IND_{R-a}(U) = IND_A(U) \; \forall a \in R
\end{aligned}
\tag{3}
$$

There are many methods as well as many software's available for computation of reducts, discussion on those is beyond the scope of this paper. We have considered Genetic Algorithm(GA) [14] for reduct computation as it can produce many reduct sets of varying length.

2.2 Cluster Description Concepts

Partitional clustering algorithm divides the data into k clusters. Pattern P of cluster C is formed by concatenating significant attribute value pairs of the form $((a_1 = v) \wedge (a_2 = v) \wedge \ldots \wedge (a_n = v))$ from that cluster. Where attribute value pair $(a_i = v)$ is defined as descriptor d, attribute $a_i \in A$ can have any value $v \in V_{ai}$. Pattern formed with the conjunction of all significant descriptors can be quite complex and for maximum comprehensibility, shorter cluster description is preferred. Hence descriptors are evaluated for their contribution in the cluster. A descriptor is said to be more contributing if, most of the entities satisfying that descriptor belongs to a single cluster. It is quite possible that some entities that satisfy the descriptor also belongs to other clusters. Therefore descriptors are measured on Precision Error (PE). Precision error for descriptor d, $PE(d)$ is defined as:

$$PE(d) = \frac{|False Positive \, C \, (d)|}{|U - C|} \tag{4}$$

where numerator defines the number of False Positive (an entity $x \in U - C$ for which $a_i = v$ is true) and denominator defines the number of entities outside C.

Problem of pattern formulation can be carried out by combining the descriptors with less PE such that this pattern distinctively describes the cluster without any error. Pattern is evaluated on Precision Error [7]. Precision Error for pattern P, $PE(P)$ is defined as:

$$PE(P) = \frac{|FalsePositive\ C\ (P)|}{|U - C|} \qquad (5)$$

Where numerator denotes the number of entities that lies outside the cluster C for which pattern P is true and denominator denotes the number of entities outside cluster C.

Pattern Length, $L(P)$ is defined as number of descriptors occurring in P.

3   Cluster Description Approaches

The field of producing patterns for individual clusters is relatively new. There are few references of cluster description approaches available in literature. Mirkin has proposed a method for cluster description applicable to only continuous attributes [7]. In Mirkin's approach attributes are normalized first and then ordered according to their contribution weights which are proportional to the squared differences between their with-in group averages and grand means. A conjunctive description of cluster is then formed by consecutively adding attributes according to the sorted order. An attribute is added to the description only if it decreases the error. This forward attribute selection process stops after the last element of attribute set is checked. Abidi et al. has proposed the rough set theory based method for rule creation for unsupervised data using dynamic reduct [1]. Dynamic reduct is defined as the frequently occurring reduct set from the samples of original decision table. However these approaches have its limitations. Mirkin's approach is applicable only to datasets having continuous attributes. Abidi in his approach has used the cluster information obtained after cluster finding and generated rules from entire data with respect to decision/cluster attribute, instead of producing description for individual clusters.

3.1   Proposed Approach (Reduct Driven Cluster Description–RCD)

Proposed approach of pattern formulation is divided into three stages. First stage deals with obtaining clusters from dataset by applying clustering algorithm. In the second stage we have computed sets of non significant and significant attributes for that cluster. Computation of reduct set (RC) provides the set of non significant attributes for a cluster. These non significant attributes (reduct) are straight away removed from the cluster. Remaining attributes then form the set of significant attributes (I) for that cluster. PE is calculated for every descriptor (d) in set I, and descriptors are arranged in ascending order of their PE score. In the third stage pattern is formulated by conjunction of descriptors in the order of minimum PE until pattern completly describes the cluster without any error.

Steps for RCD

1. Data Clustering
   Apply clustering algorithm on dataset to obtain clusters.
2. Reduct Computation.
   Compute unsupervised reduct (RC) for individual cluster C.
3. Cluster Description
   a) Computation of Descriptor Set (I)
      I=A-RC; A is attribute set and RC is reduct set of C.
   b) Ranking of Descriptors
      Calculate PE(d)(Equ.4) for every descriptor in set I.
      Arrange the set I in ascending order of PE value.
   c) Pattern Formulation P
      Take first descriptor(di)of minimum PE from I
      and assign it to P
      Compute PE(P)(Equ. 5)
      If PE(P) is zero, then output P
      Otherwise P=P^di+1
      keep on concatenating the descriptor from I in order of PE
      till PE(P)is zero.
   d) Computation of Pattern Length L(P)
      L(P)= Number of descriptor in pattern P
4. Repeat step 2 and 3 for every cluster

4   Experimental Results

4.1   Data Description

We have considered benchmarking mushroom dataset from UCI repository for demonstration of RCD approach [9]. Dataset consists of large number of records that is 8124 records. The number of edible and poisonous mushrooms in the data set is 4208 and 3916 respectively. Class attribute (edible (e) or poisonous (p)) and attribute stalk root with missing values are not considered for clustering. Details of 22 categorical attributes that describes the physical characteristics of mushrooms is given below:

1. cap-shape: bell=b, conical=c, convex=x, flat=f, knobbed=k, sunken=s;
2. cap-surface: fibrous=f, grooves=g, scaly=y, smooth=s;
3. cap-color: brown=n, buff=b, cinnamon=c, gray=g, green=r, pink=p, purple=u, red=e, white=w, yellow=y;
4. bruises: bruises=t, no=f;
5. odor: almond=a, anise=l, creosote=c, fishy=y, foul=f, musty=m, none=n, pungent=p, spicy=s;
6. gill-attachment: attached=a, descending=d, free=f, notched=n;

7. gill-spacing: close=c, crowded=w, distant=d;
8. gill-size: broad=b, narrow=n;
9. gill-color: black=k, brown=n, buff=b, chocolate=h, gray=g, green=r, orange=o, pink=p, purple=u, red=e, white=w, yellow=y;
10. stalk-shape: enlarging=e, tapering=t;
11. stalk-root: bulbous=b, club=c, cup=u, equal=e, rhizomorphs=z, rooted=r, missing=?;
12. stalk-surface-above-ring: ibrous=f, scaly=y, silky=k, smooth=s;
13. stalk-surface-below-ring: ibrous=f, scaly=y, silky=k, smooth=s;
14. stalk-color-above-ring: brown=n ,buff=b, cinnamon=c, gray=g, orange=o, pink=p, red=e, white=w, yellow=y;
15. stalk-color-below-ring: brown=n, buff=b, cinnamon=c, gray=g, orange=o, pink=p, red=e, white=w, yellow=y;
16. veil-type: partial=p, universal=u;
17. veil-color: brown=n, orange=o, white=w, yellow=y;
18. ring-number: none=n, one=o, two=t;
19. ring-type: ring-type: cobwebby=c, evanescent=e, flaring=f, large=l, none=n, pendant=p, sheathing=s, zone=z;
20. spore-print-color: black=k, brown=n, buff=b, chocolate=h, green=r, orange=o, purple=u, white=w, yellow=y;
21. population: abundant=a, clustered=c, numerous=n, scattered=s, several=v, solitary=y;
22. habitat: grasses=g, leaves=l, meadows=m, paths=p, urban=u, waste=w, woods=d.

4.2   Data Clustering

We have used Weka implementation [13] of EM algorithm for cluster finding as it can handle continuous as well as categorical attributes. EM is a mixture based algorithm that attempts to maximize the likelihood of the model [8]. By default, EM selects the number of clusters automatically by maximizing the logarithm of the likelihood of future data, estimated using cross-validation. Beginning with one cluster, it continues to add clusters until the estimated log-likelihood decreases.

When EM clustering algorithm is applied on mushroom dataset, it learned 14 numbers of clusters from the data. Table 1 shows the result obtained with EM algorithm. There is wide variance among the size of the clusters that range from

Table 1. Clustering results with EM algorithm

cluster number	1	2	3	4	5	6	7	8	9	10	11	12	13	14
poisonous	288	1728	84	0	0	0	0	256	1296	0	192	0	72	0
edible	0	0	112	192	768	96	1728	0	0	512	96	192	224	288

96 entities to 1728 entities. As shown in Table 1, except clusters 3, 11 and 13 which are mix clusters, all other clusters are pure clusters. Pure clusters in the sense that mushrooms in every cluster are either all poisonous or all edible.

4.3 Reduct Computation

We have used Rosetta software [12] for computation of reduct using GA. Unsupervised reduct are computed for individual pure poisonous and edible clusters. Table 2 shows the reduct attributes in poisonous and edible clusters. Although all the four clusters(1, 2, 8, and 9) are poisonous, yet reduct attributes are not common among these clusters. Similarly reduct attributes are not common among pure edible clusters(4, 5, 6, 7, 10, 12 and 14).

Table 2. Reduct attribute in poisonous and edible clusters

Poisonous Clusters	
Cluster1	cap-shape, cap-color, gill-color, stalk-surface-above-ring, stalk-surface-below-ring, population, habitat
Cluster2	cap-shape, cap-surface, cap-color, odor, stalk-surface-above-ring, stalk-surface-below-ring, stalk-color-above-ring, stalk-color-below-ring, habitat
Cluster8	cap-shape, cap-surface, cap-color, gill-color, spore-print-color, population, habitat
Cluster9	cap-shape, cap-surface, cap-color, gill-color, stalk-color-above-ring, stalk-color-below-ring, population, habitat
Edible Clusters	
Cluster4	cap-shape, gill-color, veil-color, spore-print-color, population
Cluster5	cap-shape, cap-surface, cap-color, gill-color, stalk-surface-above-ring, stalk-surface-below-ring, spore-print-color, population
Cluster6	cap-shape, cap-surface, cap-color, odor, gill-color, spore-print-color
Cluster7	cap-shape, cap-surface, cap-color, gill-color, stalk-color-above-ring, stalk-color-below-ring, spore-print-color, population
Cluster10	cap-shape, cap-surface, cap-color, odor, gill-color, spore-print-color, population, habitat
Cluster12	cap-shape, cap-color, odor, gill-color, spore-print-color, population, habitat
Cluster14	cap-shape, cap-surface, cap-color, gill-color, stalk-surface-above-ring, stalk-surface-below-ring, population

4.4 Cluster Description

Let us consider Cluster1 for pattern formulation. We remove the reduct attributes of Cluster1(Ref. Table 2)(cap-shape, cap-color, gill-color, stalk-surface-above-ring, stalk-surface-below-ring, population, habitat). Cluster1 is left with remaining descriptors (cap-surface=s, bruises=t, odor=f, gill-attachment=f, gill-spacing=c, gill-size=b, stalk-shape=t, stalk-color-above-ring=w, stalk-color-below-ring=w, veil-color=w, ring-number=o, ring-type=p, spore-print-color=h) having same value for all the entities within this cluster. We then calculated PE of these descriptors to find out the major contributing descriptors. Let us consider calculation of PE (Equ.4) for descriptor (cap-surface=s) in Cluster1. Descriptors cap-surface=s has support of 2556 entities in the dataset, out of which Cluster1 has support of 288 entities. PE is defined as number of false positive for that descriptor divided by the total number of entities outside that cluster.

PE (cap-surface=s) = (2556-288)/ (8124-288) = .2894

Table 3 and Table 4 shows the descriptors along with value of PE for pure edible and poisonous clusters respectively.

Pattern generation for Cluster1 involves conjunction of three descriptors spore-print-color=h ∧ odor=f ∧ cap-surface=s (Ref. Table 4) for describing the Cluster

Table 3. PE for descriptors in edible clusters

Cluster5	gill-spacing=w(.0739), habitat=g(.1876), ring-type=e(.2729), odor=n(.3752), stalk-color-below-ring=w(.4915), stalk-color-above-ring=w(.5024), stalk-shape=t(.5220), bruises=f(.5410), gill-size=b(.6585), ring-number=o(.9135), gill-attachment=f(.9714), veil-color=w(.9728)
Cluster6	gill-spacing=w(.1514), habitat=d(.3801), gill-size=n(.3009), bruises=t(.4085), ring-type=p(.4823), population=v(.4912), stalk-color-below-ring=w(.5341), stalk-color-above-ring=w(.5440), stalk-shape=t(.5620), stalk-surface-below-ring=s(.6028), stalk-surface-above-ring=s(.6327), veil-color=w(.9750), ring-number=o(.9207) , gill-attachment=f(.9738)
Cluster7	habitat=d(.2220), bruises=t(.2576), odor=n(.2814), ring-type=p(.3502), stalk-shape=t(.4502), gill-attachment=f(.9671), stalk-surface-below-ring=s(.5015), stalk-surface-above-ring=s(.5390), gill-size=b(.6072), gill-spacing=c(.7948), ring-number=o(.9005), veil-color=w(.9687)
Cluster10	bruises=t(.3763), stalk-shape=e(.3947), ring-type=p(.4540), stalk-color-below-ring=w(.5087), stalk-color-above-ring=w(.5192), stalk-surface-below-ring=s(.5812), stalk-surface-above-ring=s(.6127), gill-size=b(.6700), gill-spacing=c(.8276), ring-number=o(.9164), gill-attachment=f(.9724), veil-color=w(.9737)
Cluster12	stalk-surface-below-ring=y(.0115), cap-surface=y(.3847), bruises=t(.4014), stalk-shape=e(.4190), ring-type=p(.4760), stalk-color-below-ring=w(.5284), stalk-color-above-ring=w(.5385), stalk-surface-above-ring=s(.6283), gill-size=b(.6833), gill-spacing=c(.8345), ring-number=o(.9198), gill-attachment=f(.9735), veil-color=w(.9747)
Cluster14	ring-number=t(.0398), gill-spacing=w(.1306), habitat=g(.2373), spore-print-color=w(.2679), stalk-shape=e(.4119), odor=n(.4134), ring-type=p(.4696), stalk-color-below-ring=w(.5227), stalk-color-above-ring=w(.5329), bruises=f(.5691), gill-size=b(.6794), gill-attachment=f(.9732), veil-color=w(.9744)

Table 4. PE for descriptors in poisonous clusters

Cluster1	spore-print-color=h(.1715), odor=f(.2390), cap-surface=s(.2894), bruises=t(.3940), ring-type=p(.4696), stalk-color-below-ring=w(.5227), stalk-color-above-ring=w(.5329), stalk-shape=t(.5513), gill-size=b(.6794), gill-spacing=c(.8325), ring-number=o(.9188), gill-attachment=f(.9732), veil-color=w(.9744)
Cluster2	gill-color=b(0), spore-print-color=w(.1031), gill-size=n(.1225), ring-type=e(.1638), population=v(.3614), stalk-shape=t(.4502), bruises=f(.4721), gill-spacing=c(.7948), ring-number=o(.9005), gill-attachment=f(.9671), veil-color=w(.9687)
Cluster8	odor=p(0), gill-size=n(.2867), bruises=t(.3965), stalk-shape=e(.4143), ring-type=p(.4717), stalk-color-below-ring=w(.5246), stalk-color-above-ring=w(.5348), stalk-surface-below-ring=s(.5948), stalk-surface-above-ring=s(.6253), gill-spacing=c(.8332), ring-number=o(.9191), veil-color=w(.9745), gill-attachment=f(.9733)
Cluster9	ring-type=l(0), spore-print-color=h(.0492), odor=f(.1266), stalk-surface-below-ring=k(.1476), stalk-surface-above-ring=k(.1575), stalk-shape=e(.3251), bruises=f(.5055), gill-size=b(.6321), gill-spacing=c(.8078), ring-number=o(.9068), gill-attachment=f(.9692), veil-color=w(.9707)

with zero PE. Similarly, for Cluster2 (Ref. Table 4), only one descriptor gill-color=b generate zero PE and hence this alone describes the cluster.

4.5   Results

Cluster description with RCD approach resulted in short patterns with **zero PE** for pure edible and poisonous clusters.

Pattern obtained with RCD for poisonous clusters are:

Cluster1 (288 entities): *spore-print-color=h $\wedge$ odor=f $\wedge$ cap-surface=s*; L(P)=3.

Cluster2 (1728 entities): *gill-color=b*; L(P)=1.

Cluster8 (256 entities): *odor=p*; L(P)=1.

Cluster9 (1296 entities): *ring-type=l*; L(P)=1.

Pattern obtained with RCD for edible clusters are:

Cluster4 (192 entities): *stalk-color-above-ring=o or stalk-color-below-ring=o*; L(P)=1.

Cluster5 (768 entities): *gill-spacing =w $\wedge$ habitat=g $\wedge$ ring-type=e*; L(P)=3.

Cluster6 (96 entities): *gill-spacing=w $\wedge$ gill-size=n $\wedge$ habitat=d $\wedge$ bruises=t*; L(P)=4.

Cluster7 (1728 entities): *habitat=d $\wedge$ bruises=t $\wedge$ odor=n*; L(P)=3.

Cluster10 (511 entities): *bruises=t $\wedge$ stalk-shape=e $\wedge$ ring-type=p $\wedge$ stalk-surface-below-ring=y $\wedge$ gill-size=b $\wedge$ ring-number=o*; L(P)=6.

Cluster12 (192 entities): *stalk-surface-below-ring=y $\wedge$ cap-surface=y $\wedge$ bruises=t*; L(P)=3.

Cluster14 (288 entities): *ring-number=t $\wedge$ gill-spacing=w*; L(P)=2.

5   Conclusion

Reduct driven approach for cluster description (RCD) is presented in this paper on benchmarking dataset. Reduct along with Precision Error has resulted in formulation of significant and user understandable patterns from clusters. It is observed that patterns obtained with RCD, distinctively described the clusters with no errors. Patterns obtained is of short length hence easily understandable to users. On average two to three attributes are used to describe the clusters. To confirm the existence of relation, future research will be focused on applying the same approach on real time benchmarking datasets.

References

1. Abidi, S.S.R., Goh, A.: Applying knowledge discovery to predict infectious disease epidemics. In: Lee, H.-Y., Motoda, H. (eds.) PRICAI 1998. LNCS, vol. 1531, pp. 170–181. Springer, Heidelberg (1998)
2. Ganter, B., Wille, R.: Formal Concept Analysis: Mathematical Foundations. Springer, New York (1997)
3. Jain, A.K., Murty, M.N., Flynn, P.J.: Data Clustering: A review. ACM Computing Surveys 31(3), 264–323 (2001)
4. Jain, R.: Rough Set based Decision Tree Induction for Data Mining, Ph.D Thesis. Jawaharlal Nehru University, New Delhi, India (2004)
5. Komorowski, J., Pawlak, Z., Polkowki, L., Skowron, A.: Rough Sets: A Tutorial. In: Pal, S.K., Skowron, A. (eds.) Rough Fuzzy Hybridization, pp. 3–99. Springer, Heidelberg (1999)
6. Michalski, R.S., Stepp, R.E.: Clustering. In: Shapiro, S.C. (ed.) Encyclopedia of artificial intelligence, New York. J. Wiley & Sons, Chichester (1992)
7. Mirkin, B.: Concept Learning and Feature Selection based on Square-Error Clustering. Machine Learning 35, 25–40 (1999)
8. Mirkin, B.: Clustering for Data Mining: Data Recovery Approach. Chapman & Hall/CRC, Boca Raton (2005)
9. Murphy, P.M.: UCI repository of machine learning databases. University of California, Irvine, http://www.ics.uci.edu/~mlearn/
10. Nguyen, S.H., Nguyen, T.T., Skowron, A., Synak, P.: Knowledge discovery by Rough Set Methods. In: Proc. of the International Conference on Information Systems Analysis and Synthesis, Orlando, USA, pp. 26–33 (1996)
11. Pawlak, Z.: Rough Sets: Theoretical Aspects of Reasoning About Data. Kluwer Academic Publisher, Dordrecht (1991)
12. Rosetta, http://www.rosetta.com
13. WEKA: A Machine Learning Software, http://www.cs.waikato.ac.nz/~ml/
14. Wroblewski, J.: Finding Minimal Reduct using Genetic Algorithms. Warsaw University of Technology, Institute of Computer Science Reports-16/95 (1995)

Experiments with Rough Set Approach to Face Recognition

Xuguang Chen and Wojciech Ziarko

Department of Computer Science University of Regina
Regina, SK, S4S 0A2, Canada

Abstract. The article reports our experiences with the application of the hierarchy of probabilistic decision tables to face recognition. The methodology underlying the classifier development for our experiments is the variable precision rough sets, a probabilistic extension of the rough set theory. The soft-cut classifier method and the related theoretical background, the feature extraction technique based on the principal component analysis and the experimental results are presented.

1 Introduction

Face recognition is an important research area with numerous potential applications, most notably in security. According to [12], face recognition methods can roughly be classified into the three categories:

(1)*Feature-based matching methods*, in which local features such as based on eyes, nose, and mouth are firstly extracted and then their locations and local information such as geometric characteristics and appearance are input into a structural classifier for recognition.

(2)*Holistic matching methods*, in which the information about whole face region will be input into a recognition system (a classifier). One of the most widely used techniques is to represent the face region as eigenfaces based on principal component analysis (PCA) [5][6].

(3)*Hybrid methods*, in which both local features and the whole face region are used for face recognition.

Many face recognition techniques applying PCA have been developed in past years, but PCA cannot guarantee that the selected principal components are the most adequate features for face recognition. One of possible solution for selecting most adequate discriminative features is to apply rough set theory [1]. That is, rough set theory is applied to select the best features from the principal components generated by PCA [6].

In this paper, we present a face representation and classification methodology, called soft-cut classifier, based on merging PCA and rough sets theory. Its basic idea for feature extraction is similar with that of Turk and Pentland [5], but different techniques, especially on how to classify a test photo into an appropriate category, have been applied. The techniques, which are introduced in sections 3-4, involve developing a hierarchy of learnt decision tables based on accumulated

C.-C. Chan et al. (Eds.): RSCTC 2008, LNAI 5306, pp. 399–408, 2008.
© Springer-Verlag Berlin Heidelberg 2008

training data (pictures of faces). The process of building, analysis and evaluation of such decision tables involves rough set theory [1], in particular the variable precision model of rough sets (VPRSM) [2]. The hierarchy is subsequently used for classification of previously unseen pictures of faces. Section 5 describes the experimental procedure, and section 6 presents some experimental results.

2   Techniques for Feature Selection

In this section, we discuss the feature value acquisition methods used for forming representation of training face pictures and for recognition, adopting existing standard techniques of principal component analysis (PCA) and of Harr wavelets.

2.1   Principal Component Analysis (PCA)

Principal component analysis (PCA) is a technique that can reduce multidimensional data sets to lower dimensions for analysis. PCA can be applied to various fields including face recognition [5].

For a data set of N samples, each of which is n-dimensional and denoted as x, we assume that a training data set $T = x^1, x^2, \cdots, x^N$ can be represented as an $N \times n$ data pattern matrix $X = [x^1, x^2, \cdots, x^N]^T$. Such a training set can be characterized by the $n \times n$ dimensional covariance matrix R_x. Then, we arrange the eigenvalues of the covariance matrix R_x in the decreasing order $\lambda_1 \geq \lambda_2 \geq \cdots \geq \lambda_N \geq 0$ with the corresponding orthonormal eigenvectors $e^1, e^2, \cdots, e^n$. Using the $m \times n$ optimal Karhunen-Love transformation matrix denoted as $W_{KLT} = [e^1, e^2, \cdots, e^m]^T$, each sample x in the data set can be transformed into

$$y = W_{KLT}x \tag{1}$$

where $m \leq n$. In this way, the optimal matrix W_{KLT} will transform the original pattern matrix X into dimension-reduced feature pattern matrix Y as

$$Y = (W_{KLT}X^T)^T = XW_{KLT}^T. \tag{2}$$

PCA can extract the features and reduce the dimensions by forming the m-dimensional feature ($m \leq n$) vector y that has only the first m most dominant principal components of x.

2.2   Harr Wavelets

A wavelet is a mathematical function used to divide a given function into different frequency components so as to study each component with a resolution matching its scale. By using wavelets, a photo can be transformed from pixel space into the space of wavelet coefficients, which can have some consistency throughout the photo class and while ignoring noise [10]. The wavelet vector spaces form the foundations of the concept of a multi-resolution analysis, which can be formalized as the sequence of approximating subspaces $V^0 \subset V^1 \subset V^2 \cdots \subset V^j \subset V^{j+1} \cdots$. In the sequence, the vector space V^{j+1} describes finer details than the space V^j. As a basis for the space V^j, the following scaling functions can be used

$$\phi_i^j = \sqrt{2^j}\phi(2^j x - i) \qquad i = 0, 1, \cdots, 2^j - 1 \qquad (3)$$

where the Harr Wavelet can be expressed as:

$$\phi(x) = \begin{cases} 1 & 0 \le x < 1 \\ 0 & otherwise \end{cases} \qquad (4)$$

Correspondingly, the vector W^j describes the subspace of details in increasing refinements. It is orthogonal complement of two consecutive approximating subspaces, $V^{j+1} = V^j \oplus W^j$. As a basis for the wavelet space W^j, the following functions can be used

$$\psi_i^j = \sqrt{2^j}\psi(2^j x - i) \qquad i = 0, 1, \cdots, 2^j \qquad (5)$$

where the Harr Wavelet can be represented

$$\psi(x) = \begin{cases} 1 & 0 \le x < 0.5 \\ -1 & 0.5 \le x < 1 \\ 0 & otherwise \end{cases} \qquad (6)$$

Two-dimensional wavelet transform can be obtained by taking the tensor product of two one-dimensional wavelet transforms [10]. The results are three types of wavelet basis functions, which are $\psi(x, y) = \psi(x) \oplus \phi(y)$, $\psi(x, y) = \phi(x) \oplus \psi(y)$, and $\psi(x, y) = \psi(x) \oplus \psi(y)$.

3   Decision Table-Based Approach

Our approach to face classification and recognition involves machine learning from training data representing photographs of faces. The end-result of the learning process is a linear hierarchy of decision tables, which subsequently is used for the purpose of recognition. The automated construction of the hierarchy of decision tables is based on the VPRSM. This probabilistic approach to rough sets is also used for the evaluation of generated decision tables and their hierarchies, especially for determination of dependencies between attributes and for their optimization.

3.1   Variable Precision Rough Sets

The rough set theory was introduced by Pawlak [1], and the variable precision model of rough sets broadens its basic ideas. In the VPRSM, conditional probabilities and prior probability $P(X)$ of the set X in the universe U, are used to represent set X approximation defining criteria. Two model precision-control parameters are used, denoted as the *lower limit* l, and the *upper limit* u, respectively. With these two parameters, the original rough set theory definitions of the negative region, the positive region, and the boundary region of a set are extended as follows.

The *lower limit* l, $0 \leq P(X) < 1$ represents the highest acceptable degree of the conditional probability $P(X|E)$ to include the elementary set E in the negative region of the set X. The negative region in VPRSM is defined as

$$NEG_l(X) = \cup\{E : P(X|E) \leq l\} \qquad (7)$$

Objects are classified into the negative region of the set X if the probability of the membership in the set X is significantly lower, as expressed by the lower limit, than the prior probability $P(X)$. The *upper limit* u, $0 < P(X) < u \leq 1$, represents the least acceptable degree of the conditional probability $P(X|E)$ to include the elementary set E in the positive region of the set X. The positive region in VPRSM is defined as

$$POS_u(X) = \cup\{E : P(X|E) \geq u\} \qquad (8)$$

Objects are classified into the positive region of the set X if the probability of the membership in the set X is significantly higher, as expressed by the lower limit, than the prior probability $P(X)$. The objects that are not classified into either the positive region or the negative region are classified into the boundary region of the decision category X. The boundary region in VPRSM is defined as

$$BND_{l,u}(X) = \cup\{E : l < P(X|E) < u\} \qquad (9)$$

3.2   Hierarchies of Probabilistic Decision Tables

The probabilistic decision tables and their hierarchies extend the notion of decision table acquired from data introduced by Pawlak [1]. The probabilistic decision table approximately represents the stochastic relation between condition and decision attributes via a set of uniform size probabilistic rules. The probabilistic decision table is a mapping that assigns each vector of condition attribute values, corresponding to an elementary set E, to its unique designation of one of VPRSM approximation regions $POS_u(X)$, $NEG_l(X)$ or $BND_{l,u}(X)$, along with associated elementary set E probabilities $P(E)$ and conditional probabilities $P(X|E)$. They can be conveniently represented in a tabular form.

In the VPRSM, the boundary region is a definable subset of the universe U, that is, it can be precisely specified by its elementary sets. To construct the hierarchy of decision tables, let us denote the boundary region as $U' = BND_{l,u}(X)$. The basic idea behind the hierarchies of probabilistic decision table construction is to treat the boundary region as a sub-universe of the universe U that is completely independent from the universe U. Such a sub-universe can have its own "private" collection of condition attributes, denoted as C', to form a new approximation sub-space, from which the "child" decision table can be derived. By repeating the step of parent-child decision table formation recursively, until the boundary region is eliminated, or some other attribute-based termination criteria are satisfied, a hierarchy of probabilistic decision tables will be formed. In our experiments, a separate hierarchy of decision tables was developed for each face to be distinguished, corresponding to the decision classes X and $\neg X$. The generated decision tables and their hierarchies need to be evaluated with respect to their expected performance as classifiers. For that purpose, two dependency measures, called the γ-dependencies and the λ-dependencies [3] respectively, were adopted in our experiments.

4   Forming Hierarchies of Probabilistic Decision Tables

The attributes of the probabilistic decision tables in the hierarchy were formed with coefficients of several levels of 2-dimensional Harr wavelets transformation. In our experiments, a hierarchy having up to five probabilistic decision tables was built, and their attributes were separately formed from the coefficients of various levels of the Harr wavelet transformation of each photo in the training set.

The Harr wavelet transformation can provide many useful features, but it is still hard to tell how powerful these features are for face recognition. Moreover, these features are too numerous, and a lot of redundant information is included, so PCA was applied to choose the most useful Harr-based features. If each photo has n Harr-based coefficients, represented as $x^i_{harr,n}$ and there are N photos in the training data set, then these photos can be represented by an $N \times n$ pattern matrix

$$X = [x^1_{harr,n}, x^2_{harr,n}, \cdots, x^N_{harr,n}]^T \tag{10}$$

For each photo, formula (1) was applied to transform its original features $x^i_{harr,n}$ from n-dimensional Harr-based coefficients into m-dimensional ($m \leq n$) PCA feature patterns, and for the whole training data set, formula (2) was applied. PCA can significantly reduce the size of features to retain the most important information for face recognition, but the question which principal components are the best for face recognition remains unresolved [6]. In our experiments, we dealt with that problem by applying rough set theory.

Before a probabilistic decision table in the hierarchy was formed, the principal components selected by PCA in previous step needed to be converted from real-valued components into binary-valued components.

As the first step in the discretization, a threshold τ satisfying $0.5 < \tau \leq 1$ was defined for each probabilistic decision table in the hierarchy . Each selected principal component was then transformed according to the following *sigmoid* function formula:

$$f'(x) = \frac{1}{1 + e^{a(c-x)}} \tag{11}$$

where x is the real value of that selected principal component, a is a parameter, the values of which would be identified heuristically, and c is the arithmetic average of that selected principal component of all photos in the training set. The transformation defines a *soft cut* in place of binary cut to allow for accommodation of situations, during discretization and recognition stages, when the feature value is close the cut and possibly affected by random noise.

Each selected principal component was discretized according to the comparison result between $f'(x)$ and the threshold τ as below:

1. if $f'(x) \geq \tau$, then the selected principal component is assigned 1 as its value;
2. if $f'(x) \leq 1 - \tau$, then 0 is assigned;
3. if $1 - \tau < f'(x) < \tau$, then no assignment is made.

The basic idea of our *soft cut* is similar with the one in [7]. It splits the real axis of each dimension into three intervals. Only those principal components that can classified into two of specific intervals are discretized. On the other hand, unlike [7], it does not attempt to find a threshold value so that all of principal components can be discretized.

Our way to deal with those principal components that cannot be discretized is similar with that of support vector machine (SVM)[8]. The SVM model creates a soft margin that permits some misclassifications, and has a cost parameter, C, that controls the rate of misclassifications. In our method, those principal components that cannot be discretized are considered as misclassified points. That is, if a photo had one, or more than one selected principal components that could not be discretized, it was automatically classified into the boundary area and considered again when working on the photos in the boundary area to build next probabilistic decision table in the hierarchy. After the discretization was completed, only those photos, the selected principal components of which have been completely discretized, were evaluated by rough sets theory. The purpose was to find a group of principal components that are the most adequate for the recognition task. According to the formula in [3], a group of discretized principal components, the ones that generate the highest λ-dependency was heuristically selected and eventually used as condition attributes for the probabilistic decision table.

In order to avoid overfitting, the following strategies were applied. Firstly, we tried to limit the size of the group of discretized principal components to a reasonable number. In practice, we set that the size must be less than 50 percent of the dimension of PCA feature patterns. For example, if its dimension was 48, the size of the group must be less 24 discretized principal components. Moreover, if two groups of discretized principal components had similar λ-dependency, the group with fewer components would be selected. When training and testing the system, the corresponding data sets were only constructed based on the method of holdout validation and the method of K-fold cross-validation.

Subsequently, photos were classified into elementary sets based on the constructed attributes. The elementary sets were then assigned to rough approximation regions: the positive region, the negative region, and the boundary region based on the formulas (1-3). The above process was recursively repeated on photos classified into the boundary area to build next layer probabilistic decision table in the hierarchy, and so on. The process described above was continued until all photos were classified into either the positive area or the negative area of a probabilistic decision table or all of Harr-wavelet coefficients have been utilized. The end-result of this process was a hierarchy of probabilistic decision tables based on photos from the training set.

5   Distance-Based Classification of Objects Using Decision Tables

The process of classification of previously unseen objects (photos), for the purpose of recognition, employs a technique based on the evaluation of the distance between an object and an elementary set (see [9] for review of related methods). This is a departure from the standard technique involving exact match between the feature vectors representing objects and elementary sets. Our approach is motivated by the need to "soften" the matching procedure in order to ignore small differences between compared patterns, possibly caused by noise, which would result in many unclassified test objects.

When a test photo (an unseen object) is input, it is first transformed by the Harr wavelet transformation and the PCA. Then, each of its selected principal components i is processed by the formula (11), with the result denoted as $f_i'(x)$. The real-valued vector $(f_i'(x))_{i=1,2,...n}$ is then compared to binary-valued vectors $(att_i^E(x))_{i=1,2,...n}$ of all elementary sets E of a probabilistic decision table based on the distance function $d(x, E)$ as follows:

$$d(x, E) = MAX_{i=1,2,...n}(|f_i'(x) - att_i^E(x)|), \tag{12}$$

where $||$ is the absolute value function.

An object is classified into an elementary set E if it satisfies the following two conditions:

1. the distance between the elementary set and the tested photo is at minimum among all elementary sets;
2. the distance is no higher than the value of a predefined threshold;

The recognition is based on the rough region location of the lowest distance elementary set, starting with the top layer of the hierarchy of decision tables. For a hierarchy with k decision tables, assume that each decision table has n rows with m condition attributes, we need $O(knm)$ operations to classify unseen objects as the worst case. If such an elementary set is located in the positive area, then the positive recognition is made, i.e. it is assumed that the test object x belongs to the decision class X. Similarly, if the elementary set is located in the negative area, then the negative recognition is made, i.e. it is assumed that the test object x does not belong to the decision class X. If the closest-match elementary set is located in the boundary region or cannot be classified into any elementary set, then no recognition is made. In this case, the process passes to the next layer decision table and the procedure is repeated until either a test photo is classified into the positive area (or the negative area) of a probabilistic decision table, or all of probabilistic decision tables in the hierarchy have been checked. In the latter case, no decision is produced.

6   Experimental Results

To evaluate the performance of soft-cut classifier in different situations, a series of experiments was performed. Facial photos of 10 men and 11 women were selected from the AR Face Database [11]. The photos comprise 756 experimental sets. Each of the sets consists of 72 facial photos of two persons, 36 facial photos for each one. The experimental sets were divided into a training set(48 photos totally, 24 photos for each person) and a test set (24 photos totally, 12 photos for each person).

The first experiment was to test the overall performance of the soft-cut classifier when the number of selected principal components was varied. The objective was to check if the overall performance of the classifier would improve with the increase of the number of selected principal components. In this experiment, the soft-cut classifier was repetitively trained by the same training sets and tested by the same test sets, but each time, the number of selected principal components was varied. Totally 756 experimental sets (756 training sets and 756 test sets) were used for this experiment. Based on the experimental results, we concluded that as a general rule, the more principal components were selected, the more test sets produced the highest accuracy rate $R \geq 0.9$. For example, as demonstrated in Table 1, when 10 principal components were selected, there were 431 test sets in the accuracy range $R \geq 0.9$ among total of 756 test sets. If 24 principal components were selected, the number of test sets the accuracy rate of which was in the range $R \geq 0.9$ was 518. It should be noted, however, that when working on a specific data set, the performance of the soft-cut classier will also depend on other factors, for example on the threshold value, as described in the third and fourth experiments.

The second experiment was to compare the overall performance of soft-cut classifier with that of the nearest-neighbor (N-N) classifier [5]. For this purpose, a N-N classifier was implemented first. These two classifiers were then trained and tested on the same data sets (756 training sets and 756 test sets) and the same selected principal components.

Based on the results, we found that when working on our data sets with a fixed number of selected principal components, the soft-cut classifier performed significantly better than the N-N classifier. Moreover, when the number of principal components was increased, the performance of soft-cut classifier would improve faster than of the N-N classifier. That is, the number of test sets with a higher accuracy rate (for instance, in the range $R \geq 0.9$) would be significantly greater than that of N-N classifier. For example, if 22 principal components were selected, there were 512 test sets among total 756 test sets with an accuracy rate $R \geq 0.9$ for soft-cut classifier versus 387 test sets for N-N classifier, which constitutes 32% improvement.

The third experiment was to test the overall performance of the soft-cut classifier when varying the threshold τ and the parameter a of the formula (11). We found out that when τ is very close to 0.5, the performance of the soft-cut classier will be the best. In this case, the classifier can work with all of 756 training/test sets. When τ is far away from 0.5, the performance of the soft-cut

Table 1. Accuracy Rate Versus Number of Principal Components

Number of Selected Principal Components	Number of Test Sets			
	R≥90%	80%≤ R <90%	70%≤ R <80%	R<70%
9	413	151	111	81
10	431	153	97	75
12	464	136	90	66
15	507	96	102	51
17	479	129	94	54
20	503	121	84	48
22	512	117	73	54
24	518	112	78	48

classier deteriorates. In the worse case, for example when $\tau = 0.65$, the soft-cut classier was not able to discretize any photo for some training sets. That is, there was no photo in certain training sets, that the selected principal components of which could be completely discretized. As for the parameter a, its value had no greater impact on the performance.

The last experiment was to identify the relationship between the number of selected principal components and the threshold value when the soft-cut classifier was working on a specific training/test set. In particular, the goal of this experiment was to check if a specific threshold value could significantly reduce the number of selected principal components and while preserving the higher accuracy rate. Based on the results, we found that number of selected principal components can be reduced for some specific threshold values. However, the process of choosing the number of eigenvectors and threshold values is a heuristic one, making it difficult to arrive at any general threshold optimization rule. According to the results from the third experiment, when τ is very close to 0.5, the performance of the soft-cut classier was the highest. Thus, the following heuristic rule can be formulated:

For a specific data set, different combinations of selected principal components and the threshold values that are around 0.5 such as $0.5 < \tau \leq 0.7$ should be tried first. Then, select such a combination which results in minimum number of selected principal components and the maximum accuracy rate. For example, the accuracy rate is 100% for several different combinations of τ and principal components. Therefore, the combination of $\tau = 0.540001$ with the minimum number of selected principal components equal to nine would eventually be selected.

7   Final Remarks

The soft-cut classifier approach involves machine learning from training data representing photographs of faces to form a linear hierarchy of decision tables, which subsequently is used for the purpose of recognition.

In order to evaluate its performance in different situations, we completed a series of experiments. Based on the experimental results, we observed that the soft-cut classifier performed significantly better on our data sets than that of the N-N classifier, especially when a large number of principal components were selected for representing pictures. We also noticed that in the case of soft-cut classifier, as a general rule, the more principal components are selected, the more test sets can produce a higher accuracy rates. In addition, we observed that when working on a specific data set, the best solution for improving the performance of the soft-cut classier is to heuristically try different combinations of selected principal components and the soft cut threshold values.

Acknowledgment. The support of the Natural Sciences and Engineering Research Council of Canada in partial funding the research presented in this article is gratefully acknowledged.

References

1. Pawlak, Z.: Rough Sets: Theoretical Aspects of Reasoning About Data. Kluwer, Dordrecht (1991)
2. Ziarko, W.: Variable Precision Rough Sets Model. Journal of Computer and System Sciences 46(1), 39–59 (1993)
3. Ziarko, W.: Partition Dependencies in Hierarchies of Probabilistic Decision Tables. In: Wang, G.-Y., Peters, J.F., Skowron, A., Yao, Y. (eds.) RSKT 2006. LNCS (LNAI), vol. 4062, pp. 42–49. Springer, Heidelberg (2006)
4. Ziarko, W.: Probabilistic Rough Sets. In: Ślęzak, D., Wang, G., Szczuka, M.S., Düntsch, I., Yao, Y. (eds.) RSFDGrC 2005. LNCS (LNAI), vol. 3641, pp. 283–293. Springer, Heidelberg (2005)
5. Turk, M., Pentland, A.: Eigenfaces for Recognition. Journal of Cognitive Neuroscience 3(1), 71–86 (1991)
6. Swiniarski, R.: An Application of Rough Sets and Harr Wavelets to Face Recognition. In: Ziarko, W., Yao, Y. (eds.) RSCTC 2000. LNCS (LNAI), vol. 2005, pp. 562–568. Springer, Heidelberg (2001)
7. Nguyen, H.S.: On Exploring Soft Discretization of Continuous Attributes. Rough-Neural Computing Techniques for Computing with Words, pp. 333–350. Springer, Heidelberg (2004)
8. Burges, C.J.C.: A Tutorial on Support Vector Machines for Pattern Recognition. Data Mining and Knowledge Discovery 2(2), 121–167 (1998)
9. Ekin, O., Hammer, P.L., Kogan, A., Winter, P.: Distance-Based Classification Methods. INFOR 37, 337–352 (1999)
10. Papageorgiou, C., Poggio, T.: A Trainable System for Object Detection. International Journal of Computer Vision 38(1), 15–23 (2000)
11. Martinez, A.M., Benavente, R.: The AR Face Database. CVC Technical Report No. 24 (1998)
12. Zhao, W., Chellpappa, R., Philiips, P.J., Rosenfeld, A.: Face Recognition: A Literature Survey. ACM Computing Surveys 35(4), 399–458 (2003)

Standard and Fuzzy Rough Entropy Clustering Algorithms in Image Segmentation

Dariusz Małyszko and Jarosław Stepaniuk

Department of Computer Science
Bialystok University of Technology
Wiejska 45A, 15-351 Bialystok, Poland
{malyszko,jstepan}@wi.pb.edu.pl

Abstract. Clustering or data grouping presents fundamental initial procedure in image processing. This paper addresses the problem of combining the concept of rough sets and entropy measure in the area of image segmentation. In the present study, comprehensive investigation into rough set entropy based thresholding image segmentation techniques has been performed. Segmentation presents the low-level image transformation routine concerned with image partitioning into distinct disjoint and homogenous regions with thresholding algorithms most often applied in practical solutions when there is pressing need for simplicity and robustness. Simultaneous combining entropy based thresholding with rough sets results in rough entropy thresholding algorithm. In the present paper, new algorithmic schemes Standard $RECA$ (**R**ough **E**ntropy **C**lustering **A**lgorithm) and Fuzzy $RECA$ in the area of rough entropy based partitioning routines have been proposed. Rough entropy clustering incorporates the notion of rough entropy into clustering model taking advantage of dealing with some degree of uncertainty in analyzed data. Both Standard and Fuzzy $RECA$ algorithmic schemes performed usually equally robustly compared to standard k-means algorithm. At the same time, in many runs yielding slightly better performance making possible future implementation in clustering applications.

Keywords: Granular computing, image clustering, rough sets, entropy measure, rough entropy measure, fuzzy rough entropy measure.

1 Introduction

During last decades, growing research attention has been focused on data clustering as robust technique in data analysis. Clustering or data grouping describes important technique of unsupervised classification that arranges pattern data (most often vectors in multidimensional space) in the clusters (or groups). Patterns or vectors in the same cluster are similar according to predefined criteria, in contrast to distinct patterns from different clusters [3], [10]. Possible areas of application of clustering algorithms include data mining, statistical data analysis, compression, vector quantization and pattern recognition [3]. Image analysis is the area where grouping data into meaningful regions referred to as image

C.-C. Chan et al. (Eds.): RSCTC 2008, LNAI 5306, pp. 409–418, 2008.

segmentation) presents the first step into more detailed routines and procedures in computer vision and image understanding.

In practical applications, most often image regions do not depict well-defined homogeneous characteristics, so it seems naturally appropriate to use techniques that additionally incorporate the ambiguity in information for performing the thresholding operation. In recent years [8], the theory of rough sets has gained considerable importance with numerous applications in diverse areas of research, especially in data mining, knowledge discovery, artificial intelligence and information systems analysis. Combination of thresholding methods with rough set theory has been attempted in [7], [4]. The authors minimize the roughness value in order to perform image thresholding by optimizing an entropy measure, which they refer to as the "rough entropy of image". Incorporation of fuzzy set based methodology and rough sets based methodologies has become a technique that is attracting much research attention owing their more flexible representation of clusters. Both techniques taken separately or combined into the same algorithmic solutions should handle uncertainty related to analyzed data and their incompleteness and imprecision with more accurateness. In clustering domain much effort has been put into extensions of fuzzy and rough theory into data clustering routines. Fuzzy and rough setting has been practically employed into such algorithmic schemes as k-means clustering, Kohonen self-organizing maps, evolutionary unsupervised learning and support vector clustering.

In Section 2 introductory information of proposed solution together with explanation of basic notions is given. In Subsection 2.1, review of existing segmentation techniques has been provided. Selected rough entropy concepts are explained in Subsection 2.2. In Subsection 2.3 evolutionary algorithms are described. Rough entropy based segmentation algorithms together with proposed Standard *RECA* and Fuzzy *RECA* algorithms are outlined in Subsections 3.1 and 3.2. Experimental setup and results are given in Section 4. Conclusions and further research are finally shortly outlined.

2   Basic Notions

2.1   Image Segmentation Methods

Segmentation operation is essential and extremely important preprocessing step in the majority of image analysis based routines such as computer vision with practical applications ranging from object extraction and detection, change detection, monitoring and identification tasks. After preprocessing stage of image handling routines, with for example noise removal, smoothing, and sharpening of image contrast, follows image segmentation step, and subsequently more specific, high-level analysis is performed such as depicting objects and regions, and final interpretation of the image or scene. In almost all areas, the quality of segmentation step determines the quality of the final image analysis output. Segmentation process is defined as an operation of image partitioning into some non-overlapped regions such that each region exhibits homogeneous properties and no two adjacent regions are homogeneous. Segmentation routines present exact partitioning of input image into distinct, homogenous regions (by means of intensity, color, texture or other relevant features). Segmentation is the standard image partitioning process that results in determining and creation of disjoint and homogeneous image regions. Regions resulting from the image segmentation according to [2] should be uniform and homogeneous with respect to some characteristics, regions interiors should be simple and without many small holes, adjacent regions should be significantly different with respect to the uniformity characteristics and each segment boundary should be comparatively simple and spatially accurate.

Unsupervised segmentation and supervised segmentation (with further classification) starts by creating partition of the image data into groups by means of defining similarity measure, which values for image data are then compared and on that basis image data partitioning follows. Image segmentation routines are divided into: histogram based routines, edge-based routines, region merge routines, clustering routines and some combination of the above routines. Exhaustive overview of the segmentation methods is available in [2].

Additionally, many segmentation techniques make use of particular data analysis approaches such as neural networks, fuzzy computing, evolutionary computing, multiscale resolution techniques and morphological analysis. Into this framework-based segmentation approaches falls thresholding and clustering with rough entropy based segmentation quality measure that is the subject of this paper.

2.2   Rough Entropy

The intention of rough set theory is to approximate an imprecise concept in the domain of discourse by a pair of exact concepts, called the lower and upper approximations. The lower approximation is the set of objects definitely belonging to the vague concept, whereas the upper approximation is the set of object possibly belonging to the same. In this way, the value $roughness(X)$ of the roughness of the set X equal 0 means that X is crisp with respect to B, and conversely if the $roughness(X) > 0$ then X is rough (i.e., X is vague with respect to B). Detailed information on rough set theory is provided e.g. in [8], [9].

Entropy is a concept introduced and primarily used in the Second Law of Thermodynamics. Entropy measures the spontaneous dispersion of energy as a function of temperature. It was introduced into communications theory as the measure of the efficiency of the information transferred through a noisy communication channel. The mathematical definition of the entropy is

$$H = -\sum_{i=0}^{n} p_i \log(p_i)$$

where H is an entropy, p_i is the statistical probability density of an event i. Considering each image cluster as a set in image attribute domain, it is possible to calculate this set roughness. After calculating cluster roughness values for every cluster, each *roughness*(C) of cluster C is considered as system state and accordingly is calculated its system entropy roughness:

$$Rough\_entropy(C) = -\frac{e}{2} \times [roughness(C) \times \log(roughness(C))]$$

Resulting image rough entropy is the sum of all rough entropies of all clusters referred to further as rough entropy measure.

2.3   Evolutionary Algorithms

Investigation into application of evolutionary algorithms in k-means clustering based image segmentation routines is given in [6]. Chromosomes represent solutions consisting of centers of k clusters - each cluster center is a d-dimensional vector of values in the range between 0 and 255 representing intensity of gray or color component. Selection operation tries to choose the best suited chromosomes from parent population that come into mating pool and after cross-over and mutation operation create child chromosomes of child population. Most frequently genetic algorithms make use of tournament selection that selects into mating pool the best individual from predefined number of randomly chosen population chromosomes. This process is repeated for each parental chromosome. The crossover operation presents probabilistic process exchanging information between two parent chromosomes during formation of two child chromosomes. Typically, one-point or two-point crossover operation is used. According to [1] crossover rate 0.9 - 1.0 yields the best results. Mutation operation is applied to each created child chromosome with a given probability pm. After cross-over operation children chromosomes that undergo mutation operation flip the value of the chosen bit or change the value of the chosen byte to other in the range from 0 to 255. Typically mutation probability rate is set in the range 0.05 - 0.1 by [1]. Termination criterion determines when algorithm completes execution and final results are presented to the user. Termination criterion should take into account specific requirements. Most often termination criterion is that algorithm terminates after predefined number of iterations. Other possible conditions for termination of the k-means algorithms depend on degree of population diversity or situation when no further cluster reassignment takes place. In the present research, apart from rough entropy measure for evaluation of population solutions two additional measures are taken into account, Index-β and k-means based partition measure. These two measures are established clustering validation measures.

Quantitative Measure - Index-β

Index $-\beta$ measures the ratio of the total variation and within-class variation. Define n_i as the number of pixels in the i-th ($i = 1, 2, ..., K$) region form segmented image. Define X_{ij} as the gray value of j-th pixel ($j = 1, ..., n_i$) in the region i and $\overline{X}_i$ the mean of n_i values of the i-th region.

The index-b is defined in the following way

$$\beta = \frac{\sum_{i=1}^{k} \sum_{j=1}^{n_i} (X_{ij} - \overline{X})^2}{\sum_{i=1}^{k} \sum_{j=1}^{n_i} (X_{ij} - \overline{X}_i)^2}$$

where n is the size of the image and $\overline{X}$ represents the mean value of the image pixel attributes. This index defines the ratio of the total variation and the within-class variation. In this context, important notice is the fact that index-b value increases as the increase of k number.

Quantitative Measure - k-means partition measure

In case of k-means clustering schemes, the quality of clustering partitions is evaluated by calculating the sum of squares of distances of all points from their nearest cluster centers. This kind of measure is a good quantitative measure describing quantitatively clustering model. In this paper, this sum of squares of distances from cluster centers is further referred to as KM measure and applied in the assessment of $RECA$ and k-means experimental results. Values of KM measure should be minimized.

3   Rough Entropy Clustering Algorithms

3.1   Standard $RECA$ Algorithm

Proposed solution is an extension of rough entropy measure introduced in [7] into multiclass clustering domain. In Rough Entropy Clustering, initial cluster centers are selected. Number of clusters is given as input parameter. Algorithm does not impose any constraints on data dimensionality and is described in [5]. For each cluster center, two approximations are maintained, lower and upper approximation. For each object in universe (namely pixel data in image segmentation setting) as described in Algorithm 2, the closest cluster center is determined, and lower and upper approximations for that cluster are incremented by 1. Additionally, upper approximation of the clusters that are located within the distance not greater than threshold value from the closest cluster is incremented also by 1. After all data objects (image pixels) are processed, and lower and upper approximation for each cluster are determined, roughness value for each cluster is determined as described in Algorithm 2.

Algorithm 1. Standard $RECA$ Algorithm Flow

Data: Input Image
Result: Optimal Threshold Value
1. Create X population with $Size$ random N-level solutions (chromosomes)
repeat

> **forall** *chromosomes of X* **do**
> > calculate their rough entropy measure values $RECA$
> > Rough_Entropy_Measure;
> **end**
> create mating pool Y from parental X population ;
> apply selection, cross-over and mutation operators to Y population;
> replace X population with Y population ;

until *until termination criteria* ;

Algorithm 2. Standard *RECA* - calculation of cluster Lower and Upper Approximations, Roughness and Entropy Roughness

---

foreach *Data object D* **do**
  Determine the closest cluster center C_i for D
  Increment Lower(C_i)++ Increment Upper(C_i)++
  foreach *Cluster C_k not further then eps from D* **do**
    | Increment Upper(C_k)++
  end

for *l = 1* **to** *C(number of data clusters)* **do**
  | roughness(l) = 1 - [Lower(l) / Upper(l)];

for *l = 1* **to** *C(number of data clusters)* **do**
  | Rough_entropy = Rough_entropy $-\frac{e}{2} \times$ [roughness(l) $\times$ log(roughness(l));

Rough Entropy clustering incorporates the notion of rough entropy into clustering model. Rough entropy measure calculation of the cluster centers is based on lower and upper approximation generated by assignments data objects to the cluster centers. Roughness of the cluster center is calculated from lower and upper approximations of each cluster center. In the next step, rough entropy is calculated as sum of all entropies of cluster center roughness values. Higher roughness measure value describes the cluster model with more uncertainty at the border. Uncertainty of the class border should be possibly high as opposed to the class lower approximation. For each selected cluster centers, rough entropy measure determines quality measure or fitness value of this cluster centers.

In order to search thoroughly space of all possible class assignments for predefined number of cluster centers, evolutionary algorithm has been employed. Each solution in the evolutionary population, represented by chromosome, consists of N cluster centers. After calculation of rough entropy measure for this class centers, new mating pool of solution is selected from parental population, based on the rough entropy measure as fitness measure. Higher fitness measure makes selection of the solution into mating pool more probable. From the (parental) mating pool new child population is created by means of selection, cross-over, mutation operations. The procedure is repeated predefined number of times or stops when some other predefined criteria are met. Detailed algorithm flow with evolutionary processing is described in Algorithm 1.

Standard *RECA* algorithm presents algorithmic routine that determines cluster centers or representatives for given image data and further image clustering is based on image data assignment into their closest cluster center. In this way, in Standard *RECA* algorithm as in other partitioning data clustering algorithms performs search for optimal data cluster centers. *RECA* population consists of predefined number of solutions or chromosomes, that represent clusters centers in d-dimensional domain. Initial population of chromosomes is then iteratively evolving by means of selection, cross-over and mutation operations with clusters rough entropy measure as a fitness value for each chromosome.

Given a chromosome with N d-dimensional clusters, lower and upper approximation for each cluster is determined by means of separate analyzing distances of all data points from cluster centers. In this way, for each image data point, its closest data cluster is determined together with distances to all other cluster centers. If data point is located near to only one cluster, then this data point is assigned to lower and upper approximation for this closest data cluster. However, if data point is located close to more that one data center with the given threshold value, then this data point is assigned to upper approximations for these closest data centers. Threshold value determines maximal admissible difference of distances of data point and two clusters that is interpreted as a data point belonging to these two data clusters. After analyzing all data points and lower and upper approximation calculation for the given cluster centers, roughness for each data cluster is calculated and further roughness entropy is determined. Roughness entropy presents the measure of segmentation quality for evolutionary algorithm that manages on that base, consecutive population iterations. Algorithm flow is stopped after predefined termination criteria are met.

3.2 Fuzzy RECA Algorithm

Fuzzy $RECA$ algorithm presents fuzzy version of $RECA$ algorithm. . During computation of lower and upper approximations for the given cluster centers, fuzzy membership value is calculated and this fuzzy membership value added to lower and upper approximation. Fuzzy membership value is calculated by means of standard formula taking into account distances of the data point from each data centers.

In Fuzzy $RECA$ setting, for the given point, lower and / or lower approximation value is incremented not arbitrary by 1, but is increased by its membership value. In this way, fuzzy concept of belongings to overlapped classes has been incorporated. Taking into account fuzzy membership values during lower and upper approximation calculation, should more precisely handle imprecise information that imagery data consists of.

Experimental data for $RECA$ algorithm and Fuzzy $RECA$ algorithm are presented in subsequent section.

4 Experimental Results

Experiments have been carried out for image set of 2d *Lenna* images. The set of *Lenna* images consisted from four images standard image, and three convoluted images with the window 3x3 with operations. In this way, for *Lenna* images, six separate 2d images were created by creating pairs: Lenna $Std-Max$, $Std-Min$, $Std-Mean$, $Max-Min$ $Max-Mean$ and $Min-Mean$.

Lenna images - Standard RECA

In Table 2 index-β values of the solutions from 2d $RECA$ algorithm two runs for $R = 5$ ranges are presented together with k-means clusterings. For each

---

Algorithm 3. Fuzzy *RECA* Algorithm Flow

Data: Input Image
Result: Optimal Threshold Value
1. Create X population with *Size* random N-level solutions (chromosomes)
repeat

> **forall** *chromosomes of X* **do**
>> calculate their rough entropy measure values *RECA*
>> Fuzzy_Rough_Entropy_Measure;
>
> **end**
> create mating pool Y from parental X population ;
> apply selection, cross-over and mutation operators to Y population;
> replace X population with Y population ;

until *until termination criteria* ;

Algorithm 4. Fuzzy *RECA* - calculation of cluster Lower and Upper Approximations, Roughness and Fuzzy Entropy Roughness

---

foreach *Data object D* **do**
> Determine the closest cluster center C_i for D
> Increment Lower(C_i) by fuzzy_membership_value of D Increment Upper(C_i)
> by fuzzy_membership_value of D
> **foreach** *Cluster C_k not further then eps from D* **do**
>> Increment Upper(C_k) by fuzzy_membership_value of D
>
> **end**

for *$l = 1$* **to** *C(number of data clusters)* **do**
> roughness(l) = 1 - [Lower(l) / Upper(l)];

for *$l = 1$* **to** *C(number of data clusters)* **do**
> Fuzzy_Rough_entropy = Fuzzy_Rough_entropy $-\frac{e}{2} \times$ [roughness(l) $\times$
> log(roughness(l));

Table 1. Lenna gray 1D image - standard image and exemplary segmentation

experiment, independently measures β-index and KM are given separately for 2d k-means and 2d $RECA$. Solutions that yielded better results than 2d standard k-means algorithm based segmentations are bolded. In each experiment as input images pairs of images $Std - Min$, $Std - Max$, $Std - Mean$, $Min - Max$, $Min - Mean$, $Max - Mean$ have been taken as input.

Lenna **images - Fuzzy RECA**

In Table 3 index-β values of the solutions from Fuzzy 2d $RECA$ algorithm two runs for $R = 5$ ranges are presented together with k-means clusterings. For each experiment, independently measures β-index and KM are given separately for 2d k-means and 2d $RECA$. Solutions that yielded better results than 2d standard k-means algorithm based segmentations are bolded. In each experiment as input images pairs of images $Std - Min$, $Std - Max$, $Std - Mean$, $Min - Max$, $Min - Mean$, $Max - Mean$ have been taken as input.

Table 2. Quality Indices for 1D-1D Images *Lenna* for 2d $RECA$

Image Lenna	KM k-means	β k-means	KM RECA	β RECA
Std-Max	654471	10.32	661310	**10.33**
Std-Min	679808	11.48	658499	11.45
Std-Mean	512745	19.18	**505791**	19.16
Max-Min	828134	7.64	826195	7.49
Max-Mean	629493	11.93	**549895**	**11.95**
Min-Mean	606263	14.02	690796	13.30

Table 3. Quality Indices for 1D-1D Images *Lenna* for 2d Fuzzy $RECA$

Image Lenna	KM k-means	β k-means	KM RECA	β RECA
Std-Max	654471	10.32	710370	**10.33**
Std-Min	679808	11.48	698514	11.46
Std-Mean	512745	19.18	568150	19.18
Max-Min	828134	7.64	854481	**7.65**
Max-Mean	629493	11.93	651475	**11.95**
Min-Mean	606263	14.02	649731	14.02

5   Conclusions

In the present study, detailed investigation into standard rough entropy thresholding algorithm has been performed. In order to make rough entropy thresholding more robust, completely new approach into rough entropy computation has been elaborated. Proposed algorithm addresses the problem of extension of rough entropy thresholding into rough entropy based clustering scheme $RECA$ and Fuzzy $RECA$. In the present paper, new algorithmic scheme in the area of rough entropy based partitioning routines has been proposed. Standard $RECA$

and Fuzzy *RECA* clustering algorithm is based on the notion of rough entropy measure and fuzzy rough entropy measure of object space partitioning or universe. Rough entropy of data partitioning quantifies the measure of uncertainty generated by the clustering scheme, and this rough entropy relates to the border of the partitioning, so rough entropy measure should be as high as it is possible. Experiments on two different types of 2d images are performed and compared to standard 2d *k*-means algorithm. Results proved comparably equal or better performance of *RECA* and Fuzzy *RECA* algorithmic schemes. In this way, future research in the area of rough entropy based clustering schemes is possible with prospect of practical robust and improved clustering performance.

Acknowledgments

The research was supported by the grant N N516 069235 from Ministry of Science and Higher Education of the Republic of Poland. The present publication was also supported by the Bialystok Technical University Rector's Grant.

References

1. Hall, O., Barak, I., Bezdek, J.C.: Clustering with a genetically optimized approach. IEEE Trans. Evo. Computation 3, 103–112 (1999)
2. Haralick, R.M., Shapiro, L.G.: Image segmentation techniques. Computer Vision, Graphics, and Image Processing 29, 100–132 (1985)
3. Jain, A.K.: Fundamentals of Digital Image Processing. Prentic Hall, Upper Saddle River (1989)
4. Malyszko, D., Stepaniuk, J.: Granular Multilevel Rough Entropy Thresholding in 2D Domain, Intelligent Information Systems 2008. In: Proceedings of the International IIS 2008 Conference, pp. 151–160 (2008)
5. Malyszko, D., Stepaniuk, J.: Rough Entropy Clustering Algorithm (in review)
6. Malyszko, D., Wierzchon, S.T.: Standard and Genetic k-means Clustering Techniques in Image Segmentation. In: CISIM 2007, pp. 299–304 (2007)
7. Pal, S.K., Shankar, B.U., Mitra, P.: Granular computing, rough entropy and object extraction. Pattern Recognition Letters 26(16), 2509–2517 (2005)
8. Pawlak, Z., Skowron, A.: Rudiments of rough sets. Information Sciences 177(1), 3–27 (2007)
9. Stepaniuk, J.: Rough–Granular Computing in Knowledge Discovery and Data Mining. Springer, Heidelberg (2008)
10. Xu, R., Wunsch, D.: Survey of clustering algorithms. IEEE Transactions on Neural Networks 16, 645–678 (2005)

Efficient Mining of Jumping Emerging Patterns with Occurrence Counts for Classification*

Łukasz Kobyliński and Krzysztof Walczak

Institute of Computer Science, Warsaw University of Technology
ul. Nowowiejska 15/19, 00-665 Warszawa, Poland
{L.Kobylinski,K.Walczak}@ii.pw.edu.pl

Abstract. In this paper we propose an efficient method of discovering Jumping Emerging Patterns with Occurrence Counts for the use in classification of data with numeric or nominal attributes. This new extension of Jumping Emerging Patterns proved to perform well when classifying image data and here we experimentally compare it to other methods, by using generalized border-based pattern mining algorithm to build the classifier.

1 Introduction

Recently there has been a strong progress in the area of rule- and pattern-based classification algorithms, following the very fruitful research in the area of association rules and emerging patterns. One of the most recent and promising methods is classification using jumping emerging patterns (JEPs). It is based on the idea that JEPs, as their support changes sharply from one dataset to another, carry highly discriminative information that allows creating classifiers, which associate previously unseen records of data to one of these datasets. As JEPs have been originally conceived for transaction databases, where each data record is a set of items, a JEP-based classifier is not usually directly applicable to relational databases, i.e. containing numeric or nominal attributes. In such case an additional discretization step is required to transform the available data to transactional form.

In this article we address the problem of efficiently discovering JEPs and using them directly for supervised learning in databases, where the data can be described as multi-sets of features. This is an enhancement of the transactional database representation, where instead of a binary relation between items and database records, an occurrence count is associated with every item in a set. Example real-world problems that could be approached in this way include market-basket analysis (quantities of bought products), as well as text and multimedia data mining (numbers of occurrences of particular features). We use a new type of JEPs to accomplish this task – the jumping emerging patterns with occurrence counts (occJEPs) – show both the original semi-naïve algorithm and

---
* The research has been partially supported by grant No 3 T11C 002 29 received from Polish Ministry of Education and Science.

C.-C. Chan et al. (Eds.): RSCTC 2008, LNAI 5306, pp. 419–428, 2008.

a new border-based algorithm for finding occJEPs and compare their discriminative value with other recent classification methods.

The rest of the paper is organized as follows. Section 2 outlines previous work done in the field, while Section 3 gives an overview of the concept of emerging patterns in transaction databases. In Sections 4–7 we introduce jumping emerging patterns with occurrence counts (occJEPs), present their discovery algorithms and describe the chosen method of performing classification with a set of found occJEPs. Section 8 presents experimental results of classification and a comparison of some of the most current classifiers. Section 9 closes with a conclusion and discussion on possible future work.

2   Previous Work

The concept of discovering jumping emerging patterns and using them in classification of transactional datasets has been introduced in [1]. Such patterns proved to be a very accurate alternative to previously proposed rule- and tree-based classifiers. Efficient mining of emerging patterns has been first studied in [2,3,4] and in the context of JEP-based classification in [5]. More recently, a rough set theory approach to pattern mining has been presented in [6] and a method based on the concept of equivalence classes in [7].

The application of association rules with recurrent items to the analysis of multimedia data has been proposed in [8], while general and efficient algorithms for discovering such rules have been presented in [9] and [10]. The extension of the definition of jumping emerging patterns to include recurrent items and using them for building classifiers has been proposed in [11].

3   Emerging Patterns

We restrict further discussion on emerging patterns to transaction systems. A transaction system is a pair $(\mathcal{D}, \mathcal{I})$, where $\mathcal{D}$ is a finite sequence of transactions $(T_1, \ldots, T_n)$ (database), such that $T_i \subseteq \mathcal{I}$ for $i = 1, \ldots, n$ and $\mathcal{I}$ is a non-empty set of items (itemspace). A support of an itemset $X \subset \mathcal{I}$ in a sequence $D = (T_i)_{i \in K \subseteq \{1, \ldots, n\}} \subseteq \mathcal{D}$ is defined as $\operatorname{supp}_D(X) = \frac{|\{i \in K: \, X \subseteq T_i\}|}{|K|}$.

Given two databases $D_1, D_2 \subseteq \mathcal{D}$ we define an itemset $X \subset \mathcal{I}$ to be a jumping emerging pattern (JEP) from D_1 to D_2 if $\operatorname{supp}_{D_1}(X) = 0 \wedge \operatorname{supp}_{D_2}(X) > 0$. A set of all JEPs from D_1 to D_2 is called a JEP space and denoted by $JEP(D_1, D_2)$.

4   Jumping Emerging Patterns with Occurrence Counts

Let a transaction system with recurrent items be a pair $(\mathcal{D}^r, \mathcal{I})$, where $\mathcal{D}^r$ is a database and $\mathcal{I}$ is an itemspace (the definition of itemspace remains unchanged). We define database $\mathcal{D}^r$ as a finite sequence of transactions $(T_1^r, \ldots, T_n^r)$ for $i = 1, \ldots, n$. Each transaction is a set of pairs $T_i^r = \{(t_i, q_i); \, t_i \in \mathcal{I}\}$, where $q_i : \mathcal{I} \to \mathbb{N}$ is a function, which assigns the number of occurrences to each item of the transaction. Similarly, a multiset of items X^r is defined as a set of pairs $\{(x, p); x \in \mathcal{I}\}$, where $p : \mathcal{I} \to \mathbb{N}$. We say that $x \in X^r \iff p(x) \geq 1$ and define $X = \{x : x \in X^r\}$. We will write $X^r = (X, P)$ to distinguish X as the set of items contained in a multiset X^r and P as the set of functions, which assign occurrence counts to particular items.

The support of a multiset of items X^r in a sequence $D^r = (T_i^r)_{i \in K \subseteq \{1,\ldots,n\}} \subseteq D^r$ is defined as: $\text{supp}_D(X^r, \theta) = \frac{|\{i \in K: X^r \overset{\theta}{\subseteq} T_i^r\}|}{|K|}$, where $\overset{\theta}{\subseteq}$ is an inclusion relation between a multiset $X^r = (X, P)$ and a transaction $T^r = (T, Q)$ with an occurrence threshold $\theta \geq 1$:

$$X^r \overset{\theta}{\subseteq} T^r \iff \forall_{x \in \mathcal{I}} \, q(x) \geq \theta \cdot p(x) \tag{1}$$

The introduction of an occurrence threshold θ allows for differentiating transactions containing the same sets of items with a specified tolerance margin of occurrence counts. It is thus possible to define a difference in the number of occurrences, which is necessary to consider such a pair of transactions as distinct sets of items. We will assume that the relation $\subseteq$ is equivalent to $\overset{1}{\subseteq}$ in the context of two multisets.

Let a decision transaction system be a tuple $(D^r, \mathcal{I}, \mathcal{I}_d)$, where $(D^r, \mathcal{I} \cup \mathcal{I}_d)$ is a transaction system with recurrent items and $\forall_{T^r \in D^r} |T \cap \mathcal{I}_d| = 1$. Elements of $\mathcal{I}$ and $\mathcal{I}_d$ are called condition and decision items, respectively. A support for a decision transaction system $(D^r, \mathcal{I}, \mathcal{I}_d)$ is understood as a support in the transaction system $(D^r, \mathcal{I} \cup \mathcal{I}_d)$.

For each decision item $c \in \mathcal{I}_d$ we define a decision class sequence $C_c = (T_i^r)_{i \in K}$, where $K = \{k \in \{1,\ldots,n\} : c \in T_k\}$. Notice that each of the transactions from D^r belongs to exactly one class sequence. In addition, for a database $D = (T_i^r)_{i \in K \subseteq \{1,\ldots,n\}} \subseteq D^r$, we define a complement database $D' = (T_i^r)_{i \in \{1,\ldots,n\}-K}$.

Given two databases $D_1, D_2 \subseteq D^r$ we call a multiset of items X^r a jumping emerging pattern with occurrence counts (occJEP) from D_1 to D_2, if $\text{supp}_{D_1}(X^r, 1) = 0 \wedge \text{supp}_{D_2}(X^r, \theta) > 0$, where θ is the occurrence threshold. A set of all occJEPs with a threshold θ from D_1 to D_2 is called an occJEP space and denoted by $occJEP(D_1, D_2, \theta)$. We distinguish the set of all minimal occ-JEPs as $occJEP_m$, $occJEP_m(D_1, D_2, \theta) \subseteq occJEP(D_1, D_2, \theta)$. Notice also that $occJEP(D_1, D_2, \theta) \subseteq occJEP(D_1, D_2, \theta - 1)$ for $\theta \geq 2$. In the rest of the document we will refer to multisets of items as itemsets and use the symbol X^r to avoid confusion.

5 A Semi-naïve Mining Algorithm

Our previous method of discovering occJEPs, introduced in [11], is based on the observation that only minimal patterns need to be found to perform classification. Furthermore, it is usually not necessary to mine patterns longer than a few items, as their support is very low and thus their impact on classification accuracy is negligible. This way we can reduce the problem to: (a) finding only such occJEPs, for which no patterns with a lesser number of items and the same or lower number of item occurrences exist; (b) discovering patterns of less than δ items.

Let C_c be a decision class sequence of a database $\mathcal{D}^\mathrm{r}$ for a given decision item c and C_c' a complement sequence to C_c. We define $D_1 = C_c'$, $D_2 = C_c$ and the aim of the algorithm to discover $occJEP_\mathrm{m}(D_1, D_2, \theta)$. We begin by finding the patterns, which are not supported in D_1, as possible candidates for occJEPs. In case of multi-item patterns at least one of the item counts of the candidate pattern has to be larger than the corresponding item count in the database. We can write this as: $X^\mathrm{r} = (X, P)$ is an occJEP candidate $\iff \forall_{T^\mathrm{r}=(T,Q)\in D_1} \exists_{x\in X}\ p(x) > q(x)$.

The first step of the algorithm is then to create a set of conditions in the form of $[p(i_j) > q_1(i_j) \vee \ldots \vee p(i_k) > q_1(i_k)] \wedge \ldots \wedge [p(i_j) > q_n(i_j) \vee \ldots \vee p(i_k) > q_n(i_k)]$ for each of the candidate itemsets $X^\mathrm{r} = (X, P)$, $X \subseteq 2^\mathcal{I}$, where j and k are subscripts of items appearing in a particular X^r and n is the number of transactions in D_1. Solving this set of inequalities results in its transformation to the form of $[p(i_j) > r_j \wedge \ldots \wedge p(i_k) > r_k] \vee \ldots \vee [p(i_j) > s_j \wedge \ldots \wedge p(i_k) > s_k]$, where r and s are the occurrence counts of respective items. The counts have to be incremented by 1, to fulfill the condition of $\mathrm{supp}^\mathrm{r}_{D_1}(X^\mathrm{r}, \theta) = 0$.

Having found the minimum occurrence counts of items in the candidate itemsets, we then calculate the support of each of the itemsets in D_2 with a threshold θ. The candidates, for which $\mathrm{supp}^\mathrm{r}_{D_2}(X, \theta) > 0$ are the minimal $occJEPs$ (D_1, D_2, θ).

6 Border-Based Mining Algorithm

The border-based occJEP discovery algorithm is an extension of the EP-mining method described in [4]. Similarly, as proved in [3] for regular emerging patterns, we can use the concept of borders to represent a collection of occJEPs. This is because the occJEP space S is convex, that is it follows: $\forall X^\mathrm{r}, Z^\mathrm{r} \in S^\mathrm{r}\ \forall Y^\mathrm{r} \in 2^{S^\mathrm{r}}\ X^\mathrm{r} \subseteq Y^\mathrm{r} \subseteq Z^\mathrm{r} \Rightarrow Y^\mathrm{r} \in S^\mathrm{r}$. For the sake of readability we will now onward denote particular items with consecutive alphabet letters, with an index indicating the occurrence count, and skip individual brackets, e.g. $\{a_1 b_2, c_3\}$ instead of $\{\{1 \cdot i_1, 2 \cdot i_2\}, \{3 \cdot i_3\}\}$.

Example 1. $S = \{a_1, a_1 b_1, a_1 b_2, a_1 c_1, a_1 b_1 c_1, a_1 b_2 c_1\}$ is a convex collection of sets, but $S' = \{a_1, a_1 b_1, a_1 c_1, a_1 b_1 c_1, a_1 b_2 c_1\}$ is not convex. We can partition it into two convex collections $S_1' = \{a_1, a_1 b_1\}$ and $S_2' = \{a_1 c_1, a_1 b_1 c_1, a_1 b_2 c_1\}$.

A border is an ordered pair $< \mathcal{L}, \mathcal{R} >$ such that $\mathcal{L}$ and $\mathcal{R}$ are antichains, $\forall X^\mathrm{r} \in \mathcal{L}\ \exists Y^\mathrm{r} \in \mathcal{R}\ X^\mathrm{r} \subseteq Y^\mathrm{r}$ and $\forall X^\mathrm{r} \in \mathcal{R}\ \exists Y^\mathrm{r} \in \mathcal{L}\ Y^\mathrm{r} \subseteq X^\mathrm{r}$. The collection of sets represented by a border $< \mathcal{L}, \mathcal{R} >$ is equal to:

$$[\mathcal{L}, \mathcal{R}] = \{Y^\mathrm{r} : \exists X^\mathrm{r} \in \mathcal{L}, \exists Z^\mathrm{r} \in \mathcal{R} \text{ such that } X^\mathrm{r} \subseteq Y^\mathrm{r} \subseteq Z^\mathrm{r}\} \qquad (2)$$

Example 2. The border of collection S, introduced in earlier example, is equal to $[\mathcal{L}, \mathcal{R}] = [\{a_1\}, \{a_1 b_2 c_1\}]$.

The most basic operation involving borders is a border differential, defined as:

$$< \mathcal{L}, \mathcal{R} >=< \{\emptyset\}, \mathcal{R}_1 > - < \{\emptyset\}, \mathcal{R}_2 > \qquad (3)$$

As proven in [3] this operation may be reduced to a series of simpler operations. For $\mathcal{R}_1 = \{U_1, \ldots, U_m\}$:

$$< \mathcal{L}_i, \mathcal{R}_i > = < \{\emptyset\}, \{U_i^r\} > - < \{\emptyset\}, \mathcal{R}_2 > \qquad (4)$$

$$< \mathcal{L}, \mathcal{R} > = < \bigcup_{i=1}^{m} \mathcal{L}_i, \bigcup_{i=1}^{m} \mathcal{R}_i > \qquad (5)$$

A direct approach to calculating the border differential would be to expand the borders and compute set differences.

Example 3. The border differential between $[\{\emptyset\}, \{a_1b_2c_1\}]$ and $[\{\emptyset\}, \{a_1c_1\}]$ is equal to $[\{b_1\}, \{a_1b_2c_1\}]$. This is because:

$$[\{\emptyset\}, \{a_1b_2c_1\}] = \{a_1, b_1, b_2, c_1, a_1b_1, a_1b_2, a_1c_1, b_1c_1, b_2c_1, a_1b_1c_1, a_1b_2c_1\}$$
$$[\{\emptyset\}, \{a_1c_1\}] = \{a_1, c_1, a_1c_1\}$$
$$[\{\emptyset\}, \{a_1b_2c_1\}] - [\{\emptyset\}, \{a_1c_1\}] = \{b_1, b_2, a_1b_1, a_1b_2, b_1c_1, b_2c_1, a_1b_1c_1, a_1b_2c_1\}$$

6.1   Algorithm Optimizations

On the basis of optimizations proposed in [4], we now show the extensions necessary for discovering emerging patterns with occurrence counts. All of the ideas presented there for reducing the number of operations described in the context of regular EPs are also applicable for recurrent patterns. The first idea allows avoiding the expansion of borders when calculating the collection of minimal itemsets $\mathrm{Min}(\mathcal{S})$ in a border differential $\mathcal{S} = [\{\emptyset\}, \{U^r\}] - [\{\emptyset\}, \{S_1^r, \ldots, S_k^r\}]$. It has been proven in [4] that $\mathrm{Min}(\mathcal{S})$ is equivalent to:

$$\mathrm{Min}(\mathcal{S}) = \mathrm{Min}(\{\bigcup\{s_1, \ldots, s_k\} : s_i \in U^r - S_i^r, 1 \leq i \leq k\})$$

In the case of emerging patterns with occurrence counts we need to define the left-bound union and set theoretic difference operations between multisets of items $X^r = (X, P)$ and $Y^r = (Y, Q)$. These operations guarantee that the resulting patterns are still minimal.

Definition 1. *The left-bound union of multisets* $X^r \cup Y^r = Z^r$. $Z^r = (Z, R)$, *where:* $Z = \{z : z \in X \vee z \in Y\}$ *and* $R = \{r(z) = max(p(z), q(z))\}$.

Definition 2. *The left-bound set theoretic difference of multisets* $X^r - Y^r = Z^r$. $Z^r = (Z, R)$, *where:* $Z = \{z : z \in X \wedge p(z) > q(z)\}$ *and* $R = \{r(z) = q(z) + 1\}$.

Example 4. For the differential: $[\{\emptyset\}, \{a_1b_3c_1d_1\}] - [\{\emptyset\}, \{b_1c_1\}, \{b_3d_1\}, \{c_1d_1\}]$. $U = \{a_1b_3c_1d_1\}$, $S_1 = \{b_1c_1\}$, $S_2 = \{b_3d_1\}$, $S_3 = \{c_1d_1\}$. $U - S_1 = \{a_1b_2d_1\}$, $U - S_2 = \{a_1c_1\}$, $U - S_3 = \{a_1b_1\}$. Calculating the Min function:

$$\text{Min}([\{\emptyset\}, \{a_1 b_3 c_1 d_1\}] - [\{\emptyset\}, \{b_1 c_1\}, \{b_3 d_1\}, \{c_1 d_1\}]) =$$
$$= \text{Min}(\{a_1 a_1 a_1, a_1 a_1 b_1, a_1 c_1 a_1, a_1 c_1 b_1, b_2 a_1 a_1,$$
$$b_2 a_1 b_1, b_2 c_1 b_1, d_1 a_1 a_1, d_1 a_1 b_1, d_1 c_1 a_1, d_1 c_1 b_1\}) =$$
$$= \text{Min}(\{a_1, a_1 b_1, a_1 c_1, a_1 b_1 c_1, a_1 b_2, a_1 b_2, b_2 c_1, a_1 d_1, a_1 b_1 d_1, a_1 c_1 d_1, b_1 c_1 d_1\}) =$$
$$= \{a_1, b_2 c_1, b_1 c_1 d_1\} .$$

Similar changes are necessary when performing the border expansion in an incremental manner, which has been proposed as the second possible algorithm optimization. The union and difference operations in the following steps need to be conducted according to Definitions 1 and 2 above:

1. Incremental expansion
2. $\mathcal{L} = \{\{x\} : x \in U^r - S_1^r\}$
3. **for** $i = 2$ **to** k
4. $\mathcal{L} = \text{Min}\{X^r \cup \{x\} : X^r \in \mathcal{L}, x \in U^r - S_i^r\}$

Lastly, a few points need to be considered when performing the third optimization, namely avoiding generating nonminimal itemsets. Originally, the idea was to avoid expanding such itemsets during incremental processing, which are known to be minimal beforehand. This is the case when the same item is present both in an itemset in the old $\mathcal{L}$ and in the set difference $U - S_i$ (line 4 of the incremental expansion algorithm above). In case of recurrent patterns this condition is too weak to guarantee that all patterns are still going to be generated, as we have to deal with differences in the number of item occurrences. The modified conditions of itemset removal are thus as follows:

1. If an itemset X^r in the old $\mathcal{L}$ contains an item x from $T_i^r = U^r - S_i^r$ and its occurrence count is equal or greater than the one in T_i^r, then move X^r from $\mathcal{L}$ to $NewL$.
2. If the moved X^r is a singleton set $\{(x, p(x))\}$ and its occurrence count is the same in $\mathcal{L}$ and T_i^r, then remove x from T_i^r.

Example 5. Let $U^r = \{a_1 b_2\}$, $S_1^r = \{a_1\}$, $S_2^r = \{b_1\}$. Then $T_1^r = U^r - S_1^r = \{b_1\}$ and $T_2^r = U^r - S_2^r = \{a_1 b_2\}$. We initialize $\mathcal{L} = \{b_1\}$ and check it against T_2^r. While T_2^r contains $\{b_2\}$, $\{b_1\}$ may not be moved directly to $NewL$, as this would falsely result in returning $\{b_1\}$ as the only minimal itemset, instead of $\{a_1 b_1, b_2\}$. Suppose $S_1^r = \{a_1 b_1\}$, then initial $\mathcal{L} = \{b_2\}$ and this time we can see that $\{b_2\}$ does not have to be expanded, as the same item with at least equal occurrence count is present in T_2^r. Thus, $\{b_2\}$ is moved directly to $NewL$, removed from T_2^r and returned as a minimal itemset.

The final algorithm, consisting of all proposed modifications, is presented below.

1. Border-differential($< \{\emptyset\}, \{U^r\} >, < \{\emptyset\}, \{S_1^r, \ldots, S_k^r\} >$)
2. $T_i^r = U^r - S_i^r$ **for** $1 \leq i \leq k$

3. **if** $\exists T_i^r = \{\emptyset\}$ **then return** $< \{\}, \{\} >$
4. $\mathcal{L} = \{\{x\} : x \in T_1^r\}$
5. **for** $i = 2$ **to** k
6. $\quad NewL = \{X^r = (X, P(X)) \in \mathcal{L} : X \cap T_i \neq \emptyset \wedge \forall x \in (X \cap T_i)\, p(x) \geq t(x)\}$
7. $\quad \mathcal{L} = \mathcal{L} - NewL$
8. $\quad T_i^r = T_i^r - \{x : \{(x, p(x))\} \in NewL\}$
9. $\quad$ **for each** $X^r \in \mathcal{L}$ sorted according to increasing cardinality **do**
10. $\quad\quad$ **for each** $x \in T_i$ **do**
11. $\quad\quad\quad$ **if** $\forall Z^r \in NewL\ \mathrm{supp}_{Z^r}(X^r \cup \{x\}, 1) = 0$
12. $\quad\quad\quad\quad$ **then** $NewL = NewL \cup (X^r \cup \{x\})$
13. $\quad$ $\mathcal{L} = NewL$
14. **return** $\mathcal{L}$

6.2 Discovering occJEPs

Creating an occJEP-based classifier involves discovering all minimal occJEPs to each of the classes present in a particular decision system. We can formally define the set of patterns in a classifier $occJEP_C^\theta$ for a given occurrence threshold θ as: $occJEP_C^\theta = \bigcup_{c \in \mathcal{I}_d} occJEP_m(C_c', C_c, \theta)$, where $C_c \subseteq \mathcal{D}_L^r$ is a decision class sequence for decision item c and C_c' is a complementary sequence in a learning database $\mathcal{D}_L^r$.

To discover patterns between two dataset pairs, we first need to remove non-maximal itemsets from each them. Next, we multiply the occurrence counts of itemsets in the background dataset by the user-specified threshold. Finally, we need to iteratively call the Border-differential function and create a union of the results to find the set of all minimal jumping emerging patterns with occurrence counts from C_c' to C_c.

1. Discover-minimal-occJEPs(C_c', C_c, θ)
2. $\mathcal{L} = $ Remove-non-maximal-itemsets(C_c)
3. $\mathcal{R} = $ Remove-non-maximal-itemsets(C_c')
4. **for** $S_i^r \in \mathcal{R}$ **do**
5. $\quad S_i^r = (S_i, s(x) \cdot \theta)$
6. **end**
7. $\mathcal{J} = \{\emptyset\}$
8. **for** $L_i^r \in \mathcal{L}$ **do**
9. $\quad \mathcal{J} = \mathcal{J} \cup$ Border-differential$(< \{\emptyset\}, \{L_i^r\} >, < \{\emptyset\}, \{S_1^r, \ldots, S_k^r\} >)$
10. **end**
11. **return** $\mathcal{J}$

Example 6. Consider a learning database $\mathcal{D}_L^r$ containing transactions of three distinct classes: $C_1, C_2, C_3 \subset \mathcal{D}_L^r$. $C_1 = \{b_2, a_1c_1\}$, $C_2 = \{a_1b_1, c_3d_1\}$ and $C_3 = \{a_3, b_1c_1d_1\}$. We need to discover occJEPs to each of the decision class sequences: $occJEP_m(C_2 \cup C_3, C_1, \theta)$, $occJEP_m(C_1 \cup C_3, C_2, \theta)$ and $occJEP_m(C_1 \cup C_2, C_3, \theta)$. Suppose $\theta = 2$. Calculating the set of all minimal patterns involves invoking the Discover-minimal-occJEPs function three times, in which the base Border-differential function is called twice each time and the resulting occJEPs are as follows: $\{a_1c_1\}$ to class 1, $\{c_3, a_1b_1\}$ to class 2 and $\{a_3, b_1c_1, b_1d_1\}$ to class 3.

7   Performing Classification

Classification of a particular transaction in the testing database $\mathcal{D}_T^r$ is performed by aggregating all minimal occJEPs, which are supported by it [5]. A scoring function is calculated and a category label is chosen by finding the class with the maximum score:

$$\text{score}(T^r, c) = \sum_{X^r} \text{supp}_{C_c}^r(X^r), \tag{6}$$

where $C_c \subseteq \mathcal{D}_T^r$ and $X^r \in occJEP_m(C_c', C_c)$, such that $X^r \subseteq T^r$. It is possible to normalize the score to reduce the bias induced by unequal sizes of particular decision sequences. This is performed by dividing the calculated score by a normalization factor: norm-score$(T^r, c) = \text{score}(T^r, c)/\text{base-score}(c)$, where base-score is the median of scores of all transactions with decision item c in the learning database: base-score$(c) = \text{median}\{\text{score}(T^r, c), \text{ for each } T^r \in C_c \subseteq \mathcal{D}_L^r\}$.

8   Experimental Results

We have used two types of data with recurrent items to assess the performance of the proposed classifier. The first is a dataset used previously in [11], which consists of images, represented by texture and color features, classified into four categories: *flower*, *food*, *mountain* and *elephant*. The data contains ca. 400 instances and 16 recurrent attributes, where each instance is an image represented by 8 types of texture and 8 types of color features, possibly occurring multiple times on a single image. The accuracy achieved by applying the classifier based on jumping emerging patterns with occurrence counts for several threshold values and compared with other frequently used classification methods is presented in Table 1. All experiments have been conducted as a ten-fold cross-validation using the Weka package [12], having discretized the data into 10 equal-frequency bins for all algorithms, except the occJEP method. The parameters of all used classifiers have been left at their default values. The results are not directly comparable with those presented in [11], as currently the occJEP patterns are not limited to any specific length and the seed number for random instance selection during cross-validation was different than before.

The second dataset used for experiments represents the problem of text classification and has been generated on the basis of the Reuters-21578 collection of documents. We have used the ApteMod version of the corpus [13], which originally contains 10788 documents classified into 90 categories. As the categories are highly imbalanced (the most common class contains 3937 documents, while the least common only 1), we have presented here the results of classification of the problem reduced to differentiating between the two classes with the greatest number of documents and all other combined, i.e. the new category labels are *earn* (36.5% of all instances), *acq* (21.4%) and *other* (42.1%). Document representation has been generated by: stemming each word in the corpus using the Porter's stemmer, ignoring words, which appear on the stoplist provided with

Table 1. Classification accuracy of four image datasets. The performance of the classifier based on jumping emerging patterns with occurrence counts (occJEP) compared to: regular jumping emerging patterns (JEP), C4.5 and support vector machine (SVM), each after discretization into 10 equal-frequency bins.

method	θ	accuracy (%)					
		flower/ food	flower/ elephant	flower/ mountain	food/ elephant	food/ mountain	food/ elephant/ mountain
	1	89.50	84.38	90.63	-	73.00	-
	1.5	94.79	96.35	98.44	78.50	87.00	87.50
occJEP	2	**97.92**	**98.96**	**97.92**	88.00	91.00	**88.50**
	2.5	92.71	97.92	95.31	83.00	90.50	85.50
	3	89.06	97.92	95.31	74.00	87.00	80.50
JEP	-	95.83	91.67	96.35	**88.50**	**93.50**	83.50
C4.5	-	93.23	89.58	85.94	87.50	92.50	82.00
SVM	-	90.63	91.15	93.75	87.50	84.50	84.50

Table 2. Classification accuracy of the Reuters dataset, along with precision and recall values for each of the classes, and the number of discovered emerging patterns / C4.5 tree size

method	θ	accuracy	earn		acq		other		patterns
			precision	recall	precision	recall	precision	recall	
	1	85.12	96.2	84.7	76.6	96.1	95.5	89.4	10029
	1.5	85.58	96.2	84.7	77.8	96.1	95.5	90.4	9276
occJEP	2	84.65	96.2	87.9	78.9	95.7	**96.6**	91.5	7274
	2.5	84.19	96.2	**87.9**	77.6	95.7	96.6	90.4	7015
	10	83.72	98.00	86.2	79.3	**97.9**	95.5	91.3	3891
JEP	-	66.98	86.8	55.0	46.2	47.1	70.7	85.3	45870
C4.5	-	73.49	92.9	65.0	67.5	52.9	69.2	88.5	51
SVM	-	**86.98**	**98.1**	85.0	**85.4**	68.6	82.8	**97.1**	-

the corpus, and finally creating a vector containing the number of occurrences of words in the particular document. We have selected the 100 most relevant attributes from the resulting data, as measured by the χ^2 statistic, and sampled randomly 215 instances for cross-validation experiments, the results of which are presented in Table 2.

9 Conclusions and Future Work

We have proposed an extension of the border-based emerging patterns mining algorithm to allow discovering jumping emerging patterns with occurrence counts. Such patterns may be used to build accurate classifiers for transactional data containing recurrent attributes. By avoiding both discretization and using all values from the attribute domain, we considerably reduce the space of items and exploit the natural order of occurrence counts. We have shown that areas that could possibly benefit by using such an approach include image and text data classification. The biggest drawback of the method lies in the number of discovered patterns, which is however less than in the case of regular JEPs found in discretized data. It is thus a possible area of future work to reduce the set of discovered patterns and further limit the computational complexity without influencing the classification accuracy.

References

1. Li, J., Dong, G., Ramamohanarao, K.: Making use of the most expressive jumping emerging patterns for classification. Knowledge and Information Systems 3(2), 1–29 (2001)
2. Dong, G., Li, J.: Efficient mining of emerging patterns: Discovering trends and differences. In: KDD 1999: Proceedings of the fifth ACM SIGKDD international conference on Knowledge discovery and data mining, pp. 43–52. ACM, New York (2000)
3. Li, J., Ramamohanarao, K., Dong, G.: The space of jumping emerging patterns and its incremental maintenance algorithms. In: ICML 2000: Proceedings of the Seventeenth International Conference on Machine Learning, San Francisco, CA, USA, pp. 551–558. Morgan Kaufmann Publishers Inc., San Francisco (2000)
4. Dong, G., Li, J.: Mining border descriptions of emerging patterns from dataset pairs. Knowledge and Information Systems 8(2), 178–202 (2005)
5. Fan, H., Ramamohanarao, K.: Fast discovery and the generalization of strong jumping emerging patterns for building compact and accurate classifiers. IEEE Transactions on Knowledge and Data Engineering 18(6), 721–737 (2006)
6. Terlecki, P., Walczak, K.: On the relation between rough set reducts and jumping emerging patterns. Information Sciences 177(1), 74–83 (2007)
7. Li, J., Liu, G., Wong, L.: Mining statistically important equivalence classes and delta-discriminative emerging pattern. In: Proceedings of 13th International Conference on Knowledge Discovery and Data Mining, San Jose, California, pp. 430–439 (2007)
8. Zaïane, O.R., Han, J., Zhu, H.: Mining recurrent items in multimedia with progressive resolution refinement. In: Proceedings of the 16th International Conference on Data Engineering, San Diego, CA, USA, pp. 461–470 (2000)
9. Ong, K.L., Ng, W.K., Lim, E.P.: Mining multi-level rules with recurrent items using FP'-Tree. In: Proceedings of the Third International Conference on Information, Communications and Signal Processing (2001)
10. Rak, R., Kurgan, L.A., Reformat, M.: A tree-projection-based algorithm for multi-label recurrent-item associative-classification rule generation. Data and Knowledge Engineering 64(1), 171–197 (2008)
11. Kobyliński, Ł., Walczak, K.: Jumping emerging patterns with occurrence count in image classification. In: Washio, T., Suzuki, E., Ting, K.M., Inokuchi, A. (eds.) PAKDD 2008. LNCS (LNAI), vol. 5012, pp. 904–909. Springer, Heidelberg (2008)
12. Witten, I.H., Frank, E.: Data Mining: Practical machine learning tools and techniques, 2nd edn. Morgan Kaufmann, San Francisco (2005)
13. Lewis, D.D., Williams, K.: Reuters-21578 corpus ApteMod version

Evolutionary Algorithm for Fingerprint Images Filtration

Marcin Jędryka and Władysław Skarbek

Warsaw Technical Academy, faculty of Electronics and Information Technology
mjedryka@ire.pw.edu.pl

Abstract. Proposed filtration algorithm for fingerprints images is based
on iterative algorithm for image binarization which minimizes within
class variance for image foreground and background classes. The proposed solution was modified to fit better to local structures of fingerprint
pattern using Gabor filters and directional filtering. The algorithm was
tested on fingerprint images database, with low quality images, giving
very good results in reducing noise and disturbances.

1 Introduction

In the fingerprint verification process three important phases can be distinguished:
fingerprint preprocessing phase, feature extraction and feature verification.
Analysis of fingerprint patterns can be based on many different features, characteristic for fingerprint images, for example coefficients extracted by local filtration and local statistical parameters. Most common methods tend to use points
characteristic for a single ridge called minutiae – namely ridge endings and ridge
bifurcations. The efficiency of verification algorithm is strictly related to efficiency
of fingerprint preprocessing phase regardless of the features on which verification
is based. In the preprocessing phase image is normalized, filtered and the background of image is segmented out. Proposed algorithm concerns mostly fingerprint
filtration problem, which is very important for fingerprint images verification. Images acquired by fingerprint sensors are typically distorted because of both: the
quality of sensor and the nature of human fingers. The typical distortions are: local discontinuity of ridge structure („scratches"), local contrast deficiency (blurred
ridges) and salt'n'pepper noise. The most common methods for filtration take advantage of fingerprints local features in order to get best efficiency. Directional
filtration uses information about local ridge orientation and is mostly applied as
lowpass filtering (mean or median) on dominant directions. The most common
type of filtration of fingerprints uses Gabor filterbank. Gabor filter fits not only
local ridge direction but also to the characteristic parallel ridges pattern in image.
Gabor filtering is very efficient also when applied to regions of very low contrast.
Proposed algorithm uses both previously mentioned methods of filtration in an
iterative algorithm based on Dirac Needles algorithm for an image binarization
[1]. The algorithm was tested on fingerprint database from the Fingerprint Verification Competition [2], with images of very low quality. Next paragraphs are organized as follows: Paragraph 2 describes Dirac's needles algorithm and proposed

C.-C. Chan et al. (Eds.): RSCTC 2008, LNAI 5306, pp. 429–437, 2008.
© Springer-Verlag Berlin Heidelberg 2008

modifications. Paragraph 3 describes filtration methods and proposed algorithm for computing orientation map of fingerprint image. Paragraph 4 describes full algorithm, the results of the algorithm are also shown.

2   Dirac's Needles Algorithm and Proposed Modifications

Dirac's needles algorithm is an evolutionary algorithm for image binarization. The goal of the algorithm is to minimize the cost function given with the equation (1).

$$var(\mu_d) = \lambda_1 \Sigma_{\mu_d \leq 0} \parallel g(p) - \bar{g}^f \parallel^2 + \lambda_2 \Sigma_{\mu_d \leq 0} \parallel g(p) - \bar{g}^b \parallel^2 \qquad (1)$$

where: μ_d is a function in an image domain P, μ_d: $P \rightarrow [-1, 1]$, related to the membership of a pixel into background or foreground of an image; $g(p)$ is a function in an image domain g: $P \rightarrow [0, 255]$, related to the graylevel values of each pixel; λ_1, λ_2 are the parameters of the algorithm. Equation (1) can be rewritten in a from (2):

$$var(\mu_d) = \lambda_1 \Sigma \parallel g(p) - \bar{g}^f \parallel^2 u(\mu_d(p)) + \lambda_2 \Sigma \parallel g(p) - \bar{g}^b \parallel^2 u(-\mu_d(p)) \qquad (2)$$

The minimum of $var(\mu_d)$ function is found using gradient algorithm with a gradient given in (3):

$$\bigtriangledown var(\mu_d) = \lambda_1 \Sigma \parallel g(p) - \bar{g}^f \parallel^2 u(\delta_d(p)) - \lambda_2 \Sigma \parallel g(p) - \bar{g}^b \parallel^2 u(\delta_d(p)) \qquad (3)$$

In this algorithm image is divided into two categories: background and foreground basing on the membership function μ_d which minimizes the cost function (1). In Dirac's needles algorithm background category is defined for pixels with $\mu_d < -\alpha$ and foreground is defined for pixels with $\mu_d > \alpha$. Pixels for which $|\mu_d| < \alpha$ belong to the category „unclassified", still for μ_d which minimizes cost function (1) this category is usually empty. In the Dirac's needles algorithm binarization is applied basing on the membership function μ_d. The optimal values of this function are evaluated in order to minimize within class variance of pixels belonging to foreground and background classes. Changing values of μ_d function has no effect on the graylevel values of image pixels, which remain constant.

Proposed modifications first of all concern the output of the algorithm. Dirac's needles algorithm is used for image binarization, that is why graylevel values of each pixel are constant and only the membership function values for each pixel are modified. Proposed algorithm is used for image filtration and the membership function is not used as an output data for image thresholding, but becomes an input image for each iteration. The graylevel values of each pixel are changed in each iteration and have a similar function in proposed algorithm that membership function had in Dirac's needles algorithm. Equation (4) becomes a modified equation (2):

$$var(g) = \lambda_1 \Sigma \parallel \varphi(p) - \bar{\varphi}^f \parallel^2 (g(p)) + \lambda_2 \Sigma \parallel \varphi(p) - \bar{\varphi}^b \parallel^2 (-g(p)) \qquad (4)$$

where: g is a graylevel value of a pixel belonging to the input image for a given iteration and is related to the membership function from a Dirac's needles algorithm, $\varphi(p)$ is an input image for a given iteration after filtration and is related to the graylevel values of pixels in Dirac's needles algorithm;

In proposed algorithm the weighting function of unary step $u(\varphi_d(p))$ is also changed. The unary step function is replaced with linear function g(p), in order to use every pixel graylevel value when computing within class variation (for line and background classes) with appropriate weights. The main cause of this modification was to make the filtration faster in each iteration. The weights in this algorithm are chosen to give the biggest gradient value $\nabla(var(g))$ for pixels which graylevel value in image g differs most from graylevel value from filtered image $-\mu$. The mean values for each class are computed basing on a global mean:

$$\bar{\varphi} = \tfrac{1}{n}\Sigma\varphi$$

$$\bar{\varphi}^b = \tfrac{1}{n_1}\Sigma\varphi_{|\varphi>\bar{\varphi}}$$

$$\bar{\varphi}^f = \frac{1}{n_2}\Sigma\varphi_{|\varphi<\bar{\varphi}} \tag{5}$$

The modified version of equation (3) is given in (6):

$$\nabla var(g) = \lambda_1\Sigma \parallel \varphi(p) - \bar{\varphi}^f \parallel^2 + \lambda_2\Sigma \parallel \varphi(p) - \bar{\varphi}^b \parallel^2 (-1) \tag{6}$$

The proposed algorithm for minimizing a given cost function occurred to be not optimal. Next modification was applied to the weighting function related to the unary step from Dirac's needles algorithm. The modified cost function is given in (7):

$$var(g) = \lambda_1\Sigma \parallel \varphi(p) - \bar{\varphi}^f \parallel^2 (\tfrac{1}{2}g^2(p) - g(p)max(g(p))) - \lambda_2\Sigma \parallel \varphi(p) - \bar{\varphi}^b \parallel^2 (\tfrac{1}{2}g^2(p) - g(p)min(g(p)))$$

$$\tag{7}$$

and a gradient of (7) is given in:

$$\nabla var(g) = \lambda_1\Sigma \parallel \varphi(p) - \bar{\varphi}^f \parallel^2 (g(p) - max(g(p))) - \lambda_2\Sigma \parallel \varphi(p) - \bar{\varphi}^b \parallel^2 (g(p) - min(g(p)))$$

$$\tag{8}$$

The filtration based on a proposed iterative algorithm, which minimizes cost function from (7), gave very good results.

3   Methods for Fingerprint Images Filtration

The filtration methods were used for acquiring images that were used to evaluate the cost function (7).

3.1   Directional Filtering

The directional filtering of fingerprint images uses important local image feature – ridge orientation [3]. The analysis along ridge direction enables the reduction of distortion of ridge pattern in regions with similar local ridge orientation. The proposed filtration algorithm bases on evaluating mean graylevel value of pixels belonging to analysed region, with a region being fitted to the local ridge direction. This filtration is done using equation given in (9):

$$\varphi(p_0) = \frac{1}{m} \Sigma_{n \in N} g(p_n) \tag{9}$$

where: N is a section with a direction related to the local ridge orientation, crossing pixel p, m – is a number of pixels belonging to section N, $\varphi(p_0)$ – is a graylevel value of pixel after filtration, $g(p_n)$ – is a graylevel value of pixel belonging to N in an input image;

The directional filtration is very efficient in reduction of noise and „scratches" in fingerprint images. Using this type of filtration enabled reduction even very strong local disturbances in fingerprint images. On Fig. 2 the result of directional filtration is shown for an image from Fig.1. The directional filtering is less efficient when applied to regions of low contrast and high curvature.

Fig. 1. Fingerprint image

3.2   Gabor Filter

Gabor Filter [4] is the most common type of filter used for fingerprint images analysis. This is because of the fact that it is fitted not only to local ridge orientation but also to the characteristic „hills and valleys" structure shown on Fig. 3, which can be approximated with a sinusoidal function. The Gabor Filter can be described with the equation:

$$g(x, y) = e^{-0.5\left(\frac{x^2}{d_x^2} + \frac{y^2}{d_y^2}\right)} cos(2\pi u x) \tag{10}$$

where – $1/u$ corresponds to local frequency of fingerprint pattern, α – corresponds to local ridge orientation:

$$x = x_0 sin\alpha + y_0 cos\alpha$$

Fig. 2. Image from Fig.1 after directional filtering

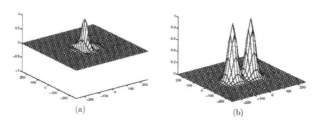

Fig. 3. The local fingerprint pattern, with a sinusoid approximating graylevel distribution

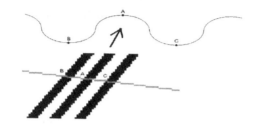

(a)   (b)

Fig. 4. Gabor filter in image domain (a) and frequency domain (b)

$$y = y_0 sin\alpha - x_0 cos\alpha \qquad (11)$$

$x0$, $y0$ correspond to coordinates of pixels, dx, dy are coefficients related to the deviation of Gauss function. On Fig.4 the example of Gabor filter is shown in the image and frequency domain. Gabor filtration is efficient in regions with quick changing local ridge orientation and in the regions with low contrast, mostly because of the mid-pass filtering. This type of filtration is not as efficient as simple directional filtering in the regions of high orientation coherence. Combining Gabor filtering with directional filtering proved to give very good results.

3.3   Evaluating of Local Ridge Orientation

The important parameter describing fingerprint ridge pattern is an orientation map. Information about local ridge orientation is used in both filtering methods as well as in whole evolutionary filtering algorithm. Correctly evaluated orientation map is crucial for the efficiency of the algorithm. Proposed method for evaluating ridge orientation enables implementation of efficient, full algorithm. Evaluation of local ridge orientation is based on the analysis of graylevel values gradients on two perpendicular axes: ∇x, ∇y, acquired with the convolution with Sobel masks. Local ridge orientation is evaluated as a division of gradients:

$$\Theta = \frac{\nabla y}{\nabla x} \tag{12}$$

Above mentioned method is especially sensitive for the distortion of fingerprint image and is insufficient for practical use. In order to improve this method in next iterations the orientation map is filtered with a low-pass filter. For this purpose orientation map is firstly transformed to the form of a vector filed, using equations:

$$\phi_x = cos(\Theta)$$

$$\phi_y = sin(\Theta) \tag{13}$$

next the filtration is done. In the process of image filtering additionally information about local orientation coherence is used. Local orientation coherence is evaluated from the equation:

$$coh = \frac{(G_{xx} - G_{yy})^2 + 4G_{xy}}{G_{xx} + G_{yy}} \tag{14}$$

where Gxx, Gyy, Gxy are local mean graylevel value gradients given from the equations:

$$G_{xx} = \Sigma_W G_x^2$$

$$G_{yy} = \Sigma_W G_y^2 \tag{15}$$

$$G_{xy} = \Sigma_W G_x G_y \tag{16}$$

Orientation coherence has 0 value for the regions with abruptly changing ridge orientation and 1 value for the regions with uniform orientation. In each successive iteration of the algorithm the condition for growing value of local coherence is tested for local windows. Only regions with improved value of orientation coherence are being filtered . Experiments proved that good results of determining orientation map can be achieved after a small number of iterations.

4 Full Algorithm of Fingerprint Image Filtration and Results

The full algorithm of image filtration is described in the following schema:

1. evaluating of fingerprint orientation map
2. evaluating images: φ_1 and φ_2, resulted from the filtration of the input image g with the directional and Gabor filters
3. evaluating of means: $\bar{\varphi}_1^f, \bar{\varphi}_1^b, \bar{\varphi}_2^f$, and $\bar{\varphi}_2^b$ for the images after filtration
4. evaluating of the gradient $\nabla var(g)$ from equation (8) separately for the images φ_1 and φ_2
5. modification of graylevel values for pixels from image g with the use of previously evaluated gradients
6. back to point 1.

Proposed algorithm was tested on fingerprint images database from the Fingerprint Verification Competition, which included images of low quality. The algorithm proved to be very efficient in the improvement of the regions heavily distorted by noise, local scratches, contrast deficiency and image blur. On Fig. 6 are shown regions of fingerprint images from Fig.5 with high distortion rate (scratches, contrasts deficiency) as a result of filtration. The proposed algorithm was compared with results for fingerprint images filtration using Gabor filtering, giving much better results when filtering fingerprint images regions with high noise level. The comparison of both algorithms using simple efficiency measure

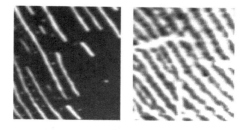

Fig. 5. Examples of fingerprint image regions with local distortion

Fig. 6. Fingerprint image regions from Fig. 5 after proposed method of filtration

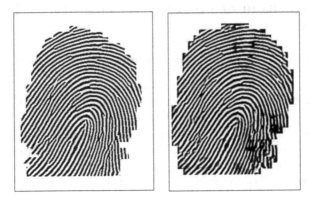

Fig. 7. Examples of fingerprint image after filtration using proposed algorithm (left image) and Gabor filtering (right image)

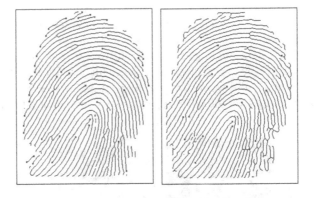

Fig. 8. Examples of minutiae extracted from fingerprint images after filtration using proposed algorithm (left image) and Gabor filtering (right image)

is not a trivial issue, since lack of the high quality images which could be use as a reference ones. The database from the Fingerprint Verification Competition contains only natural fingerprint images, mostly highly disturbed. The first proposed comparison method was simple visual comparison of results. The example of this comparison method is shown on Fig. 7, on Fig. 9 the original image is shown. Acquired filtered images using proposed filtration algorithm proved to be better than when using Gabor filter. The second method for comparison results was the analysis of minutiae detected from images filtered using both algorithms. The example of this comparison method is shown on Fig. 8, on Fig. 9 the original image is shown. In this case again the lack of information about true position of minutiae in fingerprint images makes the method less valuable. Nevertheless the detection of minutiae using proposed filtration algorithm gives in authors opinion much better results than when using Gabor filtering.

Fig. 9. The original fingerprint image from examples from Fig. 7 and 8

5 Conclusions

The algorithm proved to be very efficient, nevertheless it still gives a little worse results in the regions of quickly changing ridge frequency. Future work includes the optimization of this problem, for example by reducing directional filtering influence in these regions. The efficiency of this algorithm is related to the proper combination of the advantages of directional filtering with Gabor filtering using evolutionary algorithm. The additional advantage is the fact that the output image can be easily binarized with a simple global thresholding algorithm, with a constant threshold. This is due to the fact, that original Dirac needles algorithm was prepared mainly for image binarization. Because of its advantages the algorithm can be effectively used for fingerprint image analysis and verification, as well as in the analysis of images with a similar structure. Proposed algorithm does not stabilize quickly, with consecutive iterations oscillations occurs. Nevertheless the efficient filtration can be seen after a few steps (tests showed that 3 full steps of algorithm are sufficient) which reduces the computational complexity of the algorithm.

Acknowledgement

The work presented was developed within VISNET II, an European Network of Excellence (http://www.visnet-noe.org), funded under the European Commission IST FP6 Programme.

References

1. Skarbek, W.: Image Segmentation by Dirac Needles. Warsaw Technical Academy. Faculty of Electronics and Information Technology, Warsaw (2007)
2. http://biometrics.cse.msu.edu/fvc04db/index.html
3. Wu, C., Shi, Z., Govindaraju, V.: Fingerprint Image Enhacement Method Using Directional median Filter. In: SPIE, Defense and Security Symposium, Orlando, FL (2004)
4. Daugman, J.G.: High Confidence Recognition of Persons by a Test of Statistical Independence. IEEE Trans., Pattern Anal. Machine Intell. 15(11) (1993)

Efficient Discovery of Top-K Minimal Jumping Emerging Patterns*

Pawel Terlecki and Krzysztof Walczak

Institute of Computer Science, Warsaw University of Technology,
Nowowiejska 15/19, 00-665 Warsaw, Poland
{P.Terlecki,K.Walczak}@ii.pw.edu.pl

Abstract. Jumping emerging patterns, like other discriminative patterns, help to understand differences between decision classes and build accurate classifiers. Since their discovery is usually time-consuming and pruning with minimum support may require several adjustments, we consider the problem of finding top-k minimal jumping emerging patterns. We describe the approach based on a CP-Tree that gradually raises minimum support during mining. Also, a general strategy for pruning non-minimal patterns and their descendants is proposed. We employ the concept of attribute set dependence to test pattern minimality. A two and multiple class version of the problem is discussed. Experiments evaluate pruning capabilities and execution time.

Keywords: jumping emerging pattern, strong jumping emerging patterns, top-k most interesting patterns, CP-Tree, rough sets, attribute set dependence.

1 Introduction

Emerging patterns (EP) were introduced to capture differences between classes in classified transaction datasets ([1]). This generic idea is being constantly extended to obtain derivative pattern types with specific properties, like, recently proposed, generalized noise-tolerant ([2]) and statistically significant EPs ([3]). Various experiments prove that such patterns are accurate in classification and provide valuable knowledge in business and bioinformatics ([4]). In addition, emerging patterns appear to share several important ideas with the rough set theory ([5]) and can benefit from its achievements, e.g. reduct computation ([6]).

In this paper we bring the concept of top-k most interesting patterns to the field of emerging patterns. We consider the problem of mining top-k minimal jumping emerging patterns (JEPs), i.e. the k most supported minimal JEPs in each decision class. A JEP is a pattern that exists in one class and is absent from the others. One usually focuses on minimal JEPs due to their generality and, thus, good classification capabilities. In a classified dataset from Fig. 1, the

---

* The research has been partially supported by grant No 3 T11C 002 29 received from Polish Ministry of Education and Science.

C.-C. Chan et al. (Eds.): RSCTC 2008, LNAI 5306, pp. 438–447, 2008.

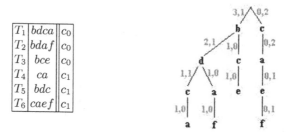

T_1	$bdca$	c_0
T_2	$bdaf$	c_0
T_3	bce	c_0
T_4	ca	c_1
T_5	bdc	c_1
T_6	$caef$	c_1

Fig. 1. Sample classified dataset with transaction items following the order $b < d < c < a < e < f$ and a respective CP-Tree

patterns bce, bdf, ab are JEPs in c_0, but only ab is minimal. At the same time, the pattern bc is not a JEP, since it is supported by transactions in both classes.

Many mining methods specify minimum support to prune infrequent patterns. However, in practice, it is hard to predict the size of the result set for a certain threshold. If its value is too high, one may obtain few patterns, if it is too low, a mining process usually becomes time-consuming or infeasible. Therefore, a specified number of top patterns may be a convenient alternative.

To provide an efficient solution, we exploit the resemblance between our problem and mining strong JEPs (SJEPs, [2]). We modify the approach based on a CP-Tree, so that it is capable of gradually raising minimum support and, thus, intensifying pattern pruning. New values of the threshold are deduced from minimal JEPs identified so far, which means that their minimality has to be verified at discovery time. For this purpose, we leverage the method from [7], which uses the notion of attribute set dependence from the rough set theory. As a general improvement for CP-Tree mining, we advocate to use our minimality test to all considered patterns and push pruning capabilities even further. The modifications are discussed for two and multiple classes. Experiments show significant savings in the number of considered patterns and total computation time.

The content is organized as follows. In Sect 2 a related work is covered. Section 3 provides preliminaries and a problem definition. Our algorithmic propositions that utilize a CP-Tree are given in Sect. 4. Section 5 contains an experimental evaluation of the presented methods. The paper is concluded in Sect. 6.

2   Related Work

Finding top-k most interesting patterns has been widely considered for frequent patterns and their derivatives. Early works look at finding k largest (most frequent) patterns ([8]) or k most frequent patterns of each length ([9]). In [10] authors consider top-k closed frequent patterns with an additional constraint on a minimum pattern length. A version for sequential patterns is presented in [11]. Two latter algorithms incorporate a support raising mechanism to mining based on a FP-Tree. In our approach, we refer to the same general strategy, however, desirable patterns, raising method and space traversal are different. To the best of our knowledge finding top-k discriminative patterns of any kind, in particular EPs, have not been analyzed in literature yet.

EPs were introduced in [1]. In our work we specifically focus on minimal JEPs, whose good classification capabilities and applications to bioinformatics are covered in [4]. A classic JEP mining approach utilizes border differentiation operation ([11]). A rough set approach that uses local projection and local reducts is presented in [6]. Moreover, many other mining strategies defined later for more general types of patterns can still be applied to this problem. In particular, an algorithm that mines a CP-Tree to find strong JEPs or generalized noise-tolerant EPs ([12,2]), or a recent method for finding equivalence pattern classes based on closed frequent pattern mining with a FP-Tree ([3]).

3   Preliminaries

This section covers theoretical background for the paper. It provides foundations for classified datasets and emerging patterns and define the problem of finding top-k minimal JEPs. Formal convention follows [6].

Let a transaction system be a pair $(\mathcal{D}, \mathcal{I})$, where $\mathcal{D}$ is a finite sequence of transactions $(T_1, .., T_n)$ (database) such as $T_i \subseteq \mathcal{I}$ for $i = 1, .., n$ and $\mathcal{I}$ is a non-empty set of items (itemspace). Let a decision transaction system be a tuple $(\mathcal{D}, \mathcal{I}, \mathcal{I}_d)$, where $(\mathcal{D}, \mathcal{I} \cup \mathcal{I}_d)$ is a transaction system and $\forall_{T \in \mathcal{D}} |T \cap \mathcal{I}_d| = 1$. Elements of $\mathcal{I}$ and $\mathcal{I}_d$ are called condition and decision items, respectively. For each $c \in \mathcal{I}_d$, we define a decision class sequence $C_c = (T_i)_{i \in K}$, where $K = \{k \in \{1, .., n\} : c \in T_k\}$. The notations C_c and $C_{\{c\}}$ are used interchangeably.

Let us consider a decision transaction system $(\mathcal{D}, \mathcal{I}, \mathcal{I}_d)$. For a database $D = (T_i)_{i \in K \subseteq \{1,...,n\}} \subseteq \mathcal{D}$, we define a complementary database $D' = (T_i)_{i \in \{1,...,n\} - K}$. The count (support) of an itemset $X \subseteq \mathcal{I} \cup \mathcal{I}_d$ in a database $D = (T_i)_{i \in K} \subseteq \mathcal{D}$ is defined as $count_D(X) = |\{i \in K : X \subseteq T_i\}|$; $(supp_D(X) = \frac{count_D(X)}{|K|})$, where $K \subseteq \{1, .., n\}$.

Let us now consider two databases $D_1, D_2 \subseteq \mathcal{D}$ referred to as negative and positive, respectively. The growth-rate of an itemset $X \in \mathcal{P}$ from D_1 to D_2 is defined as $GR_{D_1 \to D_2}(X) = 0$, for $supp_{D_1}(X), supp_{D_2}(X) = 0$; $= \infty$, for $supp_{D_1}(X) = 0$ and $supp_{D_2}(X) \neq 0$; $= \frac{supp_{D_2}(X)}{supp_{D_1}(X)}$, otherwise. Also, the support-ratio of an itemset $X \in \mathcal{P}$ is $SR(X) = max(GR_{D_1 \to D_2}(X), GR_{D_2 \to D_1}(X))$. Given ρ as minimum growth rate, a ρ-emerging pattern from D_1 to D_2 is defined as an itemset $X \subseteq \mathcal{I}$, for which $GR_{D_1 \to D_2}(X) \geq \rho$. For brevity, if D_1 is known from the context, we talk about patterns in D_2. Similarly, a jumping emerging pattern (JEP) in D_2 is an itemset $X \subseteq \mathcal{I}$ with an infinite growth-rate, $GR_{D_1 \to D_2}(X) = +\infty$. In addition, given ξ as minimum support, we define a ξ-strong jumping emerging pattern (SJEP) in D_2 as an itemset $X \subseteq \mathcal{I}$, for which $supp_{D_1}(X) = 0$ and $supp_{D_2}(X) \geq \xi$, and these not hold for its any proper subset. In other words, X can be seen as a minimal JEP in D_2 with $supp_{D_2}(X) \geq \xi$.

The set of all JEPs from D_1 to D_2 is called a JEP space and denoted by $JEP(D_1, D_2)$. JEP spaces can be described concisely by means of borders ([11]). For $c \in \mathcal{I}_d$, we use a border $< \mathcal{L}_c, \mathcal{R}_c >$ to represent a JEP space $JEP(C'_c, C_c)$.

Pattern collections $\mathcal{L}$ and $\mathcal{R}$ are called a left and a right bound, and contain minimal and maximal patterns in terms of inclusion, respectively.

Let us consider a decision transaction system $DTS = (\mathcal{D}, \mathcal{I}, \mathcal{I}_d)$. The problem of finding top-k minimal JEPs for DTS is defined as finding a collection $\{J_c\}_{c \in \mathcal{I}_d}$ such that for each $c \in \mathcal{I}_d$: $|J_c| = k$ and $\forall_{X \in J_c, Y \in \mathcal{L}_c - J_c} supp_{C_c}(X) \geq supp_{C_c}(Y)$. In other words, for each decision class $c \in \mathcal{I}_d$, one looks for a set of k minimal JEPs from C'_c to C_c with the highest support. Note that, similarly to formulations for closed frequent patterns ([10]), equally supported patterns are indiscernible, thus, the result set may not be unequivocal.

Example 1. For the decision transaction system in Fig. 1 JEP spaces for classes c_0, c_1 have the following form, respectively: $< \{df, bf, be, ad, ab\}, \{abcd, abdf, bce\} >$, $< \{ef, cf, ae\}, \{ac, bcd, acef\} >$. Top-2 minimal JEPs in the class c_0 are: ad, ab.

4   Top-K Patterns Discovery

Finding top-k minimal JEPs can be simply accomplished by selecting the k most supported elements from the left bound of a priory computed complete JEP space. This naive approach does not take advantage of the usually low value of k and unnecessarily considers numerous infrequent patterns. In order to limit a search space, we leverage the resemblance between mining SJEPs with a given support threshold and top-k minimal JEPs, where this threshold is specified implicitly by the value of k. Hereinafter, we consider our problem for a decision transaction system $DTS = (\mathcal{D}, \mathcal{I}, \mathcal{I}_d)$ and $k \in \mathbb{N}$, where $\mathcal{I}_d = \{c_0, c_1\}$.

If we assume that J_c is the result for the class $c \in \mathcal{I}_d$, we may easily calculate the best tuned minimum count as $\xi = min_{X \in J_c}(supp_{D_c}(X))$. Then, our problem can be immediately transformed to finding ξ-SJEPs and selecting k most supported patterns from the result set. Since the value of ξ is not known upfront, it has to be discovered during the mining process. Similar strategy is utilized to solve other related problems ([10,13]), although actual implementations strongly depend on properties of a particular mining algorithm.

4.1   Mining Based on CP-Tree

In order to address the problem of finding top-k minimal JEPs, we modify a popular method for mining SJEPs based on a CP-Tree ([12]). Due to space limitations, only the main sketch is provided here (details in [12,2]).

Let us assume an order $\prec$ in $\mathcal{I}$. A Contrast Pattern tree (CP-Tree) is a modified prefix tree. It is multiway and ordered. Each edge from a node N to one of its children is labeled with an item i. The set of edge labels of N is denoted by $N.items$. For $i \in N.items$, we associate its edge with a child ($N.child[i]$), positive and negative count ($N.posCount[i]$, $N.negCount[i]$). Children of each node and edges on each rooted path are ordered in the way the respective labels follow $\prec$. A set of labels on a rooted path represents a transaction prefix with counts in the positive and negative class indicated by respective counts of the last node of this path. For example, on the leftmost path bd is marked with $2, 1$, since it occurs in 2 transactions in c_0 and 1 transaction in c_1.

A CP-Tree concisely represents a two-class transaction decision system and enables efficient pattern mining. In order to construct a tree for DTS, items of each transaction from $(T_i \cap \mathcal{I} : T_i \in \mathcal{D}\}$ are sorted according to $\prec$ and the resulting strings are inserted like to a regular prefix tree. As a part of each insertion positive or negative counts are incremented based on the transaction's class (*insert-tree*, [12]). A tree for our sample system is given in Fig. 1.

SJEPs can be found by mining a CP-Tree as expressed by the *mine-tree* routine (Algorithm 1). We parameterized the original scheme to demonstrate our propositions. Two result sets for positive and negative patterns are maintained. All found JEPs are added to respective result sets based on the *accept-pattern* function. For pattern pruning two thresholds, a positive and negative minimum count, are used. The *raise-count* procedure is responsible for raising their values. Once the current pattern β is serviced, the function *visit-subtree* indicates, if its subtree should be considered. Besides traversal of nodes the algorithm performs subtree merging that allows us to consider subtrees associated with successive transaction prefixes. This step is performed by the *merge* procedure and modifies the structure of a tree. It is unrelated to our proposals, thus, omitted.

In the original algorithm for SJEP mining, all found JEPs are collected. Thus, the routine $accept\text{-}pattern(\beta, T.negCount[i], T.posCount[i], minPosCount)$ for positive patterns checks basically if $T.negCount[i] = 0$. The thresholds are constant, thus, $raise\text{-}count(minPosCount, posPatterns)$ returns $minPosCount$. Also the subtree is considered, if there is a chance to find patterns fulfilling respective thresholds, i.e. *visit-subtree* checks if $(T.posCount[i] \geq minPosCount \lor T.negCount[i] \geq minNegCount)$. In order to group JEPs closer to the top of the tree, the order $\prec$ is based on support-ratio ([12]). Since result sets contain also non-minimal JEPs, minimal ones has to be identified after mining is finished.

4.2   Minimum Support Raising

As mentioned before, our problem can be solved by finding a superset of ξ-SJEPs with a certain minimum threshold ξ. Higher values of the threshold likely lead to more efficient pruning. At the same time, one cannot exceed a certain threshold that would prune patterns from the top-k set. Our discussion refers to one class $c \in \mathcal{I}_d$. Analogous logic and structures are needed for the other.

We propose to gradually raise a minimum support threshold while a CP-Tree is being mined. In order to achieve that, minimal JEPs are collected upfront, i.e. a check for minimality is performed when a JEP is found. In consequence, inferences on minimum support can be made based on patterns identified so far. The following theorem states that whenever one knows a pattern collection of the size at least k, the minimum support ξ equal to the minimum over supports of patterns from this collection, ensures that a respective set of ξ-SJEPs contains at least k elements. The trivial proof is omitted to space limitations.

Theorem 1. $\forall_{P \in \mathcal{L}_c} |P| \geq k \land min_{X \in P}(supp_{D_c}X) \geq \xi \Rightarrow |\xi\text{-}SJEP(D'_c, D_c)| \geq k.$

Therefore, to solve our problem it is sufficient to store the current result in a priority queue, e.g. a heap, of at most k elements ordered by non-increasing

---

Algorithm 1. mine-tree (T,α)

---

1. **for all** $(i \in T.items)$ **do**
2.   **if** $T.child[i].items$ is not empty **then**
3. $merge(T.child[i], T)$
4.   **end if**
5. $\beta = \alpha \cup T.items[i]$
6.   **if** $accept\text{-}pattern(\beta, T.posCount[i], T.negCount[i], minNegCount)$ **then**
7. $negPatterns := negPatterns \cup \beta; count(\beta) := T.negCount[i]$
8. $minNegCount := raise\text{-}count(minNegCount, negPatterns)$
9.   **else**
10.   **if** $accept\text{-}pattern(\beta, T.negCount[i], T.posCount[i], minPosCount)$ **then**
11. $posPatterns := posPatterns \cup \beta; count(\beta) := T.posCount[i]$
12. $minPosCount := raise\text{-}count(minPosCount, posPatterns)$
13.   **else**
14.   **if** $visit\text{-}subtree(T, \beta)$ **then**
15. $mine\text{-}tree(T.child[i], \beta)$
16.   **end if**
17.   **end if**
18.   **end if**
19.   delete subtree i
20. **end for**

---

supports. For simplicity, we operate on counts instead of supports. At the beginning minimum count is equal to 0. Whenever a new minimal JEP is identified and fewer than k elements have been collected or its count is higher than the current minimum count, it gets inserted to the queue. If k elements are collected, one may set the minimum count to the count of the top element of the queue plus one. This way the threshold can be raised without a significant overhead.

In the original algorithm, all found JEPs in c are added to a respective result set. If one collects only minimal JEPs, their minimality has to be tested in the *accept-pattern* routine. For $X \subseteq \mathcal{I}$ and $D \subseteq \mathcal{D}$, we propose that X is D-discernibility minimal iff $\forall_{Y \subset X} supp_{D'}(X) < supp_{D'}(Y)$. In fact, it is sufficient to examine only immediate subsets of X to check this property. The proof is analogous to Theorem 5 in [7] and omitted due to space limitations.

Theorem 2. *X is D-discernibility minimal* $\Leftrightarrow \forall_{a \in X} supp_{D'} X < supp_{D'} (X - a)$

In fact, this notion is closely related to attribute set dependence from the rough set theory. Note that, if X is a JEP, we have $supp_{D'_c}(X) = 0$, thus, X is minimal, iff all its immediate subsets have non-zero negative counts.

Example 2. The following JEPs are successively discovered by the original algorithm for c_0: $abcd, abd, bdf, abc, bce, ab, be, bf, acd, ad, df$. Additional computation is required to identify minimal patterns and pick ab,ad. If we check minimality upfront, ab is the first minimal JEP, since $supp_{D'_0}(a), supp_{D'_0}(b) > 0$. For $k = 2$, one collects ab, be (counts: 2, 1), and sets minimum count to 2. Then, it prunes all patterns but ad, so that even additional minimality checks are not needed.

4.3 Pruning of Non-minimal Patterns

For a given node, associated with a certain pattern X, all its descendants refer to patterns that subsume X. The following theorem states that only supersets of a D-discernibility minimal pattern can be minimal JEPs. The proof is analogous to the one for Theorem 4 in [7] and omitted due to space limitations.

Theorem 3. $\forall_{X \subseteq \mathcal{I}} X$ is not D-discernibility minimal $\implies \forall_{Y \supseteq X} Y \notin \mathcal{L}_c$

Therefore, one may use such a minimality test in the *visit-subtree* function and prevent unnecessary traversing and merging. According to Theorem 2, it requires checking counts in the negative class for all immediate subsets of β. Although this pruning approach may be very powerful, the test is performed for every pattern that was not added to result sets and, in consequence, can impact the overall efficiency. Here, we use a classic counting method that stores a list of transaction identifiers for each individual item and computes the count of a given pattern by intersecting the lists referring to its items ([14]). Bit operations are employed to improve performance.

Example 3. Theorem 3 allows us to avoid mining subtrees of non-minimal patterns. For example, in Fig. 1, the patterns bd and bc are not minimal, since $supp_{D_0'} bd = supp_{D_0'} bc = supp_{D_0'} b$, and their subtrees can be pruned after merging.

4.4 Multiple Classes Case

Frequently, decision transaction systems contain more than two classes. In this case, it is recommended ([11,12]) to induce EPs in each class separately and treat the remaining classes as a negative class. A classic approach based on a CP-Tree searches simultaneously for patterns in a positive and negative class. Since negative ones would be discarded anyway, it is economical to modify our algorithm, so that a search space and information stored in nodes are smaller.

As it was explained before, children of each node and nodes on each path are ordered by support-ratio. This strategy makes positive and negative EPs with a high growth-rate remain closer to the root of a tree. Since we are not interested in finding negative patterns, it is better to use the ordering based solely on growth-rate to the positive class. Also, there is no need to store exact negative count in each node. A single bit that indicates if the associated pattern is a JEP or not, is sufficient. As far as mining is concerned, only one priority queue and variable for minimum count are required.

5 Experimental Evaluation

Our experiments consider finding top-k minimal JEPs for different values of k. For each dataset, two cases are examined: with a minimality check only for JEPs and with pruning of non-minimal patterns. The efficiency is contrasted with finding top-∞ JEPs in both cases and classic finding SJEPs with a CP-Tree without a minimum support pruning (Classic). Mining is performed for

Table 1. Dataset summary and JEP finding with the Classic method

No	Dataset	Objects	Items	Classes	Minimal JEPs	Classic exam. patterns	Classic comp. time
1	breast-wisc	699	29	2	499	8024	313
2	dna	500	80	3	444716	-	-
3	heart	270	22	2	151	125008	625
4	krkopt	28056	43	18	1187	146004	18250
5	kr-vs-kp	3196	73	2	-	-	-
6	lung	32	220	3	203060	-	-
7	lymn	148	59	4	1699	1670728	92188
8	mushroom	8124	117	2	1818	15279338	577254
9	tic-tac-toe	958	27	2	1429	45886	2578
10	vehicle	846	72	4	5045	19319956	2024497
11	zoo	101	35	7	103	198326	2109

Table 2. Finding top-k minimal JEPs without and with non-minimal pattern pruning

Dataset	Examined patterns				Computation time			
	10	20	50	∞	10	20	50	∞
breast-wisc	1095	1460	2864	8024	93	110	140	203
	270	443	1205	3618	94	78	109	187
dna	5182	8538	16768	-	4407	5296	6985	-
	2330	3520	6308	787714	4156	4891	6265	65235
heart	33720	52170	80474	125008	219	312	422	609
	1952	2952	4364	6523	125	171	219	281
krkopt	58982	66753	76337	146004	16109	16203	16485	19766
	54446	61586	70233	134244	20250	21204	22594	36875
kr-vs-kp	-	-	-	-	-	-	-	-
	19147911	33734042	54507469	-	2067294	3834973	6497624	-
lung	150996628	189214719	-	-	2871705	3614961	-	-
	138	252	1515	49782	578	797	662	252598
lymn	24749	44842	97792	1670728	656	984	1625	14016
	1512	2328	4342	36409	360	484	656	2438
mushroom	5591	20747	141041	15279338	3625	4672	9485	165798
	306	474	2220	64839	2969	3156	4469	17672
tic-tac-toe	4716	6535	10251	45886	563	625	672	954
	3847	5117	7335	23855	578	609	672	984
vehicle	1277105	1798589	2878406	19319956	32406	40609	56235	207830
	39374	56734	92145	418401	5844	7188	9718	27407
zoo	59976	93980	146856	198326	593	828	1234	1532
	2393	4062	7679	10851	156	188	281	359

each class separately. Each test is described by the number of examined patterns (mine-tree calls) and total computation time in ms. Results are averaged over several repetitions. In order to demonstrate efficiency of the pruning approaches, larger datasets from UCI Repository were chosen. They were transformed to transaction systems, so that an item refers to an attribute value pair. Tests were performed on a machine with 2 Quad CPU, 2.4GHz each and 4GB of RAM.

In Classic, one discovers patterns that are not necessarily minimal. Final pruning is performed at the end and may involve large collections of candidates. In order to show improvement in JEP finding, we compare it to top-∞, where pruning is performed upfront. Significant improvement can be observed for datasets with large number of patterns, like *mushroom* or *vehicle*.

Thanks to minimum count raising finding top-k minimal JEPs is usually possible in reasonable time for small k. Even without pruning of non-minimal patterns, benefits in a number of examined patterns and an execution time are significant (*dna, lymn, lung, krkopt, vehicle*).

In terms of examined patterns, efficiency of pruning of non-minimal patterns is visible for the majority of sets. Significant impact on a total execution time can be observed for datasets: *lung, vehicle, zoo*. For three datasets (*dna, lung, kr-vs-kp*) we were unable to obtain solution neither with Classic nor with top-∞ without pruning of non-minimal patterns. However, after applying the latter pruning both datasets *dna* and *lung* could be solved.

Notice that, if only a small part of all patterns can be pruned, additional overhead of checking pruning conditions decreases gains in a computation time (e.g. *breast-wisc, heart, krkopt, tic-tac-toe*).

6   Conclusions

In this paper we have considered the problem of finding top-k minimal jumping emerging patterns (JEPs). Such approach may give good insight on a classified dataset when a full computation is time-consuming or infeasible and a proper value of minimum support is hard to estimate. The classic case of two classes and its extension to multiple classes have been covered.

We have proposed an algorithm that follows the approach to finding strong JEPs based on a CP-tree. A search space is pruned with minimum support, whose value is being gradually raised as new minimal JEPs are discovered. In order to enable this, minimality of each new JEP has to be verified by the time it is found rather than at the end of the process. Our minimality test adapts the idea of set dependence from the rough set theory. In fact, it is sufficient to prove indispensability of each item of a pattern to ensure a specific support in a negative class. Independently of the main problem, we have put forward a general pruning strategy for a CP-Tree. Instead of verifying minimality only for JEPs, we apply it to each considered pattern and possibly avoid examining its descendants. Both methods may lead to significant space reductions, but they come at the price of additional pattern counting.

Experiments show significant savings both in time and the number of examined patterns, when a small number of highly supported JEPs is requested. This effect is even stronger when the general pruning approach of non-minimal patterns is applied. For datasets with large itemspaces, it was the only feasible method. For small datasets, the overhead of minimality verification is noticeable.

References

1. Dong, G., Li, J.: Efficient mining of emerging patterns: discovering trends and differences. In: KDD, pp. 43–52. ACM Press, New York (1999)
2. Fan, H., Ramamohanarao, K.: Fast discovery and the generalization of strong jumping emerging patterns for building compact and accurate classifiers. IEEE Trans. on Knowl. and Data Eng. 18(6), 721–737 (2006)
3. Li, J., Liu, G., Wong, L.: Mining statistically important equivalence classes and delta-discriminative emerging patterns. In: KDD, pp. 430–439 (2007)
4. Li, J., Dong, G., Ramamohanarao, K.: Making use of the most expressive jumping emerging patterns for classification. Knowl. Inf. Syst. 3(2), 1–29 (2001)
5. Terlecki, P., Walczak, K.: On the relation between rough set reducts and jumping emerging patterns. Information Sciences 177(1), 74–83 (2007)
6. Terlecki, P., Walczak, K.: Local projection in jumping emerging patterns discovery in transaction databases. In: Washio, T., Suzuki, E., Ting, K.M., Inokuchi, A. (eds.) PAKDD 2008. LNCS (LNAI), vol. 5012, pp. 723–730. Springer, Heidelberg (2008)
7. Terlecki, P., Walczak, K.: Attribute set dependence in apriori-like reduct computation. In: Wang, G.-Y., Peters, J.F., Skowron, A., Yao, Y. (eds.) RSKT 2006. LNCS (LNAI), vol. 4062, pp. 268–276. Springer, Heidelberg (2006)
8. Shen, L., Shen, H., Pritchard, P., Topor, R.: Finding the n largest itemsets. In: ICDM (1998)
9. Fu, A.W.-C., Kwong, R.W.-w., Tang, J.: Mining n-most interesting itemsets. In: Ohsuga, S., Raś, Z.W. (eds.) ISMIS 2000. LNCS (LNAI), vol. 1932, pp. 59–67. Springer, Heidelberg (2000)
10. Wang, J., Lu, Y., Tzvetkov, P.: Tfp: An efficient algorithm for mining top-k frequent closed itemsets. IEEE Trans. on Knowl. and Data Eng. 17(5), 652–664 (2005); Senior Member-Jiawei Han
11. Dong, G., Li, J.: Mining border descriptions of emerging patterns from dataset pairs. Knowl. Inf. Syst. 8(2), 178–202 (2005)
12. Fan, H., Ramamohanarao, K.: An efficient single-scan algorithm for mining essential jumping emerging patterns for classification. In: Chen, M.-S., Yu, P.S., Liu, B. (eds.) PAKDD 2002. LNCS (LNAI), vol. 2336, pp. 456–462. Springer, Heidelberg (2002)
13. Tzvetkov, P., Yan, X., Han, J.: Tsp: Mining top-k closed sequential patterns. Knowl. Inf. Syst. 7(4), 438–457 (2005)
14. Savasere, A., Omiecinski, E., Navathe, S.B.: An efficient algorithm for mining association rules in large databases. In: VLDB, pp. 432–444 (1995)

Hierarchical Tree for Dissemination of Polyphonic Noise

Rory Lewis[1], Amanda Cohen[2], Wenxin Jiang[2], and Zbigniew Raś[2]

[1] University of Colorado at Colorado Springs,
1420 Austin Bluffs Pkwy, Colorado Springs, CO 80918, USA
`rorlewis@uncc.edu`
[2] University of North Carolina at Charlotte,
9201 University City Blvd., Charlotte, NC 28223, USA
`{acohen24,wjiang3,ras}@uncc.edu`

Abstract. In the continuing investigation of identifying musical instruments in a polyphonic domain, we present a system that can identify an instrument in a polyphonic domain with added noise of numerous interacting and conflicting instruments in an orchestra. A hierarchical tree specifically designed for the breakdown of polyphonic sounds is used to enhance training of classifiers to correctly estimate an unknown polyphonic sound. This paper shows how goals to determine what hierarchical levels and what combination of mix levels is most effective has been achieved. Learning the correct instrument classification for creating noise together with what levels and mixed the noise optimizes training sets is crucial in the quest to discover instruments in noise. Herein we present a novel system that disseminates instruments in a polyphonic domain.

1 Introduction

The challenge for automatic indexing of instruments and Music Instrument Retrieval has moved from the monophonic domain to the polyphonic domain [15,7]. Previously we presented the rationale and need for creating a categorization system more conducive for music information retrieval, see [6]. Essentially, the Dewey-based, Hornbostel-Sachs classification system, [2,9] which classified all instruments into the four categories of Idiophones (vibrating bodies), Membranophones (vibrating membranes), Chordophones (vibrating strings), and Aerophones (vibrating air) firstly, permits instruments to fall into a more than one category and secondly, humanistic conventions of categorization of certain instruments such as a piano or tamborine are alien to machine recognition. After fine tuning our categorization, see Figure 1, we focused on solving an issue that was prevalent in MIR of polyphonic sounds: In the past, if the training data did not work, one did not know if the bug was in the classification tree or if it was in the levels used to mix noise. With the classification issue resolved we could focus on learning the optimal choice of mixes to create training noise and the optimal levels of mix ratios for noise. Knowing the aforementioned allows discovery of optimal conditions for machine learning how distortions caused by noise can be eliminated in finding an instrument in a polyphonic domain.

C.-C. Chan et al. (Eds.): RSCTC 2008, LNAI 5306, pp. 448–456, 2008.

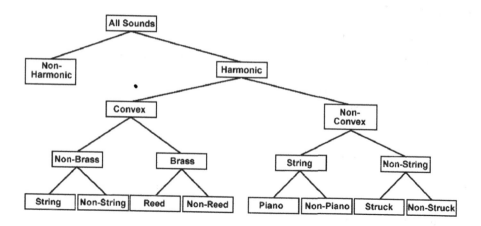

Fig. 1. Instrument Hierarchy Tree: Categorized by the MPEG-7 audio descriptor LogAttackTime. We split the Convex and Non-Convex categories to smaller, more specific groupings of instruments. We select instruments from the smaller categories and combined them to make the polyphonic sounds.

2   Creating a Controlled Noise

With the issue of hierarchical categorization of music instruments solved we decided to create the best controlled environment for training. To do this we decided to use similarly pitched notes. We randomly chose Middle-C because it was, simply put, in the middle of the spectrum. Even when certain instruments could not reach Middle-C we still used the closest C, be it up or down one octave. We also decided to create training sets consisting of both polyphonic and single instrument tuples. The test set comprised all the mixed sounds with different noise-ratios. Also, it was clear that the environment of this polyphonic domain would have to be controlled. The issue would be how one control's noise in a manner that empirical calculations can be run and tested upon the polyphonic domain? Considering our database is MPEG-7 based the authors decided to use 1) MPEG-7 descriptors [5] and five non-MPEG-7 based descriptors upon the following rationale:

In the temporal domain we differentiate between tone-like and noise-like sounds, where the center of gravity of a sound's frequency power spectrum is located, the variation and deviation, the slight variations of harmonicity of some sounds, percussive and harmonic sounds and the time averaged over the energy envelope. We also decided to incorporate descriptors that take into account how human's hear sounds in the time domain such as the ears perception to the frequency components in the mel-frequency scale, the average number of positive and negative traces of a sound wave that cross over zero, the frequencies that fall below a specific magnitude and variations of the energy through the frequency scales.

3   Descriptors

3.1   MPEG-7 Descriptors

SpectrumSpread. To differentiate between tone-like and noise-like sounds the authors used SpectrumSpread because its an economical descriptor that indicates whether the power is concentrated in the vicinity of its centroid, or else spread out over the spectrum.

$$S = \sqrt{\sum_n log_2(f(n)/1000) - C))^2 P'_x(n) \Big/ \sum_n P'_x(n)} \qquad (1)$$

where $P'_x(n)$ is the power spectrum, $f(n)$ is the corresponding frequency. C is spectrum centroid and S is the spectrum spread, in the form of RMS deviation with respect to the centroid.

SpectrumCentroid. To identify instruments with a strong or weak center of gravity of the log-frequency power spectrum we used the SpectrumCentroid descriptor which is defined as the power weighted log-frequency centroid. Here, frequencies of all coefficients are scaled to an octave scale anchored at 1 kHz

$$C = \sum_n log_2(f(n)/1000) P'_x(n) \Big/ \sum_n P'_x(n) \qquad (2)$$

where $P'_x(n)$ is the power associated with the frequency $f(n)$.

HarmonicSpectral Variation and Deviation (HSV) and (HSD). To realize shifts within a running window of the harmonic peaks we used the HSV and HSD descriptors, where the HarmonicSpectralVariation is the mean over the sound segment duration of the instantaneous HarmonicSpectralVariation. The HarmonicSpectralDeviation is the sound segment duration of the instantaneous HarmonicSpectralDeviation within a running window computed as the spectral deviation of log-amplitude components from a global spectral envelope.

$$HSV = \sum_{frame=1}^{nbframes} IHSV(frame) \Big/ nbframes \qquad (3)$$

$$HSD = \sum_{frame=1}^{nbframes} IHSD(frame) \Big/ nbframes \qquad (4)$$

where nbframes is the number of frames in the sound segment.

HarmonicPeaks. To differentiate instruments based on peaks of the spectrum located around the multiple of the fundamental frequency of the signal we used the HarmonicPeaks descriptor. The descriptor here looks for the maxima of the amplitude of the Short Time Fourier Transform (STFT) close to the multiples of the fundamental frequency. The frequencies are then estimated by the positions of these maxima while the amplitudes of these maxima determine their amplitudes.

LogAttackTime (LAT). The motivation for using the MPEG-7 temporal descriptor, LogAttackTime (LAT), is because segments containing short LAT periods cut generic percussive (and also sounds of plucked or hammered string) and harmonic (sustained) signals into two separate groups [4,5]. The LAT is the logarithm of the time duration between the point where the signal starts to the point it reaches its stable part.[10] The range of the LAT is defined as $log_{10}(\frac{1}{samplingrate})$ and is determined by the length of the signal. Struck instruments, such as most percussive instruments have a short LAT whereas blown or vibrated instruments contain LATs of a longer duration.

$$LAT = log_{10}(T1 - T0),\tag{5}$$

where $T0$ is the time the signal starts; and $T1$ reaches its sustained part (harmonic space) or maximum part (percussive space).

TemporalCentroid. To sort instruments based upon the time averaged over the energy envelope we used the TemporalCentroid descriptor which is extracted as follows:

$$TC = \sum_{n=1}^{length(SEnv)} n/sr \cdot (SEnv)(n) \Big/ \sum_{n=1}^{length(SEnv)} (SEnv)(n)\tag{6}$$

3.2   Non-MPEG-7 Descriptors

Energy MFCC. The MPEG-7 work is in the frequency domain but what about differentiating instruments in the time-domain. In other words, like we hear instruments? We know that the ears perception to the frequency components of sound do not follow the linear scale but the mel-frequency scale [3], which in the linear frequency domain below 1,000 Hz and a logarithmic spacing above 1,000 Hz [13]. To do this, filters have in the past been spaced linearly at low frequencies and logarithmically at high frequencies [12]. We chose MFCC because it can key in on the known variation of the ears critical band-widths with frequency [11]:

$$M(f) = 2595log_{10}(1 + f/700)\tag{7}$$

where f is frequency in Hertz. Based on this assumption, the mel-frequency cepstrum coefficient, once known, opens the door to computing MFCC [1].

ZeroCrossingDensity. When a pure sound, a monophonic harmonic sound is affected by noise, such as we are doing, the average number of positive and negative traces of the sound wave that cross over zero (zero-crossings) per second is affected. Using the ZeroCrossing Density descriptor allows us to consider this dimension of the experiment.

RollOff. To differentiate frequencies that fall below an experimentally chosen percentage of the accumulated magnitudes of the spectrum [14]. We chose to include the RollOff descriptor.

Flux. When non-linear sound waves are disturbed, such as being bombarded by noise, the measurement of the variations of the energy through the frequency scales is flux. Having a means to measure these changes as various levels of noise are imposed upon a sound is crucial in differentiating the noise.

4   Experiments

In our experiments we used Weka 3.5.7 to build models for each training data and chose the J48 decision tree as our classification algorithm. We had observed in previous research that the J48 decision tree had a better performance in detecting instrument timbres (see [17] and [16]). The goal is to find rules of how modification of the mix levels of various combinations of instruments influences the quality of the trained classifiers. We used the McGill University CDs, used worldwide in research on music instrument sounds [8]. To test how the accuracy of a classifier improves the estimation of the dominant instrument in a polyphonic sound, we built a training dataset comprising mixtures of single instrument sounds and polyphonic sounds. The polyphonic sounds comprised one dominant sound and a specific mix of instruments located in the leaves at the same level of the hierarchical tree with decreased amplitude which we observed as "noise." Continuing with this strategy we combined more instruments according to our hierarchical tree. However, before making the polyphonic sounds for the entire hierarchical tree, we ran experiments to determine what levels of noise would be optimal for each instrument in order to ensure a trained robust estimation of the classification model's unknown polyphonic sound. To make the size of training data reliable we used the pitch of a single tone of 4C containing 10 different dominant instruments.

Table 1. List of instruments making the basis for the noise

instrumnet	category
ElectricGuitar	string
Oboe	reed
B-flatclarinet	string
CTrumpet	brass
TenorTrombone	brass
Violin	string
Accordian	reed
TenorSaxophone	reed
DoubleBass	string
Piano	string

As shown in Table 2, we observed that the 75% mixture got the best performance in terms of dominant instrument estimation. In order to decide whether the 75% result also holds at each node of the hierarchy tree we divided the entire training group into 2 sub groups of reed and string. After dividing into the 2 sub groups we repeated the test to verify whether the 75% level was indeed the most effective level to use. Here we observed

Table 2. The results showing classification confidence from the 10-fold cross validation process

Training Set	Accuracy
mix100+single	69.90%
mix75+single	79.56%
mix50+single	73.82%
mix30+single	73.48%
mix25+single	70.26%
mix65+single	73.06%
mix80+single	73.58%

Table 3. Results after we divided the instrument sounds into 2 groups according to each category

Training Set	String Accuracy	Reed Accuracy
mmix100+single	68.14%	71.70%
mix80+single	73.76%	82.94%
mix75+single	79.50%	83.98%
mix65+single	70.63%	71.72%
mix50+single	77.66%	76.53%
mix30+single	68.05%	53.19%
mix25+single	74.34%	78.47%

As shown in Table 3, the 75% noise ratio is still the best choice for each single group of instruments, regardless the distribution of the whole tests changed a little bit when different path of hierarchy tree is followed.

Figure 2 shows a graphic representation of the performance of the polyphonic sounds with the non-dominant instrument at 80% volume, 75% volume, 65%, 50%, 30%, and 25%, and the accuracy with which the database identified the dominant instrument. It also shows the performance of the specific groups of sounds, the string group and the reed group. In all three data sets, a non-dominant instrument volume of 75% yielded the most accurate results. However, in the case of the reed group a non-dominant of 80% gave only slightly less accurate results in comparison and more accurate results than any other group at that volume. A non-dominant instrument volume of 30% yielded the least accurate results in the case of the reed group and the string group, with the reed group showing the least accurate results of any instrument group at any volume. The accuracy of the whole data set was lowest at 100%. Figure 3 shows the same results with more exact values on the left-hand vertical axis. The values on the right-hand axis show the percentile that the data falls in, in order to see how the groups compare overall to each other. Overall the reed group yielded the most accurate results. Even at its lowest accuracy, the results of the reed group didn't drop below 72%, which cannot be said for either the string group or the entire set of sounds. In fact the accuracy of the string group drops to nearly 68% at its lowest point. The string group also displayed the greatest changes in accuracy, with the largest difference between two non-dominant instrument volumes being roughly 9%.

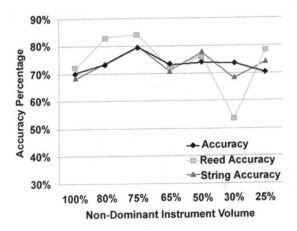

Fig. 2. Graphic Representation of the Results: This graph shows the performance of the polyphonic sounds and the accuracy with which the database identified the dominant instrument. It also shows the performance of the specific groups of sounds, the string group and the reed group.

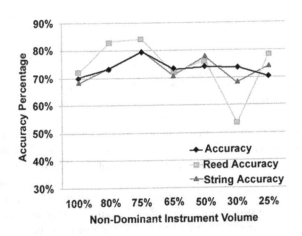

Fig. 3. Graphic Representation of the Results with Percentile Axis: This graph shows the same results with more exact values on the left-hand vertical axis. The values on the right-hand axis show the percentile that the data falls in, in order to see how the groups compare overall to each other.

5   Conclusion and Future Work

Using the new hierarchical tree in our closed domain show that a 75% volume is optimal. Knowing this we now have the tools to know that future errors in retrieving instruments in a polyphonic sound if wrong, will be because of a property inherent in the expanded domain. This is good news as it directs us to the fault. Our next domain will bear instruments playing various sets of harmonics in tex training set not all on one similar pitch. Once this is achieved, the ultimate goal of identifying instruments in a non-harmonic or harmonic noise will be a step closer.

Acknowledgment

This work is supported by the National Science Foundation under grant IIS-0414815.

References

1. Davis, S., Mermelstein, P.: Comparison of parametric representation for mono-syllabic word recognition in continuously spoken sentences. IEEE Transactions of Acoustics, Speech, and Signal Processing ASSP-28(4), 357–366 (1980)
2. Doerr, M.: Semantic problems of thesaurus mapping. Journal of Digital Information 1(8), Article No. 5: 2001-03-26:2001–03 (2001)
3. Fang, Z., Zhang, G.: Integrating the energy information into mfcc. In: International Conference on Spoken Language Processing, October 16-20, vol. 1, pp. 389–292 (2000)
4. Gomez, E., Gouyon, F., Herrera, P., Amatriain, X.: Using and enhancing the current mpeg-7 standard for a music content processing tool. In: Proceedings of the 114th Audio Engineering Society Convention, Amsterdam, The Netherlands (March 2003)
5. Martinez, J.M., Koenen, F.P.R.: Iso/iec jtc 1/sc 29. In: Information Technology–Multimedia Content Description Interface – Part 4: Audio (2001)
6. Lewis, R., Wieczorkowska, A.: Categorization of musical instrument sounds based on numerical parameters. In: Kryszkiewicz, M., Peters, J.F., Rybinski, H., Skowron, A. (eds.) RSEISP 2007. LNCS (LNAI), vol. 4585, pp. 784–792. Springer, Heidelberg (2007)
7. Lewis, R., Zhang, X., Raś, Z.: Knowledge discovery based identification of musical pitches and instruments in polyphonic sounds. International Journal of Engineering Applications of Artificial Intelligence 20(5), 637–645 (2007)
8. Opolko, F., Wapnick, J.: Mums – mcgill university master samples. cd's (1987)
9. Patel, M., Koch, T., Doerr, M., Tsinaraki, C.: Semantic interoperability in digital library systems. Technology-enhanced Learning and Access to Cultural Heritage. UKOLN, University of Bath, IST-2002-2.3.1.12 (2005)
10. Peeters, G., McAdams, S., Herrera, P.: Instrument sound description in the context of mpeg-7. In: Proceedings of the International Computer Music Conference (ICMC 2000), Berlin, Germany (2000)
11. Picone, J.: Signal modeling techniques in speech recognition. In: Bullock, T.H. (ed.) Life Science Research Report, vol. 81(9), pp. 1215–1247. IEEE, Los Alamitos (1993)
12. Schroeder, M.: Recognition of complex acoustic signals. In: Bullock, T.H. (ed.) Life Science Research Report, vol. 55, pp. 323–328. Abakon Verlag, Berlin (1977)
13. Stephens, S., Volkman, J.: The relation of pitch to frequency. American Journal of Psychology 53(3), 329–353 (1940)

14. Wieczorkowska, A., Kolczynska, E.: Quality of musical instrument sound identification for various levels of accompanying sounds. In: Kok, J.N., Koronacki, J., Lopez de Mantaras, R., Matwin, S., Mladenič, D., Skowron, A. (eds.) ECML 2007. LNCS (LNAI), vol. 4701, pp. 28–36. Springer, Heidelberg (2007)
15. Wieczorkowska, A., Synak, P., Lewis, R., Raś, Z.: Creating reliable database for experiments on extracting emotions from music. Intelligent Information Processing and Web Mining, Advances in Soft Computing, 395–404 (2005)
16. Zhang, X., Raś, Z.: Analysis of sound features for music timbre recognition. In: Proceedings of the International Conference on Multimedia and Ubiquitous Engineering (MUE 2007), Seoul, South Korea, April 26-28, 2007, pp. 3–8. IEEE Computer Society, Los Alamitos (2007)
17. Zhang, X., Raś, Z.: Differentiated harmonic feature analysis on music information retrieval for instrument recognition. In: Proc. of IEEE GrC 2006, IEEE International Conference on Granular Computing, Atlanta, Georgia (May 2006)

A Data Driven Emotion Recognition Method Based on Rough Set Theory*

Yong Yang[1,2], Guoyin Wang[2], Fei Luo[2], and Zhenjing Li[2]

[1] School of Information Science and Technology,
Southwest Jiaotong University,
Chengdou, 610031, P.R. China
[2] Institute of Computer Science & Technology,
Chongqing University of Posts and Telecommunications,
Chongqing, 400065, P.R. China
{yangyong,wanggy}@cqupt.edu.cn

Abstract. Affective computing is becoming a more and more important topic in intelligent computing technology. Emotion recognition is one of the most important topics in affective computing. It is always performed on face and voice information with such technology as ANN, fuzzy set, SVM, HMM, etc. In this paper, based on the idea of data driven data mining and rough set theory, a novel emotion recognition method is proposed. Firstly, an information system including facial features is taken as a tolerance relation in rough set, based on the idea of data driven data mining, a suitable threshold is selected for the tolerance relation. Then a reduction algorithm based on condition entropy is proposed for the tolerance relation, SVM is taken as the final classifier. Simulation experiment results show that the proposed method can use less features and get higher recognition rate, and the proposed method is proved effective and efficient.

Keywords: Affective computing, Emotion recognition, Rough set, Data driven data mining.

1 Introduction

It is always a dream that computers can simulate and communicate with a human, or have emotions that human have. A lot of research works have been done in this field in recent years. Affective computing is one of them. Affective computing is computing that relates to, arises from, or deliberately influences emotion, which is firstly proposed by Picard at MIT in 1997 [1]. Affective computing consists of recognition, expressing, modelling, communicating and responding to emotion. Emotion recognition is one of the most fundamental and

---

* This paper is partially supported by National Natural Science Foundation of China under Grant No.60773113 and No. 60573068, Natural Science Foundation of Chongqing under Grant CSTC2007BB2445.

C.-C. Chan et al. (Eds.): RSCTC 2008, LNAI 5306, pp. 457–464, 2008.

important modules in affective computing. It is always based on facial and audio information, which is in accordance with people's recognition to emotion. Its applications have reached almost every aspect of our daily life, for example, health care, children education, game software design, and human-computer interaction. Nowadays, emotion recognition is always studied using ANN, fuzzy set, SVM, HMM, Rough Set and the recognition rate often arrives at 64% to 98% [2][3]. In our previous work in [4][5][6][7], rough set has been used for feature selection and SVM is taken as the classifiers in emotion recognition with audio and visual information, and high recognition rate are resulted. But in the course, we find that facial feature and voice feature are measured, and these features may be imprecise and contain error. In traditional rough set theory, information system is taken as an equivalence relation. A process of discretization in equivalence relation is necessary since facial features and emotion voice features are both continuous value. Unfortunately, information should be lost in discretization, and the result can be impacted. To solve this question, a novel emotion recognition method based on tolerance relation is proposed in this paper, and based on the idea of data driven data mining, a method for suitable threshold is introduced, SVM is still taken as the classifier. Experiment results show the proposed method is effective and efficient. The rest of this paper is organized as follows. Basic concepts of rough set theory and proposed method are introduced in section 2, Simulation experiments and discussion are introduced in section 3. Finally, conclusion and future works are discussed in section 4.

2   Proposed Emotion Feature Selection Method

2.1   Basic Concept of Rough Set Theory

Rough set (RS) is a valid mathematical theory for dealing with imprecise, uncertain, and vague information, it was developed by Professor Z. Pawlak in 1980s [8][9]. Some basic concepts of rough set theory are introduced here for the convenience of following discussion.

Def. 1. A decision information system is a continuous value information system and it is defined as a pair $S = (U, R, V, f)$, where U is a finite set of objects and $R = C \cup D$ is a finite set of attributes, C is the condition attribute set and $D = \{d\}$ is the decision attribute set. With every attribute $a \in R$, a set of its values V_a is associated. Each attribute a determines a function $f_a : U \rightarrow V_a$.

Def. 2. For a subset of attributes $B \subseteq A$, the indiscernibility relation is defined by $Ind(B) = \{(x,y) \in U \times U : a(x) = a(y), \forall a \in B\}$.

The indiscernibility relation defined in this way is an equivalence relation. Obviously, $Ind(B) = \{(x,y) \in U \times U : a(x) = a(y), \forall a \in B\}$. By $U/Ind(B)$ we mean the set of all equivalence classes in the relation $Ind(B)$. The classical rough set theory is based on an observation that objects may be indiscernible (indistinguishable) due to limited available information, and the indiscernibility relation defined in this way is an equivalence relation indeed. The intuition behind the notion of an indiscernibility relation is that selecting a set of attribute $B \subseteq A$ effectively defines a partition of the universe into sets of objects that can not be discerned/distinguished using the attributes in B only. The equivalence classes $E_i \in U/Ind(B)$, induced by a set of attributes $B \subseteq A$, are referred to as object classes or simply classes. The classes resulted from $Ind(A)$ and $Ind(D)$ are called condition classes and decision classes respectively.

Def. 3. A decision information system is a continuous value information system and it is defined as a pair $S = (U, R, V, f)$, where U is a finite set of objects and $R = C \cup D$ is a finite set of attributes, C is the condition attribute set and $D = \{d\}$ is the decision attribute set.$\forall c \in C$, c is continuous vale attribute,$\forall d \in D$, d is continuous value attribute or discrete value attribute.

Since the attribute vales could be imprecise and contain error in continuous value information systems, the definition of equivalence relation is too rigid to be used in these systems. Therefore, the concept of tolerance relation is used for depicting continuous value information system. There are many research works about tolerance relation in rough set theory[10][11][12].

Def. 4. A binary relation $R(x, y)$ defined on a attribute set B is called a tolerance relation if it satisfies:
  1) symmetrical: $R(x, y) = R(y, x)$.
  2) reflextive: $R(x, x) = R(x, x)$.

Let an information system $S = (U, R, V, f)$ be a continuous value information system, $\forall x, y \in U, \forall a \in C$, a relation $R(x, y)$ defined on C is defined as $R(x, y) = \{(x, y) \,||a_x - a_y| < \alpha, \alpha \geq 0\}$. Apparently, $R(x, y)$ is a tolerance relation according to Def. 4 since $R(x, y)$ is symmetrical and reflexive. $R(x, y)$ is used for depicting continuous value information systems in this paper with the motivation that $\forall x, y \in U, \forall a \in C$, attribute value a_x and a_y are equal indeed while $|a_x - a_y| < \alpha$, since there are some error when a_x and a_y are measured.

In classical rough set theory, equivalence relation constitutes a partition of U, but tolerance relation constitutes a cover of U, and equivalence relation is a particular type of tolerance relation.

If $\forall x \in U$, an equivalence class of x is taken as the neighborhood of x , the amount of equivalence class of x is the number of neighborhood which x belongs to. Similarly, in tolerance relation, a tolerance class of x is taken as the neighborhood of x , the amount of tolerance class of x is the number of neighborhood which x belongs to. The bigger tolerance class is, the more uncertainty it contains, the less knowledge it contains. Therefore, concept of knowledge entropy and conditional entropy are defined as follows according to the knowledge which tolerance class contains in tolerance relation.

Def. 5. Let $U = \{x_1, x_2, ..., x_{|U|}\}$, $R(x_i, x_j)$be a tolerance relation defined on attribute set B, $n_R(x_i)$be tolerance class of x_i with respect to B, which denote neighborhood of x_i on B, then, $|x_i|_R = |\{n_R(x_j) \,|x_i \in n_R(x_j), 1 \leq j \leq |U|\}|$ so, knowledge entropy $E(R)$ under relation R is defined as $E(R) = -\frac{1}{|U|}\sum_{i=1}^{|U|}\log_2\frac{|x_i|_R}{|U|}$.

Def. 6. Let R and Q are tolerance relation defined on U, a relation satisfy R and Q simultaneous can be taken as $R \cup Q$, and it is a tolerance relation defined on U too. $\forall x_i \in U$, $n_{R\cup Q}(x_i) = n_R(x_i) \cap n_Q(x_i)$, therefore, knowledge of $R \cup Q$ can be defined as $E(R \cup Q) = -\frac{1}{|U|}\sum_{i=1}^{|U|}\log_2\frac{|x_i|_{R\cup Q}}{|U|}$.

Def. 7. Let R and Q are tolerance relation defined on U, conditional entropy of R relative to Q is defined as $E(R\,|Q) = E(R \cup Q) - E(Q)$.

Def. 8. Let R be a tolerance relation defined on attribute B, $\forall a \in B$, if $E(\{d\}\,|R) = E(\{d\}\,|R - \{a\})$, then a isn't necessary and can be reductive.

2.2 Selecting Threshold Value of Tolerance Relation Based on Idea of Data Driven Data Mining

In the face of knowledge reduction and data mining for continuous value information systems, tolerance relation could be constituted for continuous value information system, but there is an open problem should be solved, that is, threshold vale of α in tolerance relation should be made certain firstly. Traditionally, threshold vale can be gotten according to expert' experience, but there could be no expert experience in some circumstance.

According to the idea of data driven data mining [13], knowledge could be expressed in many different ways, there should be some relationship between the different formats of the same knowledge. In order to keep the knowledge unchanged in a data mining process, properties of the knowledge should remain unchanged during the knowledge transformation process. Otherwise, there should be some mistake in the process of knowledge transformation. Based on the idea, knowledge reduction can be seen a process of knowledge transformation, and in the course, properties of the knowledge should remain unchanged. Based on the idea, classification ability of conditional attribute set relative to decision attribute set can be taken as an important property of knowledge when knowledge reduction is used in classification. Therefore, classification ability of conditional attribute set relative to decision attribute set is made certain when a decision information table is given, and the ability should be unchanged in the process of attribute reduction, accordingly, conditional entropy $E(D\,|C)$ should be unchanged.

Let's constitute a tolerance relation R on the conditional attribute set C of continuous value information system S. From the standpoint of conditional entropy $E(D\,|C)$, we can found the result as follows.

1) In a continuous value information system S, each instance is different from the others, and conditional attribute values of an instance could be different with another instance', but their decision attribute may be same. If threshold vale of α in tolerance relation is taken the minimum, for example, α is taken data precision

of the decision information table, tolerance class $n_R(x_i)$ of an instance x_i only contains x_i itself. Accordingly, conditional entropy $E(D\,|C) = 0$.

2) If threshold vale of α in tolerance relation is increased from data precision, tolerance class $n_R(x)$ of any instance x contains more and more instances. At first, $n_R(x) \subset n_Q(x)$, $|x|_{R \cup Q} = n_R(x) \cap n_Q(x) = n_R(x)$, $E(D\,|C) = 0$.

3) If threshold vale of α is increased continually, tolerance class $n_R(x)$ of any instance x contains more and more instances, at a time, $n_R(x) \not\subset n_Q(x)$, $|x|_{R \cup Q} = n_R(x) \cap n_Q(x) \neq n_R(x)$, $E(D\,|C) \neq 0$.

4) If threshold vale of α is increased continually to some extent, $n_Q(x) \subset n_R(x)$, $|x|_{R \cup Q} = n_R(x) \cap n_Q(x) = n_Q(x)$, at the same time, tolerance class of condition attribute set C is bigger, that is, $|x|_R$ is bigger, then $E(C)$ is decreased, $E(D\,|C)$ is increased.

The relationship between entropy, condition entropy and $|x|_R$ can be depicted in Fig. 1.

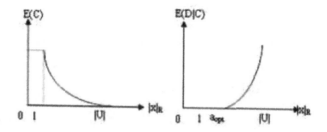

Fig. 1. The relationship between entropy, condition entropy and $|x|_R$

From Fig. 1 and discussion above, if threshold vale of α take α_{opt} , it could make $E(D\,|C) = 0$ and classification ability of conditional attribute set relative to decision attribute set is unchanged, at the same time, tolerance class of x is biggest relative to $E(D\,|C) = 0$. In a sense, knowledge granular of conditional attribute set is biggest in α_{opt}, then, generalization is the best.

In a word, in this section, parameter selection of α is discussed, based on the idea of data driven data mining, As the suitable threshold vale of α , α_{opt} is found and it can keep the classification ability of conditional attribute set relative to decision attribute set, and at the same time, it can keep generalization the most. It is predominant for the course of finding α_{opt} since the method is based on data driven and dose not need expert's experience, therefore, the method has more robustness.

2.3 Proposed Attribute Reduction Algorithm Based on Conditional Entropy in Tolerance Relation

Based on the definition about reduct in tolerance relation in Def. 8 and proposed suitable threshold vale of α_{opt}, a data driven attribute reduction algorithm based on conditional entropy named DDARACE is proposed for emotion recognition system.

Alg. 1: DDARACE

Input: a decision table $S = (U, C \cup D, V, f)$ according to a emotion recognition system, where U is a finite set of objects, C is the conditional attribute set and include 33 facial features. and $D = \{d\}$ is the decision attribute set and include 7 basic emotion.

Output: a relative reduction B of S

Step1: Compute α_{opt}, then set up a tolerance relation on conditional attribute set C.

Step2: Compute condition entropy of decision attribute set D relative to condition attribute set C in decision table S, $E(D|C)$.

Step3: $\forall a_i \in C$, compute $E(D|\{a_i\})$. Sort $E(D|\{a_i\})$ according to a_i.

Step4: Let $B = C$, deal with each a_i as the follows according to $E(D|\{a_i\})$ descendently.

Step4.1: Compute $E(D|B - \{a_i\})$;

Step4.2: If $E(D|C) = E(D|B - \{a_i\})$, attribute a_i should be reduct, $B = B - \{a_i\}$, otherwise, a_i could not be reduct, B is holding.

3   Experiment and Discussion

In this paper, three comparative experiments are done on two facial emotion data sets . One facial emotion data set comes from CMU, which include 405 expression images with 7 basic emtoions. Another facial emotion data set is extracted from 6 volunteers, in which three are female and three are male. The dataset contains 652 expression images. For both of the dataset, 33 facial features are extracted for emotion recognition accoring to [4-7].

Three compared experiments are taken to prove the effective of the proposed method. SVM is taken as classifier in the three experiments, meanwhile the training course of SVM adopts same parameters for all the three experiments. On the other hand, 4-fold cross-validation is taken for every experiment.

Proposed algorithm DDARACE is taken as attribute reduction and SVM is taken classifier in the first experiment, and it can be abbreviated DDARACE+ SVM.

CEBARKNC [14] is a reduction algorithm based on conditional entropy in equivalence relation in traditional rough set. It is taken is taken as attribute reduction and SVM is taken classifier in the second experiment, and it can be abbreviated CEBARKNC +SVM. The attribute reduction of CEBARKNC is taken by experiment tools of RIDAS [15].

In third experiment, process of attribute reduction is omitted, all the data is trained and classified by SVM.

When we compare No.1with No.2 experiment from Table 1 and Table 2, we can find that No. 1 experiment can use nearly as many features as the No.2 experiment, but the correct classification rate is a little less than No.2 experiment

Table 1. Three comparative experiments on CMU facial database

	DDARACE+SVM		CEBARKNC+SVM		SVM	
	Rate	Number	Rate	Number	Rate	Number
1	83.1	14	81.1	12	81.1	33
2	70.2	10	73.2	12	75.2	33
3	61.3	11	60.3	13	66.3	33
4	75.5	10	77.4	13	76.4	33
average	72.5	11.25	73	12.50	74.75	33

Table 2. Three comparative experiments on Self-construction facial database

	DDARACE+SVM		CEBARKNC+SVM		SVM	
	Rate	Number	Rate	Number	Rate	Number
1	77.91	10	76.69	13	90.80	33
2	83.44	11	80.98	13	90.18	33
3	91.41	25	85.28	15	90.80	33
4	81.60	10	72.39	13	91.41	33
average	83.59	14	78.83	13.50	90.80	33

in the first dataset, but the correct classification rate is much more better than No.2 experiment in the second data set. Therefore, we can draw a conclusion that DDARACE is also a useful feature selection method using in emotion recognition system compared with CEBARKNC.

When we compare No.1 with No. 3 experiment from Table 1 and Table 2, we can find that although No. 3 experiment can get much more better correct classification rate than No.1 experiment in the second dataset and a little better correct classification rate than No.1 experiment in the first dataset, but it use more features. Since No.1 experiment use less features and get a nearly correct classification rate compared with No.3 experiment, therefore, we can draw a conclusion that proposed method, DDARACE used as feature selection and SVM taken as classifier, is also an effective method for emotion recognition systems, and it is more suitable real time emotion recognition for it just use less feature for recognition.

4   Conclusion and Future Works

In this paper, based on a reduction algorithm based conditional entropy in tolerance relation, and method of finding appropriate threshold vale of α in tolerance relation, a novel emotion recognition method is proposed. From the experimental results, the proposed method is effective and efficient. Since the proposed method doesn't need expert knowledge and parameter of the method is decided according to data only, therefore, the method should be robust in real applications. In the future work, ensemble of feature selection in tolerance relation will be studied and used for emotion recognition.

References

1. Picard, R.W.: Affective Computing. MIT Press, Cambridge (1997)
2. Picard, R.W.: Affective Computing: Challenges. International Journal of Human-Computer Studies 1, 55–64 (2003)
3. Picard, R.W., Vyzas, E., Healey, J.: Toward Machine Emotional Intelligence: Analysis of Affective Physiological State. IEEE Transactions on Pattern Analysis and Machine Intelligence 10, 1175–1191 (2001)
4. Yang, Y., Wang, G.Y., Chen, P.J., Zhou, J., He, K.: Feature Selection in Audiovisual Emotion Recognition Based on Rough Set Theory. Transaction on Rough Set VII, 283–294 (2007)
5. Yang, Y., Wang, G.Y., Chen, P.J., Zhou, J.: An Emotion Recognition System Based on Rough Set Theory. In: Proceeding on Active Media Technology 2006, pp. 293–297 (2006)
6. Chen, P.J., Wang, G.Y., Yang, Y., Zhou, J.: Facial Expression Recognition Based on Rough Set Theory and SVM. In: Wang, G.-Y., Peters, J.F., Skowron, A., Yao, Y. (eds.) RSKT 2006. LNCS (LNAI), vol. 4062, pp. 772–777. Springer, Heidelberg (2006)
7. Zhou, J., Wang, G.Y., Yang, Y., Chen, P.J.: Speech Emotion Recognition Based on Rough Set and SVM. In: Proceedings of ICCI 2006, pp. 53–61 (2006)
8. Pawlak, Z.: Rough sets. International J. Comp. Inform. Science. 11, 341–356 (1982)
9. Pawlak, Z.: Rough Classification. International Journal of Man-Machine Studies 5, 469–483 (1984)
10. Kryszkiewicz, M.: Rough set approach to incomplete information systems. Information Sciences 1-4, 39–49 (1998)
11. Kryszkiewicz, M.: Rules in incomplete information system. Information Sciences 3-4, 271–292 (1999)
12. Grzymala-Busse, J.W.: Characteristic Relations for Incomplete Data A Generalization of the Indiscernibility Relation. In: Tsumoto, S., Słowiński, R., Komorowski, J., Grzymała-Busse, J.W. (eds.) RSCTC 2004. LNCS (LNAI), vol. 3066, pp. 244–253. Springer, Heidelberg (2004)
13. Wang, G.Y.: Domain-oriented Data-driven Data Mining (3DM): Simulation of Human Knowledge Understanding. Web Intelligence Meets Brain Informatics, 278–290 (2007)
14. Wang, G.Y., Yu, H., Yang, D.C.: Decision table reduction based on conditional information entropy. Journal of computers 7, 759–766 (2002) (in Chinese)
15. Wang, G.Y., Zheng, Z., Zhang, Y.: RIDAS–A Rough Set Based Intelligent Data Analysis System. In: The First Int. Conf. on Machine Learning and Cybernetics (ICMLC 2002), pp. 646–649 (2002)

Learning from Soft-Computing Methods on Abnormalities in Audio Data

Alicja Wieczorkowska

Polish-Japanese Institute of Information Technology,
Koszykowa 86, 02-008 Warsaw, Poland
alicja@pjwstk.edu.pl

Abstract. In our research we deal with polyphonic audio data, containing layered sounds of representing various timbres. Real audio recordings, musical instrument sounds of definite pitch, and artificial sounds of definite and indefinite pitch were applied in this research. Our experiments included preparing training and testing data, as well as classification of these data. In this paper we describe how results obtained from classification allowed us to discover abnormalities in the data, then adjust the data accordingly, and improve the classification results.

Keywords: Music Information Retrieval, sound recognition.

1 Introduction

Audio data are difficult to deal with, since we have to work with samples representing amplitude of the recorded complex sound wave. Therefore, parameterization is needed before further processing, in order to replace a long sequence of amplitude values for a given audio channel with a relatively compact feature vector. However, identification of instruments present in a given piece of music (if we are interested in more information than just a tune), still poses a big problem. Since all users of Internet and even stand-alone computers have access to large amount of music data in digital form, it is desirable to have the possibility to automatically search through such data, in order to find the favorite tune, played by favorite instrument, etc.

Decomposing complex audio wave into source waves representing particular instruments or voices is hardly feasible (unless spatial cues are used, but the position of sources may overlap anyway). Sound engineers usually record each instrument or vocal on separate tracks, but the user listening to the CD only gets the final result, which is down-mixed into 2 stereo channels. Still, extraction of the main, dominating sounds in the mix recorded in any audio track/channel, is possible, although the more sources, the more difficult the task, especially if spectra overlap, and the overlapping sounds are of the same length.

In our research, we deal with polyphonic, multi-timbral recordings, mainly representing musical instrument sounds of definite pitch, since they are crucial to recognize tunes (maybe parallel) that can be present in a given audio recording. We used singular isolated sounds of musical instruments of definite

C.-C. Chan et al. (Eds.): RSCTC 2008, LNAI 5306, pp. 465–474, 2008.

pitch, accompanied by various sounds: complex orchestral sounds, sounds of other instruments of definite pitch, and artificial sounds of definite pitch (harmonic waves) and indefinite pitch (noises). Our purpose was to recognize the instrument dominating in a given recording, i.e. the loudest one, using a limited set of exemplary instruments and accompanying (mixed) sounds, and also possibly obtaining the generalization property of the classifier, i.e. recognition of the specified instruments in the audio environment different than mixes used for training purposes. It is a difficult task, because there is no standard sound parameterization, data set, nor experiment set-up, and it can be difficult even for humans. Moreover, the sound data change significantly when a different articulation (the way the sound is played) is applied, or even when a different specimen of the instrument is used. Sounds of different instruments can be more similar to each other than sounds of the same instrument played in a different way. Consequently, the classification cannot rigidly fit the data, to obtain classifiers with ability to recognize new data with possibly high accuracy. This is why soft computing methods seem to be good tools to deal with such data. Also, sounds are described by musicians using subjective descriptors, and no clear dependencies between numerical and subjective properties of sounds can be easily found. Therefore, observing any numerically definable sound properties is also desirable, which is another advantage to be gained from these experiments.

This paper presents and discusses results of experiments on recognition of the dominating instrument in sound mix, which was already investigated by the author [17], [18]. The focus of this paper is to find abnormalities in the analyzed data, and conclude with an improved set-up for further experiments in the future.

1.1   Literature Review

Automatic recognition of musical instrument is not a new topic and has been already investigated by many researchers. A broad review of parameterization and classification methods used in the research on recognition of isolated singular sounds is given in [3]. Sound parameterization techniques include various spectral domain, time domain, and timbral-spectral methods, based on Fourier transform, wavelet analysis, cepstral coefficients, constant-Q coefficients, and so on [1], [2], [4], [16]. MPEG-7 based features are applied for sound parameterization purposes [10], [12]. Research on instrument identification in mixes has also been performed, with and without using spacial clues [6], [14], [15]. Various classifiers have been applied, including k-nearest neighbor, decision trees, support vector machines, artificial neural networks, rough set based classifiers, hidden Markov models, and other techniques [5], [3], [7], [9]. Obviously, the papers mentioned here represent only part of the research performed in this area, still they picture the main trends and techniques applied.

Identification of particular sounds in mixes is especially important for supporting automatic transcription of musical recordings, and although we should not expect full automatic transcription, the main notes still can be extracted. In our research performed in the area of instrument recognition in sound mixes, we were dealing with audio data parameterized mainly based on MPEG-7 parameters, and applying various classifiers, including Bayesian networks, decision trees, support vector machines, and other methods, using WEKA classification software [13], [17], [18]. Some experiments were performed using cross-validation for the training data, but later we decided to use different data for training and testing, in order to test generalization abilities of the classifiers used, and to obtain more reliable evaluation of our methodology [8], [18].

2   Audio Parameterization

In this research, audio data parameterization was mainly based on MPEG-7 parameters, and other features already applied for the recognition of musical instruments [19]. The parameterization was performed using 120 ms analyzing frame, sliding along the entire parameterized sound, with Hamming window and hop size 40 ms. Such a long frame was used in order to parameterize low sounds, if needed. Most of the calculated parameters represent average value of parameters calculated through consecutive frames of a sound. Since some of the features are multi-dimensional, statistical descriptions of those features were extracted in order to avoid too high dimensionality of the data. The feature set consists of the following 219 parameters [18], [19]:

- MPEG-7 audio descriptors:

    - basic spectral descriptors, using 32 frequency bins, with values averaged through frames for the entire sound: *AudioSpectrumSpread*, *AudioSpectrumCentroid*, *AudioSpectrumFlatness* - 25 out of 32 frequency bands were used in this case;
    - spectral basis - *AudioSpectrumBasis*; minimum, maximum, mean, distance, and standard deviation were extracted, for 33 subspaces;
    - timbral temporal - *LogAttackTime*, *TemporalCentroid*;
    - timbral spectral, averaged through the entire sound - *HarmonicSpectralCentroid*, *HarmonicSpectralSpread*, *HarmonicSpectralVariation*, *HarmonicSpectralDeviation*;

- other descriptors:

    - *Energy* - average energy of spectrum in the parameterized sound;
    - MFCC - min, max, mean, distance, and standard deviation of the MFCC vector through the entire sound;
    - *ZeroCrossingDensity*, averaged through the given sound;
    - *RollOff* - the frequency below which a chosen percentage of the accumulated magnitudes of the spectrum is concentrated (averaged over all frames);
    - *Flux* - the difference between the magnitude of the DFT points in a given frame and its successive frame, averaged through the entire sound (value multiplied by 10^7 to comply with the requirements of the WEKA classifiers);

- *AverageFundamentalFrequency*, with maximum likelihood algorithm applied for pitch estimation;
- *Ratio* $r_1, \ldots, r_{11}$ - ratio of the amplitude of a harmonic partial to the total harmonic partials.

Detailed description of sound parameters is behind the scope of this paper, and it can be found in [10], [12].

Since there is no one standard parameter set that can be applied for all sound-related research purposes, we use the parameterization which has already been applied in similar research. Still, the experiments allow checking if this parameter set performs satisfactorily, or should be modified.

3   Classification Set-Up

The experiments were performed using audio samples from McGill University Master Samples CDs [11], containing singular sounds of musical instruments, recorded with 44.1 kHz sampling rate and with 16-bit resolution. Data from the left channel of stereo recordings were taken for further processing. Initial experiments were focused on 4 instruments of definite pitch:
B-flat clarinet, trumpet, violin, and cello.

This set was later replaced by the following 8 instruments:

B-flat clarinet, cello, trumpet, flute, oboe, tenor trombone, viola, and violin.

Some of the sounds were played with vibration, which makes recognition more difficult because of changes in pitch, amplitude, and timbre of such sounds.

WEKA [13] software was used for classification experiments. Initially, Bayesian Network, decision trees (Tree J48), Logistic Regression Model (LRM), and Locally Weighted Learning (LWL) were applied. However, the obtained results did not show consistency between those classifiers. Later, we decided to apply Support Vector Machine (SMO) classifier, since it is suitable for multi-dimensional data, as it aims at finding the hyperplane that best separates observations belonging to different classes in multi-dimensional feature space. Also, such a classifier was already reported successful in case of musical instrument sound identification.

Initial experiments were performed using cross-validation, but in further experiments, different data were prepared for training and testing purposes, to allow better evaluation of the classifier. The details of the experiments are given in the next section.

4   Experiments and Results

In our experiments, we were using singular musical instrument sounds, mixed with other sounds, of level lower than the main sound. We experimented with various levels of added (mixed) sounds, in order to check how accuracy of the recognition of the main sound changes depending on the level of accompanying sound. Initially we planned to perform ten-fold or three-fold cross-validation, but finally different training and testing data were used in experiments.

4.1   Initial Experiments - 4 Instruments

To start with, we decided to use the following musical instrument sounds as the recognition goal:

1. B-flat clarinet - 37 sound objects,
2. C-trumpet (also trumpet muted, mute Harmon with stem out) - 65 objects,
3. violin vibrato - 42 objects
4. cello vibrato - 43 objects.

These sounds were mixed with orchestral recordings for training purposes. Adagio from Symphony No. 6 in B minor, Op. 74, Pathetique by P. Tchaikovsky was used, choosing 4 short excerpts representing data based on a chord, but changing in time (with fast string passages). The level of added sounds was diminished to 10, 20, 30, 40, and 50% of the original amplitude.

For testing, the singular sounds were mixed with singular sounds of the same instrument (440 Hz sounds, i.e. A4 were chosen for this purpose).

The following classifiers from WEKA have been applied:

- Bayesian network,
- Logistic Regression Model (LRM),
- Locally Weighted Learning (LWL),
- decision tree J48.

The results obtained from these experiments are shown in Table 1.

Table 1. Results for initial experiments with classifiers from WEKA, for 4 instruments

Mix level	Bayesian network	LRM	LWL	J48
10%	80.98%	85.11%	67.42%	76.46%
20%	76.33%	89.36%	66.36%	79.65%
30%	77.39%	85.90%	62.63%	91.62%
40%	76.73%	84.18%	55.85%	66.36%
50%	75.13%	82.98%	53.86	71.94%

As we can see, no clear dependency can generally be observed between the accuracy and the level of mixed sounds - at least different trends are visible for various classifiers, but generally the obtained accuracy was relatively high, so we can conclude that the parameterization used is satisfying. As we have expected, lower levels of added (mixed) sounds generally result in higher accuracy.

Local maxima of classification accuracy oscillated around 10-30% (with another maximum around 50% for decision trees). This result was quite a bit surprising, since we hoped to observe more clear dependency. This result suggested that there might be some abnormality in the data regarding levels of added (mixed) sounds. Because of big amount of audio data, we did not check all samples before experiments, but the results showed that there are abnormalities in the data, and more careful elaboration of the audio data for experiments is needed. First of all, since the levels of the main sounds and mixed sounds were not normalized, in further experiments we decided to apply normalization of the mixed (added) sound with respect to the RMS level of the main sound, before further processing of the sound loudness.

Also, more steps in added levels could show more details, especially with denser steps for lower levels of added sounds. Therefore, we decided to improve the experiment set-up, this time choosing more thresholds of levels of mixed sounds, and not in linear, but in geometrical way instead, since this is more suitable for human hearing.

4.2   Extended Experiments - 8 Instruments

Continuing our experiments, we decided to use a larger set of decision classes, including the following 8 instruments:

1. B-flat clarinet,
2. cello - bowed, played vibrato,
3. trumpet,
4. flute played vibrato,
5. oboe,
6. tenor trombone,
7. viola - bowed, played vibrato,
8. violin - bowed, played vibrato.

For each instrument, 12 sounds representing octave no. 4 (in MIDI notation) were chosen - we wanted to avoid a huge data set. To maintain reasonable size of the data sets, and avoid all possible combinations of sounds, we decided to perform experiments on the most difficult case, i.e. when spectra of mixed sounds fully overlap. Therefore, these data were mixed for training with artificially generated harmonic waves, triangular and saw-tooth, of always of the same pitch as the main sound, and also with noises, white and pink. The following levels of mixed, accompanying sounds were chosen:

- 50%,
- $50\%/\sqrt{2} \approx 35.36\%$,
- 25%,
- $25\%/\sqrt{2} \approx 17.68\%$
- 12.5%,
- $12.5\%/\sqrt{2} \approx 8.84\%$,
- 6.25%.

The level of each added (mixed) sound was first normalized with respect to the RMS of the main sound, then silence replaced the beginning and the end of each added sound, and fade-in and fade-out effects were applied. Next, the level was diminished, according to the desired percentage level. Therefore, we made sure that the main sound is actually always dominating, i.e. louder than any accompanying sound, which was not previously assured.

Testing was performed on mixes of the main sound with the sum of other 7 instrument sounds of the same pitch, with the RMS adjusted to the level of the main sound, with silence at the beginning and the end, and fade in and fade out effects applied, to assure that all the time, including transients, the main sound is still louder, and then processed the level similarly as during training.

Since previously we did not observe consistency between results obtained from different classifiers, we decided to use one classifier only. This time, support vector machine was chosen, as reported successful in similar research and suitable to multidimensional data.

These experiments yielded the following results:

- 50.00% level: 81.25% correctness,
- 35.36% level: 90.63% correctness,
- 25.00% level: 87.50% correctness,
- 17.68% level: 94.79% correctness,
- 12.50% level: 81.25% correctness,
- 08.84% level: 100% correctness,
- 06.25% level: 100% correctness.

As we can see, the recognition rate has significantly improved, and the lowest levels of mixed sounds did not influence the recognition accuracy, which is the expected result. Other results oscillate around 80-90%, so still there are some difficult data to classify in our data set. This suggests having a closer look into the details, with hope to find the most problematic sounds. The contingency table for this experiment is shown in Table 2.

As we can observe, especially viola and violin pose difficulties in recognition. The percentage of confusion is quite high, showing again some abnormality in the data. This result suggests that the sounds used in experiments should be more carefully checked (also by human experts). However, on the other hand, those instruments are difficult to distinguish even for humans, so problems with their separation are not so surprising.

Therefore, this experiment shows the general results as quite high, but the difficulties with recognition of violin and viola suggests some abnormalities in the data. Again, this is a hint that maybe the data need further improvement. Indeed, when listening to the recorded samples afterwards, we discovered that mixes suffer from some imperfections. Actually, the sounds were not perfectly in tune, and also vibration of mixed sounds was different, which can be problematic in experiments. This suggests that maybe new samples should be prepared for experiments, with musicians playing together rather than independently (and then mixed), playing in sync, and more in tune.

5   Summary and Conclusions

The performed experiments aimed at recognition of musical instrument sound, dominating in mixes of instrumental sounds of definite pitch as the main sounds,

Table 2. Contingency table for the support vector machine classifiers, trained and tested for 8 instruments

Classified as ->	clarinet	cello	trumpet	flute	oboe	trombone	viola	violin
clarinet+6.25%	12							
clarinet+8.84%	12							
clarinet+12.5%	10	1			1			
clarinet+17.68%	12							
clarinet+25%	12							
clarinet+35.36%	12							
clarinet+50%	10	1			1			
cello+6.25%		12						
cello+8.84%		12						
cello+12.5%		12						
cello+17.68%		12						
cello+25%		12			.			
cello+35.36%		12						
cello+50%		12						
trumpet+6.25%			12					
trumpet+8.84%			12					
trumpet+12.5%			12					
trumpet+17.68%			12					
trumpet+25%			12					
trumpet+35.36%			12					
trumpet+50%			12					
flute+6.25%				12				
flute+8.84%				12				
flute+12.5%	1	1		8	2			
flute+17.68%				12				
flute+25%				12				
flute+35.36%				12				
flute+50%	1	1		8	2			
oboe+6.25%					12			
oboe+8.84%					12			
oboe+12.5%					12			
oboe+17.68%					12			
oboe+25%					12			
oboe+35.36%					12			
oboe+50%					12			
trombone+6.25%						12		
trombone+8.84%						12		
trombone+12.5%						12		
trombone+17.68%						12		
trombone+25%						12		
trombone+35.36%						12		
trombone+50%						12		
viola+6.25%							12	
viola+8.84%							12	
viola+12.5%		3					9	
viola+17.68%		2					10	
viola+25%		3					9	
viola+35.36%		2					10	
viola+50%		3					9	
violin+6.25%								12
violin+8.84%								12
violin+12.5%	1				1		7	3
violin+17.68%							3	9
violin+25%					1		8	3
violin+35.36%					2		5	5
violin+50%	1				1		7	3

and added other sounds. The feature vector was based on parameterization already applied in similar research, and also in this research the parameter set ued yielded good results.

The initial experiments showed that no clear dependency could be found between the level of mixed sounds and the correctness of recognition for all classifiers applied. Therefore, further experiments focused on one classifier, and more steps used, especially for lower levels. Since linearly chosen steps of added unprocessed sounds did not work as expected, data were considered abnormal and checked. Therefore, level processing was applied in further experiments, to assure that the main sound is actually the loudest one, and geometrical step was selected in further research, with denser steps for lower levels. The obtained results showed significant improvement of accuracy, even though the most difficult case of the same pitch of sounds was chosen.

The results again showed abnormality in the data with some instruments far too difficult to distinguish, thus being considered again as abnormality in the data, and suggesting further work with those sounds. Although these instruments, violin and viola, have very similar timbre, we believe that more careful preparation of the data may improve the recognition rate.

As a result, further experiments are planned as a follow-up, to assure removal of probably still existing abnormalities in the data, obtaining sounds yielding consistent results, and conforming to general standards of such a recognition, even in case of data difficult to recognize for humans.

Acknowledgments. This work was supported by the National Science Foundation under grant IIS-0414815, and also by the Research Center of PJIIT, supported by the Polish National Committee for Scientific Research (KBN).

The author would like to express thanks to Dr. Xin Zhang for her help with data parameterization, to Zbigniew W. Raś from the University of North Carolina at Charlotte for fruitful discussions, and to Elżbieta Kubera from the University of Life Sciences in Lublin for help with experiments.

References

1. Aniola, P., Lukasik, E.: JAVA Library for Automatic Musical Instruments Recognition. In: AES 122 Convention, Vienna, Austria (2007)
2. Brown, J.C.: Computer identification of musical instruments using pattern recognition with cepstral coefficients as features. J. Acoust. Soc. Am. 105, 1933–1941 (1999)
3. Herrera, P., Amatriain, X., Batlle, E., Serra, X.: Towards instrument segmentation for music content description: a critical review of instrument classification techniques. In: International Symposium on Music Information Retrieval ISMIR (2000)
4. Kaminskyj, I.: Multi-feature Musical Instrument Sound Classifier w/user determined generalisation performance. In: Australasian Computer Music Association Conference ACMC, pp. 53–62 (2002)

5. Kitahara, T., Goto, M., Okuno, H.G.: Pitch-Dependent Identification of Musical Instrument Sounds. Applied Intelligence 23, 267–275 (2005)
6. Klapuri, A., Virtanen, T., Eronen, A., Seppanen, J.: Automatic transcription of musical recordings. In: Consistent and Reliable Acoustic Cues for sound analysis CRAC Workshop (2001)
7. Kostek, B., Dziubinski, M., Dalka, P.: Estimation of Musical Sound Separation Algorithm Efectiveness Employing Neural Networks. Journal of Intelligent Information Systems 24, 133–157 (2005)
8. Livshin, A., Rodet, X.: The importance of cross database evaluation in musical instrument sound classification: A critical approach. In: International Symposium on Music Information Retrieval ISMIR (2003)
9. Martin, K.D., Kim, Y.E.: Musical instrument identification: A pattern-recognition approach. In: 136th meeting of the Acoustical Society of America, Norfolk, VA (1998)
10. ISO/IEC JTC1/SC29/WG11: MPEG-7 Overview,
    http://www.chiariglione.org/mpeg/standards/mpeg-7/mpeg-7.htm
11. Opolko, F., Wapnick, J.: MUMS - McGill University Master Samples. CD's (1987)
12. Peeters, G., McAdams, S., Herrera, P.: Instrument Sound Description in the Context of MPEG-7. In: International Computer Music Conference ICMC 2000 (2000)
13. The University of Waikato: Weka Machine Learning Project,
    http://www.cs.waikato.ac.nz/~ml/
14. Virtanen, T.: Algorithm for the separation of harmonic sounds with time-frequency smoothness constraint. In: 6th International Conference on Digital Audio Effects DAFX (2003)
15. Viste, H., Evangelista, G.: Separation of Harmonic Instruments with Overlapping Partials in Multi-Channel Mixtures. In: IEEE Workshop on Applications of Signal Processing to Audio and Acoustics WASPAA 2003 (2003)
16. Wieczorkowska, A.: Wavelet Based Analysis and Parameterization of Musical Instrument Sounds. In: ISSEM 1999, pp. 219–224 (1999)
17. Wieczorkowska, A., Kolczyńska, E.: Quality of Musical Instrument Sound Identification for Various Levels of Accompanying Sounds. In: Raś, Z.W., Tsumoto, S., Zighed, D.A. (eds.) MCD 2007. LNCS (LNAI), vol. 4944, pp. 93–103. Springer, Heidelberg (2008)
18. Wieczorkowska, A., Kolczyńska, E.: Identification of Dominating Instrument in Mixes of Sounds of the Same Pitch. In: Ann, A., Matwin, S., Raś, Ś.D. (eds.) ISMIS 2008. LNCS (LNAI), vol. 4994, pp. 455–464. Springer, Heidelberg (2008)
19. Zhang, X., Ras, Z.: Discriminant feature analysis for music timbre recognition. In: ECML/PKDD Third International Workshop on Mining Complex Data (MCD 2007), pp. 59–70 (2007)

Shadowed Clustering for Speech Data and Medical Image Segmentation

Bishal Barman[1], Sushmita Mitra[2], and Witold Pedrycz[3]

[1] Electrical Engineering Department
S. V. National Institute of Technology, Surat - 395 007, Gujarat, India
bishalbarman@gmail.com
[2] Center for Soft Computing Research
Indian Statistical Institute, Kolkata - 700 108, India
sushmita@isical.ac.in
[3] Electrical and Computer Engineering Department
University of Alberta, Edmonton, Canada - T6G 2G7
pedrycz@ece.ualberta.ca

Abstract. The paper presents a novel application of the shadowed clustering algorithm for uncertainty modeling and CT scan image segmentation. The core, shadowed and the exclusion regions, generated via shadowed c-means (SCM), quantize the ambiguity into three zones. This leads to faster convergence and reduced computational complexity. It is observed that SCM generates the best prototypes even in the presence of noise, thereby producing the best approximation of a structure in the unsupervised mode. A comparison with rough-fuzzy clustering algorithm reveals the automatic determination of the threshold and absence of externally tuned parameters in SCM. Experiments suggest that SCM is better suited for extraction of regions under vascular insult in the brain via pixel clustering. The relative efficacy of SCM in brain infarction diagnosis is validated by expert radiologists.

Keywords: Shadowed clustering, three-valued logic, rough-fuzzy clustering, image segmentation, CT scan imaging.

1 Introduction

Shadowed set theory aims at ambiguity demarcation. Gradual distribution of membership values have known to work reasonably well in the fuzzy framework. Though complete, the membership values often represent excessive detail in the form of precise numeric values. Shadowed set, proposed by Pedrycz [1], provides an alternate mechanism for handling uncertainty. Together with fuzzy logic, neural network, genetic algorithms and rough sets, shadowed sets could be considered the next extension to broaden the paradigm of *Soft computing* methodologies [2]. Fuzzy and rough clustering have been well documented in the literature [3] [4] [5]. The development of fuzzy c-means (FCM) [3], rough c-means (RCM) [4] and the hybridized rough-fuzzy c-means (RFCM) [6] [7] serve as specific instances of the extension of the c-means framework [K-means or

C.-C. Chan et al. (Eds.): RSCTC 2008, LNAI 5306, pp. 475–484, 2008.
© Springer-Verlag Berlin Heidelberg 2008

hard c-means (HCM)]. Fuzzy clustering, through membership values, seeks to handle overlapping clusters. Rough clustering, via the notion of approximation spaces, deals with vagueness and ambiguity in data in terms of upper and lower approximation.

Shadowed clustering, called shadowed c-means (SCM), was developed [8] to connect the ideas of FCM, RCM and RFCM. It incorporated the membership concept from fuzzy sets, while simultaneously avoiding the use of too many external parameters as in rough clustering. In this article, we present the application of shadowed clustering for segmentation of regions under vascular insult in brain images. Clustering of the highly overlapped vowels in Indian Telegu speech data [9] is also investigated. SCM provides maximum importance to the core members during clustering, threby increasing the robustness and reliability of the algorithm. We use the membership based Xie-Beni validity index [10] to optimize the number of clusters for the speech data.

The paper is organized into five sections. Section 2 leads the reader into the basic notions about shadowed sets, while Section 3 discusses the rough-fuzzy and the shadowed clustering algorithms. Cluster validation via Xie-Beni index is also described in this section. Section 4 presents the results on the speech data as well as the segmentation of the CT scan imagery. Section 5 draws the conclusion.

2   Shadowed Sets

Shadowed sets looks to answer the question of optimum level of resolution required in precision. Traditional methods, such as fuzzy sets, tend to capture ambiguity exclusively through membership values. Naturally, this leads to the problem of e*xcessive precision in describing imprecise phenomenon* [1] [11]. There is hardly any problem in assigning membership values close to 0 or 1, but a lot of confusion does exist regarding the assignment of a grade of 0.5. To solve the above issues, Pedrycz [1] proposed the idea of shadowed sets to improve the observability of imprecise phenomenon.

Consider a fuzzy set, $\mathbb{G}$, as shown in Fig. 1. Rather than having uniform membership values, defined by the membership function, we go for quantization of the same on the lines of three valued logic. In doing so, some membership values are markedly reduced to zero, while some are elevated to one. To compensate for the ambiguity thus introduced, we declare a particular region as the zone of uncertainty. This area of the universe of discourse has intermediate membership values on a unit interval between [0 1], but is left undefined. In order to induce a shadowed set, a fuzzy set must accept a specific threshold. This facilitates the transformation of the domain of discourse into clearly marked zones of vagueness. We call this mapping, a shadowed set.

$$\mathbb{G} : \mathbf{X} \rightarrow \{0, 1, [0\ 1]\} \tag{1}$$

Elements with grade equal to one form the core, while the elements with $\mathbb{G}(x) = [0, 1]$ lie in the shadow of the mapping; the rest forms the exclusion.

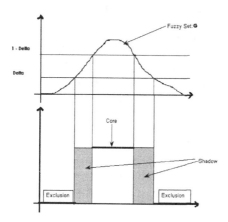

Fig. 1. The fuzzy set G inducing a shadowed set via a threshold, δ

To obtain the threshold, Pedrycz proposed an optimization based on balance of vagueness. A particular threshold, δ, is selected for the quantization process and is expressed in terms of the relationship

$$\mathcal{P}(\delta) = \left| \int_{-\infty}^{L_1} \mathbb{G}(x)\,dx + \int_{L_2}^{\infty} (1 - \mathbb{G}(x))\,dx - \int_{L_1}^{L_2} dx \right|, \qquad (2)$$

where $\delta \in (0, \; 1/2)$ such that $\mathcal{P}(\delta) = 0$. L_1 and L_2 are points where $\mathbb{G}$ is thresholded [11].

Shadowed sets reveal very interesting relationship with rough sets. Although conceptually quite similar, the mathematical foundations of rough sets is very different. We must remember that in rough sets, we define the approximation space in advance and the equivalent classes are kept fixed. In shadowed set theory, on the other hand, the class assignment is dynamic.

Computation of the threshold δ for common membership functions, such as triangular and Gaussian, can be found in [1] [11]. The minima of $\mathcal{P}(\delta)$ gives the δ_{opt} [11].

3   Rough and Shadowed Clustering with Validation

Rough sets [5] are used to model clusters in terms of upper and lower approximations, that are weighted by a pair of parameters while computing cluster proto-
types [4]. We observe that the rough set theory assigns objects into two distinct regions, *viz.*, lower and upper approximations, such that objects in lower approx-
imation indicate definite inclusion in the concept under discussion while those in the upper approximation correspond to possible inclusion in it [4]. Shadowed clustering attempts to overcome the problem of assigning weighting parameters in RFCM. This increases the stability of the algorithm and minimizes data-
dependency [8].

3.1   Rough-Fuzzy C-Means (RFCM)

A rough-fuzzy c-means (RFCM) algorithm, involving an integration of fuzzy and rough sets, has been developed [6]. This allows one to incorporate fuzzy membership value u_{ik} of a sample $\mathbf{x}_k$ to a cluster mean $\mathbf{v}_i$, relative to all other means $\mathbf{v}_j \ \forall \ j \neq i$, instead of the absolute individual distance d_{ik} from the centroid. The major steps of the algorithm are provided below.

1. Assign initial means $\mathbf{v}_i$ for the c clusters.
2. Compute membership u_{ik} for c clusters and N data objects as

$$u_{ik} = \frac{1}{\sum_{j=1}^{c} \left(\frac{d_{ik}}{d_{jk}} \right)^{\frac{2}{m-1}}}, \tag{3}$$

   where m is the fuzzifier.
3. Assign each data object (pattern) $\mathbf{x}_k$ to the lower approximation $\underline{B}U_i$ or upper approximation $\overline{B}U_i$, $\overline{B}U_j$ of cluster pairs U_i, U_j by computing the difference in its membership $u_{ik} - u_{jk}$ to cluster centroid pairs $\mathbf{v}_i$ and $\mathbf{v}_j$.
4. Let u_{ik} be maximum and u_{jk} be the next to maximum.
   If $u_{ik} - u_{jk}$ is less than some *threshold*
      then $\mathbf{x}_k \in \overline{B}U_i$ and $\mathbf{x}_k \in \overline{B}U_j$ and $\mathbf{x}_k$ cannot be a member of any lower approximation,
      else $\mathbf{x}_k \in \underline{B}U_i$ such that membership u_{ik} is maximum over the c clusters.
5. Compute new mean for each cluster U_i as

$$\mathbf{v}_i = \begin{cases} w_{low} \dfrac{\sum_{\mathbf{x}_k \in \underline{B}U_i} u_{ik}^m \mathbf{x}_k}{\sum_{\mathbf{x}_k \in \underline{B}U_i} u_{ik}^m} + w_{up} \dfrac{\sum_{\mathbf{x}_k \in (\overline{B}U_i - \underline{B}U_i)} u_{ik}^m \mathbf{x}_k}{\sum_{\mathbf{x}_k \in (\overline{B}U_i - \underline{B}U_i)} u_{ik}^m} & \text{if } \underline{B}U_i \neq \emptyset \wedge \overline{B}U_i - \underline{B}U_i \neq \emptyset, \\[3ex] \dfrac{\sum_{\mathbf{x}_k \in (\overline{B}U_i - \underline{B}U_i)} u_{ik}^m \mathbf{x}_k}{\sum_{\mathbf{x}_k \in (\overline{B}U_i - \underline{B}U_i)} u_{ik}^m} & \text{if } \underline{B}U_i = \emptyset \wedge \overline{B}U_i - \underline{B}U_i \neq \emptyset, \\[3ex] \dfrac{\sum_{\mathbf{x}_k \in \underline{B}U_i} u_{ik}^m \mathbf{x}_k}{\sum_{\mathbf{x}_k \in \underline{B}U_i} u_{ik}^m} & \text{otherwise.} \end{cases}$$

$$\tag{4}$$

6. **Repeat** Steps 2-5 **until** convergence, *i.e.*, there are no more new assignments.

In rough clustering algorithms, we commonly use a number of parameters, *viz.* $w_{up} = 1 - w_{low}$, $0.5 < w_{low} < 1$, $m = 2$, and $0 < threshold < 0.5$.

3.2   Shadowed C-Means (SCM)

Based on the concept of shadowed sets, we delineate here the shadowed c-means (SCM) clustering algorithm [8]. The quantization of the membership values into core, shadowed and exclusion region permit reduced computational complexity. The elements corresponding to the core should not have any fuzzy weight factor in terms of its membership values. In other words, unlike uniform computation of u_{ik} as in FCM, here the u_{ik} should be unity for core patterns while calculating the centroid. The elements corresponding to the shadowed region lie in the zone of uncertainty, and are treated as in FCM. However, the members of the exclusion region are incorporated in a slightly different manner. Here the fuzzy weight factor for the exclusion is designed to have the fuzzifier raised to itself in the form of a double exponential. The centroid for the ith class is evaluated as $v_i =$

$$\frac{\sum_{x_k|u_{ik}\geq(u_{i_{max}}-\delta_i)} x_k + \sum_{x_k|\delta_i<u_{ik}<(u_{i_{max}}-\delta_i)} (u_{ik})^m x_k + \sum_{x_k|u_{ik}\leq\delta_i} (u_{ik})^{m^m} x_k}{\phi_i + \eta_i + \psi_i},$$

(5)

where

$$\phi_i = \mathbf{card}\{x_k|u_{ik} \geq (u_{i_{max}} - \delta_i)\},$$

(6)

$$\eta_i = \sum_{x_k|\delta_i<u_{ik}<(u_{i_{max}}-\delta_i)} (u_{ik})^m,$$

(7)

$$\psi_i = \sum_{x_k|u_{ik}\leq\delta_i} (u_{ik})^{m^m},$$

(8)

and δ_i is the threshold for the ith class. This arrangement causes a much wider dispersion and a very low bias factor for elements which can generally be considered outside the class under discussion or most definitely, the exclusion members. This prevents the mean from getting drifted from its true value. It also minimizes the effect of noise and outliers. The threshold to induce the core, shadowed and exclusion region is automatically calculated through functional optimization using eqn. (2).

The mean in eqn. (5) basically tries to first get a coarse idea regarding the cluster prototype (using the first term in the numerator and denominator, respectively) and then proceeds to tune and refine this value using data from the shadowed and exclusion region. This enables a better estimation of the actual cluster prototypes. The major steps of the algorithm are outlined below [8].

1. Assign initial means, v_i, $i = 1, \ldots, c$. Choose values for fuzzifier m, and t_{max}. Set iteration counter $t = 1$.
2. **Repeat** steps (3) - (5) by incrementing t **until** no new assignment is made and $t < t_{max}$.
3. Compute u_{ik} by eqn. (3) for c clusters and N data objects.
4. Compute threshold δ_i for the ith class, in terms of eqn. (2), as $\mathcal{P}(\delta_i) =$

$$\left| \sum_{x_k|u_{ik}\leq\delta_i} u_{ik} + \sum_{x_k|u_{ik}\geq u_{i_{max}}-\delta_i} (u_{i_{max}} - \delta_i) - \mathbf{card}\{x_k|\delta_i < u_{ik} < (u_{i_{max}} - \delta_i)\} \right|$$

(9)

such that

$$\delta_i = \delta_{opt} = \arg\min_{\delta_i} \mathcal{P}(\delta_i).$$

(10)

5. Update mean, v_i, using eqn. (5).

The range of feasible values of δ_i could be taken as $[u_{i_{min}}, \frac{u_{i_{min}}+u_{i_{max}}}{2}]$.

3.3   Cluster Validation

Prespecification of the number of clusters in partitive algorithms is a necessity. Hence the results are dependent on the choice of c. In this article, we compute the optimal number of clusters c_0 in terms of the Xie-Beni index [10]. As the Xie-Beni index is designed to work in a fuzzy framework, therefore it was employed to validate our results. The fuzzy validity function identifies overall compact and separate fuzzy c-partitions. This function depends upon the data set, geometric distance measure, distance between cluster centroids, as well as the fuzzy partitions, irrespective of the algorithm used. We define ρ as a fuzzy clustering validity function

$$\rho = \frac{\sum_{i=1}^{c} \sum_{j=1}^{N} u_{ij}^2 ||\mathbf{v_i} - \mathbf{x_j}||^2}{N \min_{i,j} ||\mathbf{v_i} - \mathbf{v_j}||^2}. \tag{11}$$

In case of FCM and RFCM algorithms, with $m = 2$, eqn. (11) reduces to

$$\rho = \frac{J_2}{N * (d_{min})^2}, \tag{12}$$

where J_2 is the fuzzy objective function with Euclidean norm and $d_{min} = \min_{i,j} ||\mathbf{v_i} - \mathbf{v_j}||$. The more separate the clusters, the larger $(d_{min})^2$ and the smaller ρ. Thus the smallest ρ, corresponding to $c = c_0$, indeed indicates a valid optimal partitioning.

4   Results

The performance of the rough-fuzzy and the shadowed clustering algorithms is presented in this section. Two real life data sets, involving vowel sounds and CT scan images of the brain, are explored.

The *Telegu* speech data, *Vowel*, is a set of 871 vowel sounds from the Indian Telegu language, obtained by the utterance of three male speakers in the age group of 30-35 years, in a Consonant-Vowel-Consonant context [9]. The three input features correspond to the first, second and third vowel format frequencies obtained through spectrum analysis of the speech data. Fig. 2(a) shows the six highly overlapped vowel classes ∂, a, i, u, e, o, marked with symbols 'star', 'plus', 'decagon', 'circle', 'upper triangle' and 'cross', respectively.

Siemens Emotion-Duo model was the clinical instrument for the acquisition of the CT scan imagery. The images, obtained in the *DICOM* format, were

Table 1. Xie-Beni index on Speech data, *Vowel*

Index	c	HCM	FCM	RCM	RFCM	SCM
Xie	5	0.2074	0.2378	0.2503	0.2142	0.3268
Beni	6	0.1665	0.1893	0.1795	0.1625	**0.1496**
	7	0.1774	0.1913	0.2049	0.1598	0.3466

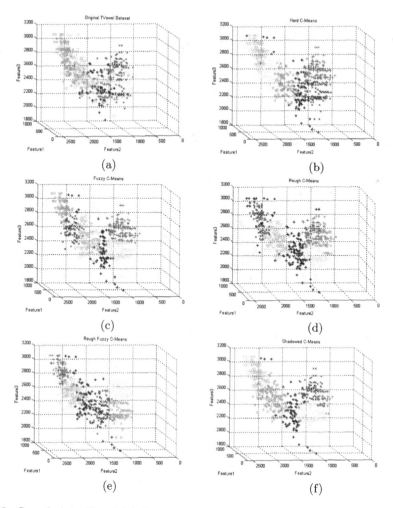

Fig. 2. Speech data *Vowel*. (a) Original, and after clustering with (b) HCM, (c) FCM, (d) RCM, (e) RFCM, (f) SCM algorithms for $c=6$.

converted to *RAW* as part of pre-processing. The images were of size 512 x 512 pixels with 16-bit gray levels. The brain images were of patients in an age-range of 30-65 years, and exhibit different cases of brain infarction. Fig. 3(a) illustrates a sample image for patient P45 indicating fresh vascular insult. We also present the segmented image for a patient, P135, via SCM, indicating chronic case of infarction. (Fig. 4)

4.1   Speech Data

The boundaries portrayed in the scatter plot of *Vowel*, as observed from Figs. 2(b)-(f), are highly fuzzy. The validity indices in Table 1 demonstrate the best

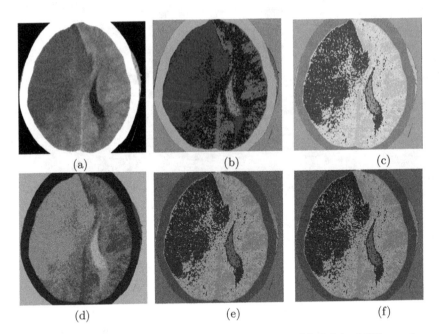

Fig. 3. Sample case of Fresh Infarction for patient, P45. (a) Original CT scan image, and the corresponding segmented versions for (b) HCM (c) FCM (d) RCM (e) RFCM (f) SCM clustering.

results with SCM for $c=6$. This corresponds to the actual number of vowel categories under consideration. For example, in case of RFCM, we observe that XB is indicative of incorrect optimization at seven partitions. On the other hand, SCM provides better modeling of the uncertainty in the overlapped data. The comparative study involving algorithms, HCM, FCM and RCM give higher values of XB index, as compared to SCM. All the algorithms were randomly initialized and the average over nine runs was computed.

4.2   Medical Image

Segmentation partitions an image into some non-overlapping meaningful regions [12]. Pixel clustering is one of the faster and efficient techniques of constituting homogeneous regions for segmentation. Here, we present sample results of different members of the family of c-means algorithms on segmentation of the infarcted region in CT scan images of the brain.

The patients under study, P45 and P135, were suffering from fresh and chronic vascular insult, respectively. The fresh infarction [Fig. 3(a)] is observable on the left, with the left side compressing the right side such that the third ventricle is not visible due to this severe edema. Dilation of the blood ventricles is the main cause of the edema here. In case of chronic infarction, the symmetry of the brain

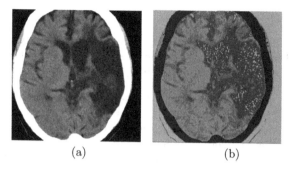

(a)   (b)

Fig. 4. Sample case of Chronic Infarction for patient, P135. (a) Original CT scan image, and the corresponding segmented version for (b) SCM clustering.

in Fig. 4(a) is not as distorted as in the previous case. Cholesterol deposit due to old age is among the main causes of such an infarction.

The problem at hand is modeled as the task of segmenting six regions comprising the gray matter (GM), the white matter (WM), the infarcted region, the skull and the background. Figs. 3(b)-(f) depict the results of segmentation under HCM, FCM, RCM, RFCM and SCM. In the absence of an accurate index to test the accuracy of segmentation in CT scan images, we resorted to expert domain knowledge. In all, 36 frames of the patient P45 and 58 frames of the P135 were studied, and the ground-truth regarding the best segmentation was validated by an experienced radiologist. As before, the SCM algorithm produced the best results. The superiority of SCM based segmentation was readily evident over RFCM based partition [7], and was confirmed by the experts.

5   Conclusion

Applications of the family of c-means algorithms to real life speech and medical imagery were described. The superiority of shadowed clustering, over algorithms like HCM, FCM, RCM, RFCM, was established here. Modulation of the membership function helped achieve a quantization in SCM. This in turn leads to reduced computational complexity, faster convergence and low memory usage. The contrast-enhancement paradigm in SCM enabled knowledge discovery in unlabeled data in a more effective manner. The CT scan image segmentation was viewed from this novel angle, for the efficient extraction of vascular infarction. In the next phase, we aim to establish an inventory, so that a second opinion is readily available to the radiologist.

Acknowledgement

Mr. Bishal Barman would like to thank the Indian National Academy of Engineering (INAE) for a fellowship to carry out this work at the Center for Soft Computing Research, Indian Statistical Institute, Kolkata, India.

References

1. Pedrycz, W.: Shadowed sets: Representing and processing fuzzy sets. IEEE Transactions on Systems, Man, and Cybernetics - B 28, 103–109 (1998)
2. Zadeh, L.A.: Fuzzy logic, neural networks, and soft computing. Communications of the ACM 37, 77–84 (1994)
3. Bezdek, J.C.: Pattern Recognition with Fuzzy Objective Function Algorithms. Plenum Press, New York (1981)
4. Lingras, P., West, C.: Interval set clustering of Web users with rough k-means. Journal of Intelligent Information Systems 23, 5–16 (2004)
5. Pawlak, Z.: Rough Sets, Theoretical Aspects of Reasoning about Data. Kluwer Academic, Dordrecht (1991)
6. Mitra, S., Banka, H., Pedrycz, W.: Rough-fuzzy collaborative clustering. IEEE Transactions on Systems, Man, and Cybernetics, Part-B 36, 795–805 (2006)
7. Mitra, S., Barman, B.: Rough-fuzzy clustering: An application to medical imagery. In: Wang, G., Li, T., Grzymała-Busse, J.W., Miao, D., Skowron, A., Yao, Y. (eds.) RSKT 2008. LNCS (LNAI), vol. 5009, pp. 300–307. Springer, Heidelberg (2008)
8. Mitra, S., Barman, B., Pedrycz, W.: Shadowed C-Means: Bridging fuzzy and rough clustering. IEEE Transactions on Pattern Analysis and Machine Intelligence, Communicated (2008)
9. Pal, S.K., Dutta Majumder, D.: Fuzzy sets and decision making approaches in vowel and speaker recognition. IEEE Transactions on Systems, Man, and Cybernetics 7, 625–629 (1977)
10. Xie, X.L., Beni, G.: A validity measure for fuzzy clustering. IEEE Transactions on Pattern Analysis and Machine Intelligence 13, 841–847 (1991)
11. Pedrycz, W.: Interpretation of clusters in the framework of shadowed sets. Pattern Recognition Letters 26, 2439–2449 (2005)
12. Gonzalez, R.C., Woods, R.E.: Digital Image Processing. Prentice Hall, Upper Saddle River (2002)

Computational Intelligence Techniques Applied to Magnetic Resonance Spectroscopy Data of Human Brain Cancers

Alan J. Barton and Julio J. Valdes

National Research Council Canada, Institute for Information Technology,
M50, 1200 Montreal Rd., Ottawa, ON K1A 0R6
alan.barton@nrc-cnrc.gc.ca, julio.valdes@nrc-cnrc.gc.ca
http://iit-iti.nrc-cnrc.gc.ca

Abstract. Computational intelligence techniques were applied to human brain cancer magnetic resonance spectral data. In particular, two approaches, Rough Sets and a Genetic Programming-based Neural Network were investigated and then confirmed via a systematic Individual Dichotomization algorithm. Good preliminary results were obtained with 100% training and 100% testing accuracy that differentiate normal versus malignant samples.

1 Introduction

Magnetic resonance spectroscopy (MRS) and magnetic resonance imaging (MRI) are two non-invasive and harmless clinical techniques that can provide useful biochemical information about a region of interest in the body. They can be particularly helpful when the organ under investigation is difficult or dangerous to reach (e.g. the brain) where direct inspection and surgery should be avoided as much as possible.

Both techniques are based on magnetic resonance (MR), which is related to the physical property called quantum spin. The MRI technique reveals water concentration levels and is used in routine examinations by clinicians; whereas the MRS technique is not used as frequently as MRI (despite its great potential). MRS information consists of a signal, possibly noisy, composed of peaks whose location and height correspond to different metabolites and their relative concentrations. Reading the most frequent chemical in an MR spectrum is relatively straightforward, but the complete interpretation of a spectrum or the comparison between two spectra usually requires an expert [14]. This reliance on specialized expertise may be one of the reasons why it has been more difficult to introduce MRS into routine medical practice.

An international project, INTERPRET http://azizu.uab.es/INTERPRET, gathered the efforts of 5 centers across Europe with the long term goal of generalizing the use of MRS. During this project, a large database of 1HMR spectra was built in order to develop an automatic MRS-based system to aid clinicians to diagnose brain tumors. Each spectrum in the database was acquired according to a pre-defined protocol and formally validated by clinicians and pathologists [9].

This paper has a preliminary character and will focus on the study of the tumor vs normal differentiation (i.e. $\{G1, G2, G3\} vs \{normal\}$), with 204 and 15 cases respectively. Future studies will cover the distinction between the different types of tumors.

C.-C. Chan et al. (Eds.): RSCTC 2008, LNAI 5306, pp. 485–494, 2008.

2   Rough Sets

The Rough Set Theory [17], [16] bears on the assumption that in order to define a set, some knowledge about the elements is needed. This is in contrast to the classical approach where a set is uniquely defined by its elements. In the Rough Set Theory, some elements may be indiscernible from the point of view of the available information and it turns out that vagueness and uncertainty are strongly related to indiscernibility.

Reducts and Minimum Reducts. Let $O = \{o_1, o_2, \cdots, o_m\}$ be a set of m objects and $A = \{a_1, a_2, \cdots, a_N\}$ a set of N attributes. Let d be a special attribute called the decision attribute. O is consistent if $\forall k, n, \forall i \in [1, N], a_i(o_k) = a_i(o_n) \rightarrow d(o_k) = d(o_n)$. A reduct is a subset $R \subseteq A$ so that $\forall k, n, \forall a \in R, a(o_k) = a(o_n) \rightarrow d(o_k) = d(o_n)$. Minimal reducts are those for which no proper subset is a reduct and are extremely important, as decision rules can be constructed from them [3]. However, the problem of reduct computation is NP-hard, and several heuristics have been proposed [21].

Reduct Computation. Genetic algorithms are the most popular representative of the evolutionary computation family of algorithms [5], [1].They have been used as an approach to reduct computation by [20], which proposed several methods based on the notion of a distinction table; which is a $(m^2 - m)/2 \times (N + 1)$ matrix B where columns i are attributes (the last one is the decision attribute d) and the rows are pairs of objects k, n. For every row $i \in [1, N]$ and every $k, n \in [1, m]$ the values of B are constructed as follows: $B[(k, n), i] = 1$ if $a_i(o_k) \neq a_i(o_n)$ and 0 otherwise. For the last row $B[(k, n), N + 1] = 1$ if $d(o_k) = d(o_n)$ and 0 otherwise. In terms of B, a reduct is a subset of columns R with the property [20] $\forall k, n, \exists i \in R, (B[(k, n), i] = 1) \vee (B[(k, n), N + 1] = 1)$. In its simplest representation, a GA with binary chromosomes of length N encodes subsets of attributes (the indices of the chromosomes for which the value is 1). The evolution is guided by a fitness function given by: $F(r) = ((N - L_r)/N) + C_r/K$, where r is a chromosome, L_r is the cardinality of the set of attributes (given by the number of 1s in the chromosome, C_r is the number of object pairs (with different values of the decision attribute) which are discerned by the attributes in R. $K = (m(m - 1))/2$ is the number of object pairs.

3   Genetic Programming

Analytic functions are among the most important building blocks for modeling, and are a classical way of expressing knowledge and have a long history of usage in science. From a data mining perspective, direct discovery of general analytic functions poses enormous challenges because of the (in principle) infinite size of the search space. Within computational intelligence, genetic programming techniques aim at evolving computer programs, which ultimately are functions. Genetic Programming (GP) introduced in [10] and further elaborated in [11], [12] and [13], is an extension of the Genetic Algorithm. The algorithm starts with a set of randomly created computer programs and this initial population goes through a domain-independent breeding process over a series of generations. It employs the Darwinian principle of survival of the fittest with operations similar to those occurring naturally, like sexual recombination of entities (crossover), occasional mutation, duplication and gene deletion.

3.1 Gene Expression Programming

There are many approaches to GP leading to a plethora of variants (and implementations). A discussion about their relative merits, drawbacks and properties is beyond the scope of this paper. One of these GP techniques is the so-called Gene Expression Programming (GEP) [7], [8]. GEP individuals are nonlinear entities of different sizes and shapes (expression trees) encoded as strings of fixed length. For the interplay of the GEP chromosomes and the expression trees (ET), GEP uses an unambiguous translation system to transfer the language of chromosomes into the language of expression trees and vise versa. The structural organization of GEP chromosomes allows a functional genotype/phenotype relationship, as any modification made in the genome always results in a syntactically correct ET or program. The set of genetic operators applied to GEP chromosomes always produces valid ETs.

3.2 Neural Networks Constructed Via Genetic Programming (NN-GP)

A general extension to GEP for vector valued functions was previously introduced [19], whereby GEP individuals consist of multiple chromosomes. Such an extension was the starting point for the construction of a technique to evolve explicit neural networks. Figure 1 shows an example of an explicit neural network consisting of $(n + m + c)$ neurons and (3) layers (other topologies are also possible), where each neuron is a chromosome in an individual. For this example, n neurons in the input layer are determined by the number of variables in the input data set; m neurons in the hidden layer determine the dimension of the non-linear space to be constructed (in this paper, $m = 1$); and c determines the number of classes that need to be discriminated. In general, c neurons in the output layer may be used, but other approaches exist. For example, this paper uses $c = 2$ and uses 1 output neuron in order to construct explicit classifiers. Future studies will investigate these issues more deeply, for example, when determining class discrimination between $c > 2$ classes.

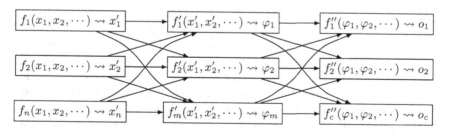

Fig. 1. Neural network representation of one specific topology containing (3) layers and $(n + m + c)$ neurons. Each box is a neuron in the network where all activity occurs (e.g. activation, aggregation, etc). Weights are learned within the neuron by NN-GP.

4 Individual Dichotomization

This is a simple screening algorithm used with the purpose of finding individual attributes that are relevant from the point of view of their ability to differentiate the classes (in a binary problem), when their values are dichotomized. The inputs for the algorithm are: *i*) the values of a given attribute A for all the objects , *ii*) the classes C_1, C_2 associated with each sample (Cancer vs Normal in this case), and *iii*) a probability threshold p_T. The algorithm proceeds as follows: *(1)* construction of the set of distinct values of A (call it Δ). If O is the set of objects and $A(o)$ is the value of the attribute for any object $o \in O$, $\Delta = \{\delta_1, \delta_2, \cdots, \delta_k\}, (k \in [1, card(O)])$ with the following properties: ($\forall \delta_i, \delta_j \in \Delta$, $\delta_i \neq \delta_j$), ($\forall o \in O, \exists \delta \in \Delta$ s.t. $A(o) = \delta$) and ($\forall \delta \in \Delta, \exists o \in O$ s.t. $A(o) = \delta$). *(2)* sort Δ in increasing order. (3) construct the set $\hat{\Delta}$ composed by the mean of all consecutive values of Δ. That is, for every pair $\delta_i, \delta_{i+1} \in \Delta$ compute $(\hat{\delta}_i = (\delta_i + \delta_{i+1})/2$. Clearly, $\hat{\Delta}$ has one element less than Δ. *(4)* use each $(\hat{\delta}_i \in \hat{\Delta}$ as a binary threshold for the values of attribute A. This divides the set of objects into two disjointed classes A_1, A_2 . *(5)* compute the contingency table of A_1, A_2 vs C_1, C_2 (6) on the table, compute the conditional probabilities $p_1 = p(C_1/A_1)$, $p_2 = p(C_1/A_2)$and retain $p_{max} = \max(p_1, p_2)$. *(7)* if $p_{max} \geq p_T$ select the attribute as relevant, and discard it otherwise. The process is repeated for all attributes and the resulting set of selected attributes gives an indication on how many of them contain a differentiation power equal or better than the pre-set probability threshold p_T . Specifically, if $p_T = 1$ the algorithm will give a set of attributes such that each of them (*individually*) will perfectly differentiate the classes $\{C_1, C_2\}$.

5 Experimental Settings

The height and shape of each resonance in the MR spectrum is determined by several parameters related to the way in which signal produced by the exited proton spin decays by a relaxation process. One of them, called the echo time (TE) is very important. The longer the TE, the more the signal has attenuated before acquisition. Hence, a short echo time spectrum (TE $\leq$ 50ms) has larger peaks than a long echo time spectrum (TE $\geq$ 130ms). A short echo time spectrum also contains more peaks, as resonances with a small relaxation value or complex coupling pattern, like mI (myo-Inositol), Glu (glutamate) and Gln (glutamine) are less pronounced at longer echo times. At short echo time signals, macromolecules are prominent; originating from proteins and membrane components. They have very broad peaks with a large contribution to an underlying and partially unknown baseline [14], [6]. The data used in this study consist of 219 long-echo MR spectra (echo time TE $\geq$ 130ms). The data acquisition protocol and the signal processing procedure is described in [18]. Each spectrum covers a range between [4.23 .. 0.45] parts per million (ppm) along the x-axis, where 200 equally spaced samples were taken. The available validated set represents different types of tumors and normal cases grouped into 4 main classes: G1: astrocytome, oligoastrocytome and oligodendrogliome, G2: glioblastome and metastasis and G3: meningiomes. This paper has a preliminary character and so will focus on the study of the tumor vs normal differentiation (i.e. $\{G1, G2, G3\} vs \{normal\}$), with 204 and 15 cases respectively. In order

Table 1. Experimental settings for the two series of experiments involving NN-GP

	Series 1 (240)	Series 2 (2250)
GEP Max. Num. Generations	50	same
GEP Population Size	5, 10, 15	10
GEP Num. Elite Individuals	1	same
GEP Inversion Rate	0.1	same
GEP Mutation Rate	0.044	same
GEP IS Transposition Rate	0.1	same
GEP RIS Transposition Rate	0.1	same
GEP One Point Recomb. Rate	0.3	same
GEP Two Point Recomb. Rate	0.3	same
GEP Gene Recombination Rate	0.1	same
GEP Gene Transposition Rate	0.1	same
GEP Num. Genes Per Chromosome	1	same
GEP Gene Headsize	2	same
GEP Gene Linking Function	Addition	same
GEP Num. Real Constants Per Gene	2, 4, 8, 200	1, 2, 3, 4, 5
GEP Constants Limits	$[-100.0, 100.0]$	same
GEP Seeds	5 unique seeds	Series 1 and 45 more
GEP Species RNC Mutation Rate	0.01	same
GEP Species DC Mutation Rate	0.044	same
GEP Species DC Inversion Rate	0.1	same
GEP Species DC IS Transposition Rate	0.1	same
GEP Functions For All Symbol Sets	Addition, Subtraction, Multiplication	
GEP Number of Symbol Sets	Determined by NN topology: 3 (one/layer)	
GEP Number of Chromosomes	Determined by NN topology: 202	
Neural Network (NN) Topology	200 Input Nodes, 1 Hidden, 1 Output	
NN Input Layer Constant Weights	1, 200	1, 100, 200
NN Input Layer Terminal Weights	1	same
NN Hidden Layer Constant Weights	1, 200	1, 100, 200
NN Hidden Layer Terminal Weights	1	same
NN Output Layer Constant Weights	1	same
NN Output Layer Terminal Weights	1	same

to simplify the application of some procedures, in particular genetic programming, the dataset (219 individuals and 200 predictive variables) was linearly re-scaled from its original range $[-44.850571, 56.267685]$ to the $[1, 100]$ range. The purpose was to work with strictly positive values and since the target range is almost the same as the original (99 vs 101.118256), the re-scaling operation is essentially a shifting. The re-scaled data was divided into a training and a test set using random stratified sampling so that class proportions were preserved. The training set contained 80% of the data (175 objects) and the test set the remaining 20% (44 objects). The NN-GP approach was investigated within a series of two experiments (See Table nn-gep-experimental-settings). The first series of 240 attempted to broadly sweep the parameter space; with the second series of 2250 being used to more closely investigate the parameter space around the good solution obtained within the first series.

6   Results

Results from Rough Sets and NN-GP are reported, along with validation via the individual dichotomization approach.

Rough Sets Results. Rough sets analysis was conducted as follows: *i)* the training set was discretized according to the global method described in [2], [4], *ii)* reducts (see Section 2) were computed using exhaustive and genetic algorithms [2], [20], *iii)* classification rules were generated from the reducts, *iv)* the test set was discretized using the same cuts produced by the discretization of the training set, and finally, *v)* the set was classified using the rules obtained for the training set. Remarkably, both reduct computation algorithms found a single reduct on the training set. Moreover, it was a simple reduct composed of a singleton attribute ($\{V270\}$). Accordingly, both sets of classification rules consist of the common single rule:

$$\text{IF } V_{270} \begin{cases} \geq 69.374496 \Rightarrow C_1 \text{ (i.e. Normal)} \\ < 69.374496 \Rightarrow C_2 \text{ (i.e. Diseased)} \end{cases}$$

which classifies the training set with 100% accuracy. When applied to the test set, it turned out that it also classifies with 100% accuracy. This is very interesting, as it shows that a single attribute (V_{270}) (out of the original 200) is capable of discriminating the spectra from normal cases from those of the malignant class. It corresponds to a concentration of approx. 1.969 ppm.

NN-GP Results. Two series of experiments, one of size 240, and the other size 2250 led to 26 explicit neural networks that, when interpreted as classifiers, had 100% training and 100% testing error; a very interesting preliminary result. In order to study the properties of these high performing solutions, the space constructed from the mapping function associated with each of the 26 networks is summarized in Fig.2. It can be seen that all 26 spaces (horizontal lines in Fig.2) perfectly separate the 2 classes and that the 26 solutions can be divided into 4 equivalence classes based on constructed space magnitude: *i)* extra large magnitude $[-150000, 200000]$ (1 solution), *ii)* large magnitude $[-4000, 6000]$ (14 solutions), *iii)* medium magnitude $[-1000, 2000]$ (2 solutions), and *iv)* small magnitude $[-200, 100]$ (9 solutions); with the small magnitude solutions lying closest to the magnitude of the training and testing data. The 26 spaces shown in Fig.2 may also be analyzed in terms of their associated mapping functions. In particular, the 26 equations contain only 50 of the 200 attributes present within the input data; with 43 attributes occurring in exactly one equation, 3 attributes occurring in two equations and 2 attributes occurring in exactly three equations. The two most frequent attributes are V_{270} occurring in exactly eleven equations and V_{271} occurring most frequently, and in sixteen equations. In addition, it is observed that V_{271} was more frequently used than V_{270} within good solution networks and that it was not discovered by the Rough Sets approaches that were investigated, which only discovered attribute V_{270}. Of the 26 good solution results (100% training and 100% testing accuracy), 3 are now highlighted that show use of the 2 most frequent variables (as both independent and joint usage) in the

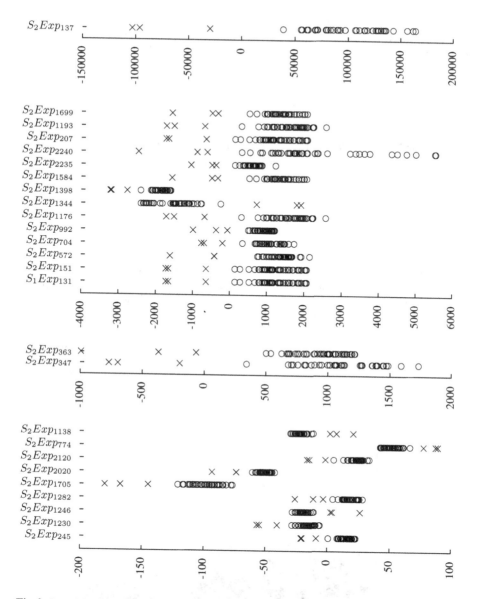

Fig. 2. Best 26 mapped 1D spaces (varying orders of magnitude) from nonlinear discriminant analysis of neural network (NN-GP) solutions having 200 input variables. All 26 spaces have an associated classifier (not shown) with 0.00 training and validation error. X = healthy class. O = diseased patient samples.

mapping and classifier results. It can be observed from Fig.2, that the mapping results may be converted into the good classifiers through rescaling (and possibly reflection about a point) of the constructed spaces. An example NDA and classifier result involving V_{270} was discovered in experiment S_2 Exp_{207} and resulted in the construction of a 200D to 1D mapping function $\varphi_1(\cdot) = 66.86 - V_{270}$ and the following classifier (with 100% training and testing accuracy):

$$\text{IF } (66.86 - V_{270})^3 \begin{cases} < 0.5 \Rightarrow C_1 \text{ (i.e. Normal)} \\ = 0.5 \Rightarrow \text{Undecidable} \\ > 0.5 \Rightarrow C_2 \text{ (i.e. Diseased)} \end{cases}$$

An example NDA and classifier result involving V_{271} was discovered in experiment S_2 Exp_{347} and resulted in the construction of a 200D to 1D mapping function $\varphi_1(\cdot) = V_{271} - V_{234} - 27.69$ and the following classifier (with 100% train/test accuracy):

$$\text{IF } -28.75(V_{271} - V_{234} - 27.69) - 50.78 \begin{cases} < 0.5 \Rightarrow C_1 \text{ (i.e. Normal)} \\ = 0.5 \Rightarrow \text{Undecidable} \\ > 0.5 \Rightarrow C_2 \text{ (i.e. Diseased)} \end{cases}$$

An example NDA and classifier result involving both V_{270} and V_{271} was discovered in experiment S_2 Exp_{1699} and resulted in the construction of a 200D to 1D mapping function $\varphi_1(\cdot) = V_{331} - V_{295} - V_{271} - V_{270} - V_{195} + V_{179}$ and the following classifier (with 100% train/test accuracy):

$$\text{IF } V_{331} - V_{295} - V_{271} - V_{270} - V_{195} + V_{179} + 137.40 \begin{cases} < 0.5 \Rightarrow C_1 \text{ (i.e. Normal)} \\ = 0.5 \Rightarrow \text{Undecidable} \\ > 0.5 \Rightarrow C_2 \text{ (i.e. Diseased)} \end{cases}$$

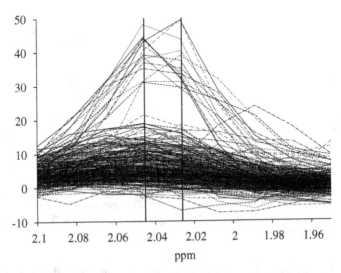

Fig. 3. All 285 MR spectra. 2 out of 200 variables may be used (independently or jointly) for discrimination. Larger values ([31.075169..48.118134] for V_{270} and [29.067427..49.497776] for V_{271}) are normal samples.

Individual Dichotomization Results. A systematic exploration of each single attribute in the training set was made with the individual dichotomization algorithm (see Section 4). The probability threshold was set to 1 ($p_T = 1$) in order to find the highest conditional probabilities of the classes given the attribute dichotomization. It was found that $P(class = normal/(V_{270} \geq 69.375)) = 1$ and that $P(class = normal/(V_{271} \geq 68.257)) = 1$. When these probabilities are computed on the test set using the same conditionals, the result was the same, showing that both V_{270} and V_{271} (spectral peaks at 1.969 and 1.95 ppm respectively), can individually discriminate the normal from the malignant cases, thus confirming the results found with rough sets and especially with the NN-GP network. Rough sets found V_{270} but not V_{271}, whereas NN-GP detected V_{270} and V_{271} as the two most important attributes, confirmed by individual dichotomization.

7 Conclusions

Computational intelligence techniques were applied to brain cancer data. Good preliminary results were obtained with 100% training and testing accuracy that differentiate normal versus malignant samples. Two out of 200 attributes were found to be most important. Rough Sets found one; whereas the NN-GP experiments found both. The results were confirmed via a systematic algorithm, which disregards attribute interactions; something that cannot (in general) be assumed *a priori*. The NN-GP approach, which, although more complex, did not miss a relevant attribute as did the Rough Sets approach. Future studies will focus on differentiation of the different cancers.

Acknowledgments

This paper is a cooperation between the SOCO Group of the Polytechnic University of Catalonia (UPC, Spain) and the Knowledge Discovery Group (National Research Council Canada). The authors thank Alfredo Vellido (UPC, Spain) for his support of the research presented in this paper. Authors gratefully acknowledge the former INTER-PRET (EU-IST-1999-10310) European project partners. Data providers: Dr. C. Majós (IDI), Dr. À. Moreno-Torres (CDP), Dr. F.A. Howe and Prof. J. Griffiths (SGUL), Prof. A. Heerschap (RU), Dr. W. Gajewicz (MUL) and Dr. J. Calvar (FLENI); data curators: Dr. A.P. Candiota, Ms. T. Delgado, Ms. J. Martín, Mr. I. Olier and Mr. A. Pérez (all from GABRMN-UAB). Prof. Carles Arús, GABRMN group leader.

References

1. Bäck, T., Fogel, D.B., Michalewicz, Z.: Evolutionary Computation 1 and 2. Institute of Physics Publishing, Bristol (2000)
2. Bazan, J.G., Nguyen, H.S., Nguyen, S.H., Synak, P., Wróblewski, J.: Rough set algorithms in classification problem. In: Polkowski, L., Tsumoto, S., Lin, T.Y. (eds.) Rough Set Methods and Applications, pp. 49–88. Physica-Verlag, Heidelberg (2000)
3. Bazan, J.G., Skowron, A., Synak, P.: Dynamic reducts as a tool for extracting laws from decision tables. In: Proc. of the Symp. on Methodologies for Intelligent Systems. LNCS (LNAI), vol. 869, pp. 346–355. Springer, Heidelberg (1994)

4. Bazan, J.G., Szczuka, M.S., Wróblewski, J.: A new version of rough set exploration system. In: Alpigini, J.J., Peters, J.F., Skowron, A., Zhong, N. (eds.) RSCTC 2002. LNCS (LNAI), vol. 2475, pp. 397–404. Springer, Heidelberg (2002)
5. Bäck, T., Fogel, D.B., Michalewicz, Z.: Handbook of Evolutionary Computation. Institute of Physics Publishing and Oxford Univ. Press, New York, Oxford (1997)
6. Devos, A.: Quantification and classification of magnetic resonance spectroscopy data and applications to brain tumour recognition. Technical Report U.D.C. 616.831-073, Katholieke Universiteit Leuven. Dept. of Electronics (2005)
7. Ferreira, C.: Gene expression programming: A new adaptive algorithm for problem solving. Journal of Complex Systems 13 (2001)
8. Ferreira, C.: Gene Expression Programming: Mathematical Modeling by an Artificial Intelligence. Springer, Heidelberg (2006)
9. Juliá-Sapé, M., Acosta, D., Mier, M., Arús, C., Watson, D.: A multi-centre, web-accessible and quality control-checked database of in vivo mr spectra of brain tumour patients. Magn Reson Mater Phy. 19 (2006)
10. Koza, J.: Hierarchical genetic algorithms operating on populations of computer programs. In: Proc. of the 11th International Joint Conf. on Artificial Intelligence, San Mateo, CA (1989)
11. Koza, J.: Genetic programming: On the programming of computers by means of natural selection. MIT Press, Cambridge (1992)
12. Koza, J.: Genetic programming II: Automatic discovery of reusable programs. MIT Press, Cambridge (1994)
13. Koza, J.R., Bennett III, F.H., Andre, D., Keane, M.A.: Genetic Programming III: Darwinian Invention and Problem Solving. Morgan Kaufmann, San Francisco (1999)
14. Ladroue, C.L.C.: Pattern Recognition Techniques for the Study of Magnetic Resonance Spectra of Brain Tumours. Ph.D thesis, St George's Hospital Medical School. London (2004)
15. Øhrn, A.: Discernibility and rough sets in medicine: Tools and applications. Technical Report NTNU report 1999:133, Norwegian University of Science and Technology, Department of Computer and Information Science (1999)
16. Pawlak, Z.: Rough Sets, Theoretical Aspects of Reasoning about Data. Kluwer Academic Publishers, New York (1991)
17. Pawlaw, Z.: Rough Sets. International Journal of Information and Computer Sciences 11, 341–356 (1982)
18. Tate, A.R., Underwood, J., Acosta, D.M., Juliá-Sapé, M., Majo, C., Moreno-Torres, A., Howe, F.A., van der Graaf, M., Lefournier, V., Murphy, M.M., Loosemore, A., Ladroue, C., Wesseling, P., Bosson, J.L.,, M.E.C., as, M.E.C.n., Simonetti, A.W., Gajewicz, W., Calvar, J., Capdevila, A., Wilkins, P.R., Bell, B.A., Rémy, C., Heerschap, A., Watson, D., Griffiths1, J.R., Arús, C.: Development of a decision support system for diagnosis and grading of brain tumours using in vivo magnetic resonance single voxel spectra. NMR Biomed. 19 (2006)
19. Valdés, J.J., Orchard, R., Barton, A.J.: Exploring Medical Data using Visual Spaces with Genetic Programming and Implicit Functional Mappings. In: GECCO Workshop on Medical Applications of Genetic and Evolutionary Computation. The Genetic and Evolutionary Computation Conference, London, England (2007)
20. Wróblewski, J.: Finding minimal reducts using genetic algorithm. In: Proc. of the Second Annual Join Conference on Information Sciences (1995)
21. Wróblewski, J.: Ensembles of classifiers based on approximate reducts. Fundamenta Informaticae 47 (2001)

A Hybrid Model for Aiding in Decision Making for the Neuropsychological Diagnosis of Alzheimer's Disease

Ana Karoline Araújo de Castro, Plácido Rogério Pinheiro,
and Mirian Calíope Dantas Pinheiro

Master Degree in Applied Computer Sciences
University of Fortaleza
Av. Washington Soares, 1321 - Bloco J sala 30,
CEP: 60811-905, Fortaleza, Ceará, Brazil
akcastro@gmail.com,
{placido,caliope}@unifor.br

Abstract. This work presents a hybrid model, combining Bayesian Networks and the Multicriteria Method, for aiding in decision making for the neuropsychological diagnosis of Alzheimer's disease. Due to the increase in life expectancy there is higher incidence of dementias. Alzheimer's disease is the most common dementia (alone or together with other dementias), accounting for 50% of the cases. Because of this and due to limitations in treatment at late stages of the disease early neuropsychological diagnosis is fundamental because it improves quality of life for patients and theirs families. Bayesian Networks are implemented using NETICA tool. Next, the judgment matrixes are constructed to obtain cardinal value scales which are implemented through MACBETH Multicriteria Methodology. The modeling and evaluation processes were carried out with the aid of a health specialist, bibliographic data and through of neuropsychological battery of standardized assessments.

Keywords: Diagnosis, neuropsychological, CERAD, alzheimer's, MACBETH, multicriteria.

1 Introduction

The World Health Organization [22] estimates that in 2025 the population over age 65 will be 800 million, with 2/3 in developed countries. It is expected that in some countries, especially in Latin America and Southeast Asia there will be an increase in the elderly population of 300% in the next 30 years.

With the increase in life expectancy health problems among the elderly population also increase and these complications tend to be of long duration, requiring qualified personnel, multi-disciplinary teams, and high cost extra exams and equipment.

Health care systems will have to confront the challenge of aiding patients and those responsible for them. The costs will be enormous the whole world over.

C.-C. Chan et al. (Eds.): RSCTC 2008, LNAI 5306, pp. 495–504, 2008.

As the population increases the number of dementias increases as a consequence. There are numerous causes of dementias and specific diagnosis depends on knowledge of different clinical manifestations and a specific and obligatory sequence of complementary exams [7].

The initial symptoms of dementia can vary, but the loss of short term memory is usually the main or only characteristic to be brought to the attention of the doctor in the first appointment. Even so, not all cognitive problems in elderly people are due to dementia. Careful questioning of patients and family members can help to determine the nature of cognitive damage and narrow the choices for diagnosis [21].

Alzheimer's disease (AD) is the most frequent cause of dementia and makes up 50% of the cases in the 65+ age group [8].

The main focus of this work is to develop a multicriteria model for aiding in decision making for the neuropsychological diagnosis of Alzheimer's disease, using Bayesian networks as a modeling tool. The processes of problem definition, qualitative and quantitative modeling, and evaluation are presented here. In this work, the modeling and evaluation processes have been conducted with the aid of a medical expert and bibliographic sources. Batteries of standardized assessments which help in the neuropsychological diagnosis of Alzheimer's disease were used for the application of the model.

The battery of tests used in this work is from the Consortium to Establish a Registry for Alzheimer's disease (CERAD). We have sought to discover which questions are most relevant for neuropsychological diagnosis of Alzheimer's disease by using this battery of tests.

The work has produced a Bayesian network and a ranking with the classification of these questions. This ranking is composed of the construction of judgment matrixes and constructing value scales for each Fundamental Point of View already defined. The construction of cardinal value scales was implemented through MACBETH.

2   Diagnosis of Alzheimer's Disease

Alzheimer's disease is characterized by a progressive and irreversible decline in some mental functions, such as memory, time and space orientation, abstract thinking, learning, the incapacity to carry out simple calculations, language disturbances, communication and the capacity to go about daily activities [21].

Diagnosis of Alzheimer's disease [1,6,7,8,11,13,16,17,19,20,21,22] is based on the observation of compatible clinical symptoms and the exclusion of other causes of dementia by means of laboratory exams and structural neuro-imagery. A variety of clinical instruments are used to come to a diagnosis such as a complete medical history, tests to evaluate memory and mental state, evaluation of the degree of attention and concentration of abilities in solving problems and level of communication.

3   CERAD - An Overview

The original mandate of the Consortium to Establish a Registry for Alzheimer's Disease (CERAD) in 1986 was to develop a battery of standardized assessments for the evaluation of cases with Alzheimer's disease who were enrolled in NIA-sponsored Alzheimer's Disease Centers (ADCs) or in other dementia research programs [15]. Despite the growing interest in clinical investigations of this illness at that time, uniform guidelines were lacking as to diagnostic criteria, testing procedures, and staging of severity. This lack of consistency in diagnosis and classification created confusion in interpreting various research findings. CERAD was designed to create uniformity in enrollment criteria and methods of assessment in clinical studies of Alzheimer's Disease and to pool information collected from sites joining the Consortium.

CERAD developed the following standardized instruments to assess the various manifestations of Alzheimer's disease: Clinical Neuropsychology, Neuropathology, Behavior Rating Scale for Dementia, Family History Interviews and Assessment of Service Needs.

4   Model Construction

4.1   Definition of Problem

In studies developed by [9] and [10] the application of the multicriteria model for aiding in diagnosis of Alzheimer's disease was presented. These models were initially validated using two patients and later, with a group of 235 patients who had not yet been diagnosed with Alzheimer's disease. In the validation with the bigger group of people we used other data set that was obtained through of study realized in 2005 with 235 elderly people in the city of So Jos dos Campos, SP, Brazil. We used in this study a questionnaire with 120 questions that supply demographic-social data, analyze the subjective perception of the elderly, the mental and physical health (aspects cognitive and emotional), independency in day-by-day, in addition to familiar and social support and the use of services.

In the present study, we sought to validate the model in a group of patients who had already been diagnosed. So, in this validation we used a neuropsychological battery of assessments which has been applied all over the world.

In the next sections the structures of the model are shown. Initially, we defined the mapping of the questions from the neuropsychological part of the battery of assessments. As a result of this mapping a Bayesian network was created, based on in the opinion of specialists. After that, with the Bayesian network, a multicriteria model was structured that indicated the questions which had the most decisive impact for the diagnosis. In addition to that, the questions of the biggest impact for the diagnosis were shown for the percentage of elderly people that presented the neuropsychological symptoms of Alzheimer's disease.

4.2 Bayesian Network Model

The Bayesian network is a graphic model that has an acyclic directed graph. The nodes and arcs of the model represent, respectively, the universal variables U=(A1, A2,...,An) and the dependencies among the variables. In the network that was constructed for the medical problem modeled, the direction of the arcs represents the relations of consequence-cause among the variables. For example, we have an arc between an A node to a B node, we say that an A node represents, semantically, a cause of B and we use the name A as one of the parents of B [12]. There are other works with applications using Bayesian Network in diagnosis of Alzheimer's disease [5].

During the construction of the Bayesian network, we sought to use the bibliographic data with a health professional. For this reason, meetings were held with a specialist nurse that helped in the structure of the network and the subsequent quantification.

In obtaining the structural model of the network, sought to identify what information relating to the problem of diagnosis, which were present in the neuropsychological part of the battery of assessment which could be represented as variables of the network, as well as the causal relationships between these variables.

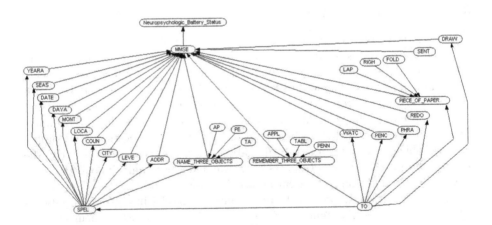

Fig. 1. Network for the diagnosis of Alzheimer's disease in relation to the criteria Mini-Mental State Examination

In the present model, the nodes of the network can be divided in the following way:

1. Main objective of the study: Definition of the neuropsychological diagnosis of Alzheimer's disease.
2. Create uniformity in enrollment criteria and methods of assessment in neuropsychological studies of Alzheimer's diagnosis.

3. Areas evaluated by the neuropsychological battery of assessments and used in the network construction [15]: Verbal Fluency, Boston Naming, Mini-Mental State Exam, Word List Memory, Constructional Praxis, Word List Recall, Word List Recognition and Constructional Praxis (Recall).

The network structure for Mini-Mental State Exam is presented in figure 1. We used Netica Software (http://www.norsys.com) for the construction of the Bayesian network.

The use of the network occurs during the definition of the descriptors and in the evaluation of the final results obtained in the multicriteria model that will be shown in the next section.

4.3   Multicriteria Model

According to [2], in decision making it is necessary to look for elements that can answer the questions raised in order to clarify and make recommendations or increase the coherency between the evolution of the process and the objectives and values considered in the environment.

In this study we used the Multi-Criteria Decision Analysis (MCDA) that is a way of looking at complex problems that are characterized by any mixture of objectives, to present a coherent overall picture to decision makers.

As a set of techniques, MCDA provides different ways of measuring the extent to which options achieve objectives. A substantial reading on MCDA methods can be found in [2,3,4,14,18], where the authors address the definitions and the problems that are involved in the decision making process.

Although it is not simple, the task of constructing the value tree is greatly facilitated with the aid of the Bayesian network. A great volume of information and inter-relations of the raised concepts are provided through the network.

In this study, we used M-MACBETH for the MCDA tool (http://www.m-macbeth.com) to help in the resolution of the problem.

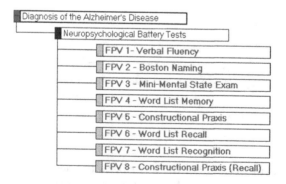

Fig. 2. Problem value tree

The evaluation process is composed of the construction of judgment matrixes and constructing value scales for each Fundamental point of view (FPV) already defined. The construction of cardinal value scales will be implemented through the MACBETH methodology developed by [4].

Figure 2 shows the tree corresponding to the FPVs. The tree represents the definitive structure of the problem that evaluates the neuropsychological diagnosis of Alzheimer's disease.

From the family of FPVs it is possible to carry out the evaluation of the attractiveness of the options for each interest. Although the definition of the describers of impact is a difficult task, it decisively contributes to a good formation of judgments and a just and transparent evaluation [3].

4.4   Describers

An FPV is operational in the moment that has a set of levels of associated impacts (describers). These impacts are defined for Nj, that can be ordered in decreasing form according to the describers [18].

Table 1. Describer for the FPV - Mini-Mental State Exam

NI	Description	Order
N16	NOT ABLE TO DRAW	1°
N15	NOT ABLE TO SENT	2°
N14	NOT ABLE TO PIECE_OF_PAPER	3°
N13	NOT ABLE TO REDO	4°
N12	NOT ABLE TO PHRA	5°
N11	NOT ABLE TO PENC	6°
N10	NOT ABLE TO WATC	7°
N9	NOT ABLE TO YEARA or NOT ABLE TO SEAS or NOT ABLE TO DATE or NOT ABLE TO DAYA or NOT ABLE TO MONT or NOT ABLE TO LOCA or NOT ABLE TO COUN or NOT ABLE TO CITY or NOT ABLE TO LEVE or NOT ABLE TO ADDR or NOT ABLE TO NAME_THREE_OBJECTS or NOT ABLE TO REMEMBER_THREE_OBJECTS	8°
N8	ABLE TO DRAW	9°
N7	ABLE TO SENT	10°
N6	ABLE TO PIECE_OF_PAPER	11°
N5	ABLE TO REDO	12°
N4	ABLE TO PHRA	13°
N3	ABLE TO PENC	14°
N2	ABLE TO WATC	15°
N1	ABLE TO YEARA or ABLE TO SEAS or ABLE TO DATE or ABLE TO DAYA or ABLE TO MONT or ABLE TO LOCA or ABLE TO COUN or ABLE TO CITY or ABLE TO LEVE or ABLE TO ADDR or ABLE TO NAME_THREE_OBJECTS or ABLE TO REMEMBER_THREE_OBJECTS	16°

In this step of construction of the describers, the decisions were made during the meetings with the health professional involved in the process.

Each FPV was operationalized in such a way as to evaluate the influence of the questions evaluated in the elderly patients that correspond to each criterion during the definition of the neuropsychological diagnosis of Alzheimer's disease.

For the evaluation of each FPV, the possible states were defined. Each FPV has a different quantity of states. These states were defined according to the exams or questions involved for each describer.

Its important remember that the describers has a structure of complete pre-order, otherwise, a superior level is always preferable a least level.

For the evaluation of the FPV Mini-Mental State Exam were defined 38 states possible. Table 1 shows the describer of the Mini-Mental State Exam with 16 levels of impact.

4.5   Analysis of Impacts

In this step, the analysis of impacts is carried out, according to each FPV: (i) the lowest and highest values of the impacts; and (ii) the relevant aspects of the distribution of the impacts in each one.

In this work, for each describer, the same values were considered to get the value function for each FPV. Therefore, scores higher than 60, obtained through the judgments matrixes were considered risk describers during the evaluation of diagnosis, in other words, the elderly person that has a great number of answers considered right in the definition of the diagnosis, becomes part of the group of people with a great probability of developing Alzheimer's disease. This perception was defined by the health professional.

4.6   Evaluation

After the definition of the FPVs, family and the construction of the describers, the next step is the construction of the cardinal value scales for each FPV. The evaluations of the judgments matrixes were made according to the opinion of the decision maker, the health area professional.

	[FPV 3]	[FPV 7]	[FPV 4]	[FPV 6]	[FPV 8]	[FPV 1]	[FPV 2]	[FPV 5]	[all lower]	Current scale	extreme
[FPV 3]	no	very weak	weak	weak	strong	strong	weak-str	extreme	positive	17.45	v. strong
[FPV 7]		no	very weak	very weak	strong	strong	strong	extreme	positive	16.59	strong
[FPV 4]			no	vweak-weak	moderate	weak-str	strong	extreme	positive	16.17	moderate
[FPV 6]				no	moderate	weak-str	strg-extr	extreme	positive	15.74	weak
[FPV 8]					no	strg-extr	strong	extreme	positive	13.62	very weak
[FPV 1]						no	weak	extreme	positive	10.64	no
[FPV 2]							no	extreme	positive	9.36	
[FPV 5]								no	positive	0.43	
[all lower]									no	0.00	

Fig. 3. Judgment of all the FPVs

After evaluating the alternatives of all the FPVs individually, an evaluation of the FPVs in one matrix only was carried out. For this, a judgment matrix was created in which the decision maker's orders are defined according to the preference of the decision maker. The decision maker defined the order based on what he judged to be more important in deciding on a diagnosis. Figure 3 presents the judgment matrix of the FPVs.

4.7 Results

In this step, we show the final result of the model - the contribution of the criteria for the neuropsychological diagnosis of Alzheimer's disease. We can see the describer values for each criterion. These values show the importance of choosing these questions that are part of the describers during the definition of the diagnosis.

Analyzing the FPV1 (Verbal Fluency), two describers achieved a value above that which was defined in the impact analysis. They are describers N2 and N3 with values of 77.78 and 88.89 respectively.

In FPV2 (Boston Naming), three describers achieved a value above that which was defined in the impact analysis. They are describers N4, N5 and N6 with values of 82.35, 88.24 and 94.12 respectively.

In FPV3 (Mini-Mental State Exam), there were 10 describers which achieved the minimum value in impact analysis. They were describers N7 to N16, with values of 67.41, 78.57, 89.73, 92.41, 95.09, 97.77, 98.21, 98.66, 99.11 and 99.55.

In FPV4 (Word List Memory), two describers achieved a value above that which was defined in the impact analysis. They are describers N2 and N3 with values of 77.78 and 88.89 respectively.

In FPV5 (Constructional Praxis), four describers achieved a value above that which was defined in the impact analysis. They are describers N5 to N8 with values of 68.42, 84.21, 91.23 and 98.25 respectively.

In FPV6 (Word List Recall) only one describer achieved the minimum value. Its value is 87.50.

In FPV7 (Word List Recognition), only one describer achieved the minimum value. Its value is 87.50.

In FPV8 (Constructional Praxis (Recall)), six describers achieved a value above that which was defined in the impact analysis. They are describers N5 to N10 with values of 61.90, 84.13, 90.48, 93.65, 96.83 and 98.41 respectively.

With this result, we can conclude that the questions that are part of these describers should be preferentially applied during the definition of neuropsychological diagnosis of Alzheimer's disease.

Another important factor to be analyzed is the great quantity of describers which achieved the minimum value in FPV3. Many studies show the importance of the definition of the diagnosis of dementia that should be carried out before the definition of the diagnosis of Alzheimer's disease [6], because many diseases can be confused with dementias and as a consequence, be confused with Alzheimer's disease. This is merely to underline the importance of these criteria for the solution of this problem.

With these results we can carry out a probabilistic analysis with the objective of producing the profiles of the elderly patients that were diagnosed by using this battery of assessments.

5 Conclusion

The diagnosis of Alzheimer's disease is made up of many steps. The first step is to discover if the patient has dementia and then the patient is assessed to see if he has Alzheimer's.

Due to these limitations, this study sought to find the best way possible in the decision making process of defining this diagnosis. By using the neuropsychological part of the battery of assessments adopted by CERAD we attempted to select the main questions involved in diagnosis of Alzheimer's. This battery of assessments was chosen because it uses all the steps of the diagnosis process, and has been used all over the world.

The MACBETH multicriteria method was used to aid in decision making with the help of the Bayesian network during the mapping of the variables involved in the problem. The criteria were defined according to the neuropsychological CERAD areas of assessment.

The questions that make up the battery of assessments were defined as the describers of the problem. With this information, the judgement matrixes were constructed using MACBETH software.

After evaluating the matrixes, a ranking was obtained showing all the questions, from most important to least important with respect to the diagnosis of Alzheimer's.

With these results we can carry out a probabilistic analysis with the objective of producing the profiles of the elderly patients that were diagnosed by this assessment.

As a future project, this model can be extended with the inclusion of new criteria or new models which can be developed using other batteries of assessments.

Acknowledgments. The authors thank the Consortium to Establish a Registry for Alzheimer's Disease (CERAD) for the divulgation of the data utilized in this case study. Ana Karoline Arajo de Castro is thankful to FUNCAP for the support she has received for this project.

References

1. American Psychiatric Association. Diagnostic and Statistical Manual of Mental Disorders, 41th edn. American Psychiatric Association, Washington (1994)
2. Costa, C.A.B., Beinat, E., Vickerman, R.: Introduction and Problem Definition, CEG-IST Working Paper (2001)
3. Costa, C.A.B., Correa, E.C., Corte, J.M.D., Vansnick, J.C.: Facilitating Bid Evaluation in Public Call for tenders: A Social-Technical Approach. OMEGA 30, 227–242 (2002)
4. Costa, C.A.B., Corte, J.M.D., Vansnick, J.C.: Macbeth, LSE-OR Working Paper (2003)
5. Belisle, P., Joseph, L., Wolson, D.B., Zhou, X.: Bayesian Estimation of Cognitive Decline in Patients with AlzheimerŠs Disease. The Canadian Journal of Statistics/La Revue Canadienne de Statistique 30, 37–54 (2002)

6. Braak, H., Braak, E.: Neuropathological Stageing of Alzheimer- Related Changes. Acta Neuropathol. 82, 239–259 (1991)
7. Buschke, H., Sliwinsky, M.J., Kuslansky, G., Lipton, R.B.: Diagnosis of Early Dementia by the Double Memory Test: Encoding Specificity Improves Diagnostic Sensitivity and Specificity. Neurology 48, 989–997 (1997)
8. Caramelli, P., Barbosa, M.T.: How to diagnose the four most frequent causes of dementia? Rev Bras Psiquiatr 24(supl. I), 7–10 (2002)
9. Castro, A.K.A., Pinheiro, P.R., Pinheiro, M.C.D.: Applying a Decision Making Model in the Early Diagnosis of Alzheimer's Disease. In: Yao, J., Lingras, P., Wu, W.-Z., Szczuka, M.S., Cercone, N.J., Ślęzak, D. (eds.) RSKT 2007. LNCS (LNAI), vol. 4481, pp. 149–156. Springer, Heidelberg (2007)
10. Castro, A.K.A., Pinheiro, P.R., Pinheiro, M.C.D.: A Multicriteria Model Applied in the Early Diagnosis of Alzheimer's Disease: A Bayesian Approach. In: International Conference on Operational Research for Development (ICORDVI), Brazil (2007)
11. Clauss, J.J., Strijers, R.L., Jonkman, E.J.: The Diagnostic Value of EEG in Mild Senile AD. Clin. Neurophysiol. 110, 825–832 (1999)
12. Hall, C.B., et al.: Bayesian and profile likelihood change point methods for modeling cognitive function over time. Computational Statistics Analysis 42, 91–109 (2003)
13. Hughes, C.P., Berg, L., Danzinger, W.L., Coben, L.A., Martin, R.L.: A New Clinical Scale for the Staging of Dementia. British Journal of Psychiatry 140, 566–572 (1982)
14. Keysalis: Hiview for Windows. Krysalis, London (1995)
15. Morris, J.C., Heyman, A., Mohs, R.C., et al.: The Consortium to Establish a Registry for Alzheimer's Disease (CERAD): Part 1. Clinical and Neuropsychological Assessment of Alzheimer's Disease. Neurology 39, 1159–1165 (1989)
16. Ostrosky-Solis, Ardila-Roselli, M.: NEUROPSI: a Brief Neuropsychoogical Test Battery in Spanish with Norms by Age and Educational Level. J. Int. Neuropsychol. Soc. 5, 413–433 (1999)
17. Petersen, R.C., Smith, G.E., Waring, S.C., et al.: Mild Cognitive Impairment: Clinical Characterization and Outcome. Archives Neurology 56, 303–308 (1999)
18. Pinheiro, P.R., Souza, G.G.C.: A Multicriteria Model for Production of a Newspaper. In: Proc: The 17th International Conference on Multiple Criteria Decision Analysis, Canada, pp. 315–325 (2004)
19. Podgorski, C.A., Lanning, B.D., Casaceli, C.J., Nardi, A.L., Cox, C.: Dementia Consults: Predictors of Compliance by Primary Care Physicians. Am. J. Alzheimers Dis Other Demen 17(1), 44–50 (2002)
20. Porto, C.S., Fichman, H.C., Caramelli, P., Bahia, V.S., Nitrini, R.: Brazilian Version of the Mattis Dementia Rating Scale Diagnosis of Mild Dementia in Alzheimere's Disease. Arq. Neuropsiquiatr. 61(2-B), 339–345 (2003)
21. Rosen, W.G., Mohs, R.C., Davis, K.L.: A New Rating Scale for Alzheimer' s Disease. Am. J. Psychiatry 141, 1356–1364 (1984)
22. World Health Organization, http://www.who.int/whr/1998/media_centre/executive_summary1/en/index5.html

Rough Sets in Data Warehousing
Extended Abstract

Dominik Ślęzak, Jakub Wróblewski, Victoria Eastwood, and Piotr Synak

Infobright Inc., Poland & Canada
www.infobright.com,
{slezak,jakubw,victoriae,synak}@infobright.com

The theory of rough sets [15,16], based on the universal framework of information systems, provides a powerful model for representing patterns and dependencies both in databases and in data mining. On the one hand, although there are numerous rough set applications to data mining and knowledge discovery [10,18], the usage of rough sets inside the database engines is still quite an uncharted territory. On the other hand, however, this situation is not so exceptional given that even the most well-known paradigms of machine learning, soft computing, artificial intelligence, and approximate reasoning are still waiting for more recognition in the database research, with huge potential in such areas as, e.g., physical data model tuning or adaptive query optimization [2,3].

Rough set-based algorithms and similar techniques can be applied to improve database performance by employing the automatically discovered dependencies to better deal with query conditions [5,9]. Another idea is to use available information to calculate rough approximations of data needed to resolve queries and to assist the database engine in accessing relevant data [20,24]. In our approach, we partition data onto *rough rows*, each consisting of 64K of original rows. We automatically label rough rows with compact information about their values on particular columns, often involving multi-table cross-relationships. One may say that we create a new information system where objects take the form of rough rows and attributes correspond to various flavors of rough row information. A number of database operations can be fully or partially processed within such a new system, with an access to the original data pieces still available, whenever required on top of rough row information. Such a framework seems to actually fit the paradigms of rough and granular computing [1,17], where calculations on granules are additionally allowed to interact with those on single items.

The above ideas guided us towards implementing the fully functional database product, with interfaces provided via integration with MySQL [13,14] and with internals based on such trends in database research as columnar stores [8,11] and adaptive compression [6,22]. Relying on relatively small, flexible rough row information enabled us to become especially competitive in the field of analytical data warehousing, where users want to analyze terabytes of data in a complex, dynamically changing fashion. We realize though that we should keep comparing ourselves against other strategies of using data about data [4,12] and redesigning various dependency/pattern/metadata/index structures originally defined over single rows to let them work at our rough row level [7,19]. In particular, searching

C.-C. Chan et al. (Eds.): RSCTC 2008, LNAI 5306, pp. 505–507, 2008.

for most efficient though reasonably compact types of rough row information can be interpreted by means of feature extraction and selection [21,23], which will additionally inspire us to refer to the rough set methods in future.

References

1. Bargiela, A., Pedrycz, W.: Granular Computing: An Introduction. Springer, Heidelberg (2003)
2. Chaudhuri, S., Narasayya, V.R.: Self-Tuning Database Systems: A Decade of Progress. In: VLDB 2007, pp. 3–14 (2007)
3. Deshpande, A., Ives, Z.G., Raman, V.: Adaptive Query Processing. Foundations and Trends in Databases 1(1), 1–140 (2007)
4. Grondin, R., Fadeitchev, E., Zarouba, V.: Searchable archive. US Patent 7, 243, 110 (2007)
5. Haas, P.J., Hueske, F., Markl, V.: Detecting Attribute Dependencies from Query Feedback. In: VLDB 2007, pp. 830–841 (2007)
6. Holloway, A.L., Raman, V., Swart, G., DeWitt, D.J.: How to barter bits for chronons: compression and bandwidth trade offs for database scans. In: SIGMOD 2007, pp. 389–400 (2007)
7. Ioannidis, Y.E.: The History of Histograms (abridged). In: VLDB 2003, pp. 19–30 (2003)
8. Kersten, M.L.: The Database Architecture Jigsaw Puzzle. In: ICDE 2008, pp. 3–4 (2008)
9. Kerdprasop, N., Kerdprasop, K.: Semantic Knowledge Integration to Support Inductive Query Optimization. In: Song, I.-Y., Eder, J., Nguyen, T.M. (eds.) DaWaK 2007. LNCS, vol. 4654, pp. 157–169. Springer, Heidelberg (2007)
10. Lin, T.Y., Cercone, N. (eds.): Rough Sets and Data Mining. Kluwer, Dordrecht (1996)
11. MacNicol, R., French, B.: Sybase IQ multiplex - designed for analytics. In: VLDB 2004, pp. 1227–1230 (2004)
12. Metzger, J.K., Zane, B.M., Hinshaw, F.D.: Limiting scans of loosely ordered and/or grouped relations using nearly ordered maps. US Patent 6 973, 452 (2005)
13. MySQL 5.1 Reference Manual: Storage Engines, dev.mysql.com/doc/refman/5.1/en/storage-engines.html
14. MySQL Business White Papers: Enterprise Data Warehousing with MySQL, www.scribd.com/doc/3003152/Enterprise-Data-Warehousing-with-MySQL
15. Pawlak, Z.: Rough sets: Theoretical aspects of reasoning about data. Kluwer, Dordrecht (1991)
16. Pawlak, Z., Skowron, A.: Rudiments of rough sets. Inf. Sci. 177(1), 3–27 (2007)
17. Pedrycz, W., Skowron, A., Kreinovich, V. (eds.): Handbook of Granular Computing. Wiley, Chichester (2008)
18. Polkowski, L., Skowron, A. (eds.): Rough Sets in Knowledge Discovery. Parts 1 & 2. Physica-Verlag (1998)
19. Ślęzak, D.: Searching for dynamic reducts in inconsistent decision table. In: IPMU 1998, vol. 2, pp. 1362–1369 (1998)
20. Ślęzak, D., Wróblewski, J., Eastwood, V., Synak, P.: Brighthouse: An Analytic Data Warehouse for Ad-hoc Queries. In: VLDB 2008 (2008)
21. Świniarski, R.W., Skowron, A.: Rough set methods in feature selection and recognition. Pattern Recognition Letters 24(6), 833–849 (2003)

22. Wojnarski, M., Apanowicz, C., Eastwood, V., Ślęzak, D., Synak, P., Wojna, A., Wróblewski, J.: Method and System for Data Compression in a Relational Database. US Patent Application, 2008/0071818 A1 (2008)
23. Wróblewski, J.: Analyzing relational databases using rough set based methods. In: IPMU 2000, vol. 1, pp. 256–262 (2000)
24. Wróblewski, J., Apanowicz, C., Eastwood, V., Ślęzak, D., Synak, P., Wojna, A., Wojnarski, M.: Method and System for Storing, Organizing and Processing Data in a Relational Database. US Patent Application, 2008/0071748 A1 (2008)

Classification Challenges in Email Archiving

Arvind Srinivasan and Gaurav Baone

ZL Technologies, Inc.
2000 Concourse Drive, San Jose, CA 95131, USA
{arvind,gbaone}@zlti.com

Abstract. This paper focuses on the technology of Email Archiving and how it has changed the way emails and other such communications are being handled in corporations of the world. In today's world, email finds itself at the top of the preferred modes of communications list. Emails are being increasingly recognized as an acceptable form of evidence in legal disputes and are treated as the most important wealth of information in a company. Every year governments in countries like Unites States of America introduce new laws governing the usage and handling of emails in the corporate worlds. In addition, industry standards are being made that require companies to retain their email communications as a mandatory requirement. These laws and regulations impose huge fines on corporations that fail to comply with these rules. In addition to the fines, any loss in email data makes the corporations susceptible to law suits and losses. To adhere to all the laws and regulations and also to make use of the wealth of information in emails companies have requirement for retention of email data, search and discovery and surveillance of this data. The Email Archiving technology has made it possible for companies to meet these challenges. In this paper, we will elaborate on the key aspects and the challenges that are faced in making an Email Archiving solution. The paper will show how most of the challenges that are faced in this area are closely related to classification and matching of data and how better and advanced techniques need to be devised to improve the email archiving technology.

1 Introduction

Electronic mail (Email) has emerged as the largest modes of communications in today's world. According to a survey performed by the Radicati Group Inc. [1], in 2005 the total number of corporate email users around the world was predicted around 571 million, with an average of 133 emails sent and received by the user in a single day. The survey predicted that the number would rise to an average of about 160 emails per day per user. Other than the speed, ease and comfort provided by this mode of communication it is also the variety and the improvement that has contributed to the rise of email usage. Email communication has grown from its conventional form to other variations such as Instant Messages, SMS logs, Phone logs and Faxes. The amount and the nature of the information that is contained in all these forms of communications makes

C.-C. Chan et al. (Eds.): RSCTC 2008, LNAI 5306, pp. 508–519, 2008.

it imperative for the corporations to capture, preserve and analyze this information for several purposes. There are a growing number of workplaces in America and in other parts of the world that are building repositories of these communications and using the repositories to incorporate industry and government regulations of retention of data, discovery of emails to handle law suits, management of storage and increase work productivity A single mid-size enterprise generates email and messaging related information that is comparable to the size of the information on the internet. The primary problem with captured archives which are such in size is their housekeeping or management. Enterprises have several requirements for the archive such as retention, discovery for satisfying legal requirements, employee productivity, storage optimization etc. Enterprises have been under pressure and have faced several penalties including high-profile lawsuits for losing email data. One classic example of such a scenario is when the White House lost email data during an upgrade of email system and was hence sued for mismanagement [3]. Also, since the 2001 stock crash several companies have faced a tough time in cases where they have lost email data. Some include Morgan Stanley, Enron, etc. This paper will elaborate more on these challenges and requirements for the enterprise and how these are met today and will identify areas where classification technologies and knowledge discovery can be used to meet the requirement and increase productivity.

2   Email Archive Data Sources

Different enterprises based on their industry have varying requirements. Fundamentally, Information capture can distinguished into two – Active Captures and Historical Captures.

2.1   Active Captures

The acquirement of real-time messaging data is classified as Active capture.

1. Real-Time Email Capture via *Journaling*. Journaling of email is a process of capturing every email communications that transpires in a company network. This process ensures that 100% of messaging information is captured for further analysis or for storage purposes. Mails are captured at the Mail Transfer Agent (MTA) level so as to ensure every mail that comes into the company network and goes out of the network and also ones that stay and rotate within the network. To enable what journaling offers, it becomes mandatory of the journaling block to exist at the company's gateway through which information enters the company. Other factors of emails like group lists in the email and blind carbon copy also need to be captured during the Journaling operation. Email servers like Microsoft Exchange, Lotus Notes, etc. which are present at the email gateway of the company to send and receive email are in the best position to journal these emails. Therefore, all major email systems like SendMail, Microsoft Exchange, Lotus Notes etc. have mechanisms in place to enable Journaling of email. Email Archives are

built by this journaling data. Email archiving solutions like ZL Technologies' Unified Archival offer several methodologies to capture journaling data from the email gateway or directly from the company email servers' journaling repositories. However, the drawback of this kind of capture is that due to its real-time nature, only current emails can be captured by this method. Any data from the past has no way to show up in the email archives of the company.

2. Mailbox Crawling. Email archiving system use Mail Server API or protocol to crawl through individual mailboxes and capture legacy information from these mailboxes. This method enables companies to track and captures human behavioral patterns as in organization of these emails in folders by end-users and other such patterns. However, the biggest differentiator between the crawling method and the journaling method is that the archive formed by the crawling method is susceptible to the user behavior. For example, a user deleting an email will prevent the crawling method from discovering the email and hence will have no way find if such an email every existed. The Mailbox Crawling methodology of capturing email overcomes the problem of Journaling wherein past emails cannot be captured. But this method is more susceptible to user actions on his mailbox.

3. Other Messaging data. Messaging data which present themselves in other formats than email also have a need to be archived. This data is usually in the form of Instant Messaging from various sources such as MSN, Yahoo Messenger, GTALK, ICQ etc. and also internal enterprise instant messaging systems such as Sametime, Microsoft Live Communication server. SMS and Phone logs from Blackberry enterprise server, dedicated chats such as Bloomberg, Reuters etc. There are several companies that convert faxes into email and send it out to employees. All these data sources also are taken into consideration while designing an email archiving system.

2.2 Historical Capture

A historical capture of messaging data is usually performed at start of a proactive archive project or more like for a reactive response to a lawsuit, subpoena, and other discovery requests.

1. Old backed up databases. For example, tape backups of Microsoft Exchange databases, Lotus Mail files or SMTP transactions from the past.
2. Mail Archives. End User Email files such as PST, NSF file, individual MIME files, .msg files or other proprietary messaging formats.
3. Other outdated legacy archives.

Fundamentally, any organization will have one or all the forms of email data that have been elaborated in Sections 2.1 and 2.2. Traditionally, the Historical Capture of email data has been the most prominent way to capture email data, but in recent times due to the rise in the demand for archiving more and more organizations are moving towards Active capture of their email data.

3   Business Drivers for Archive Creation

This section will focus on the main business requirements for having an email archiving solution for maintaining an enterprise wide repository or archive of all the messaging communications in a company.

1. Industry Specific Regulations
   Corporations have to abide by several industry specific regulations related to internal and external communications in order to be in business. A failure to adhere to these regulation results in severe monetary penalties and sometimes even the closure of business. The regulations usually demand retaining certain period worth of mail of the company's employees for a particular period of time. SEC-17a-4, NASD 3010 [2] are some such examples for industry regulations for financial companies. The Health Insurance Portability and Accountability Act (HIPPA) is one such regulation in the healthcare industry. In addition to industry regulations, there are laws in every country that require organizations to store emails and make them available when requested.
2. Laws [2]
   There are several United States federal and state security and privacy laws that require companies to preserve their email communications. Federal Rules of Civil Procedure (FRCP), Florida Government-in-the-Sunshine Law and California Security Breach Notification Act (SB 1386) are some such examples in the United States of America. Other countries like Japan also have similar laws like Japan's Personal Information Protection Act (PIPA) and United Kingdom's UK Freedom of Information Act.
3. Corporate Governance
   Corporations feel the need to monitor the flow of information through the messaging channels inside the company. The protection of Intellectual property, proactive action against internal and external fraud, and litigation requests are some reasons that require a complete repository of messaging communications from where companies can extract information. This is one reason companies turn towards email archiving solutions as they provide an unified approach towards all the message communications in the company environment.
4. Storage Management
   For the ease of administration, management and control of mail servers and the data corporations turn towards email archiving products that provide a feasible way to control and handle these problems. Companies find it more convenient to transfer huge storage requirements and expensive operations off of the primary mail servers to the email archiving solutions with a hope to reduce the stress and work load on the mail servers.
5. Employee Productivity
   Email archives can be used by employees in a company to find information from past communications in a fast manner. The archiving solutions also offer techniques to reduce stress of actual mail servers thus increasing their performance and making the email interaction of employees more productive, fast and reliable.

4   Email Archiving Challenges

In previous section, we saw a high-level picture of the business requirements for a need of an email archiving solution. This section will translate these high-level needs to actual challenges.

4.1   Retention

Retention in simple words is the period for which an email should be retained in the email archive. Retention must be driven by corporate policy to meet several laws and regulations. The retention policies are variable in nature as they undergo changes depending on changes made to policies and regulations on which they are based. The law or regulation requires companies to keep records of certain type of messaging transaction for certain amount of period. At the same time to avoid increasing costs for storage and management, the companies want to retain just enough to satisfy the legal requirement and reduce the liability of carrying such data.

Typically the retention policy is set by a set of stake-holders which includes compliance, legal and records management groups. The requirements are typically set by:

1. The user or a group to which a user adheres to, for example a department in a company.
2. The category of record (for example, financial records, customer transaction records, Spam mails, etc.)

Deletion of old mails from the repository of emails is controlled by the retention that is set for the particular mail. There are cases where a mail may be required to stay beyond its set retention period for legal or other reasons. This special category wherein mails are kept beyond their life cycle or retention periods is termed as *Legal-Hold*. It's usually the case that an email is referred by more than one user and so the conflict is resolved by the highest priority retention (longer or shorter as per the requirement). The determination of which retention period must be chosen for a particular email requires recognizing and establishing categories and classification of an email (or record) to associate it with the appropriate categories. The categories are human-based such as based on which folder the mail was found, etc. or automated or a combination of both human-based and automated categories. The categorization is performed based on the content, headers and other attributes of an email and then a decision is made to determine the closest matching retention type. This categorization and decision making process plays a very crucial role in email archiving.

4.2   Surveillance or Supervision

As described in Section 3, enterprises are required to abide by several federal, state and industry wide regulations which require them to monitor their employees' communications. The types and nature of these regulations vary from the type of business that a company operates. For example, financial companies have to supervise communication (SEC 17a-4, NASD-3010) so that they don't over-promise stock returns, commit fraud, insider trading activity and other suspicious. Similarly, Human resources departments of companies proactively monitor emails for inappropriate language, sexual harassment and other such violation. Likewise, companies need to prevent the leaking of critical company information like IP or Earnings Reports etc., to ensure their place in the competitive world of industries. There are basically four types of supervisions that can be performed by a company.

1. Preventive Supervision or *Pre-Review*. In this form of supervision, the incoming and outgoing mails are blocked from reaching their destination and undergo supervision. If the supervision reveals that the mail does not comply with the company's standards, it can be prevented from delivery until being approved by a human supervisor or reviewer. Extreme care is taken to avoid false positives in case of categorizing a mail as *Pre-Review* as this adds to the latency in communication and adds to the workload of the supervisors.
2. Reactive Supervision or *Post-Review*. This approach is taken by companies to proactively monitor employee communication to avoid forced or unforced violations within the company. The mail is captured and flagged based on its content and presented to the supervisors as a potential violation.
3. *Random-Sampling*. Some percentage of every user's mail is picked at random and presented for supervision. This approach can prove extremely useful in cases of employees trying to beat the automated review system by generating communication which may be incorrect to look at from a machine's perspective but may contain sensitive information in the form of misspelled words and forced grammatical errors. The random sampling of the employees' mail can ensure that a general communication behavior of the employee can be determined and forced or unforced misinterpretations can be alerted.
4. Targeted –Review – After fact, investigator are looking for certain type of violation (for example, finding users who may be involved in trading a particular stock during a lock-out period).

Pre-Review and *Post-Review* requires technology to detect potential frauds by analyzing the mail and attachment contents. Several methods have been used including

- Simple Keyword Detections in the scope of certain part of the mail.
- Word Association techniques to determine the context of a sentence. For example, in a financial company the words *park* and *stock* within 10 words in a sentence can assumed to be a sentence that may be revealing or advising on stocks and their prices.
- Natural Language Processing may be used to detect the language structure of the content. (For example, the word *Sue* used as a proper noun as opposed to the word used as a common noun.)

- Reducing false positives by ignoring certain sections (for example, Disclaimer messages in emails.) The challenge in this case is that it may be easy and direct to filter known sections of the emails as false positives but detecting unknown sections of an email that fall in the same category is a challenge. For example, a company's own disclaimer that is attached at the end of every outgoing email from the company email is a known section that may be treated as a section that may trigger a false alarm. However, for incoming emails from other companies those carry similar sections but with completely different content in the disclaimers also need to be triggered as false alarms.
- Automated mails such as spam emails, newsletters from Wall street Journal, analyst reports etc. require to be treated specially. This is done using mail header analysis and content analysis. However, this process can prove to be laborious depending on the size and the format of the content. Classification techniques can be of great help here to optimize and speed up the process.
- Other standard mail types such as Read-Receipt, Out of Office Response, Delivery Status Notification etc. also fall into a special category. These types are usually auto-generated based triggered based on some preset events. Simple techniques like header analysis are used to determine such emails.
- Classification Mail direction also plays an important role in determining the action that needs to be taken on the mail. For example, a company may choose to be extremely critical over mails that generate inside the company and go out of the company than mails that come into the company from outside.
- Forced or unforced changes to emails either in their entirety or in part may also make a supervision system behave differently than expected. For example, if the supervision system is configured remove attachments of a certain type to not enter into the company network due to security and virus protection reasons the mail could then be changed to send the same exact attachment with a different extension in order for the passage of the email to be successfully. Thus compromising
- Duplicate mails need to be identified and handled appropriately to avoid unwanted work on the part of supervisors who supervise mail in the company.

4.3   Search and Discovery

Search and Discovery of mails form an integral part of any email archiving system. Being the single largest compilation of all the messaging communications of a company, an email archive is expected to be the best place to search and find mails for any given range of period. Mails are free text indexed using the keyword, which means, the indexing process stores unique words in the search engine associated with the email archive. Although, a simple search and discovery of these mails using keyword searches may not be very efficient way of finding a mail as the information returned from the search may be too generic in nature and not specific to the search that is being performed. To understand the classification challenges in this area, we will list the several types of search requirements and their possible solutions of an email archive.

1. Internal Risk Assessment / mounting defense. Internal Legal counsel may anticipate and assess risk or collection information for defense. Concept searches, Document Ranking etc are the important techniques that can be employed in this case. Concept searches are searches that can be customized based on the results received from a simple keyword search and the searching process may be optimized to return better results in future searches. Employing synonyms of words, phonetically similar words (homophones and homonyms), etc. may be some ways to enhance the searching experience and improve search results.

2. Responding to Discovery request. During a subpoena when mail information is requested, the two counsels or the court negotiate for mails that belong to certain people for certain period of time for a certain *concept*. Usually a *concept* is defined by a keyword. Counsels from the two sides involved end up negotiating keywords that can be searched on and mail related to these keywords could be traded or presented as evidence in courts. In such cases, Proximity search request are becoming the order of the day. Proximity searching uses concepts of *distance* between words to identify the usage of these words in various contexts of the email. To make these proximity searches possible and a practically viable and reliable option, *ranking* of keywords plays a very important role. However, these approaches are prone to lot of false-positives results that can present themselves in the search results. Another important concept is an *Attorney Client Privilege* which allows certain communications between the attorney and the client to remain confidential. This set of communication may not be produced as evidence during a legal battle.

3. Productivity / Internal Investigation. Internal company wide investigations for proactive monitoring or for reactive purposes like investigating illegal activity and fraud inside the company forms another major part of search and discovery. Building Concepts or categories in the search engine and overlaying taxonomy could be the most useful way to go about this search approach. For example, internal supervisors of a financial company may want to find all the users' mails that have involved in internal trading in the past one year. From the example, *Internal trading* could be treated as a concept that is a combination of several keywords and their specific usage patterns. If there could be ways of defining these patterns and analyzing the content to fit these patterns using classification approaches, then the searches could be made to return more focused results.

4.4   Storage Management

The amount of storage required to store this goliath of email communications is the most challenging financial and logistic problem that enterprises face. With the rise in the usage of email and the rise in the average size of the email storage management of these emails becomes an important aspect of email archiving. Some challenges in this respect are listed in this section.

Single Instancing of Mails: This is the most important concept in storage management. Email data as a whole may be considered to have a huge amount of redundancy. In terms of the content, attachments, and other parts of the email, there is a good enough possibility to find shared aspects between two separate emails. This opens up a scope for taking advantage of this redundancy and using compression techniques on the email archive to save storage of duplicate emails. However, one point to note is that no matter how similar two emails are, they will always be unique and distinguishable either in their creation time or the source where they were generated or the destination that they are supposed to reach. Identifying these similarities and analyzing the possibilities of avoiding storage of duplicate data of emails is known as the single instancing feature of an email archiving solution. Mail servers like Microsoft Exchange also employ techniques to enable single instancing of email. Another factor to consider in email archiving products is that since the archive receives mail from several data sources (as listed in Section 2), the possibility of duplication is increased, however separation of the mails based on their source of generation also remains a challenge. To solve the problem better, an empirical analysis of email data may give a clearer picture at the constraints that can be used to design a single instancing store for an email archive. It may be observed, for example, that the attachments contribute to most of the storage when an email is stored and are also the most potential parts of an email that may be duplicated in a corporate environment. Thus, separating out an attachment and storing them separated and forming a single instance store of attachments would be an effective approach towards this problem.

5   Research Challenges

This section will focus on some research challenges in the area of email archiving. Most of these problems share their origins to several other fields of Computer Science.

5.1   Mail Classification

As elaborated in previous sections of this paper, the primary function of an email archiving solution revolves around the category of the mail. The mail category is responsible and eventually determinates action that needs to be taken on the mail, the way it needs to be presented, the purpose of the mail and its lifecycle. For example, in the insurance industry as per regulation documents related to asbestos litigations are to be retained for 10 years or more as per laws. In such a case, if there a process is able to exactly pin-point the difference between a mail dealing with stocks and their prices and a fixed income record related to asbestos litigations the application of retention on these documents can be performed easily and more effectively. Some key challenges in this area are as follows:

- The possibilities of false positives occurring during this classification process are great in number and their effect can prove to be disastrous. Reducing the occurrences of these false positives should be one of the primary goals while solving the mail classification problem.

- The process must be real-time, precise and has to account for the heavy email volumes of a corporation.
- The process should be robust yet flexible. It should be able to allow human input in order to form categories that are customized to an organization.
- The process should be able to give back feedback to the organization dealing with the types and behavior of its email content so that organizations can proactively take actions and improve the classification process by adding better rules and policies.

5.2 Surveillance Techniques

Development of better surveillance techniques to help detect fraud, usage of inappropriate language, policy violations like sexual harassment, etc. is a constant challenge in the area of surveillance of emails using an email archiving product. The problem can be treated as a segmentation problem wherein the mail is divided into several parts and analyzed based on several rules and a decision is made. Some key challenges in this area are as follows:

- The process needs to be intelligent enough to be invariant to attempts to bypass the rules. A practical example is when a user, after having detected over a certain period of time will learn what kind of mail content is acceptable and what is not. The user will then try to find ways to beat the system by manipulating emails by changing the order of words, forcing grammatical errors that may be readable by a human viewer but may be impossible to interpret for a machine, changing attachment names to bypass file-extension based email rules, etc.
- The system can be made to learn over time based on specific user activity and the nature of the emails in general.
- Other than detecting potential cases for surveillance, the system must also be effective in discarding unwanted information with great accuracy and speed. Detection of spam emails, automated mails such as read receipts, vacation responses, etc. can be made faster and easier.
- The process must be able to detect and skip certain sections of the email like disclaimers which are repeatable and do not constitute as scan-able portion of the email.
- The process must be prone to false positives and must be scalable in terms of size of data and ease of use.

5.3 Understanding Non-textual Data

In today's world, email is not just a medium of textual data. With companies offering virtually unlimited storage for their employee's emails and web-mail companies competing based on the storage options that are offered to free and paid email users, email communication is slowly moving towards being a storage system of its own. Several heavy email users prefer storing documents as attachments in their emails. More so mail servers have expanded their roles and provide users easy and effective options of storing and sharing files using their email clients. This trend makes non-textual data in emails an important component when it looked upon in the light of retention, surveillance etc. Better understanding of image, audio and video will help companies improve their use of an email archiving solution. Companies can be enabled to effectively track and apply policies over inappropriate images, scanned copies of documents, inappropriate video content, sensitive audio records of meetings, etc. In this respect, there is a need for:

- Optical Character Recognition (OCR) technologies to detect image and video content and translate the content into a deterministic data. This data can then be used to act upon these formats of data. Some examples where companies can use such a technology to monitor such content is the prevention of inappropriate images and videos like pornography in company network, extracting text out of images to detect if the content abides by company policies, etc.
- Classification of audio content to extract out information regarding the nature of content. Confidential and internal presentations or meeting recording being inappropriately distributed is one example of an application to this problem.
- Extraction and proper definition of non-textual data can lead to creating advanced classes of documents wherein a category of documents can have an ensemble of both text and non-text data. Such an approach can be used to create advanced analysis on the behavior of the information in the enterprise.
- Digital rights management algorithms can be incorporate to detect unauthorized software or media in email.

5.4   Search and Discovery

Searching for email data can be made more effective by using techniques such as page ranking to improve finding similar documents. The challenge also lies in determining the similarity criteria and how it can be controlled by a user to make it more usable. Search techniques can be further improved to provide more information that will help analyze the nature of the data. Some key challenges in this area are as follows:

- Searches can be made event driven. Techniques need to be devised to categorize huge amounts of searches so as to make them fall into several events or classes. This approach will make the searching experience more productive and effective.
- Techniques for finding similar documents, both text and non-text, which will require coming up with similarity criteria. The similarity criteria should then be made configurable and easily interpretable by a user.
- Internationalization also must play a role in designing systems for search and discovery. Consideration of several languages and creating event based search results that comprise of documents that transcend language can then be made possible.

5.5   Single Instancing

As described in Section 4.4, single instancing is the sole factor that makes storage management an important and effective feature of email archiving solutions. Traditionally, single instancing has made use *hashing* techniques to compare content and detect similarity in emails. This determination decides if an email is a duplicate of another and if it should be stored again or not. There is a need for:

- Better hashing technologies to improve determination of uniqueness in documents that include both textual as well as non-textual data. For example, a renamed attachment file must still be treated as the original file and stored in addition to a simple pointer to the new file name so as to preserve the email's integrity and at the same time, saving space.
- Better and new techniques that will rely on better matching technologies. The techniques must be smart and must be able to take several factors like mail direction, effects of mail supervision and retention, etc. to determine whether to store an email or not.
- Techniques that are robust, simple to understand, easy to configure and most importantly adaptive to changes.

6   Conclusion

We have made an effort in this paper to bring to light the technology involved in email archiving and the challenges that are faced in making an email archiving solution. Most of the problems that have been listed boil down to problems in the area of classification, detection and pattern matching. There are several methodologies that have been developed over the years in these areas, however efforts on fitting them in the context of an email archiving solution is an exercise that has gathered momentum only in the recent past. The authors believe that technologies such as Bayes' classification, ID3, Rough sets, concepts of nearest neighborhood, Fuzzy logic, neural networks can be effectively employed to solve the various problems in email archiving. Techniques to incorporate results human visualization and analysis can be developed to make the technology more intelligent and attain its goals.

References

1. Taming the Growth of Email – An ROI analysis. The Radicati Group, Inc. (March 2005)
2. Rules & Regulations related to Email Archiving, ZL Technologies, Inc., http://www.zlti.com/resources/index.html#rule_overviews
3. White House loses e-mail during 'upgrade', gets sued (April 30, 2008), http://news.cnet.com/8301-10784_3-9932956-7.html

Approximation Theories: Granular Computing vs Rough Sets

Tsau Young ('T. Y.') Lin

Department of Computer Science, San Jose State University
San Jose, CA 95192-0249
tylin@cs.sjsu.edu

Abstract. The goal of approximation in granular computing (GrC), in this paper, is to learn/approximate/express an unknown concept (a subset of the universe) in terms of a collection of available knowledge granules. So the natural operations are "and" and "or". Approximation theory for five GrC models is introduced. Note that GrC approximation theory is different from that of rough set theory (RST), since RST uses "or" only. The notion of universal approximation theory (UAT) is introduced in GrC. This is important since the learning capability of fuzzy control and neural networks is based on UAT. Z. Pawlak had introduced point based and set based approximations. We use an example to illustrate the weakness of set based approximations in GrC.

1 Introduction

Granular Computing (GrC) can be .interpreted from three semantic views, namely, uncertainty theory, knowledge engineering (**KE**) and how-to-solve/compute-it. In this paper, we concentrate on the KE views: the primary goal of this paper is to develop and investigate the approximation theory that can *approximate/learn/express an unknown concept (represented by an arbitrary subset of the universe)* in terms of a set of basic units of available knowledge (represented by granules.)

2 Category Theory Based GrC Models

It is important to note that the following definition is basically the same as the category model of relational databases that we proposed in 1990 [4]. This observation seems to say that the abstract structures of knowledge and data are the same. Let CAT be a given category.

Definition 1. *Category Theory Based GrC Model:*

1. $\mathcal{C} = \{C_j^h, h, j, = 1, 2, \ldots\}$ *is a family of objects in the Category CAT, it is called the universe (of discourse).*
2. *There is a family of Cartesian products, $C_1^j \times C_2^j \times \ldots$ of objects, $j = 1, 2, \ldots$ of various lengths.*

C.-C. Chan et al. (Eds.): RSCTC 2008, LNAI 5306, pp. 520–529, 2008.

3. *Each n-ary relation object R^j, which is a sub-object of $C_1^j \times C_2^j \times \ldots C_n^j$, represents some constraint.*

4. *$\beta = \{R^1, R^2, \ldots\}$ is a family of n-ary relations (n could vary); so β is a family of constraints.*

The pair $(\mathcal{C}, \beta)$, called *Category Theory Based GrC Model*, is a formal definition of *Eighth GrC Model*.

To specify the general category to the categories of functions, Turing machines and crisp/fuzzy sets, we have **Sixth**, **Seventh** and **Fifth GrC models**, respectively. In Fifth GrC model, by limiting n-ary relations to $n = 2$, we have **Fourth GrC Models**, which are information tables based on binary relations [8], [9]. By restricting the number of relations to one, we have **Third GrC Model (Binary GrC Model)**.

Again from Fifth GrC Model, we have Second GrC Model (Global GrC Model) by requiring all n-ary relations to be symmetric.

Note that a binary relation B defines a binary (right) neighborhood system as follows:

$$p \longrightarrow B(p) = \{y \ mid \ (p.y) \in B\},$$

By considering the collection of $B(p)$ for all binary relations in the Fourth GrC Model, we have **First** GrC Model (Local GrC Model).

3   Approximations: RST vs GrC

In this paper, we are taking the following view: A granule represents a (basic unit) of available knowledge. Based on this view, what should be the admissible operations? We believe "and" and "or" so we take intersection and union as basic operations. For technical considerations (for the infinitesimal granules, equivalently, topological spaces), we take finite intersections and unions of any subfamily as acceptable knowledge operations. We do believe a negation of a piece of available knowledge is not necessary a piece of available knowledge; so negation is not an acceptable operation.

Note that the approximation in this sense is different from that of generalized RST, which is based on the sole operation "or". (Is this a miss-interpretation of Pawlak's idea by the RST community? In classical RST, the intersections are not needed, since they are always empty.) Anyway, in practice generalized RST does not regard the "and" of two known concepts as a known concept. So strictly speaking, generalized rough set approximations are not concept approximations.

GrC has eight models; each has its own approximation theory. We will provide a set of generic definitions here, then discuss the details for individual models.

Definition 2. *Three (point based) approximations (we will suppress "point based" in future discussions).*

Let $\beta = \mathcal{C}_1$ be a granular structure (the collection of granules). Let $\mathcal{G}_1$ be the collection of all possible finite operations of $\mathcal{C}_1$; note that operations are either *finite intersections or point-wise finite intersections. Let G be a variable that varies through the collection $\mathcal{G}_1$, then we define*

1. *Upper approximation:*

   $C[X] = \overline{\beta}[X] = \{p : \forall\ G,\ such\ that,\ p \in G\ \&\ G \cap X \neq \emptyset\}.$

2. *Lower approximation:*

   $I[X] = \underline{\beta}[X] = \{p : \exists\ a\ G,\ such\ that,\ p \in G\ \&G \subseteq X\}.$

3. *Closed set based upper approximation:*

   [17] used closed closure operator. It applies closure operator repeatedly (for transfinite times) until the resultants stop growing. The space is called Frechet (V)-space or (V)-space.
   $Cl[X] = X \cup C[X] \cup C[C[X]] \cup C[C[C[X]]] \ldots$ (transfinite). For such a closure, it is a closed set.

The concept of approximations just defined is derived from topological spaces.

For RST, they can also be defined as follows:

Definition 3. *Set based approximations*

1. *Upper approximation:*

   $C[X] = \overline{\beta}[X] = \bigcup\{G : \forall\ G,\ such\ that,\ G \cap X \neq \emptyset\}.$

2. *Lower approximation:*

   $I[X] = \underline{\beta}[X] = \bigcup\{G : \exists\ a\ G,\ such\ that,\ G \subseteq X\}.$

These definitions do not work as well for many GrC models.

4   Rough Set Theory (RST)

Let us consider the easiest case first. Let U be a classical set, called the universe.

Definition 4. *Let β be a partition, namely, a family of subsets, called equivalence classes, that are mutually disjoint and their union is the whole universe U. Then the pair (U, β) is called RST GrC Model (0th GrC Model or Pawlak Model)*

The two definitions of approximations agree in RST, in this case $\mathcal{G}_1$ is the partition plus $\emptyset$.

5   Topological GrC Model

Next, we consider the approximation theory of a special case in the First GrC Model (Local GrC Model), namely the classical topological space (U, τ), where τ is the topology.

Definition 5. *Topological GrC Model (0.5th GrC Model) is (U, τ). A topology τ is a family of subsets, called open sets, that satisfies the following (global version) axioms of topology: The union of any family of open sets is open and a finite intersections of open sets is open [1].*

A subset $N(p)$ in a topological space U is a neighborhood of p if $N(p)$ contains an open set that contains p. Note that every point in this open set has regarded $N(p)$ as its neighborhood. Such a point will be called the center of $N(p)$ in First GrC Model. The union of all such open sets is $O(p)$ is the maximal open set in $N(p)$. It is clear every point in $O(p)$ regards $N(p)$ as its neighborhood. So $O(p)$ is the collection of center set. In First GrC, it is denoted by $C(p)$. The topology can also be defined by neighborhood system.

Definition 6. *Topological GrC Model (0.5th GrC Model) is (U, TNS). Topological neighborhood system (TNS) is an assignment that associates each point p a family of subsets, TNS(p), that satisfies the (local version) axioms of topology; see [1]. In this case topology is the family $\{TNS(p) \; \forall \; p \; U\}$.*

6   Second GrC Model

Definition 7. *Let $\beta = \mathcal{C}_1 = \{F^1, F^2, \ldots\}$ be a family of subsets. Then the pair (U, β) is called Global GrC Model (2.nd GrC Model or Partial Covering Model).*

In this case $\mathcal{G}_1$ is the family of all possible finite intersections of $\mathcal{C}_1$.

Theorem 1. *The approximation space of Full Covering Model (point based) is a topological space. However, under rough set approximation, it is not a topological space.*

Let τ be the collection of all possible unions of $\mathcal{G}_1$ (when $\mathcal{C}_1$ is a full covering), then τ is a topology.

Proposition 1. *$\mathcal{G}_1$ is a semi-group under intersection.*

The set based definitions may not be useful, for example, $C[X]$ may always be U if β is a topology.

7   First GrC Model

Now, we will generalize this idea to First GrC Model. Let U and V be two classical sets. Let NS be a mapping, called neighborhood system(NS)

$$NS : V \longrightarrow 2^{(P(U))},$$

where $P(X)$ is the family of all crisp/fuzzy subsets of X. 2^Y is the family of all crisp subsets of Y, where $Y = P(U)$. In other words, NS associates each point p in V, a (classical) set $NS(p)$ of crisp/fuzzy subsets of U. Such a crisp/fuzzy subset is called a neighborhood (granule) at p, and the set $NS(p)$ is called a neighborhood system at p; note that $NS(p)$ could be a collection of crisp sets or fuzzy sets.

Definition 8. *First GrC Model:* The 3-tuple (V, U, β) *is called* **Local GrC Model**, *where* β *is a neighborhood system (NS). If* $V = U$, *the 3-tuple is reduced to a pair* (U, β). *In addition, if we require NS to satisfy the topological axioms, then it becomes a TNS.*

Let $NS(p)$ be the neighborhood system at p. Let $G(p)$ be the collection of all finite intersections of all neighborhoods in $NS(p)$. Let G be a variable that varies through $G(p)$.

Definition 9. *With such a* G, *the previous equations given above do define the appropriate notions of* $C[X]$, $I[X]$, $Cl[X]$ *for First GrC Models.*

7.1   Algebraic Structure of GrS

Let $N(p)$ represent an arbitrary neighborhood of $NS(p)$. Let $C_N(p)$, called the center set of $N(p)$, consists of all those points that have $N(p)$ as its neighborhood. (Note that $C_N(p)$ is the maximal open set $O(p)$ in $N(p)$).

Now we will observe something deeper: The finite intersections of all neighborhoods in $NS(p)$ is $G(p)$. A hard question is: Do the intersections of neighborhoods at distinct points belong to $G(p)$?

Proposition 2. *The theorem of intersections*

1. $N(p) \cap N(q)$ *is in* $G(p) = G(q)$, *iff* $C_N(p) \cap C_N(q) \neq \emptyset$.
2. $N(p) \cap N(q)$ *is not in any* $G(p) \; \forall \; p$, *iff* $C_N(p) \cap C_N(q) = \emptyset$.

If we regard $N(p)$ as a known basic knowledge, then we should define the knowledge operations: Let $\circ$ be the "and" of basic knowledge (a neighborhood). For technical reasons, $\emptyset$ is regarded as a piece of the given basic knowledge.

Definition 10. *New operations*

1. $N(p) \circ N(q) = N(p) \cap N(q)$, *iff* $C_N(p) \cap C_N(q) \neq \emptyset$.
2. $N(p) \circ N(q) = \emptyset$, *iff* $C_N(p) \cap C_N(q) = \emptyset$.

The second property says that even though $N(p) \cap N(q)$ may not be equal to $\emptyset$, it does *not* form a neighborhood, hence not a knowledge.

8   Third and Fourth GrC Model

Let U and V be two classical sets. Each $p \in V$ is assigned a subset $B(p)$; intuitively, it is a "basic knowledge" (a set of friends or a "neighborhood" of positions as in quantum mechanics).

$$p \longrightarrow B(p) = \{Y_i, \; i = 1, \ldots\} \subseteq U$$

Such a set $B(p)$ is called a (right) binary neighborhood and the collection $\{B(p) \mid \forall p \in V\}$ is called the binary neighborhood system (BNS).

Definition 11. *Third GrC Model: The 3-tuple* (U, V, β), *where* β *is a BNS, is called a* **Binary GrC Model**. *If* $U = V$, *then the 3-tuple is reduced to a pair* (U, β).

Observe that BNS is equivalent to a binary relation(BR):

$$BR = \{(p, Y) \mid Y \in B(p) \text{ and } p \in V\}.$$

Conversely, a binary relation defines a (right) BNS as follows:

$$p \longrightarrow B(p) = \{Y \mid (p, Y) \in BR\}.$$

So both modern examples give rise to BNS, which was called a binary granular structure in [8]. We would like to note that based on this (right) BNS, the (left) BNS can also be defined:

$$D(p) = \{Y \mid p \in B(Y)\} \text{ for all } p \in V\}.$$

Note that BNS is a special case of NS, namely, it is the case when the collection NS(p) is a singleton B(p). So the Third GrC Model is a special case of First GrC Model.

The algebraic notion, binary relations, in computer science, is often represented geometrically as graphs, networks, forest and etc. So Third GrC Model has captured most of the mathematical structure in computer science.

Observe that BNS is a special cases of NS. So we have

Definition 12. *Let B be a BNS, then*

1. $B(p) \circ B(q) = B(p) = B(q)$, *iff* $C_B(p) \cap C_B(q) \neq \emptyset$.
2. $B(p) \circ B(q) = \emptyset$, *iff* $C_B(p) \cap C_B(q) = \emptyset$. *Note that* $B(p) \cap B(q)$ *may not be empty, but it is not a neighborhood of any point.*

Observe that in Binary GrC Model, two basic pieces of knowledge are either the same or the set theoretical intersection does not represent any basic knowledge.

Next, instead of a single binary relation, we consider the case: β is a set of binary relations. It was called a [binary] knowledge base [8]. Such a collection naturally defines a NS.

Definition 13. *Fourth GrC Model: the pair* (U, β), *where* β *is a set of binary relations, is called a* **Multi-Binary GrC Model**. *This model is most useful in databases; hence it has been called Binary Granular Data Model(BGDM), in the case of equivalence relations, it is called Granular Data Model(GDM).*

Observe that Fourth GrC Model can be converted by a mapping say $First - Four$, to First Model, and First GrC Model induces, say by $Four - First$, to Fourth Model. So First and Fourth models are equivalent, however, the conversions are not natural, because, the two maps are not the inverse of each other.

9   Fifth GrC Model

Definition 14. *Fifth GrC Model:*

1. *Let $\mathcal{U} = \{U_j^h, h, j, = 1, 2, \ldots\}$ be a given family of classical sets, called the universe. Note that distinct indices do not imply the sets are distinct.*
2. *Let $U_1^j \times U_2^j \times \ldots$ be a family of Cartesian products of various length.*
3. *A constraint is expressed by an n-ary relation, which is a subset $R^j \subseteq U_1^j \times U_2^j \times \ldots U_n^j$.*
4. *The constraints are the collection $\beta = \{R^1, R^2, \ldots\}$ of n-ary relations for various n.*

The pair $(\mathcal{U}, \beta)$, called Relational GrC Model, is a formal definition of Fifth GrC Model.

Note that this granular structure is the relational structure (without functions) in the First Order Logic, if n only varies through finite cardinal number.

Higher Order Concept Approximations (**HOCA**).

In Fifth GrC model, we consider the relations (subsets of product space) as basic knowledge, and any subset in a product space as a new concept. We will illustrate the idea in the following case: U^j is either a copy of V or U. Moreover, in each product space, there is at most one copy of V, but no restrictions on the number of copies of U. If a Cartesian product has no V component, it is called U-product space. If there is one and only copy of V, it is called a product space with unique V.

1. u and u^1 is said to be directly related, if u and u^1 are in the same tuple (of a relation in β), where u^1 could be an element of U or V.
2. u and u^2 is said to be indirectly related, if there is a finite sequence $u_i, i = 1, 2, \ldots, t$ such that (1) u_i and u_{i+1} are directly related for every i, and (2) $u = u_1$ and $u^2 = u_t$.
3. An element $u \in U$ is said to be v-related $(v \in V)$, if u and v are directly or indirectly related.
4. v-neighborhood, U_v, consists of all the $u \in U$ that is v-related.

Such a relational granular model (with unique V) induces a map:

$$B : V \longrightarrow 2^U; v \longrightarrow U_v,$$

which is a binary neighborhood system(BNS), where U_v is a v-neighborhood in U, and hence induces a *binary granular model* (U, V, B). Next, we will consider the case $U = V$, then

Definition 15. *The high order approximations of Fifth GrC model is the approximations based on the v-neighborhood system.*

[Digression] In the case $n=2$, depending on the given relation is either $V \times U$ or $U \times V$, the neighborhood systems so obtained is left neighborhood system or right neighborhood system.

The algebraic notion, n-ary relations, in computer science, is often represented geometrically as hypergraphs, hyper-networks, simplicial complexes and etc.

10   Models in Other Categories

Let us consider the category of differentiable, continuous or discrete real-valued functions (think of them as generalized fuzzy sets) on some underlying spaces (these spaces can be differentiable manifolds, topological spaces, or classical sets). In general any collection of functions can be a granular structure, but we will be more interested in those collection that have universal approximation property (for example, a Schauder base in a Banach space). In such case, the approximations are done under appropriate topology on functions spaces.

For a category of Turing machines (algorithms), it is still unclear as how to define the concept approximations.

11   Future Directions

1) Higher Order Concept Approximations,

we may consider v-direct-related neighborhood system.

2) Admissible operations in Granular Structure.

For simplicity, we will consider the Global GrC Model (2nd GrC Model). In other words, β is a partial/full covering. In this paper, we have not introduced the admissible operations into GrS; GrS is represented by $\mathcal{C}_1$, the admissible operations are carried in $\mathcal{G}_1$. In this section, we will include the admissible operations into granular structure. Let $A(GrS)$ be the algebraic structure generated by GrS using the admissible operations. The three approximations can be stated as follows: Let G be a variable that varies through $A(GrS)$, then the same equations given in Section 3 will be used to define C[X], I[X] and Cl[X]. Based on this terminology, we say

1. RST-Based View. $A(GrS)$ is a complete semi-group under union.
2. Topology Based View. $A(GrS)$ is a topology (closed under finite intersection and general unions). This view is what we have adopted in this paper. Actually, what we hope is a bit more general. We would require only that $A(GrS)$ to be a topology on a subset($=\bigcup(A(GrS))$), not necessarily the whole universe.
3. Complete Boolean Ring Based View. $A(GrS)$ is an algebraic structure that is closed under intersections and unions (of any sub-family), but not the complement.
4. σ-ring based view. This is similar to the previous item, except that we restrict it to countable intersections and countable unions.

3) Set based Approximation Theory:

Pawlak offered a set based approximation theory; in RST, both theories are the same. However, for other GrC model, they are different. In the case of Topological

GrC Model $C[X]$ is always equal to U; uninteresting. Now, we have noticed that some information will be lost in the set based approximation theory. Let us consider the *non-reflexive* and *symmetric* binary relational GrC Model (Second GrC Model). To be specific, we consider a finite universe $U = \{a, b, c\}$. Let Δ_1 be a binary neighborhood system defined by

$$\Delta_1 : U \longrightarrow 2^U : a \longrightarrow B_1; \quad b \longrightarrow B_2; \quad c \longrightarrow B_3,$$

where $\{B_1, B_2, B_3\}$ are three distinct fixed subsets of U. Now, let us consider a new BNS, denoted by Δ_2,

$$\Delta_2 : U \longrightarrow 2^U : b \longrightarrow B_1; \quad a \longrightarrow B_2; \quad c \longrightarrow B_3.$$

In fact, we could consider 6 (= 3!) cases. All these six BNS have the same covering. So these six BNS have the *same* set based approximation, and hence the same $C[X]$, $I[X]$, and $Cl[X]$. In other words, the set base approximation cannot reflect the SIX differences. Nieminen considered such approximation for tolerance relations [13], we have used point based notion [6]. It is implicitly in [3] as we have treated it as a generalization of topology.

4) Numerical Measure Based Approximation Theory:

We will illustrate the idea from "infinite RST:" Let us consider a family of partitions on the real line.

1. The first partition P_1 consists of unit closed- open interval [n, n+1),where $-\infty < n < \infty$. They form a partition of real line,
2. The second partition P_2 consists of $1/2$ unit half closed-open intervals, $[n.n + (1/2)), \ldots$
3. The m-th partition P_m consists of $1/(2^m)$ unit half closed-open intervals, $[n.n + (1/(2^m))$.

Let β be the family of the union of these families ($m = 1, 2, \ldots$). It is important to observe that β is a covering but not a partition, though every $P_m, m = 1, \ldots$ is a partition. Now we have the following universal approximation theorem: Let μ be the Lebesgue measure of real line.

Definition 16. *A subset X is a good concept, if for every given ϵ, we can find a finite set of granules such that $|\mu(C[X] - \mu(I[X])| \leq \epsilon$.*

Note that this is a variation of Pawlak's accuracy measure.

Theorem 2. *Every measurable set is a good concept.*

This is a universal approximation theorem. The learning capability of fuzzy control and neural network is based on such a theorem, see for exmaple [15].

References

1. Kelley, J.: General Topology. American Book-Van Nostrand-Reinholdm (1955)
2. Lin, T.Y.: Neighborhood systems and relational database. In: Proceedings of CSC 1988, p. 725 (1988)
3. Lin, T.Y.: Neighbourhood systems and approximation in database and knowledge base systems. In: Proceedings of the Fourth International Symposium on Methodologies of Intelligent Systems (Poster Session), pp. 75–86 (1989b)
4. Lin, T.Y.: Relational Data Models and Category Theory (Abstract). In: ACM Conference on Computer Science 1990, p. 424 (1990)
5. Lin, T.Y.: Peterson Conjecture and Extended Petri nets. In: Fourth Annual Symposium of Parallel Processing, April, pp. 812–823 (1990)
6. Lin, T.Y.: Topological and Fuzzy Rough Sets. In: Slowinski, R. (ed.) Decision Support by Experience - Application of the Rough Sets Theory, pp. 287–304. Kluwer Academic Publishers, Dordrecht (1992)
7. Lin, T.Y.: Neighborhood Systems -A Qualitative Theory for Fuzzy and Rough Sets. In: Wang, P. (ed.) Advances in Machine Intelligence and Soft Computing, Duke University, North Carolina, vol. IV, pp. 132–155 (1997) ISBN:0-9643454-3-3
8. Lin, T.Y.: Granular Computing on binary relations I: data mining and neighborhood systems. In: Skoworn, A., Polkowski, L. (eds.) Rough Sets In Knowledge Discovery, pp. 107–121. Physica-Verlag (1998a)
9. Lin, T.Y.: Granular Computing on Binary Relations II: Rough set representations and belief functions. In: Skoworn, A., Polkowski, L. (eds.) Rough Sets In Knowledge Discovery, pp. 121–140. Physica-Verlag (1998b)
10. Lin, T.Y.: Granular Computing: Fuzzy logic and rough sets. In: Zadeh, L., Kacprzyk, J. (eds.) Computing with Words in Information/Intelligent Systems, pp. 183–200. Physica-Verlag (1999)
11. Lin, T.Y.: Chinese Wall Security Policy Models: Information Flows and Confining Trojan Horses. In: DBSec 2003, pp. 275–287 (2003)
12. Lin, T.Y., Sutojo, A., Hsu, J.-D.: Concept Analysis andWeb Clustering using Combinatorial Topology. In: ICDM Workshops 2006, pp. 412–416 (2006)
13. Nieminen, J.: Rough Tolerance Equality and Tolerance Black Boxes. Fundamenta Informaticae 11, 289–296 (1988)
14. Pawlak, Z.: Rough sets. Theoretical Aspects of Reasoning about Data. Kluwer Academic Publishers, Dordrecht (1991)
15. Park, J.W., Sandberg, I.W.: Universal Approximation Using Radial-Basis-Function Networks. Neural Computation 3, 246–257 (1991)
16. Polya, G.: How to Solve It, 2nd edn. Princeton University Press, Princeton (1957)
17. Sierpenski, W., Krieger, C.: General Topology (Mathematical Exposition No 7). University of Toronto Press (1952)
18. Spanier, E.H.: Algebraic Topology. McGraw Hill, New York (1966); (Paperback, Springer, December 6, 1994)
19. Zimmerman, H.: Fuzzy Set Theory –and its Applications. Kluwer Academic Publisher, Dordrecht (1991)
20. Zadeh, L.A.: Some reflections on soft computing, granular computing and their roles in the conception, design and utilization of information/ intelligent systems. Soft Computing 2, 23–25 (1998)
21. Zadeh, L.A.: Toward a theory of fuzzy information granulation and its centrality in human reasoning and fuzzy logic. Fuzzy Sets and Systems 90, 111–127 (1997)

Author Index

Lecture Notes in Artificial Intelligence (LNAI)